CODES ON EUCLIDEAN SPHERES

North-Holland Mathematical Library

VOLUME 63

ELSEVIER
Amsterdam – London – New York – Oxford – Paris – Shannon – Tokyo

CODES ON EUCLIDEAN SPHERES

Thomas ERICSON
Department of Electrical Engineering
Linköping University
Linköping, Sweden

Victor ZINOVIEV
Institute for Problems of Information Transmission
Russian Academy of Sciences
Moscow, Russia

2001
ELSEVIER
Amsterdam – London – New York – Oxford – Paris – Shannon – Tokyo

ELSEVIER SCIENCE B.V.
Sara Burgerhartstraat 25
P.O. Box 211, 1000 AE Amsterdam, The Netherlands

First edition 2001

Library of Congress Cataloging in Publication Data
A catalog record from the Library of Congress has been applied for.

ISBN: 0 444 50329 3
ISSN: 0924 6509

⊗The paper used in this publication meets the requirements of ANSI/NISO Z39.48-1992 (Permanence of Paper).

Printed in The Netherlands.

Preface

Packing of spheres involves a number of problems that have interested mathematicians since antiquity. The fascination of the topic is largely due to the fact that the problems are usually very easy to formulate and understand, and still in most cases extremely difficult to solve. Kepler formulated his famous conjecture regarding packing of equal spheres in Euclidean space in 1611. Hilbert included that problem in his famous list at the International Congress of Mathematics in Paris, 1900. The problem is still unsolved. The famous discussion between Newton and Gregory regarding the number of surrounding spheres that can touch a single sphere took place in 1694. The question remained unanswered until 1874, when a proof in favor of Newton was suggested. However, this proof turned out to be incomplete. A full solution was not obtained until 1953 in a paper by Schütte–van der Werden.

In recent time the view has widened. In particular problems involving packing of spheres in spaces of higher dimension have been considered. The first steps were taken by Minkowski around 1900 and by Fejes Tóth around 1940. Later important contributions were made by Rankin, Rogers, Leech, Delsarte, Levenshtein, Kabatianskii and many others.

Apart from its pure mathematical interest the topic is of great importance in various branches of physics, biology and engineering. Recently there has been a lot of activity in connection with error control in telecommunication systems. Since the pioneering work by Shannon, Hamming, and Golay around 1950 an extensive theory of error-correcting codes has evolved. The major part of this theory is formulated in terms of discrete mathematical systems such as finite fields and combinatorial designs. However, any practical implementation involves a representation using signals which can naturally be interpreted as points in Euclidean space. In this way many problems in modern telecommunication theory can be seen as natural generalizations of classical packing problems.

Our ambitions in this book is to offer a reasonably complete and self-contained account of the current state of art within the special part of the theory which is concerned with codes on *spheres* in Euclidean space, often referred to as *spherical codes*. The reason for our interest in this special field is its great importance in modern telecommunication. Although we emphasize the mathematical theory we always have the engineering applications in mind. In particular we discuss to some degree such engineering aspects as decoding of spherical codes. Therefore we hope that this book will be of interest both to mathematicians and to engineers, especially to those within the field of telecommunication.

The book is the result of more than ten years of cooperation. During this period—and long before—our work has been largely inspired and influenced by a large number of colleagues and friends. It will not be possible to mention all of them, but we feel that a few of them deserve to be explicitly recognized.

The writing of the book started during a very pleasant joint stay of both authors at INRIA, Paris during the academic year 1993-1994. For the kind invitation and for providing a very creative environment during this stay we both like to express our sincere gratitude to Paul Camion and Pascale Charpin. Both of these people have greatly influenced the content of this book.

We also like to express our special thanks to Peter Boyvalenkov for great help in the computations of the improvements on the Levenshtein bound, to Danyo Danev for helping with computations of bounds and for careful reading of several parts of the manuscript, to Dmitrii Zinoviev for helping with computer search for codes from packings and lattices, and to Martin Bossert, Ezio Biglieri, Ingemar Ingemarsson, Grigory Kabatianskii, Vladimir Levenshtein, Vladimir Sidelnikov, and Vladimir Yudin for many useful remarks.

Very special thanks go to Jean-Jacques Moulis for his extensive help with all the difficulties involved with the Latex typesetting program and for helping us to produce all the figures in electronic form. We also like to express our gratitude to Mrs. Gunilla Svahn Roming for her skillful typing of early versions of the manuscript and to Mrs. Marianne Anse-Lundberg for drawing early versions of many complicated figures.

Finally both of us want to express our deep gratitude to our families who could bear with us during all these years when a large fraction of our spare time was devoted to our work on this book.

Linköping and Moscow, January 2001.

Thomas Ericson   Victor Zinoviev

Contents

Chapter 1

Introduction

1.1 Definitions and basic properties

The *unit sphere* Ω_n in Euclidean n-space $\mathbb{R}^n$ is the set of all unit norm vectors:

$$\Omega_n \triangleq \{x = (x_1, x_2, \cdots, x_n) \in \mathbb{R}^n : \|x\| = 1\},$$

where

$$\|x\| = \sqrt{x_1^2 + x_2^2 + \cdots + x_n^2}.$$

A *spherical code* is a finite subset $\mathcal{X}$ in Ω_n. The special term "spherical code" is employed in order to announce that there are certain special aspects that are of interest. The main focus is the relation between *cardinality* and *internal distances.*

Points in a code are usually referred to as *codewords*. In order to avoid trivial cases we always assume that the number of codewords is at least two.

The three most important parameters characterizing a spherical code $\mathcal{X} \subseteq \Omega_n$ are *size* $M = |\mathcal{X}|$, *dimension* $n = \dim(\mathcal{X})$, and *squared minimum distance.* The size is simply the number of codewords in the code. The dimension is defined as the smallest dimension of any

Euclidean space containing the code. The squared minimum distance is the smallest squared distance between pairs of distinct codewords:

$$\rho \triangleq \min\{||x-y||^2 : x, y \in \mathcal{X},\ x \neq y\}.$$

We usually require that the codewords are *distinct.*

Generally speaking we are interested in finding codes with *large* size, *large* minimum distance and *small* dimension. Obviously these ambitions are in conflict. A core problem is to characterize the set of all triples (n, ρ, M) for which spherical codes exist. In general this is a very difficult problem. Therefore, instead of offering optimal solutions we must—in most cases—content ourselves with providing upper and lower bounds. Usually we do this by fixing the dimension n and then investigating the trade-off between size M and squared minimum distance ρ. Upper bounds to these parameters—to be discussed in Chapters 2 and 3 and in Sections 4 and 6 of the present chapter—are usually quite general and indicate limits beyond which performance cannot reach. Lower bounds, in contrast, are normally very special and are usually given in terms of explicit constructions. Most of our interest will be devoted to such explicit constructions.

Although the limits of performance—in terms of size, minimum distance and dimension—are our main concern, there are also many other aspects that are of interest and that will be discussed at various depths. Foremost among those is the *decoding problem.* This problem occurs in practical applications, usually in the context of telecommunication. Mathematically the problem is this: given a code $\mathcal{X} \subseteq \Omega_n$ and an arbitrary point $y \in \mathbb{R}^n$ find a point $\hat{x} \in \mathcal{X}$ on minimum distance from y. The point $\hat{x}$ is the receiver's guess about the transmitted codeword $x \in \mathcal{X}$, which during transmission has been subject to distortion and has been transformed to a received signal represented by the point $y \in \mathbb{R}$. In principle a solution can always be found by exhaustive search. However, if the code is large—and for good performance large codes are imperative—such a search is *practically* impossible. Therefore the decoding problem is one of *algorithmic complexity*: to design an algorithm which in an efficient way for any $y \in \mathbb{R}^n$—or at least for all y in a sufficiently large subset of $\mathbb{R}^n$—finds an $\hat{x} \in \mathcal{X}$ on smallest possible distance.

For communication applications codes are often evaluated by probabilistic methods and performance is often characterized in terms of error probability. Any such analysis is necessarily an analysis of a *complete system.* As a consequence the result is not a characterization of the code alone, but rather a characterization of the *behaviour* of the code in a specified *environment*, including source, channel, decoding algorithm and possibly still further things. No such analysis will be performed here. Our interest will be devoted to codes only and any analysis or characterization will be in terms of entities derived from the codes alone. Our approach will be *geometric* rather than probabilistic.

Let $\mathcal{X} = \{x_1, x_2, \ldots, x_M\} \subseteq \Omega_n$ be a spherical code. A *codeword matrix* is a matrix X having the codewords $x_i = (x_{i1}, x_{i2}, \ldots, x_{in})$ in $\mathcal{X}$ as its rows. A *Gram matrix* is a matrix G given by

$$G = XX^T,$$

where the exponent T indicates transponate. Clearly G is a square matrix of order M, in which the ij-th component G_{ij} equals the inner product $\langle x_i, x_j \rangle$ between the i-th and the j-th codeword. The squared distance between these codewords is

$$\|x_i - x_j\|^2 = 2 - 2\langle x_i, x_j \rangle,$$

taking into account the fact that all codewords have unit norm. The *Gram matrix* contains complete information about the internal distances in the code, as well as of the dimension n and the size M.

Let $G = XX^T$ be a Gram matrix to a spherical code $\mathcal{X}$ with code matrix X. The following properties are evident:

(*i*) The matrix G is *symmetric* and *positive definite.*

(*ii*) The diagonal elements G_{ii} are all equal to unity, while all non-diagonal elements G_{ij} satisfy $-1 \leq G_{ij} < 1$.

Conversely, to any matrix G satisfying all of these properties we can always find a spherical code $\mathcal{X}$ having G as its Gram matrix. This is a simple consequence of the well known fact that any positive definite matrix G can be represented as a product $G = XX^T$. The fact that all of the diagonal elements in G are equal to unity ensures that all rows in X have unit norm, which is precisely the condition needed in order that they shall define a spherical code.

The above observations imply that we can specify a spherical code either by its codeword matrix X or by its Gram matrix G. Both possibilities will be employed in the sequel.

The enumeration of the codewords is inessential: we insist that a code is an *unordered* set of codewords. Therefore we must agree that two matrices X and Y differing only in the ordering of the rows define the same code.

Also the *basis* used in the description of the code is often—although not always—inessential. More precisely: as we are mainly interested in the distance structure we usually like to identify codes which can be obtained from each other by *distance preserving transformations.* Recalling that distance preserving transformations on the unit sphere are exactly those produced by *orthogonal matrices* we are led to the following definition.

Definition 1.1.1 *Two code matrices X and Y are equivalent if and only if there is a permutation matrix P and an orthogonal matrix A such that the code matrix Y can be expressed as a product of the form $Y = PXA$. Two codes $\mathcal{X}$ and $\mathcal{Y}$ are equivalent if and only if they can be represented by equivalent code matrices.*

We notice that if X and Y are equivalent code matrices their Gram matrices are related as follows:

$$G_{YY} = YY^T = PXAA^TX^TP^T = PG_{XX}P^T.$$

Sometimes it is convenient to view the code $\mathcal{X}$ as a subset in a sphere Ω_N where N is larger than the dimension of $\mathcal{X}$, i.e. to use a coordinate description with an excessive number of coordinates. The dimension of the *code*, however, is independent of the basis used in the description of it. Let G be a Gram matrix to the code $\mathcal{X}$. The following relation is always valid, and can in fact be taken as a definition:

$$n = \dim(\mathcal{X}) = rank(G).$$

Suppose $\mathcal{X} = \{x_1, x_2, \ldots, x_M\} \subseteq \Omega_n$ is a spherical code with the property that for some unit norm vector $a \notin \mathcal{X}$ we have $\langle x_i, a\rangle = \theta$ for all i, with $\theta \neq 0$. A code with this property is said to be *planar*. The following argument shows that such a code cannot be optimal.

The possibilities $\theta = \pm 1$ are ruled out by the requirement that the code should contain at least two codewords, so we may assume $\theta^2 \neq 1$. Define

$$y_i \triangleq \frac{1}{\sqrt{1-\theta^2}}\,(x_i - \theta a), \qquad i = 1, 2, \ldots, M.$$

The set $\mathcal{Y} = \{y_1, y_2, \ldots, y_M\}$ is a new spherical code. Denote by $\rho_{\mathcal{X}}$ and $\rho_{\mathcal{Y}}$ the squared minimum distances in $\mathcal{X}$ and $\mathcal{Y}$. A simple calculation reveals

$$\rho_{\mathcal{Y}} = \frac{\rho_{\mathcal{X}}}{1-\theta^2} > \rho_{\mathcal{X}}.$$

As a rule we avoid constructions leading to planar codes.

The basic parameters introduced above—dimension n, minimum distance ρ, and size M—are all invariant under distance preserving transformations. We hasten to emphasize, however, that there are also properties of interest that are *not* invariant. Foremost among those is the *alphabet* used to describe the code.

Definition 1.1.2 *Let $\mathcal{X}$ be a spherical code in Ω_N, where N is a composite integer, $N = km$. A set $\mathcal{L} = \{a_1, a_2, \ldots, a_q\} \subseteq \mathbb{R}^m$ is an* alphabet *for $\mathcal{X}$ if $\mathcal{X} \subseteq \mathcal{L}^k \subseteq \mathbb{R}^N$.*

Equivalent codes can have very different alphabets. For reasons of implementation it is usually desirable to have an alphabet of as small a size as possible. Sometimes this requires describing the code with an excessive number of coordinates. It is a challenging problem—about which very little is known—to find the smallest possible alphabet by which a given code can be generated.

1.2 Examples of spherical codes

Spherical codes can be constructed in many different ways. Before going into any detailed discussion regarding possibilities and limitations it might be useful to consider a few illustrative examples. The presentation is tutorial at this point and by no means complete. More details will be offered in subsequent chapters.

Example 1.2.1: Let $\mathcal{A}$ be a binary code with parameters (n, d, M). Consider the mapping $\varphi : \{0,1\} \to \mathbb{R}$ defined by $\varphi(\theta) = (-1)^\theta$; $\theta = 0, 1$. By applying this mapping on each coordinate any codeword $u = (u_1, u_2, \ldots, u_n) \in \mathcal{A}$ is mapped on a vector $x = (x_1, x_2, \ldots, x_n) \in \mathbb{R}^n$ such that the norm $\|x\|$ satisfies $E \triangleq \|x\|^2 = n$. The code $\mathcal{A} \subseteq \mathbb{F}_2^n$ is mapped on a code $\mathcal{X} \subseteq \mathbb{R}^n$ with minimum *squared Euclidean* distance $d_E^2 = 4d$. The quantity E is called *energy*.

The codewords in $\mathcal{X}$ are not located on the unit sphere in $\mathbb{R}^n$ but on a sphere with squared radius n. By proper normalization we obtain a spherical code with parameters $(n, \rho, M) = (n, 4d/n, M)$. We notice that the spherical code is generated by the binary alphabet $\mathcal{L} = \{\pm 1/\sqrt{n}\}$.

For a numerical example take $\mathcal{A}$ to be the simple parity check code with parameters $(n, d, M) = (5, 2, 16)$. The corresponding spherical code has parameters $(n, \rho, M) = (5, 8/5, 16)$. □

Example 1.2.2: Let $\mathcal{C}$ be a binary constant weight code with parameters (N, w, d, M), where N is the length, w is the weight, d is the minimum distance, and M is the size. Let a mapping from $\{0,1\}$ to $\mathbb{R}$ be defined by $\varphi(\theta) = w - \theta N$. Applying this mapping componentwise to the codewords $u = (u_1, u_2, \ldots, u_N)$ in $\mathcal{C}$ we obtain vectors $x = (x_1, x_2, \ldots, x_N)$ in $\mathbb{R}^N$ satisfying the following equations:

$$x_1 + x_2 + \cdots + x_N = 0,$$

$$x_1^2 + x_2^2 + \cdots + x_N^2 = Nw(N - w).$$

The first identity implies that the code is located in an $(N-1)$-dimensional subspace of $\mathbb{R}^N$. The second identity says that all codewords have the same norm. The squared minimum distance is $d_E^2 = N^2 d$. By proper normalization we obtain a spherical code with parameters

$$(n, \rho, M) = (N - 1, \frac{Nd}{Nw - w^2}, M).$$

In principle the dimension could of course be smaller than $N - 1$, but for codes of interest that is usually not the case.

This construction also employs a binary alphabet, but in this case it is non-symmetric: $\mathcal{L} = \{w, w - N\}$ (for the non-normalized code).

For a *numerical* example, let the constant weight code C have parameters $(N, w, d, M) = (10, 4, 4, 30)$, (this corresponds to a Steiner system $S(10, 4, 3)$, see Appendix A). The resulting spherical code has parameters $(n, \rho, M) = (9, 5/3, 30)$. □

Example 1.2.3: Consider the binary vector $(1,1,0,1,0,0,0)$. Taking all cyclic shifts of this vector we obtain a constant weight binary code with parameters $(n, w, d, M) = (7, 3, 4, 7)$, (corresponding to the Steiner system $S(7, 3, 2)$). It is known as the *Simplex* code of length 7. It can also be characterized as the codewords of weight 3 in the binary Hamming code of length $n = 7$. If we interpret the zeros and ones as real numbers, and if we also take all possible sign changes, we get a (non-normalized) code with parameters $n = 7$, $E = 3$, $d_E^2 = 4$, $M = 7{\cdot}8 = 56$. We can characterize it symbolically using the notation

$$\mathcal{X} = (\pm1, \pm1, 0, \pm1, 0, 0, 0)^C,$$

where " $\pm$ " indicates that we should take all possible combinations of signs, and where the exponent C indicates that we should take all possible cyclic shifts. The alphabet (for the non-normalized code) is $\mathcal{L} = \{0, \pm1\}$. The *normalized* code has parameters $(n, \rho, M) = (7, 4/3, 56)$. □

Example 1.2.4: Let $\mathcal{X}$ be the spherical code obtained from a binary code with parameters $(n, d, M) = (8, 2, 128)$ using the componentwise antipodal mapping described in Example 1.2.1. The code $\mathcal{X}$ has parameters $(n, \rho, M) = (8, 1, 128)$. Let $\mathcal{Y}$ be the spherical code obtained from a binary constant weight code with parameters $(N, w, d, M) = (8, 2, 2, 28)$ using the construction described in Example 1.2.3. The spherical code has parameters $(n, \rho, M) = (8, 1, 112)$. Symbolically we may write

$$\mathcal{X} = \frac{1}{2\sqrt{2}}(\pm1^7, \mp1), \qquad \mathcal{Y} = \frac{1}{\sqrt{2}}(0^6, \pm1^2)^S,$$

where the exponents inside the parentheses indicate the number of occurrences of the corresponding symbols, the " $\pm$ " indicates that we should take all possible sign changes, and the exponent S outside the second parenthesis indicates that we should take all possible permutations (S stands for "symmetric group"). In the words of the code $\mathcal{X}$ the seven first signs can be selected arbitrarily, while the last sign must always be chosen so as to maintain even parity in the number of minus signs. This is indicated by the sign "$\mp$".

We notice that the first spherical code is binary with alphabet $\mathcal{L} = \{\pm 1/(2\sqrt{2})\}$, while the second is a ternary code with alphabet $\mathcal{L} = \{0, \pm 1/\sqrt{2}\}$. It follows that any pair $(x, y) \in \mathcal{X} \times \mathcal{Y}$ will satisfy $\|x - y\|^2 = 1$. This in turn means that the union $\mathcal{Z} = \mathcal{X} \cup \mathcal{Y}$ has the same minimum distance as $\mathcal{X}$ and $\mathcal{Y}$ individually and that $\mathcal{Z}$ is a spherical code with parameters $(n, \rho, M) = (8, 1, 240)$. □

Example 1.2.5: Consider Figure 1.1. It displays seven points in Euclidean space $\mathbb{R}^2$ in the form of a regular hexagon and its center. It also indicates a possible partition in three subsets, which we indicate in an obvious way by the symbols " $\cdot$ ", "$\triangle$" and "∇". Suppose the hexagon is centered at the origin. Using a symbolism of the same kind as in the previous examples we can define a code $\mathcal{X}$ by the formula

$$\mathcal{X} = (\cdot, \triangle, \nabla)^C,$$

where the triangles indicate that all different combinations of letters from the two triangular configurations should be employed and where the exponent C indicates that all cyclic shifts should be taken. Clearly in this way we obtain $3^3 = 27$ points in Euclidean 6-space $\mathbb{R}^6$. By proper normalization this transforms into a spherical code with parameters $(n, \rho, M) = (6, 3/2, 27)$. Our description uses an alphabet $\mathcal{L}$ of size 7 (the hexagon and its center). The convex set spanned by the codewords is sometimes called the *Hessian polytope*. □

Example 1.2.6: The integer solutions to the equation

$$u + v + x + y \equiv 0 \mod 2$$

form a *lattice* in $\mathbb{R}^4$ called D_4. If (u, v, x, y) is a solution then the sum of squares $u^2 + v^2 + x^2 + y^2$ is even. Conversely, for each even number e there is a *set* of solutions in D_4 such that the sum of the squares

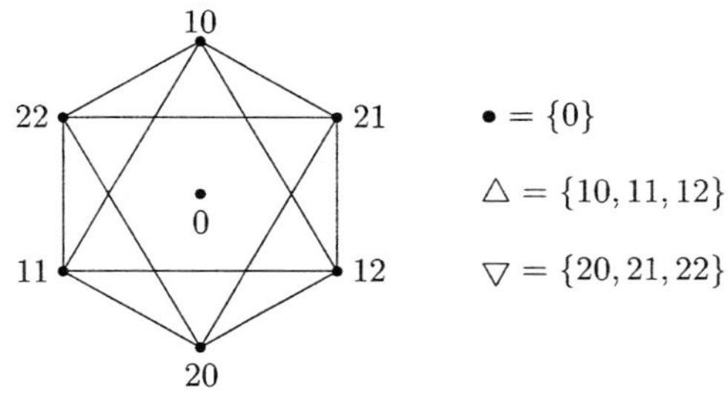

Figure 1.1: Construction of the Hessian polytope.

equals e. Such a set of solutions is called a *shell*. Any shell defines—upon proper normalization—a spherical code. Let us consider the first two of these codes.

Solutions with $e = 2$ have the form $(\pm 1^2, 0^2)^S$. This gives us a spherical code $\mathcal{X}$ with parameters $(n, \rho, M) = (4, 1, 24)$. Solutions with $e = 4$ have either the form $(\pm 1^4)$ or the form $(\pm 2, 0^3)^S$. There are 16 solutions of the first kind and 8 of the second. Together they form again a spherical code with the parameters $(n, \rho, M) = (4, 1, 24)$.

It can be shown that these two codes—obtained from the first two shells in the D_4 lattice—are equivalent. Moreover, taking the *union* of them we get a code with parameters $(n, \rho, M) = (4, 2(1 - 1/\sqrt{2}), 48)$. □

Example 1.2.7: Choose $\sigma = 1/2 - \sqrt{5}/2$ and consider the code

$$\mathcal{X} = (0, \pm\sigma, \pm 1)^C.$$

It has 12 codewords located at the corners of a regular icosahedron, see Figure 1.2. Each codeword has squared norm $1 + \sigma^2$ and the distance between adjacent codewords is $4\sigma^2$. The normalized spherical code has minimum distance $\rho = 4\sigma^2/(1 + \sigma^2) = 2 - 2/\sqrt{5} \approx 1.1056$, and hence parameters $(n, \rho, M) = (3, 2 - 2/\sqrt{5}, 12)$. □

Example 1.2.8: As a final example we consider the following three codes:

$$\mathcal{X} = \frac{1}{2}(\pm 2, 0^3)^C, \qquad \mathcal{Y} = \frac{1}{2}(\pm 1^4), \qquad \mathcal{Z} = \frac{1}{2}(0, \pm 1, \pm \sigma, \pm \tau)^A,$$

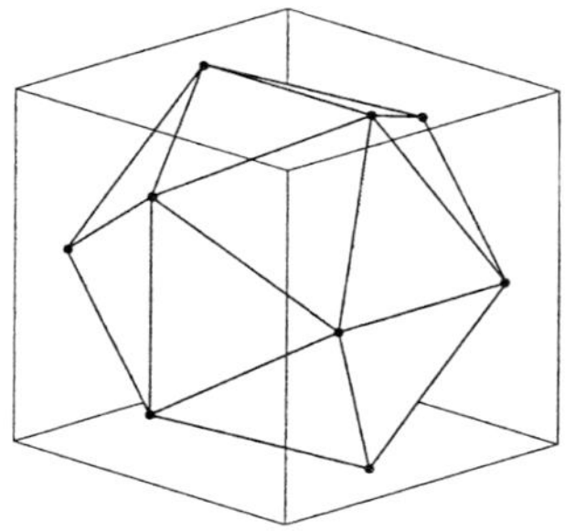

Figure 1.2: The icosahedron. The edges of the cube are parallel to the axis in the coordinate system used in the description suggested in Example 1.2.7.

where σ is defined as in Example 1.2.7 and where $\tau = 1/2 + \sqrt{5}/2$. The exponent A indicates that all *even* permutations of coordinates should be taken (the exponent A stands for the *alternant group*). We obtain codes with parameters $(n, \rho, M) = (4, 2, 8)$, $(4, 1/2, 16)$, $(4, (3-\sqrt{5})/2, 96)$ respectively. Taking the union of the three codes we get a spherical code with parameters $(n, \rho, M) = (4, (3-\sqrt{5})/2, 120)$. The four-dimensional polytope spanned by this code is known as 600-*cell*, see Coxeter, [75]. □

Three main principles have been employed in the above examples:

- *Code concatenation:* In several of our examples the spherical codes were obtained by coordinate-wise mappings of codes defined in various Hamming spaces. We will refer to constructions of this kind as *code concatenations.* The mapping—as we saw—can be obtained in many different ways. A main problem is to find efficient such mappings. The process of representing information in Euclidean space is often referred to as *modulation.* When codes are involved one often talks about *coded modulation.*

- *Lattices:* In Example 1.2.6 we pointed out the obvious fact that spherical codes can be obtained from the shells in a lattice. Good codes are often obtained from *unions of shells.*

- *Direct constructions:* In two cases (Examples 1.2.7 and 1.2.8) we constructed our spherical codes directly, without relying upon any code in Hamming space or any lattice. A few such constructions are available, most of them employing various results and ideas from *group theory* and from the theory of *association schemes.*

Of course there are no strict borders between these methods, and the classification is somewhat arbitrary. Often one and the same code can be constructed in many different ways and often the best results are obtained by combining various constructions. Nevertheless, in our subsequent presentation we will largely follow the classification indicated above and discuss in separate chapters constructions based on concatenation, codes obtained from lattices and codes obtained by special direct constructions.

The simplest description of a code is often obtained in a scale where the code is not located on the *unit* sphere but on some other sphere. In the sequel we shall often take advantage of this fact and describe codes in the scale we find most convenient. A code such that all codewords have the same norm is called a *constant energy code.* A *spherical* code is the special case when the norm is *unity.*

It turns out that it is often convenient to specify first an *alphabet* as a subset in $\mathbb{R}$ or $\mathbb{R}^2$ (sometimes some even larger subspace of $\mathbb{R}^n$) and then characterize the code in terms of this alphabet. In this way we usually obtain a constant energy code for which we estimate first the minimum squared Euclidean distance d_E^2 and then the squared norm E—usually called the *energy* of the code. We then compute the normalized distance

$$\rho \triangleq d_E^2/E,$$

which of course is the squared minimum distance in the corresponding normalized code. In the end it is always the parameters of the *normalized* code that are of interest.

A small remark is motivated regarding the informal notation we have employed. We have used "$\pm$" to indicate sign-changes and the letters S, A, C for indicating various permutations of coordinates. This notation and variations of it (like the one used in Example 1.2.5) will be used throughout the text whenever applicable. Unfortunately, however, many codes are too complicated to be described in this simple manner. Therefore the notation will mainly be used when discussing simple illustrative examples, as we did above.

1.3 Two basic functions

Let $A_n(\rho)$ denote the largest possible size for a spherical code with dimension n and minimum squared distance at least ρ. Formally we may write

$$A_n(\rho) \triangleq \max\{|\mathcal{X}| : \rho(\mathcal{X}) \geq \rho, \ \dim(\mathcal{X}) = n\},$$

where $\mathcal{X} = \{x_1, x_2, \ldots, x_M\}$ is a spherical code and where $\rho(\mathcal{X})$ denotes the minimum squared Euclidean distance in $\mathcal{X}$.

Although in general this function is not known exactly, there are quite a few things we can say about it. In fact, the study of spherical codes is largely a study of this function. We begin with a few simple basic facts.

Theorem 1.3.1 *Let $n \geq 1$ be fixed. The function $A_n(\rho)$ is monotonically non-increasing as a function of the parameter ρ, $0 < \rho \leq 4$. Moreover, it is a step function, everywhere continuous from left.*

Proof: Only the statement on left continuity requires an argument. Consider the function $f : \Omega_n^M \longrightarrow \mathbb{R}$ defined by

$$f(x_1, x_2, \ldots, x_M) \triangleq \min_{1 \leq i < j \leq M} ||x_i - x_j||^2 .$$

It is continuous and defined over a compact domain. As a consequence it has a maximum. The maximum is the largest ρ such that $A_n(\rho) = M$. It follows that $A_n(\rho)$ is left-continuous. □

The maximum of the function f denotes the largest possible squared minimum distance for any spherical code $\mathcal{X}$ of dimension n and size M. We denote this quantity by $\rho_n(M)$. Formally we have

$$\rho_n(M) = \max\{\rho(\mathcal{X}) : |\mathcal{X}| \geq M, \ \dim(\mathcal{X}) = n\}.$$

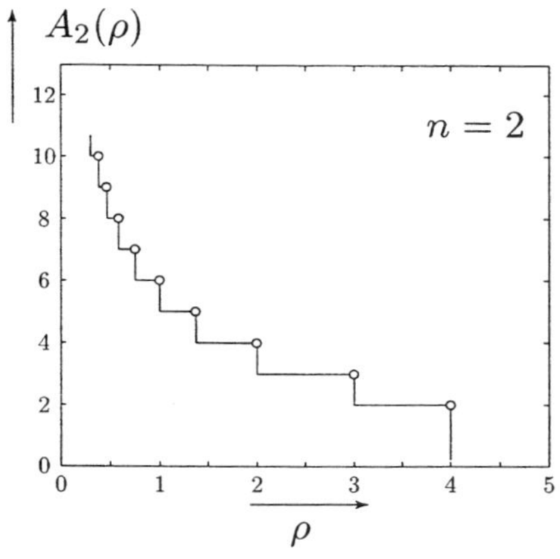

Figure 1.3: The function $A_n(\rho)$ in the cases $n = 2, 3$. Optimal codes are indicated by small circles.

The functions $A_n(\rho)$ and $\rho_n(M)$ are each other's inverses. In principle it would be enough to study just one of them, but it turns out to be convenient to use both. We will freely use one or the other depending on what appears as most convenient at a given occasion.

Dimensions $n = 1$ and $n = 2$ are trivial. For $n = 1$ the space Ω_n contains only two points, $\Omega_1 = \{\pm 1\}$, so we have

$$\begin{aligned} A_1(\rho) &= 2, \quad 0 < \rho \leq 4, \\ \rho_1(M) &= 4, \quad M = 2, \end{aligned}$$

while $\rho_1(M)$ is undefined for $M > 2$. For $n = 2$ we have

$$\rho_2(M) = 2 - 2\cos(2\pi/M) = 4\sin^2(\pi/M), \qquad M = 2, 3, 4, \ldots .$$

In terms of the inverse function $A_n(\rho)$ the same result takes the form

$$A_2(\rho) = M, \quad \rho_2(M+1) < \rho \leq \rho_2(M), \quad M = 2, 3, 4, \ldots ,$$

or

$$A_2(\rho) = \left\lfloor \frac{\pi}{\arcsin\sqrt{\rho/4}} \right\rfloor, \quad 0 < \rho \leq 4,$$

where $\lfloor x \rfloor$ denotes the integer part of x.

The function $A_n(\rho)$ is displayed in Figure 1.3 for the cases $n = 2, 3$. A detailed discussion of the case $n = 3$ will appear in Chapter 3. For the moment we will only draw attention to the interesting fact that the inverse function $\rho_3(M)$ takes the same value for $M = 5$ and $M = 6$: in both cases we have $\rho_3(M) = 2$. The geometrical meaning is the remarkable fact that the largest minimum distance that can be achieved for 5 points on a 3-dimensional sphere is no larger than the largest minimum distance that can be established among 6 points. Although relatively rare, similar phenomena occur in any dimension $n \geq 3$.

We define a spherical code $\mathcal{X} \subseteq \Omega_n$ to be *optimal* if its parameters (n, ρ, M) are such that *both* of the conditions $M = A_n(\rho)$, and $\rho = \rho_n(M)$ are satisfied. As a consequence there is no optimal code of size $M = 5$ in dimension $n = 3$.

For dimensions n larger than 2 the functions $A_n(\rho)$ and $\rho_n(M)$ are only partially known. In short it can be said that they are known only in the interval $2 \leq \rho \leq 4$, while for the interval $0 < \rho < 2$ exact values are available only in a very limited number of special cases. We proceed to derive the exact results for the interval $2 \leq \rho \leq 4$.

1.4 The Rankin bounds

Any explicit code offers a lower bound on the functions $A_n(\rho)$ and $\rho_n(M)$, while an upper bound requires a different kind of argument. We shall determine the functions $A_n(\rho)$ and $\rho_n(M)$ in the interval $2 \leq \rho \leq 4$ by first deriving three upper bounds originally obtained by Rankin [183], and then demonstrating that the bounds are actually tight by giving some very simple explicit codes that satisfy the bounds with equality.

Theorem 1.4.1 (Rankin's first bound) *Any spherical code $\mathcal{X}$ has parameters (n, ρ, M) satisfying the following inequality*

$$\rho \leq \frac{2M}{M-1}.$$

Remark 1.4.1: Notice that the bound does not depend on the dimension n.

Remark 1.4.2: Alternatively the result can also be stated as an inequality for the function $\rho_n(M)$, namely

$$\rho_n(M) \leq \frac{2M}{M-1}.$$

In terms of the inverse function the same result takes the following form

$$A_n(\rho) \leq \left\lfloor \frac{\rho}{\rho - 2} \right\rfloor, \quad 2 < \rho \leq 4.$$

Proof: Let $\mathcal{X} = \{x_1, x_2, \ldots, x_M\}$ be a spherical code with parameters (n, ρ, M). For $i \neq j$ we have $\langle x_i, x_j \rangle \leq 1 - \rho/2$. From this observation we obtain.

$$\sum_{i,j} \langle x_i, x_j \rangle \leq M + M(M-1)(1 - \rho/2).$$

On the other hand we also have

$$\begin{aligned}\sum_{i,j} \langle x_i, x_j \rangle &= \sum_{i,j} \sum_{k=1}^{n} x_{ik} x_{jk} \\ &= \sum_{k=1}^{n} \sum_{i,j} x_{ik} x_{jk} = \sum_{k=1}^{n} (\sum_{i=1}^{M} x_{ik})^2 \geq 0.\end{aligned}$$

Combining these two inequalities we obtain the result by a simple reordering of the terms. □

Theorem 1.4.2 (Rankin's second bound) *The size $A_n(\rho)$ of a largest possible spherical code with squared minimum distance ρ in the unit Euclidean sphere Ω_n satisfies the following inequality:*

$$A_n(\rho) \leq n + 1, \quad 2 < \rho \leq 4.$$

Proof: Again let $\mathcal{X} = \{x_1, x_2, \ldots, x_M\}$ be a spherical code with parameters (n, ρ, M). Notice that the assumption $2 < \rho \leq 4$ is equivalent with the condition $\langle x_i, x_j \rangle < 0$, $i \neq j$, $i, j, = 1, 2, \ldots, M$. In particular we have $\langle x_i, x_M \rangle < 0$, $i = 1, 2, \ldots, M-1$. But we also have $\langle x_i, x_M \rangle > -1$, $i = 1, 2, \ldots, M-1$. For suppose this last inequality is violated for some index, say for $i = M-1$. Then we would have $x_{M-1} = -x_M$, and consequently for any i, $1 \leq i \leq M-2$, we would have $\langle x_i, x_{M-1} \rangle = -\langle x_i, x_M \rangle > 0$, contrary to the original assumption. Thus we have

$$\gamma_i \triangleq 1 - \langle x_i, x_M \rangle^2 > 0, \qquad i = 1, 2, \ldots, M-1,$$

and therefore the following definition is valid:

$$y_i \triangleq \sqrt{\frac{1}{\gamma_i}}\,(x_i - \langle x_i, x_M \rangle x_M), \qquad i = 1, 2, \ldots, M-1.$$

The following relations follow:

$$\begin{aligned}\sqrt{\gamma_i \gamma_j}\,\langle y_i, y_j \rangle &= \langle x_i - \langle x_i, x_M \rangle x_M, x_j - \langle x_j, x_M \rangle x_M \rangle = \\ &= \langle x_i, x_j \rangle - \langle x_i, x_M \rangle \langle x_j, x_M \rangle, \quad 1 \leq i, j \leq M-1.\end{aligned}$$

For $i = j$ this gives us $\|y_i\| = 1$, while for $i \neq j$ we obtain $\langle y_i, y_j \rangle < 0$, $1 \leq i, j \leq M-1$. We notice that the points y_i are all located in the hyperspace orthogonal to x_M. It follows that $\mathcal{Y} = \{y_1, y_2, \ldots, y_{M-1}\}$ is a spherical code with parameters $(n_\mathcal{Y}, \rho_\mathcal{Y}, M_\mathcal{Y})$ satisfying $n_\mathcal{Y} = n-1$, $\rho_\mathcal{Y} > 2$, $M_\mathcal{Y} = M-1$.

As $\rho_\mathcal{Y} > 2$ the procedure can be repeated. That means that we can generate a sequence $\mathcal{X}_n, \mathcal{X}_{n-1}, \ldots, \mathcal{X}_1$ of codes with $\mathcal{X}_n = \mathcal{X}$, $\mathcal{X}_{n-1} = \mathcal{Y}$, and such that the code $\mathcal{X}_k$ has parameters (n_k, ρ_k, M_k) satisfying

$$n_k = k, \quad \rho_k > 2, \quad M_k = M - n + k, \quad k = 1, 2, \ldots, n.$$

For $k = 1$ we obtain $M_1 = M - n + 1 \leq 2$, which gives us $M \leq n+1$. $\square$

Theorem 1.4.3 (Rankin's third bound) *The largest possible size $A_n(\rho)$ of a spherical code with squared minimum distance $\rho = 2$ in the sphere Ω_n satisfies the following bound:*

$$A_n(2) \leq 2n.$$

Proof: The proof is just a minor modification of the proof of Theorem 1.4.2. In the present case the possibility $\langle x_i, x_j \rangle = -1$ is *not* excluded, with the result that the sizes of successive codes in the sequence $\mathcal{X}_n, \mathcal{X}_{n-1}, \ldots, \mathcal{X}_1$ decrease with either 1 (as in the previous case) or by 2. Consequently the estimate for M_k must be modified to the inequality

$$M_k \geq M - 2n + 2k.$$

For $k = 1$ we get $2 \geq M_1 \geq M - 2n + 2$, which gives us $M \leq 2n$. □

Theorem 1.4.1 is the counterpart for spherical codes to the well known *Plotkin* bound for codes in Hamming spaces. Indeed, denote by $A_n^{(2)}(d)$ the maximum size of a code with minimum Hamming distance d in the binary space $\mathbb{F}_2^n$. Using the binary antipodal construction described in Example 1.2.1 we obtain from Theorem 1.4.1:

$$A_n^{(2)}(d) \leq A_n(4d/n) \leq \left\lfloor \frac{2d}{2d-n} \right\rfloor, \qquad n < 2d \leq 2n,$$

which is just slightly weaker than the Plotkin bound ([167], p. 41). With Theorem 1.4.3 the same construction gives us

$$A_n^{(2)}(n/2) \leq 2n; \quad n \text{ even}.$$

Theorems 1.4.1 and 1.4.2 offer different bounds in the same domain, while Theorem 1.4.3 gives a bound at a point just outside the domain of applicability of the two other bounds. It should be mentioned that Theorem 1.4.3 (Rankin's third bound) was known already in 1951 by Davenport–Hajos [78], see also [92], [2], [188].

We conclude this section by noticing that Theorems 1.4.1 and 1.4.2 can be jointly expressed in the following form:

$$A_n(\rho) \leq \min \left\{ \left\lfloor \frac{\rho}{\rho - 2} \right\rfloor, \; n+1 \right\}, \qquad 2 < \rho \leq 4.$$

1.5 The Simplex and the Biorthogonal codes

There are two very simple constructions which show that the Rankin bounds are actually *tight* in the range $2 \leq \rho \leq 4$. The first is the

Simplex code $\mathcal{S}_n$, which in our informal notation can be defined as

$$\mathcal{S}_n \triangleq (1^n, -n)^C.$$

Notice that with this definition each codeword $x = (x_1, x_2, \ldots, x_n)$ satisfies

$$x_1 + x_2 + \cdots + x_{n+1} = 0; \quad x_1^2 + x_2^2 + \cdots + x_{n+1}^2 = n + n^2.$$

The first identity shows that the code is described in an excessive number of coordinates and the second one shows that the code has constant energy. It is easy to see that the dimension of the code is n. The normalized code has parameters $(n, \rho, M) = (n, 2 + 2/n, n + 1)$. We will use the notation $\mathcal{S}_n$ and we will refer to the code as the *Simplex* code whether it is normalized or not. We notice that the code has the property that any pair of distinct codewords have the same squared distance $\rho = 2 + 2/n$.

Now suppose n and ρ are given. All *Simplex* codes $\mathcal{S}_i$ with $i \leq n$ and such that $2+2/i \geq \rho$ generate lower bounds on $A_n(\rho)$. For $2 < \rho \leq 4$ we get

$$A_n(\rho) \geq \max\{i + 1 : i \leq n,\ 2 + 2/i \geq \rho\} = \min\left\{\left\lfloor \frac{\rho}{\rho - 2} \right\rfloor, n + 1\right\}.$$

This means that the combination of Rankin's first and second bounds actually gives the optimum performance.

We noticed in Section 1.4 that Rankin's first bound does not depend on the dimension n. From the formula above follows that if dimension is not a restriction—i.e. if we consider the problem of finding the best configuration of M points on a sphere in *any* dimension—the solution is given by the *Simplex* code. The solution is unique up to equivalence.

The *Biorthogonal* code $\mathcal{B}_n$ can be defined as follows:

$$\mathcal{B}_n \triangleq (0^{n-1}, \pm 1)^C.$$

This code is already normalized. The parameters are $(n, \rho, M) = (n, 2, 2n)$. A comparison with Rankin's third bound reveals that the

Biorthogonal code satisfies this bound with equality. This shows both that the bound is tight and the code is optimal. Again the code is unique up to equivalence. For each codeword there is one other codeword in antipodal position on squared distance 4; all other codewords are orthogonal to the given one, which means that they are on distance 2.

In summary we have solved the coding problem in the range $2 \leq \rho \leq 4$. The remaining part—the range $0 < \rho < 2$—is considerably more difficult. As we have already mentioned there are only a few isolated cases where optimal solutions are known. A few additional upper bounds on the function $A_n(\rho)$ are known. By far the most important of these is the *Levenshtein* bound. A major part of Chapter 2 will be devoted to that bound, while a much simpler bound—the sphere packing bound—will be stated in the next section. Several chapters will be devoted to various constructions.

1.6 The Chabauty–Shannon–Wyner bound

A general *lower* bound for spherical codes was obtained independently by Chabauty [66], Shannon [203], and Wyner [228]. We will refer to this bound as the Chabauty–Shannon–Wyner bound. It is based on the same kind of argument as the famous Gilbert–Varshamov bound for codes in Hamming space. It is closely related to the sphere packing *upper* bound.

In order to state and prove the Chabauty–Shannon–Wyner bound we need a few geometrical concepts and results regarding spheres in Euclidean n-space. The material is standard and can be found in any textbook, for instance Coxeter [75], see also Appendix B.

Let the dimension satisfy $n \geq 2$. The set of points y in Ω_n located within distance $\rho = 2 - 2\cos\beta$ from a fixed point x is called a *spherical cap*:

$$Cap_n(x, \beta) \triangleq \{y \in \Omega_n : \langle x, y \rangle > \cos\beta\}.$$

The main thing we need is a formula for the size of a cap as a function of the angle β. It is clear that the size is independent of the point x. It will be denoted $C_n(\beta)$.

By Appendix B we generally have

$$C_n(\beta) \triangleq k_{n-1} \int_0^{\beta} \sin^{n-2} \alpha \, d\alpha.$$

As a special case we have

$$|\Omega_n| = C_n(\pi),$$

where $|\Omega_n|$ denotes the Lebesgue measure of the unit sphere in $\mathbb{R}^n$.

Denote by $S_n(x,t)$ the open sphere of radius t around x:

$$S_n(x,t) \triangleq \{y \in \mathbb{R}^n : \|x-y\| < t\}.$$

Notice that we have $Cap_n(x,\beta) = \Omega_n \cap S_n(x,t)$, with $1-t^2/2 = \cos\beta$. The following fundamental observation is crucial.

Lemma 1.6.1 *For any points x, y in $\mathbb{R}^n$ the following equivalence holds:*

$$\|x-y\| \geq 2t \;\Leftrightarrow\; S_n(x,t) \cap S_n(y,t) = \emptyset.$$

Proof: First suppose $\|x-y\| \geq 2t$ and let z be any point in $S_n(x,t)$. By the triangle inequality we have

$$\|y-z\| \geq \|x-y\| - \|x-z\| > t,$$

which means that z is outside $S_n(y,t)$. In the same way it follows that any point in $S_n(y,t)$ is outside $S_n(x,t)$. It follows that $S_n(x,t)$ and $S_n(y,t)$ are disjoint.

Next suppose that $S_n(x,t)$ and $S_n(y,t)$ are disjoint. Let z be of the form $z = \theta x + (1-\theta)y$, where θ is chosen so that $\|x-z\| = t$. We have

$$\|x-y\| = \|x-z\| + \|z-y\|,$$

and as $S_n(x,t)$ and $S_n(y,t)$ are disjoint we also have $\|z-y\| \geq t$. The inequality $\|x-y\| \geq 2t$ follows. □

An immediate consequence of this lemma is the so called sphere packing bound. It can be stated as follows.

Theorem 1.6.1 *Let $\mathcal{X}$ be any spherical code with parameters (n, ρ, M). Then the following inequality holds*

$$M \leq A_{SP}^{(n)}(\rho) \triangleq \left\lfloor \frac{C_n(\pi)}{C_n(\theta/2)} \right\rfloor,$$

where $\rho = 2 - 2\cos\theta$.

Proof: By Lemma 1.6.1 the codewords in $\mathcal{X}$ can be surrounded by disjoint caps with angle $\theta/2$. This fact implies $C_n(\pi) \geq M \cdot C_n(\theta/2)$. The final result follows by observing that M is an integer. □

A spherical code with the property that it cannot be extended without decreasing the minimum distance is said to be *maximal*. Any maximal code $\mathcal{X}$ with minimal squared Euclidean distance $\rho = 2-2\cos\theta$ must satisfy the relation

$$\bigcup_{x \in \mathcal{X}} \overline{Cap\,(x, \theta)} = \Omega_n,$$

where the overbar indicates set closure. By the sphere packing bound it is clear that maximal codes with parameters (n, ρ) exist for any dimension $n \geq 2$ and any value of the squared minimum distance ρ. The Chabauty–Shannon–Wyner bound is a simple consequence of this fact.

Theorem 1.6.2 (Chabauty–Shannon–Wyner bound) *There exists, for any $n \geq 2$ and for any $\rho \leq 2$, a code $\mathcal{X}$ in Ω_n with parameters (n, ρ, M) satisfying the inequality*

$$M \geq A_{CSW}^{(n)}(\rho) \triangleq \left\lceil \frac{C_n(\pi)}{C_n(\theta)} \right\rceil,$$

where $\rho = 2 - 2\cos\theta$.

Proof: Let $\mathcal{X}$ be a code with parameters (n, ρ, M) and suppose it is maximal. By the union bound we have

$$|\Omega_n| \leq \sum_{x \in \mathcal{X}} |\overline{Cap\,(x, \theta)}| = M \cdot C_n(\theta).$$

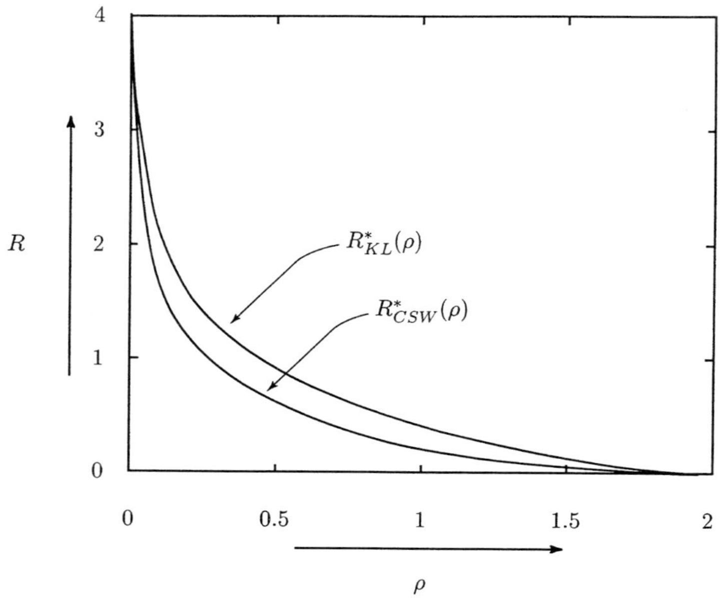

Figure 1.4: The asymptotic lower bound by Chabauty–Shannon–Wyner along with the asymptotic upper bound by Kabatianskii–Levenshtein.

The final result follows by observing that M is an integer. □

Remark 1.3: Notice that both bounds can be evaluated without knowing the coefficients k_n.

In small dimensions the Chabauty–Shannon–Wyner bound is usually superseded by various more explicit constructions. However, for large dimensions it offers the best bound known. We proceed to derive the asymptotic form.

Let us define

$$R^*(\rho) \triangleq \limsup_{n\to\infty} \frac{1}{n} \log A_n(\rho).$$

From the Chabauty–Shannon–Wyner bound we obtain, using standard arguments for exponential sums, the following asymptotic bound:

$$R^*(\rho) \geq R^*_{CSW}(\rho),$$

where

$$R^*_{CSW}(\rho) \triangleq \limsup_{n\to\infty} \frac{1}{n} \log A^{(n)}_{CSW}(\rho) = \begin{cases} -\log \sin\theta, & 0 < \rho < 2, \\ 0, & 2 \le \rho \le 4, \end{cases}$$

and where $\rho = 2 - 2\cos\theta$. The limit is obtained by observing that for large n both integrals in the formula for $A^{(n)}_{CSW}(\rho)$ are dominated by the largest parts of the integrands. Therefore the quotient between the integrals is—for large n—approximated by the quotient between the maximal values of the two integrands. Clearly, for θ smaller than $\pi/2$ the maximum of the denominator occurs at $\alpha = \theta$, while for θ larger than $\pi/2$ the maximum occurs at $\alpha = \pi/2$. As a consequence the normalized logarithm is zero in the angular range $\pi/2 \le \theta \le \pi$, which in terms of squared Euclidean distance reads $2 \le \rho \le 4$.

The fact that $R^*(\rho)$ is positive in the range $0 < \rho < 2$ means that the size M of the best spherical codes increases exponentially with the dimension n for fixed normalized distance ρ. It follows from the Rankin bounds that there is no exponential increase in the range $2 \le \rho \le 4$.

The *Levenshtein* bound [156], [157]—to be discussed in the next chapter—provides the best general *upper* bound for spherical codes of finite length. The best known *asymptotic* form is known as the Kabatianskii–Levenshtein bound [140]. It is given by the following expression:

$$R^*(\rho) \le R^*_{KL}(\rho),$$

where

$$R^*_{KL}(\rho) \triangleq \frac{1+\sin\theta}{2\sin\theta} \log \frac{1+\sin\theta}{2\sin\theta} - \frac{1-\sin\theta}{2\sin\theta} \log \frac{1-\sin\theta}{2\sin\theta},$$

and where

$$\rho = 2 - 2\cos\theta, \quad 0 < \theta \le \pi/2.$$

The Kabatianskii–Levenshtein bound is displayed in Figure 1.4, along with the Chabauty–Shannon–Wyner bound.

1.7 The direct sum

Let $\mathcal{X}$ and $\mathcal{Y}$ be two spherical codes with parameters $(n_\mathcal{X}, \rho_\mathcal{X}, M_\mathcal{X})$ and $(n_\mathcal{Y}, \rho_\mathcal{Y}, M_\mathcal{Y})$ respectively. We define the *direct sum* $\mathcal{Z} = \mathcal{X} \oplus \mathcal{Y}$ as the set of all vectors z of the form $z = (x\cos\theta, y\sin\theta)$, where the parameter θ is selected so as to maximize the normalized squared distance $\rho_\mathcal{Z} = \min\{\rho_\mathcal{X}\cos^2\theta, \rho_\mathcal{Y}\sin^2\theta\}$ of the resulting spherical code $\mathcal{Z}$. The maximum occurs if we choose

$$\cos\theta = \sqrt{\frac{\rho_\mathcal{Y}}{\rho_\mathcal{X}+\rho_\mathcal{Y}}}, \qquad \sin\theta = \sqrt{\frac{\rho_\mathcal{X}}{\rho_\mathcal{X}+\rho_\mathcal{Y}}}.$$

Formally we write

$$\mathcal{Z} = \mathcal{X} \oplus \mathcal{Y} \triangleq \left(\mathcal{X}\sqrt{\frac{\rho_\mathcal{Y}}{\rho_\mathcal{X}+\rho_\mathcal{Y}}}, \mathcal{Y}\sqrt{\frac{\rho_\mathcal{X}}{\rho_\mathcal{X}+\rho_\mathcal{Y}}}\right).$$

A simple computation reveals the following relations.

Proposition 1.7.1 *Let $\mathcal{Z} = \mathcal{X} \oplus \mathcal{Y}$ be the direct sum of the spherical codes $\mathcal{X}$ and $\mathcal{Y}$. The parameters of the codes $\mathcal{X}$, $\mathcal{Y}$, $\mathcal{Z}$ are related as follows:*

$$n_\mathcal{Z} = n_\mathcal{X} + n_\mathcal{Y},$$

$$\frac{1}{\rho_\mathcal{Z}} = \frac{1}{\rho_\mathcal{X}} + \frac{1}{\rho_\mathcal{Y}}$$

$$M_\mathcal{Z} = M_\mathcal{X} \cdot M_\mathcal{Y}.$$

The direct sum represents a trivial combination of spherical codes. Using a codeword from a direct sum $\mathcal{X} \oplus \mathcal{X} \oplus \cdots \oplus \mathcal{X}$ is the same thing as using an appropriate sequence of successive codewords from the simple code $\mathcal{X}$. As an example consider the binary antipodal code $\mathcal{X} = \Omega_1$ with parameters $(n, \rho, M) = (1, 4, 2)$. The n-fold direct sum is the hypercube in $\mathbb{R}^n$, which is a code with parameters $(n, \rho, M) = (n, 4/n, 2^n)$. As the dimension increases the size of this code increases exponentially while the normalized distance decreases and ultimately approaches zero.

The same kind of behaviour is obtained also for any other sequence of codes obtained as direct sums of some fixed code. By the Chabauty–Shannon–Wyner bound we know, however, that it is possible to have an *exponential* increase in size while still having the distance staying strictly away from zero. In fact, the distance could approach any value in the interval $0 < \rho < 2$. Thus, the performance of codes generated by direct sums is usually very far from optimal. In a sense this is a main reason why there is a theory of codes: in each dimension new codes have to be considered.

Chapter 2

The linear programming bound

2.1 Introduction

The best upper bounds for codes are usually those obtained from linear programming techniques. The basic idea belongs to Delsarte [79]. The particular case of spherical codes was developed in successive steps by Delsarte–Goethals–Seidel [81], [82], Kabatianskii–Levenshtein [140], Levenshtein [155], [156], [157], Boyvalenkov [42], [43] and Boyvalenkov–Danev–Bumova [45]. The most explicit result, which we shall refer to as the Levenshtein bound, was obtained by Levenshtein [155], [156], [157].

The linear programming bound is largely based on the theory of *orthogonal polynomials*. The bound is expressed in terms of so called *Gegenbauer polynomials*. These polynomials are orthogonal with respect to a certain inner product, which is defined in terms of a weighted integral over the interval $[-1, 1]$. They are also related to so called *spherical polynomials*, which are multi-variate polynomials defined on the *unit sphere* Ω_n. The key feature for our purposes is the fact that the Gegenbauer polynomials are *positive definite*. This property is a simple consequence of the *addition formula*, which is the bridge between Gegenbauer polynomials and spherical polynomials.

In order to derive the linear programming bound we need first to offer precise definitions to the above concepts and to establish a few facts about the polynomials occurring in the theory. All these results are classical, [1], [7], [17], [37], [195], [191].

2.2 Spherical polynomials

Let $f(x)$ be a polynomial of the real variables $x_1, x_2, \ldots, x_n$. We usually collect the variables x_i in a vector $x = (x_1, x_2, \ldots, x_n)$. We will be interested in polynomials evaluated on the unit sphere Ω_n. This means that two polynomials $f(x)$ and $g(x)$ such that $f(x) = g(x)$ for all $x \in \Omega_n$ will be regarded as identical elements. The equivalence class of all polynomials taking the same value for all $x \in \Omega_n$ will be referred to as a *spherical polynomial.* The set of all spherical polynomials over Ω_n will be denoted $Pol(n)$.

For any integer vector $u = (u_1, u_2, \ldots, u_n)$ with nonnegative components u_i the special polynomial x^u is defined as

$$x^u \triangleq \prod_{i=1}^{n} x_i{}^{u_i}.$$

A polynomial of this form is called a *monomial.* The sum $\deg(x^u) \triangleq u_1 + u_2 + \cdots + u_n$ is called the *degree* of the monomial x^u.

A *polynomial* $f(x)$ is a linear combination of monomials:

$$f(x) = \sum_u f_u x^u. \tag{2.2.1}$$

An expression of the form (2.2.1) will also be referred to as a *monomial expansion.* If $f(x)$ is interpreted as a polynomial over $\mathbb{R}^n$ its monomial expansion is unique. In that case the degree of $f(x)$ is defined as the largest of the degrees of the monomials in its monomial expansion. For spherical polynomials, however, the monomial expansion is *not* unique. For a *spherical polynomial* we define the degree as the *smallest* of the degrees in sense of polynomials over $\mathbb{R}^n$ of any of the polynomials in the corresponding equivalence class.

We denote by $Pol(n,k)$ the set of all spherical polynomials $f(x)$ of degree k or less. It is clear that $Pol(n,k)$ is a linear space. Our first concern will be to determine the dimension of this space.

A spherical polynomial $f(x) \in Pol(n,k)$ is said to be *homogeneous* if it can be expressed as a sum of monomials, all of the same degree. We denote by $Hom(n,i)$ the set of all spherical polynomials which can be represented as homogeneous polynomials of degree i.

We notice that $Hom(n,i-2)$ is a subspace in $Hom(n,i)$. Moreover, $Hom(n,i)$ is a subspace in $Pol(n,k)$ for any $i \leq k$. The following lemma holds.

Lemma 2.2.1 $Pol(n,k) = Hom(n,k) \oplus Hom(n,k-1)$.

Proof: Consider the general expansion (2.2.1). Any monomial x^u is an element in $Hom(n,i)$, where $i = \deg(x^u)$. If $i < k-1$, let us replace x^u with $x^u\|x\|^{2s}$, where the nonnegative integer s is selected such that $k-1 \leq 2s+i \leq k$. Regarded as functions over Ω_n the polynomials x^u and $x^u\|x\|^{2s}$ are the same. It follows that it is always possible to express $f(x)$ as a linear combination of elements in $Hom(n,k)$ and $Hom(n,k-1)$. Conversely, any such linear combination is obviously an element in $Pol(n,k)$. Moreover, it is also easy to see that any polynomial $f(x) \in Pol(n,k)$ has a unique representation in the form $f(x) = g(x) + h(x)$, where $g(x) \in Hom(n,k)$ and $h(x) \in Hom(n,k-1)$. This completes the proof. □

It is now easy to determine the dimension of the space $Pol(n,k)$. We first observe that the monomials x^u of degree k are linearly independent and span $Hom(n,k)$. Thus, the dimension of the linear space $Hom(n,k)$ equals the number of vectors $u = (u_1, u_2, \ldots, u_n)$ with nonnegative integer components adding to k. This number is well known and is given by $\binom{n-1+k}{k}$. Combining this observation with Lemma 2.2.1 we obtain

Lemma 2.2.2

$$\dim Pol(n,k) = \binom{n-1+k}{k} + \binom{n-2+k}{k-1}. \tag{2.2.2}$$

For $k \geq 2$ let $\varphi : Hom(n,k) \rightarrow Hom(n,k-2)$ be a surjective homomorphism with kernel $\ker(\varphi)$. Notice that the image $\mathrm{Im}\,(\varphi) = Hom(n,k-2)$ is also a subspace in $Hom(n,k)$. It follows that the kernel is independent of φ and that we always have $Hom(n,k) \cong \ker(\varphi) \oplus Hom(n,k-2)$. It can be shown that the *laplacian operator* is a surjective homomorphism of this kind. Therefore the elements in the kernel are usually referred to as *harmonic polynomials*, and therefore the kernel is usually denoted $Harm(n,k)$. Although the laplacian operator itself plays no role in our theory we will follow the conventional terminology. For $k=0$ we define $Harm(n,0) = <1>$ and for $k=1$ we define $Harm(n,1) = < x_1, x_2, \ldots, x_n >$. We have

$$Hom(n,k) = Harm(n,k) \oplus Hom(n,k-2), \quad k \geq 2. \tag{2.2.3}$$

It follows that any spherical polynomial $f(x) \in Pol(n,k)$ has a unique representation as a sum of harmonic polynomials. We state this important result as a theorem.

Theorem 2.2.1 *The polynomial space $Pol(n,k)$ is a direct sum of the polynomial spaces $Harm(n,i)$, $i = 0,1,\ldots,k$:*

$$Pol(n,k) = \bigoplus_{i=0}^{k} Harm(n,i). \tag{2.2.4}$$

Let r_k denote the dimension of the linear space $Harm(n,k)$. We have $r_0 = 1$ and $r_1 = n$. For $k \geq 2$ we obtain as an immediate consequence of (2.2.3) the following explicit formula:

$$\begin{aligned} r_k &= \binom{n+k-1}{k} - \binom{n+k-3}{k-2} \\ &= \binom{n+k-3}{k-1} \cdot \frac{n+2k-2}{k}, \quad k \geq 2. \end{aligned} \tag{2.2.5}$$

For $f, g \in Pol(n)$ we define the following inner product:

$$(f,g) \triangleq \int_{\Omega_n} f(x) g(x) d\mu(x), \tag{2.2.6}$$

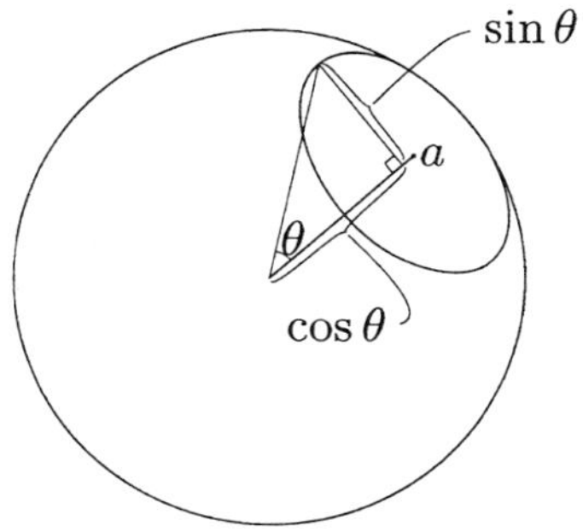

Figure 2.1: An $(n-1)$-dimensional sphere on the surface of an n-dimensional sphere.

where μ is the uniform Lebesgue measure over Ω_n, normalized so that

$$\int_{\Omega_n} d\mu(x) = 1.$$

Let a be a fixed vector on the unit sphere Ω_n and let $F(t)$ be a polynomial of degree k of the real variable t. The fixed vector $a \in \Omega_n$ and the real polynomial $F(t)$ define a spherical polynomial $f(x) \in Pol(n,k)$ according to the formula $f(x) = F(\langle x, a \rangle)$, where $\langle x, a \rangle$ denotes the usual inner product between vectors in $\mathbb{R}^n$. Spherical polynomials of this kind are called *zonal spherical polynomials.* They are characterized by the property of being constant for all $x \in \Omega_n$ such that $\langle x, a \rangle = \cos\theta$, where θ is a fixed angle, $0 \le \theta \le \pi$. We notice that all such vectors are located on an $(n-1)$-dimensional sphere of radius $\sin\theta$ with center in the point $a\cos\theta$, see Figure 2.1.

Let $F(t)$ and $G(t)$ be two polynomials of the real variable t and define $f(x) = F(\langle x, a \rangle)$, $g(x) = G(\langle x, a \rangle)$. Consider the inner product between $f(x)$ and $g(x)$ as defined in (2.2.6). Integrating first over the $(n-1)$-dimensional sphere centered at $a\cos\theta$ and then substituting $\cos\theta$ with the variable t we obtain

$$(f,g) = k_{n-1} \int_{-1}^{1} F(t)G(t)(1-t^2)^{\frac{n-3}{2}}\, dt, \tag{2.2.7}$$

where k_{n-1} is a constant as defined in Appendix B.

The integral in (2.2.7) induces an inner product between the real polynomials $F(t)$ and $G(t)$. We simply define

$$(F, G)_a = (f, g), \tag{2.2.8}$$

where $f(x) = F(\langle x, a \rangle)$, $g(x) = G(\langle x, a \rangle)$. We will see in a moment that this inner product is actually independent of the fixed vector $a \in \Omega_n$. The Gegenbauer polynomials constitute a family of orthogonal polynomials with respect to this inner product.

In Section 2.4 we will see that to any inner product there is a unique sequence of orthogonal polynomials. The Gegenbauer polynomials could be defined directly from this fact. However, for the present discussion we prefer another more explicit definition. The reason is that by this alternative definition we are able to derive very easily one of the fundamental results upon which the linear programming bound is based, namely the so called *addition formula.*

Consider the spaces $Harm(n, i)$. For each i let $\{v_{ij} : 1 \leq j \leq r_i\}$ be an orthonormal basis spanning $Harm(n, i)$. Theorem 2.2.1 implies that the union $\{v_{ij} : 1 \leq j \leq r_i,\ 0 \leq i \leq k\}$ of all these bases is an orthogonal set of polynomials spanning the space $Pol(n, k)$. Thus we have

$$(v_{ij}, v_{lm}) = \int_{\Omega_n} v_{ij}(x) v_{lm}(x) d\mu(x) = \delta_{il}\delta_{jm},$$

where δ_{il} and δ_{jm} are Kronecker symbols. In particular, as $r_0 = 1$ and $v_{01}(x) = 1$ we obtain for any $j = 1, 2, \ldots, r_i$

$$(v_{ij}, v_{01}) = \int_{\Omega_n} v_{ij}(x) d\mu(x) = \begin{cases} 1, & \text{if } \ i = 0, \\ 0, & \text{if } \ i \neq 0. \end{cases} \tag{2.2.9}$$

This implies, among other things, that the integral over Ω_n of any polynomial $f(x) \in Pol(n)$ equals the coefficient f_{01} of the constant polynomial $v_{01}(x) = 1$ in its expansion in the basis $\{v_{ij} : 1 \leq j \leq r_i, 0 \leq i \leq k\}$.

Let the orthonormal basis $\{v_{ij}(x) : 1 \leq j \leq r_i\}$ be given and let $U \in \mathcal{O}_n$ be an arbitrary orthogonal matrix. It is easy to see that the family $\{v_{ij}(xU) : 1 \le j \le r_i\}$ is another orthonormal basis in $Harm(n,i)$. It follows that we have

$$v_{ij}(xU) = \sum_{m=1}^{r_i} v_{im}(x)V_{mj}, \quad j = 1, 2, \ldots, r_i,$$

where $V = [V_{mj}]$ is an $r_i \times r_i$ orthogonal matrix. Using this observation it is easy to show that the sum

$$\sum_{j=1}^{r_i} v_{ij}(xU)v_{ij}(yU) \tag{2.2.10}$$

is independent of the orthogonal matrix U and so depends only on the inner product $\langle x, y\rangle$. Indeed, Ω_n is a so called 2-point homogeneous space. This means that to any two pairs $x, y \in \Omega_n$ and $x', y' \in \Omega_n$ such that $\langle x, y\rangle = \langle x', y'\rangle$ there exists an orthogonal matrix U such that $x' = xU$ and $y' = yU$. Therefore the sum (2.2.10) is uniquely determined by the inner product $\langle x, y\rangle$. We notice that this sum is a polynomial of degree i. This can be seen for instance by choosing $x = (t, \sqrt{1-t^2}, 0, \ldots, 0)$ and $y = (1, 0, 0, \ldots, 0)$ in (2.2.10). Thus the following definition is valid.

Definition 2.2.1 *For* $i = 0, 1, \ldots,$ *let* $\{v_{ij}(x),\ j = 1, 2, \ldots, r_i\}$ *be an orthonormal basis in* $Harm(n,i)$. *We define the polynomials* $Q_i(t)$ *by the formula*

$$Q_i(\langle x, y\rangle) \triangleq \frac{1}{r_i} \sum_{j=1}^{r_i} v_{ij}(x)\, v_{ij}(y). \tag{2.2.11}$$

The real polynomials $Q_i(t)$ so obtained are referred to as *Gegenbauer polynomials*. Notice that they are always defined with respect to a family of spaces $Harm(n,i)$, where the parameter n is fixed while the parameter i runs through all natural numbers, $i = 0, 1, 2, \ldots$. The dependence upon the parameter n is usually suppressed in the notation.

Corresponding definitions can be made in many different polynomial spaces. They are generally referred to as *addition formulas*. The formula (2.2.11) is therefore referred to as the addition formula for the *Gegenbauer polynomials.*

We notice that the polynomials $Q_i(t)$ are subject to the following normalization:

$$Q_i(1) = \int_{\Omega_n} Q_i(1) d\mu(x) = \frac{1}{r_i} \sum_{j=1}^{r_i} \int_{\Omega_n} v_{ij}^2(x) d\mu(x) = 1.$$

Using this fact a simple computation reveals that the following orthogonality condition prevails:

$$(Q_i, Q_k) = \frac{1}{r_i}\, \delta_{ik}. \tag{2.2.12}$$

We notice that $Q_i(t)$ is orthogonal to all polynomials of degree less than i. Therefore the inner product $(t\,Q_i, Q_k) = (Q_i, t\,Q_k)$ is nonzero only if $k-1 \le i \le k+1$. It is also easy to establish the identity $Q_k(-t) = (-1)^k Q_k(t)$, which holds for any $t \in \mathbb{R}$ and any $k \in \mathbb{N}$. We conclude that the polynomials $Q_k(t)$ satisfy a recursion of the form

$$t\,Q_k(t) = a_k\, Q_{k+1}(t) + b_k\, Q_{k-1}(t), \quad k = 1, 2, \dots\ . \tag{2.2.13}$$

Moreover, inserting $t = 1$ and recalling the normalization $Q_k(1) = 1$, we conclude that $a_k + b_k = 1$ holds for all $k \ge 1$. Observing the equalities $Q_0(t) = 1$ and $Q_1(t) = t$ we see that it is natural to define $a_0 = 1$, $b_0 = 0$. For $k \ge 1$ the coefficients a_k and $b_k = 1 - a_k$ can be obtained by the recursion (2.2.13). The following theorem summarizes our observations.

Theorem 2.2.2 *The Gegenbauer polynomials $Q_k(t)$ satisfy the recursion*

$$t\,Q_k(t) = a_k Q_{k+1}(t) + b_k Q_{k-1}(t), \quad k = 1, 2, \dots \tag{2.2.14}$$

where $a_0 = 1$, $b_0 = 0$, and where for $k = 1, 2, \dots$ the coefficients a_k, b_k satisfy

$$(i) \quad a_k + b_k = 1,$$

$$(ii) \quad b_k = \frac{r_{k-1}}{r_k}\, a_{k-1}.$$

Proof: For $k \geq 1$ we have

$$0 = (a_k Q_{k+1}, Q_{k-1}) = (t\, Q_k - b_k Q_{k-1}, Q_{k-1}) =$$
$$= (t\, Q_k, Q_{k-1}) - b_k \|Q_{k-1}\|^2.$$

But we also have

$$(t\, Q_k, Q_{k-1}) = (Q_k, t\, Q_{k-1}) =$$
$$= (Q_k, a_{k-1} Q_k + b_{k-1} Q_{k-2}) = a_{k-1} \|Q_k\|^2.$$

By (2.2.12) we have $\|Q_k\|^2 = 1/r_k$. The theorem follows. □

A straightforward computation gives

$$\frac{r_{k-1}}{r_k} = \frac{k}{n+k-3} \cdot \frac{n+2k-4}{n+2k-2} \, .$$

Inserting this in the formulas given in Theorem 2.2.2 we obtain the following explicit expressions:

$$a_k = \frac{n-2+k}{n-2+2k}, \quad b_k = \frac{k}{n-2+2k}, \quad k = 0, 1, \dots \, . \tag{2.2.15}$$

This, in turn, gives us the following recursion for the Gegenbauer polynomials:

$$Q_0(t) = 1, \quad Q_1(t) = t,$$
$$(k+n-2)\, Q_{k+1}(t) = (2k+n-2)\, t\, Q_k(t) - k\, Q_{k-1}(t), \tag{2.2.16}$$
$$k = 1, 2, \dots \, .$$

The first few polynomials are

$$Q_0(t) = 1,$$

$$Q_1(t) = t,$$

$$Q_2(t) = \frac{nt^2 - 1}{n-1},$$

$$Q_3(t) = \frac{(n+2)t^3 - 3t}{n-1},$$

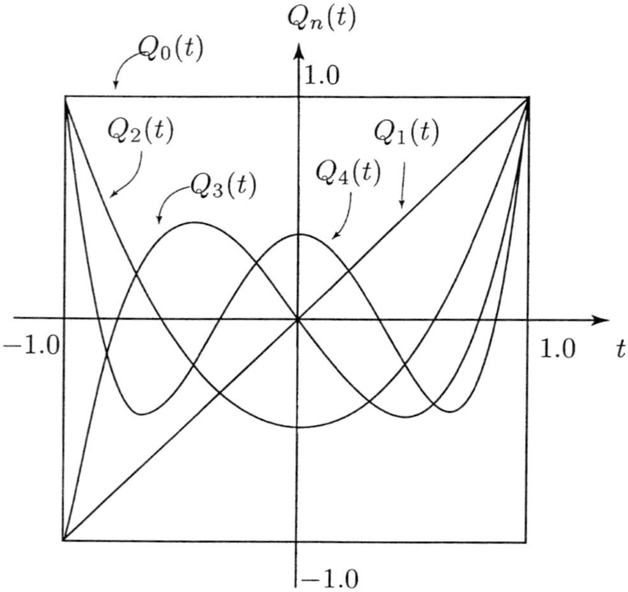

Figure 2.2: The first few Gegenbauer polynomials, $Q_i(t)$, $0 \leq i \leq 4$. The dimension is $n = 3$.

$$Q_4(t) = \frac{(n+2)(n+4)t^4 - 6(n+2)t^2 + 3}{n^2 - 1},$$

$$Q_5(t) = \frac{1}{(n^2-1)}[(n+4)(n+6)t^5 - 10(n+4)t^3 + 15t],$$

$$Q_6(t) = \frac{1}{(n^2-1)(n+3)}[(n+4)(n+6)(n+8)t^6$$

$$-15(n+4)(n+6)t^4 + 45(n+4)t^2 - 15].$$

The first few graphs are illustrated in Figure 2.2. We notice that the i-th polynomial $Q_i(t)$ has precisely i zeros, all of them in the interval $(-1, 1)$.

We are now able to establish the fundamental property upon which the use of the Gegenbauer polynomials in connection with the linear programming bound is crucially dependent, namely the fact that the Gegenbauer polynomials are *positive definite.* We start with a definition.

Definition 2.2.2 *A real-valued function $F(t)$ is positive definite if for any set $\mathcal{X} = \{x_1, x_2, \cdots, x_M\}$ of vectors x_i in Ω_n the $M \times M$ matrix $A(F) = [A_{ij}]$ with elements $A_{ij} = F(\langle x_i, x_j \rangle)$ is positive definite in the usual sense of matrices.*

The definition implies that $F(t)$ is positive definite if for any real constants $\alpha_1, \alpha_2, \ldots, \alpha_M$ and for any set $\mathcal{X} = \{x_1, x_2, \ldots, x_M\}$ of vectors x_i in Ω_n the inequality

$$\sum_{i,j=1}^{M} \alpha_i \alpha_j F(\langle x_i, x_j \rangle) \geq 0$$

holds.

The crucial result we need in connection with the linear programming bound is the following.

Theorem 2.2.3 *All Gegenbauer polynomials are positive definite.*

Proof: The result follows immediately from the definition (2.2.11). With $\alpha = \{\alpha_1, \alpha_2, \ldots, \alpha_M\}$ and $\mathcal{X} = \{x_1, x_2, \ldots, x_M\}$ as above we obtain

$$\begin{aligned} \sum_{i,j=1}^{M} \alpha_i \alpha_j Q_k(\langle x_i, x_j \rangle) &= \sum_{i,j=1}^{M} \alpha_i \alpha_j \frac{1}{r_k} \sum_{\ell=1}^{r_k} v_{k\ell}(x_i) v_{k\ell}(x_j) \\ &= \sum_{\ell=1}^{r_k} \frac{1}{r_k} \sum_{i,j=1}^{M} \alpha_i \alpha_j v_{k\ell}(x_i) v_{k\ell}(x_j) \\ &= \frac{1}{r_k} \sum_{\ell=1}^{r_k} \left(\sum_{i=1}^{M} \alpha_i v_{k\ell}(x_i) \right)^2 \geq 0. \end{aligned}$$

□

Theorem 2.2.4 *A real-valued polynomial*

$$F(t) = F_0 Q_0(t) + F_1 Q_1(t) + \cdots + F_k Q_k(t)$$

is positive definite if and only if all Gegenbauer coefficients F_i are nonnegative.

Proof: It is obvious that $F(t)$ is positive definite if all coefficients F_i are nonnegative. The main content of the theorem is the converse. In order to prove this result we first notice that—by a limiting argument—the function $F(t)$ is positive definite if and only if for any function $a(x) \in Pol(n)$ the following inequality holds:

$$\int\int F(\langle x, y\rangle)\, a(x)\, a(y)\, d\mu(x)\, d\mu(y) \geq 0.$$

Next we notice that F_i can be computed as follows:

$$F_i = r_i(F, Q_i) = r_i \int F(\langle x, y\rangle)\, Q_i(\langle x, y\rangle)\, d\mu(x).$$

Integrating also with respect to the variable y we obtain

$$F_i = \int F_i d\mu(y) = r_i \int\int F(\langle x, y\rangle)\; Q_i(\langle x, y\rangle)\, d\mu(x)\, d\mu(y) =$$

$$= \sum_{j=1}^{r_i} \int\int F(\langle x, y\rangle)\, v_{ij}(x)\, v_{ij}(y)\, d\mu(x)\, d\mu(y) \geq 0.$$

□

Theorem 2.2.5 *The product $F(t)\, G(t)$ of two positive definite functions $F(t)$ and $G(t)$ is positive definite.*

Proof: From the defining formula (2.2.11) it is easily shown that the product $Q_i(t)\, Q_j(t)$ of two Gegenbauer polynomials is positive definite. After this observation Theorem 2.2.5 follows as a simple corollary to Theorem 2.2.4. □

By Theorem 2.2.5 follows that the product $Q_i(t)\, Q_j(t)$ has a Gegenbauer expansion of the form

$$Q_i(t)\, Q_j(t) = \sum_{k=0}^{i+j} p_{ij}^k\, Q_k(t)\,, \tag{2.2.17}$$

where all coefficients p_{ij}^k are nonnegative. In particular we notice the following equalities:

$$p_{ij}^0 = r_0(Q_i\, Q_j, Q_0) = (Q_i, Q_j) = (1/r_i)\, \delta_{ij}\ . \tag{2.2.18}$$

2.3 The linear programming bound

The problem of determining the function $A_n(\rho)$ can be viewed as a maximization problem, where we try to maximize the size $M = |\mathcal{X}|$ over all spherical codes $\mathcal{X}$ with dimension n and minimum squared distance $\rho = 2-2s$. The linear programming bound relates this *maximization* problem to a *minimization* problem involving certain real functions $f(t)$, which we call *test functions*. Any such test function $f(t)$ satisfying certain constraints provides an upper bound on the function $A(n,\rho)$. The precise relation is as follows.

Theorem 2.3.1 *Let $f(t)$ be a polynomial with Gegenbauer expansion $f(t) = f_0 + f_1Q_1(t) + \cdots + f_kQ_k(t)$, where the Gegenbauer polynomials $Q_i(t)$ are defined with respect to the family $Harm(n,i)$, $i = 0,1,2,\ldots$ Suppose the polynomial $f(t)$ satisfies the following conditions:*

(*i*) $f(t) \leq 0, \quad -1 \leq t \leq s,$

(*ii*) $f_0 > 0, \quad f_1, f_2, \ldots, f_k \geq 0.$

Then any spherical code $\mathcal{X} \subset \Omega_n$ with minimal squared distance $\rho = 2-2s$ has size M satisfying the following bound:

$$M \leq \frac{f(1)}{f_0}\,. \tag{2.3.1}$$

Proof: The result follows by evaluating the sum

$$\sum_{x,y\in\mathcal{X}} f(\langle x,y\rangle)$$

in two different ways. We obviously have

$$\sum_{x,y\in\mathcal{X}} f(\langle x,y\rangle) = Mf(1) + \sum_{x\neq y} f(\langle x,y\rangle).$$

This expression is upper bounded by $Mf(1)$, because by assumption we have $f(\langle x,y\rangle) \leq 0$ whenever $x \neq y$; $x,y \in \mathcal{X}$.

From the Gegenbauer expansion using (2.2.11) we obtain

$$\sum_{x,y\in\mathcal{X}} f(\langle x,y\rangle) = M^2 f_0 + \sum_{i=1}^{k} \frac{f_i}{r_i} \sum_{j=1}^{r_i} \left(\sum_{x\in\mathcal{X}} v_{ij}(x) \right)^2 .$$

By assumption we have $f_i \geq 0$ for all $i = 1, 2, \ldots, k$, and by definition $r_i > 0$. Therefore $M^2 f_0$ is a lower bound. This proves the theorem. □

The conditions for equality in (2.3.1) are of interest. Let the inner products in $\mathcal{X}$ be $s_0 > s_1 > \ldots > s_d$, with $s_0 = 1$, $s_1 = s$. Then equality holds if and only if both of the following conditions are satisfied:

$$(i) \quad f(s_i) = 0, \quad i = 1, 2, \ldots, d,$$

$$(ii) \quad f_i \sum_{x\in\mathcal{X}} v_{ij}(x) = 0, \quad j = 1, 2, \ldots, r_i, \ i = 1, 2, \ldots, k.$$

As a first example choose $f(t) = t - s$. The conditions of the theorem are satisfied for $s < 0$. We get

$$A_n(\rho) \leq \frac{1-s}{-s} = \frac{\rho}{\rho - 2}, \quad \rho = 2 - 2s > 2,$$

which we recognize as Rankin's first bound .

As a second example take $f(t) = (t+1)(t-s)$. By (2.2.16) we have

$$Q_2(t) = \frac{nt^2 - 1}{n - 1},$$

and so we have $f_0 = 1/n - s$, $f_1 = 1 - s$, $f_2 = 1 - 1/n$. The conditions of the theorem are satisfied for $s < 1/n$ and we get

$$A_n(\rho) \leq \frac{2n(1-s)}{1 - ns} = \frac{2\rho}{\rho - 2(1 - 1/n)},$$

$$\rho = 2 - 2s > 2(1 - 1/n).$$

Both bounds are displayed in Figure 2.3. We notice that the new bound is better in the range $2 - 2/n < \rho < 2 + 2/n$. In particular it contains the third Rankin bound.

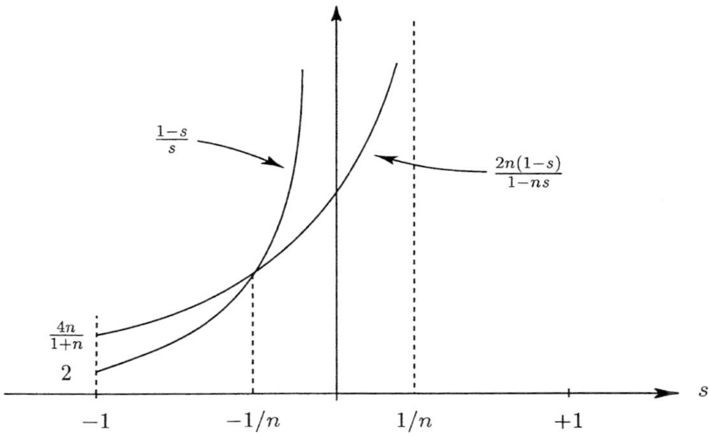

Figure 2.3: A first improvement upon the Rankin bound

We notice that by increasing the degree we were able to extend the range of the bound to smaller squared distances ρ. It is natural to try polynomials of still higher degree in an attempt to obtain further improvements for even smaller ρ. The problem is how to choose the test polynomials $f(t)$ so as to obtain the best possible bound. In general this problem is still unsolved. However, Levenshtein [157] suggested a sequence of test functions which in various ranges provide very good results. The resulting bound is referred to as the *Levenshtein bound.* It turns out that in some cases the Levenshtein bound is actually the best bound possible of the linear programming type, while in other cases improvements are possible. Boyvalenkov–Danev–Bumova [45] gave a criterion which uniquely characterizes those cases. Even more: they also indicated a method to obtain a better bound—still of the linear programming type—for all cases where the Levenshtein bound is not the best bound possible.

The Levenshtein bound—as well as the improvements suggested by Boyvalenkov et al [45]—is expressed in terms of several different orthogonal polynomials derived from the Gegenbauer polynomials. We proceed to derive a few of the properties satisfied by orthogonal polynomials in general.

2.4 Orthogonal polynomials

The Gegenbauer polynomials constitute a basis in the linear space $\mathbb{R}[t]$ of polynomials $f(t)$ of a real variable t. They are orthogonal with respect to a specific inner product, namely the one defined by (2.2.8). However, other bases and other inner products are possible and sometimes more useful.

An *inner product* in $\mathbb{R}[t]$ is a symmetric bilinear form which to any pair $\{f, g\}$ in $\mathbb{R}[t]$ assigns a real number (f, g) such that the Cauchy inequality

$$(f, g)^2 \leq (f, f)\,(g, g)$$

holds. The square root of the inner product of an element f with itself is usually called a *norm*, and the special notation $\|f\|^2 = (f, f)$ is often employed.

Let an inner product be given. A set of polynomials $Q_i(t)$ such that

$$\begin{aligned} (i) \quad & \deg(Q_i) = i, \\ (ii) \quad & (Q_i, Q_j) = (1/r_i)\,\delta_{ij}, \\ & i, j = 0, 1, 2, \dots\ , \end{aligned}$$

is called an *orthogonal* set of polynomials. The inner product in use is part of the definition: different inner products give rise to different families of orthogonal polynomials. For our purposes the normalization $Q_i(1) = 1$ will always be assumed. This condition determines the constants r_i. In fact, by Gram–Schmidt orthogonalization the whole sequence $\{Q_i(t)\}$ is actually determined by this condition.

However, Gram–Schmidt orthogonalization is not the only way of determining the orthogonal polynomials corresponding to a given inner product. For computational purposes the so called *three term recursion* is often more useful. Notice that the polynomial $t\,Q_k(t)$ has degree $k+1$ and so has an expansion

$$t\,Q_k(t) = \sum_{i=0}^{k+1} \alpha_i Q_i(t),$$

where the coefficients are given by

$$\alpha_i = r_i(t\,Q_k, Q_i), \quad i = 0, 1, \ldots, k+1.$$

By definition $Q_k(t)$ is orthogonal to all $Q_i(t)$ with $i < k$. As the equality $(t\,Q_k, Q_i) = (Q_k, t\,Q_i)$ prevails we conclude that α_i must be zero for all i except $i = k-1, k, k+1$. This observation gives us the following recursion:

$$Q_0(t) = 1, \qquad Q_1(t) = \frac{(1,1)\,t - (1,t)}{(1,1) - (1,t)},$$

$$t\,Q_k(t) = a_k\,Q_{k+1}(t) + c_k Q_k(t) + b_k\,Q_{k-1}(t), \tag{2.4.1}$$

where for $k = 1, 2, \ldots$

$$a_k = r_{k+1}(t\,Q_k, Q_{k+1}),$$
$$c_k = r_k(t\,Q_k, Q_k),$$
$$b_k = r_{k-1}(t\,Q_k, Q_{k-1}).$$

As $t\,Q_k(t)$ is outside the space spanned by $Q_i(t),\ 0 \le i \le k$ it follows that a_k is non-zero. It also follows that both a_k and $Q_{k+1}(t)$ can be computed from $Q_k(t)$ and $Q_{k-1}(t)$. We notice that the normalization condition $Q_i(1) = 1$ implies the identity

$$a_k + c_k + b_k = 1, \quad k = 1, 2, \ldots\,. \tag{2.4.2}$$

The recursion (2.2.14) for the Gegenbauer polynomials is, of course, just a special case of the more general recursion (2.4.1), which alternatively can be stated in the following more explicit form:

$$Q_0(t) = 1, \quad Q_1(t) = \frac{(1,1)\,t - (1,t)}{(1,1) - (1,t)},$$

$$Q_{k+1}(t) = \frac{t - c_k}{a_k}\,Q_k(t) - \frac{b_k}{a_k}\,Q_{k-1}(t), \tag{2.4.3}$$

where for $k = 1, 2, \ldots$

$$b_k = \frac{(t\,Q_k, Q_{k-1})}{(Q_{k-1}, Q_{k-1})}, \qquad c_k = \frac{(t\,Q_k, Q_k)}{(Q_k, Q_k)},$$

and where a_k is given by (2.4.2).

One consequence of (2.4.1) is that two successive polynomials $Q_i(t)$ and $Q_{i+1}(t)$ have no common non-trivial factor. Indeed, suppose $q(t)$ is a common factor in $Q_{k+1}(t)$ and $Q_k(t)$. By (2.4.1) follows that $q(t)$ must be a factor also in $Q_{k-1}(t)$. But then it is a common factor in $Q_k(t)$ and $Q_{k-1}(t)$ and by the same argument it is also a factor in $Q_{k-2}(t)$. Continuing this argument we finally conclude that $q(t)$ must also be a factor in $Q_0(t) = 1$, which implies that $q(t)$ must be a constant.

We also have the following important result.

Theorem 2.4.1 *Each polynomial $Q_k(t)$, $k \geq 0$, has precisely k distinct real zeros, all located in the open interval $(-1, 1)$.*

Proof: The proposition is trivially true for $Q_0(t) = 1$. For $k \geq 1$ the polynomial $Q_k(t)$ factors into linear and possibly quadratic factors. If $Q_k(t)$ has a multiple real zero or a pair of conjugate complex zeros then it contains a nonnegative factor which is not a constant. We will show that this is impossible. Indeed, assume $Q_k(t) = a(t)b(t)$, where $a(t) \geq 0$ for all t. We get

$$(Q_k, b) = (ab, b) = (1, ab^2).$$

Now if $a(t)$ is not a constant then it has positive degree and then $b(t)$ has degree less than k. But $Q_k(t)$ is orthogonal to all polynomials of degree less than k. It follows that the inner product $(Q_k, b) = (1, ab^2)$ must be zero, which is obviously impossible. The conclusion is that there cannot be any complex zeros or multiple real zeros in $Q_k(t)$. This proves the first part of the statement. The fact that all zeros are located in the interval $(-1, 1)$ can easily be established by an induction argument using the three-term formula. □

From the recursion (2.4.1) we also obtain the *Christoffel–Darboux* identities:

$$\sum_{i=0}^{k} r_i Q_i(s) Q_i(t) = r_k a_k \frac{Q_{k+1}(s)Q_k(t) - Q_k(s)Q_{k+1}(t)}{s - t}, \qquad (2.4.4)$$

$$\sum_{i=0}^{k} r_i(Q_i(t))^2 = r_k a_k \left(Q'_{k+1}(t)Q_k(t) - Q'_k(t)Q_{k+1}(t)\right), \qquad (2.4.5)$$

where the prime indicates differentiation. The first identity follows directly from the iteration (2.4.1), while the second identity follows as a limit of (2.4.4) as s tends to t.

From the second identity we conclude that successive polynomials have *interlacing zeros*. We will explain what this means. First let θ be a zero of the polynomial $Q_{k+1}(t)$. By (2.4.5) we get

$$\sum_{i=0}^{k} r_i(Q_i(\theta))^2 = r_k a_k Q'_{k+1}(\theta)Q_k(\theta).$$

The left-hand member of this equality is positive, because all terms are certainly nonnegative, and at least the k-th term, as we have just seen, is strictly positive. This implies that $Q'_{k+1}(\theta)$ and $Q_k(\theta)$ are non-zero and in addition that they also have the same sign.

Now let θ and τ be two adjacent zeros of the polynomial $Q_{k+1}(t)$. That means that the evaluations $Q'_{k+1}(\theta)$ and $Q'_{k+1}(\tau)$ of the derivative Q'_{k+1} have opposite signs, and consequently the same holds also for $Q_k(\theta)$ and $Q_k(\tau)$. Then by continuity $Q_k(t)$ must have an odd number of zeros in the open interval between θ and τ. As the same holds for all of the k open intervals between the $k+1$ zeros of $Q_{k+1}(t)$, and as $Q_k(t)$ has exactly k zeros, it also follows that there is precisely one zero in each one of these intervals. This is what we mean by saying that successive polynomials have interlacing zeros.

Let the zeros of $Q_{k+1}(t)$ be $\theta_0, \theta_1, \ldots, \theta_k$. Define

$$\rho_j = \left(r_k a_k Q'_{k+1}(\theta_j)Q_k(\theta_j)\right)^{-1}.$$

Notice that ρ_j is always positive. From the Christoffel–Darboux identities we obtain the following orthogonality relations:

$$\rho_j \sum_{i=0}^{k} r_i Q_i(\theta_j)Q_i(\theta_m) = \delta_{jm}\,. \qquad (2.4.6)$$

Let $U = [U_{ij}]$ be a square matrix with components

$$U_{ij} = \sqrt{\rho_i r_j}\, Q_j(\theta_i), \quad i,j = 0,1,\ldots,k.$$

We see that (2.4.6) implies

$$\begin{aligned}(UU^T)_{ij} &= \sum_{m=0}^{k} U_{im}U_{jm} \\ &= \sum_{m=0}^{k} \sqrt{\rho_i r_m}\, Q_m(\theta_i)\sqrt{\rho_j r_m}\, Q_m(\theta_j) \\ &= \sqrt{\rho_i\rho_j}\sum_{m=0}^{k} r_m Q_m(\theta_i)Q_m(\theta_j) = \delta_{ij}.\end{aligned}$$

Thus $UU^T = I$, where I is an identity matrix. Therefore U is an *orthogonal* matrix. But then we also have $U^TU = I$, which is the same as

$$\delta_{ij} = \sqrt{r_i r_j}\sum_{m=0}^{k} \rho_m Q_i(\theta_m)Q_j(\theta_m). \tag{2.4.7}$$

This identity gives rise to the following interesting fact. Let $f(t)$ and $g(t)$ be two arbitrary polynomials, both of degree at most k. We obtain

$$(f,g) = \sum_{i=0}^{k} \frac{f_i g_i}{r_i} = \sum_{i,j} \frac{f_i g_j}{\sqrt{r_i r_j}}\delta_{ij},$$

where f_i and g_j are the coefficients in the expansion of the polynomials $f(t)$ and $g(t)$ in the orthogonal basis $Q_0(t), Q_1(t), \ldots, Q_k(t)$. Now replacing δ_{ij} with the right-hand member of (2.4.7) we obtain

$$\begin{aligned}(f,g) &= \sum_{i,j} \frac{f_i g_j}{\sqrt{r_i r_j}}\sqrt{r_i r_j}\sum_{m=0}^{k} \rho_m Q_i(\theta_m)Q_j(\theta_m) \\ &= \sum_{m=0}^{k} \rho_m \sum_{i=0}^{k} f_i Q_i(\theta_m)\sum_{j=0}^{k} g_j Q_j(\theta_m) \\ &= \sum_{m=0}^{k} \rho_m f(\theta_m)g(\theta_m).\end{aligned}$$

Actually, for the special case $g(t) = 1$ this result can be strengthened to the following result, known as the *Gauss–Jacobi* formula.

Theorem 2.4.2 *Let $f(t)$ be a polynomial of degree at most $2k+1$. Then the following identity holds:*

$$(1, f) = \sum_{m=0}^{k} \rho_m f(\theta_m). \tag{2.4.8}$$

Proof: For given polynomials $f(t)$ of degree at most $2k+1$ and $Q_{k+1}(t)$ of degree $k+1$, there exist two unique polynomials, $a(t)$ and $b(t)$, both of degree at most k, such that

$$f(t) = Q_{k+1}(t)\, b(t) + a(t).$$

If $\theta_0, \theta_1, \ldots, \theta_k$ are the $k+1$ zeros of $Q_{k+1}(t)$, then we clearly have $f(\theta_i) = a(\theta_i),\ i = 0, 1, \ldots, k$. As $b(t)$ has degree at most k it follows that we have $0 = (Q_{k+1}, b) = (1, Q_{k+1}b) = (1, f - a)$, and so

$$(1, f) = (1, a) = \sum_{m=0}^{k} \rho_m a(\theta_m) = \sum_{m=0}^{k} \rho_m f(\theta_m),$$

where

$$\rho_m = [r_k a_k Q'_{k+1}(\theta_m) Q_k(\theta_m)]^{-1} > 0, \quad m = 0, 1, \ldots, k.$$

□

As we have already noticed the coefficients ρ_m—which of course depend on the inner product—are always positive. A slight variation of this formula will be very useful in the study of the Levenshtein bound.

2.5 The Levenshtein bound

The key problem in applying the linear programming bound is to find an efficient polynomial $f(t) = f_0 + f_1 Q_1(t) + \cdots + f_k Q_k(t)$ satisfying the following two properties:

$$(i) \quad f(t) \leq 0, \quad -1 \leq t \leq s,$$

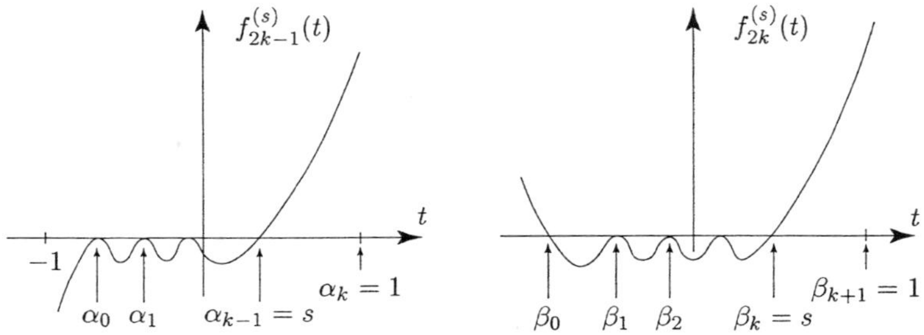

Figure 2.4: Test polynomials as suggested by Levenshtein

$$(ii) \quad f_0 > 0, \quad f_1, f_2, \ldots, f_k \geq 0.$$

We denote by $D(n, s)$ the set of all polynomials satisfying (i) and (ii) for a given $s = 1-\rho/2$. The parameter n denotes the dimension of the space under consideration. Levenshtein proposed using polynomials of the forms

$$f(t) = (t - s)A(t)^2,$$
$$f(t) = (t + 1)(t - s)B(t)^2,$$

where $A(t)$ and $B(t)$ are polynomials with simple roots, all in the open interval $(-1, s)$:

$$A(t) = \prod_{i=0}^{k-2}(t - \alpha_i), \quad -1 < \alpha_0 < \alpha_1 < \ldots < \alpha_{k-2} < s, \tag{2.5.1}$$

$$B(t) = \prod_{i=1}^{k-1}(t - \beta_i), \quad -1 < \beta_1 < \beta_2 < \ldots < \beta_{k-1} < s. \tag{2.5.2}$$

The test function $f(t)$ will have zeros of multiplicity two in all the points α_i, $i = 0, 1, \ldots, k-2$ or in the points β_i, $i = 1, 2, \ldots, k-1$ as the case may be. In addition it will have zeros of multiplicity one at $t = s$ and—in the second case—also at the point $t = -1$. Following Levenshtein we define $\alpha_{k-1} = s$, $\alpha_k = 1$, $\beta_0 = -1$, $\beta_k = s$, $\beta_{k+1} = 1$. We notice that we have $\deg f(t) = 2k - 1$ in the first case and $\deg f(t) = 2k$ in the second case. The two cases are illustrated in Figure 2.4.

Levenshtein proposed a sequence of polynomials of increasing degrees to be used for different values of the parameter s in the interval $\mathcal{I} =$

$[-1, 1)$. More precisely a partitioning $\mathcal{I} = \bigcup_{m \geq 1} \mathcal{I}_m$ was proposed such that $f(t)$ is of degree m for $s \in \mathcal{I}_m$. The form (2.5.1) is used whenever m is odd and the form (2.5.2) is used whenever m is even. A good choice of the intervals $\mathcal{I}_m$ and the functions $A(t)$ and $B(t)$ is in essence the main achievement obtained by Levenshtein. We proceed to discuss this solution in detail.

We will need several families of orthogonal polynomials closely related to the Gegenbauer family. Levenshtein refers to them as "adjacent" polynomials. Recall that the Gegenbauer polynomials are uniquely determined by the inner product

$$(f, g) = (1, fg) = k_{n-1} \int_{-1}^{1} f(t)\, g(t)(1 - t^2)^{\frac{n-3}{2}}\, dt,$$

and the normalization condition $Q_i(1) = 1$, $i \in \mathbb{N}$. For any nonnegative integers p, q we define a new inner product $(f, g)_{p,q}$ according to the following rule:

$$(f, g)_{p,q} = c^{p,q} \left((1 - t)^p (1 + t)^q, fg \right),$$

where the constant $c^{p,q}$ is chosen such that $(1, 1)_{p,q} = 1$. It is easy to check that this definition yields a valid inner product, i.e. one that is bilinear and satisfies the Cauchy inequality. To any such inner product we associate a sequence of orthogonal polynomials $Q_i^{p,q}(t)$. The normalization $Q_i^{p,q}(1) = 1$ will still be assumed, and by defining

$$r_i^{p,q} = (Q_i^{p,q}, Q_i^{p,q})^{-1},$$

we obtain

$$(Q_i^{p,q}, Q_j^{p,q}) = \delta_{ij} / r_i^{p,q}.$$

We notice that with $p = q = 0$ we obtain the usual Gegenbauer polynomials. Define

$$K_i^{p,q}(t, s) = \sum_{j=0}^{i} r_j^{p,q} Q_j^{p,q}(t) Q_j^{p,q}(s).$$

The Christoffel–Darboux identities take the forms

$$(t - s) K_i^{p,q}(t, s) = r_i^{p,q} a_i^{p,q} [\, Q_{i+1}^{p,q}(t) Q_i^{p,q}(s) - Q_i^{p,q}(t) Q_{i+1}^{p,q}(s) \,], \qquad (2.5.3)$$

$$K_i^{p,q}(t,t) =$$
$$= r_i^{p,q} a_i^{p,q} [\, Q'^{p,q}_{i+1}(t) Q_i^{p,q}(t) - Q'^{\,p,q}_i(t) Q_{i+1}^{p,q}(t)\,], \tag{2.5.4}$$

where $Q'^{\,p,q}_i(t)$ means the derivative of $Q_i^{p,q}(t)$, and where the coefficients $a_i^{p,q}$ are defined in the same way as for the Gegenbauer polynomials (see (2.2.14)).

The following relations can easily be proved:

$$Q_i^{p,q+1}(t) = \frac{K_i^{p,q}(t,-1)}{K_i^{p,q}(1,-1)}\,, \tag{2.5.5}$$

$$Q_i^{p+1,q}(t) = \frac{K_i^{p,q}(t,1)}{K_i^{p,q}(1,1)}\,. \tag{2.5.6}$$

We will mainly be concerned with the cases $(p,q) = (0,1), (1,0)$ and $(1,1)$.

The adjacent polynomials satisfy all the general properties of orthogonal polynomials discussed in Section 2.4.

Denote by $t_{i,j}^{p,q}$, $j = 1,2,\ldots,i$ the i zeros of the polynomial $Q_i^{p,q}(t)$ in increasing order. For the largest zero $t_{i,i}^{p,q}$ we employ the simplified notation $t_i^{p,q}$ and for the Gegenbauer polynomials we omit the upper indices. In addition to the usual interlacing properties the following separation lemma can be proved.

Lemma 2.5.1 *The zeros of the adjacent polynomials $Q_i^{1,0}(t)$ and $Q_i^{1,1}(t)$ satisfy the following separation conditions:*

$$t_k^{1,0} < t_k^{1,1} < t_{k+1}^{1,0}.$$

Proof: Both inequalities can be proved using the Christoffel–Darboux identities and by slight modifications of the arguments used in Section 2.4. For details we refer to Levenshtein [159]. □

We are now ready to describe formally the Levenshtein bound. We first define the half-open intervals $\mathcal{I}_m = [\tau_{m-1}, \tau_m)$, $m = 1,2,\ldots$, where

$$\tau_m \triangleq \begin{cases} t_k^{1,0}, & if \quad m = 2k-1, \\ t_k^{1,1}, & if \quad m = 2k. \end{cases}$$

The first few values are

$$\tau_0 = -1, \quad \tau_1 = -1/n, \quad \tau_2 = 0,$$

$$\tau_3 = \frac{1}{\sqrt{n+3}+1}, \quad \tau_4 = \frac{1}{\sqrt{n+2}}.$$

We then define the test functions $f_m^{(s)}(t)$—where $f_m^{(s)}(t)$ should be used for $s \in \mathcal{I}_m$—as follows:

$$f_m^{(s)}(t) = \begin{cases} (t-s)(K_{k-1}^{1,0}(t,s))^2, & m = 2k-1, \\ (t+1)(t-s)(K_{k-1}^{1,1}(t,s))^2, & m = 2k. \end{cases}$$

By Lemma 2.5.1 follows that the intervals $\mathcal{I}_m$ constitute a partition of the half-open interval $\mathcal{I} = [-1, 1)$. We notice that the degree of the polynomial $f_m^{(s)}(t)$ equals m, so as the parameter s increases from -1 to 1 the degree of the test polynomials increases monotonically through all positive integers.

We need to show that the polynomials $f_m^{(s)}(t)$ are valid test functions. Condition (i) is obviously satisfied by the very construction. The difficult part of the argument is to establish the validity of condition (ii). We proceed in successive steps.

Consider first the polynomials $Q_i^{1,0}(t)$, $i = 0, 1, \ldots, k-1$. By (2.5.6) we have

$$Q_i^{1,0}(t) = \frac{K_i(t,1)}{K_i(1,1)}.$$

Recall the identity

$$K_i(t,1) = \sum_{j=0}^{i} r_j Q_j(t),$$

where all coefficients r_j are positive. We have

$$K_i(1,1) = \sum_{j=0}^{i} r_j > 0.$$

It follows that $Q_i^{1,0}(t)$ has a Gegenbauer expansion with positive coefficients. By Theorem 2.2.4 it is positive definite.

Next consider the polynomial $K_{k-1}^{1,0}(t,s)$. We have

$$K_{k-1}^{1,0}(t,s) = \sum_{j=0}^{k-1} r_j^{1,0} Q_j^{1,0}(t) Q_j^{1,0}(s) \ .$$

Recall that $f_{2k-1}^{(s)}(t)$ is employed as a test function in the interval

$$\mathcal{I}_{2k-1} = \left[\tau_{2k-2}, \tau_{2k-1}\right) = \left[t_{k-1}^{1,1}, t_k^{1,0}\right).$$

By Lemma 2.5.1 we have $Q_i^{1,0}(s) > 0$ for $i = 0, 1, \ldots, k-1$, while $Q_k^{1,0}(s) \leq 0$. As all coefficients $r_j^{1,0}$ are positive we conclude that $K_{k-1}^{1,0}(t,s)$ is positive definite.

Consider now the function $\psi(t) = (t-s)K_{k-1}^{1,0}(t,s)$. By applying the Christoffel–Darboux formula we get

$$(t-s)K_{k-1}^{1,0}(t,s) \ = \ r_{k-1}^{1,0} a_{k-1}^{1,0} \left(Q_k^{1,0}(t) Q_{k-1}^{1,0}(s) - Q_{k-1}^{1,0}(t) Q_k^{1,0}(s) \right) .$$

Clearly $r_{k-1}^{1,0} a_{k-1}^{1,0} > 0$. We just noticed that $Q_{k-1}^{1,0}(s) > 0$ while $Q_k^{1,0}(s) \leq 0$. It follows that $(t-s)K_{k-1}^{1,0}(t,s)$ is also positive definite.

We finally notice that the test function $f_{2k-1}^{(s)}(t)$ is a product of two factors, namely $K_{k-1}^{1,0}(t,s)$ and $(t-s)K_{k-1}^{1,0}(t,s)$. Both of them are positive definite for $s \in \mathcal{I}_{2k-1}$. By Theorem 2.2.5 follows that $f_{2k-1}^{(s)}(t)$ is positive definite and so belongs to the set $D(s,n)$ of valid test functions.

By similar arguments it can also be shown that the function

$$f_{2k}^{(s)}(t) = (t+1)(t-s)(K_{k-1}^{1,1}(t,s))^2$$

is a valid test function in the interval $\mathcal{I}_{2k}$. For the details we refer to Levenshtein [157], [158].

Now that we have established that the polynomials $f_m^{(s)}(t)$ are valid test functions the natural question is: how good are they?

We consider first the case $m = 2k-1$. Let us return to the expression (2.5.1). A key result is a lemma, which is a slight modification of the Gauss–Jacobi formula (2.4.8). Let the real numbers $\alpha_0 < \alpha_1 < \ldots < \alpha_k$ be given. Define

$$\varphi(t) = \prod_{j=0}^{k} (t - \alpha_j),$$

and notice that the polynomial $\varphi(t)$ has degree $k+1$. For $i = 0, 1, \ldots, k$ define

$$g_i(t) = \frac{\varphi(t)}{t - \alpha_i}\, \gamma_i\,,$$

where the constants γ_i are chosen such that $g_i(\alpha_j) = \delta_{ij}$. We notice that all the polynomials $g_i(t)$ have degree k. Now let $f(t)$ be an arbitrary polynomial of degree deg $f(t) \leq 2k-1$ and define

$$a(t) = \sum_{i=0}^{k} f(\alpha_i)\, g_i(t).$$

We notice that $a(t) = f(t)$ for $t = \alpha_i$, $i = 0, 1, \ldots, k$. It follows that we have

$$f(t) - a(t) = \varphi(t)\, \eta(t)$$

for some polynomial $\eta(t)$ of degree at most $k-2$. Suppose we have been able to choose the numbers $\alpha_0 < \alpha_1 < \ldots < \alpha_k$ such that $\varphi(t)$ is orthogonal to all polynomials of degree $k-2$ or less. We will see in a moment that such a choice is indeed possible. For such a $\varphi(t)$ we have

$$0 = (\varphi, \eta) = (1, \varphi\, \eta) = (1, f - a),$$

which gives us

$$(1, f) = (1, a) = \sum_{i=0}^{k} f(\alpha_i)(1, g_i) = \sum_{i=0}^{k} \rho_i f(\alpha_i),$$

where

$$\rho_i = (1, g_i), \quad i = 0, 1, \ldots, k.$$

How to choose $\varphi(t)$ such that it is orthogonal to all polynomials of degree at most $k-2$? The following choice will do:

$$\varphi(t) = (1-t)\psi(t),$$

where

$$\psi(t) = (t-s)K_{k-1}^{1,0}(t,s).$$

We have already noticed above that the function $(t-s)K_{k-1}^{1,0}(t,s)$ is a linear combination of the polynomials $Q_k^{1,0}(t)$ and $Q_{k-1}^{1,0}(t)$. With respect to the inner product $(f, g)_{1,0}$ both of them are orthogonal to all polynomials of degree at most $k-2$. It follows that for any $\eta(t)$ of degree at most $k-2$ we have

$$(\varphi, \eta) = ((1-t)\psi, \eta) = (\psi, \eta)_{1,0} = 0.$$

In fact, we are now able to prove the following lemma.

Lemma 2.5.2 *For fixed $s \in \mathcal{I}_{2k-1}$ let $\alpha_0 < \alpha_1 < \ldots < \alpha_{k-2}$ be the zeros of $K_{k-1}^{1,0}(t,s)$ viewed as a polynomial in t. Define $\alpha_{k-1} = s$, $\alpha_k = 1$ and let us define*

$$\psi(t) \triangleq (t-s)K_{k-1}^{1,0}(t,s), \quad \varphi(t) \triangleq (1-t)\psi(t).$$

For any polynomial $f(t)$ with $\deg f(t) \leq 2k-1$ we have

$$f_0 = (1, f) = \sum_{i=0}^{k} \rho_i f(\alpha_i),$$

where for $i = 0, 1, \ldots, k-1$

$$\rho_i = (1, g_i) = (c^{1,0}(1-\alpha_i)K_{k-1}^{1,0}(\alpha_i, \alpha_i))^{-1} > 0,$$

while for $i = k$ we have

$$\rho_k = \frac{-Q_k(s)}{K_{k-1}(s,1) - Q_k(s)K_{k-1}(1,1)}.$$

Proof: The only things we have not yet established are the formulas for the coefficients ρ_i. They follow easily, however, from the following expressions, which yield polynomials with exactly the properties assumed in the derivation:

$$g_i(t) = \frac{(1-t)K_{k-1}^{1,0}(t,\alpha_i)}{(1-\alpha_i)K_{k-1}^{1,0}(\alpha_i,\alpha_i)}, \quad i = 0, 1, \ldots, k-1,$$

$$g_k(t) = \frac{(t-s)K_{k-1}^{1,0}(t,s)}{(1-s)K_{k-1}^{1,0}(1,s)}.$$

□

Corollary 2.5.1 *Let s belong to the interval $\mathcal{I}_{2k-1}$. For any polynomial $f(t)$ with degree $\deg f(t) \leq 2k-1$ such that $f(t) \leq 0$ in $[-1, s]$ we have*

$$f_0 \leq \rho_k f(1),$$

with equality if and only if $f(t)$ contains $\psi(t) = (t-s)K_{k-1}^{1,0}(t,s)$ as a factor.

Proof: The inequality holds because each one of the coefficients ρ_i, $i \leq k-1$ is positive and $\alpha_k = 1$. The condition for equality is precisely the condition under which $f(\alpha_i) = 0$ for $i = 0, 1, \ldots, k-1$. □

We notice now that the function

$$f_{2k-1}^{(s)}(t) = (t-s)\left(K_{k-1}^{1,0}(t,s)\right)^2$$

satisfies all the conditions in Corollary 2.5.1. It follows that the bound $1/\rho_k$ is achieved by $f_{2k-1}^{(s)}(t)$.

For the case $m = 2k$ we need the following result.

Lemma 2.5.3 *For fixed $s \in \mathcal{I}_{2k} = [t_k^{1,0}, t_k^{1,1})$ let $\beta_1 < \beta_2 < \ldots < \beta_{k-1}$ be the zeros of the function $K_{k-1}^{1,1}(t,s)$. Define $\beta_0 = 1$, $\beta_k = s$, $\beta_{k+1} = 1$ and let the polynomials $\psi(t)$ and $\varphi(t)$ be as follows:*

$$\psi(t) = (t+1)(t-s)K_{k-1}^{1,1}(t,s), \quad \varphi(t) = (1-t)\psi(t).$$

For any polynomial $f(t)$ of degree at most $2k$ we have

$$f_0 = \sum_{i=0}^{k+1} \gamma_i f(\beta_i),$$

where $\gamma_i > 0$, $i = 1, 2, \ldots k$, for any $s \in \mathcal{I}_{2k}$ and $\gamma_0 > 0$ for $t_k^{1,0} < s < t_k^{1,1}$, while $\gamma_0 = 0$ for $s = t_k^{1,0}$, and where

$$\gamma_{k+1} = \frac{K_k(s,-1)}{K_k(1,1)K_k(s,-1) - K_k(1,-1)K_k(s,1)}.$$

Proof: See Levenshtein [158],[159]. □

Corollary 2.5.2 *Suppose $s \in \mathcal{I}_{2k}$. For any polynomial $f(t)$ with degree at most $2k$ such that $f(t) \leq 0$ for $t \in [-1, s]$ the inequality $f_0 \leq \gamma_{k+1} f(1)$ prevails. If $t_k^{1,0} < s < t_k^{1,1}$ equality holds if and only if $f(t)$ contains*

$$\psi(t) = (t+1)(t-s)K_{k-1}^{1,1}(t,s)$$

as a factor. If $s = t_k^{1,0}$ equality holds if and only if $f(t)$ contains

$$(t-s)K_{k-1}^{1,1}(t,s)$$

as a factor.

Proof: See Levenshtein [158], [159]. □

We say that a test function $f(t) \in D(n,s)$ is *extreme* if it minimizes the quotient $f(1)/f_0$ among all functions in $D(n,s)$ of degree less than or equal to the degree of $f(t)$. By Corollaries 2.5.1 and 2.5.2 we see that the polynomial $f_m^{(s)}(t)$ is extreme for $s \in \mathcal{I}_m$. This does *not* mean that the bound produced by these test functions is best possible. As we will see in the next section functions of *higher* degree sometimes give better bounds. But it *does* mean that no polynomial of the *same degree or lower* can produce a better bound. Extremality in this sense of the Levenshtein test functions was first shown by Sidelnikov [205].

We now define the *Levenshtein functions* $L_m(s)$ as follows:

$$L_{2k-1}(s) = \frac{1}{\rho_k} = K_{k-1}(1,1)\left(1 - \frac{Q_{k-1}^{1,0}(s)}{Q_k(s)}\right), \tag{2.5.7}$$

$$L_{2k}(s) = \frac{1}{\gamma_{k+1}} = K_k(1,1)\left(1 - \frac{Q_k^{1,0}(s)}{Q_k^{0,1}(s)}\right). \tag{2.5.8}$$

The Levenshtein bound is given by the following formula:

$$A_n(\rho) \leq L(s) \triangleq L_m(s), \quad s \in \mathcal{I}_m, \quad m = 1, 2, \ldots$$

where $\rho = 2 - 2s$. We formulate this important result in the following theorem.

Theorem 2.5.1 (Levenshtein bound) *Let $\mathcal{X}$ be a spherical code with parameters (n, ρ, M) and let $s = 1 - \rho/2$. Then the size M is necessarily bounded according to $M \leq L(s) = L_m(s)$, $s \in \mathcal{I}_m$, $m = 1, 2, \ldots$.*

Using the Christoffel–Darboux identities and the formulas (2.5.5) and (2.5.6) we can express $Q_k^{1,0}(t)$ and $Q_k^{0,1}(t)$ in terms of Gegenbauer polynomials $Q_k(t)$ as follows

$$\begin{aligned} Q_k^{1,0}(t) &= \frac{(n-1)(Q_k(t) - Q_{k+1}(t))}{(2k+n-1)(1-t)}, \\ Q_k^{0,1}(t) &= \frac{Q_k(t) + Q_{k+1}(t)}{1+t}. \end{aligned}$$

Using these expressions we can rewrite the Levenshtein functions. If $t_{k-1}^{1,1} \leq s < t_k^{1,0}$ we have

$$L_{2k-1}^{(n)}(s) = \binom{k+n-3}{k-1}$$

$$= \left(\frac{2k+n-3}{n-1} - \frac{Q_{k-1}(s) - Q_k(s)}{(1-s)Q_k(s)} \right), \tag{2.5.9}$$

while if $t_{k-1}^{1,0} \leq s < t_k^{1,1}$ we have

$$L_{2k}^{(n)}(s) = \binom{k+n-2}{k}$$

$$= \left(\frac{2k+n-1}{n-1} - \frac{(1+s)(Q_k(s) - Q_{k+1}(s))}{(1-s)(Q_k(s) + Q_{k+1}(s))} \right). \tag{2.5.10}$$

In the first few cases the following explicit expressions are valid (we suppress the dependence on the dimension n for the Levenshtein functions $L_i(s)$):

$$L_1(s) = \frac{1-s}{-s}, \qquad s \in \mathcal{I}_1,$$

$$L_2(s) = \frac{2n(1-s)}{1-ns}, \qquad s \in \mathcal{I}_2,$$

$$L_3(s) = \frac{n(2+(n+1)s)(1-s)}{1-ns^2}, \qquad s \in \mathcal{I}_3,$$

$$L_4(s) = \frac{2n(1+(n+2)s)(1-s)}{1+2s-(n+2)s^2}, \qquad s \in \mathcal{I}_4,$$

$$L_5(s) = \frac{n(1-s)((n+3)(n+2)s^2+4s(n+2)-(n-1))}{2s(3-(n+2)s^2)}, \qquad s \in \mathcal{I}_5,$$

$$L_6(s) = \frac{n(1-s)(n+1)(n+2)(1-2s-(n+4)s^2)}{s(n+2)(s^2(n+4)-3s-3)+3}, \qquad s \in \mathcal{I}_6,$$

where the intervals I_i are given as follows:

$$\mathcal{I}_1 = [-1, -1/n),$$

$$\mathcal{I}_2 = [-1/n, 0),$$

$$\mathcal{I}_3 = [0, 1/(\sqrt{n+3}+1)\,),$$

$$\mathcal{I}_4 = [1/(\sqrt{n+3}+1), 1/\sqrt{n+2}\,),$$

$$\mathcal{I}_5 = [1/\sqrt{n+2}, t_3^{1,0}\,),$$

$$\mathcal{I}_6 = [t_3^{1,0}, \sqrt{3/(n+4)}\,).$$

2.6 The Boyvalenkov–Danev–Bumova criterion

A natural question to ask is whether the Levenshtein bound is the best bound possible of the linear programming type. Extremality means that no polynomial of degree less than or equal to m could outperform $f_m^{(s)}(t)$ in the interval I_m. However, there might be polynomials of higher degree which give better bounds. This question was addressed by Boyvalenkov–Danev–Bumova [45]. They found a criterion by which it can be tested. Moreover, they also developed a method to improve the Levenshtein bound whenever an improvement is possible. Improved bounds have been obtained in a large number of cases. Unfortunately in those cases the possibility of further improvements cannot be excluded.

Notice that the parameters α_i and ρ_i occurring in Lemma 2.5.2 as well as the parameters β_i and γ_i occurring in Lemma 2.5.3 are functions of the parameter s. For any positive integer j we define a function $P_j(s)$ as follows

$$P_j(s) = \begin{cases} \sum_{i=0}^{k} \rho_i \, Q_j(\alpha_i), & if \quad s \in I_{2k-1}, \\ \sum_{i=0}^{k} \gamma_i \, Q_j(\beta_i), & if \quad s \in I_{2k}. \end{cases} \tag{2.6.1}$$

Suppose $s \in I_m$. According to the modified Gauss–Jacobi formula given by Lemma 2.5.2 and by Lemma 2.5.3 we have $P_j(s) = 0$ for $j \le m$. The interesting observation by Boyvalenkov–Danev–Bumova is that the values taken by the function $P_j(s)$ for $j > m$ can be used in order to determine whether the Levenshtein bound can be improved or not.

Theorem 2.6.1 (Boyvalenkov–Danev–Bumova, [45])
Improvement to the bound given by Levenshtein, $L(s) = L_m(s)$, $s \in I_m$, can be obtained if and only if $P_j(s) < 0$ for some $j > m$.

Proof: We consider only the case $m = 2k-1$. It is obvious that exactly the same argument applies also to the case $m = 2k$.

Let $s \in I_{2k-1}$. First suppose $P_j(s) \ge 0$ for all $j > 2k$. Let $f(t)$ be an arbitrary test function of degree $\tau \ge 2k$. We represent it in the form $f(t) = g(t) + h(t)$, with

$$g(t) = \sum_{i=0}^{2k-1} g_i \, Q_i(t), \quad h(t) = \sum_{i=2k}^{\tau} h_i \, Q_i(t). \tag{2.6.2}$$

Notice that $\deg g(t) \leq 2k-1$, so the modified *Gauss–Jacobi* formula is applicable on $g(t)$. We have

$$f_0 = g_0 = \sum_{i=0}^{k} \rho_i \, g(\alpha_i) = \sum_{i=0}^{k} \rho_i \left(f(\alpha_i) - h(\alpha_i) \right). \tag{2.6.3}$$

By (2.6.1) and (2.6.2) we have

$$\sum_{i=0}^{k} \rho_i \, h(\alpha_i) = \sum_{j=2k}^{\tau} h_j \, P_j(s).$$

Inserting this in (2.6.3) we obtain

$$f_0 = \sum_{i=0}^{k} \rho_i \, f(\alpha_i) - \sum_{j=2k}^{\tau} h_j \, P_j(s).$$

For $f(t) \in D(n,s)$ we have $f(\alpha_i) \leq 0$, $i = 0, 1, \ldots, k-1$. By assumption we have $P_j(s) \geq 0$, $j \geq 2k$. It follows that we have $f_0 \leq \rho_k f(1)$ or—what is the same

$$\frac{1}{\rho_k} = L_{2k-1}(s) \;\leq \frac{f(1)}{f_0}.$$

This means that whatever test function $f(t)$ is employed the bound $L_{2k-1}(s)$ cannot be improved by any polynomial of degree $\tau \geq 2k$. We already know that $L_{2k-1}(s)$ is achievable by a polynomial of degree $2k-1$ and that no better bound can be obtained by any polynomial of this degree or lower. It follows that $L_{2k-1}(s)$ is the best possible linear programming bound whenever the condition $P_j(s) \geq 0$ is fulfilled for all $j \geq 2k$.

Now suppose the opposite, i.e. suppose $P_j(s) < 0$ for some $j \geq 2k$. We claim that an improvement of the Levenshtein bound is possible in this case. In order to prove this we need to come up with a test function $f(t) \in D(n,s)$ which produces such an improvement.

It is clear that the degree must exceed $2k-1$. We try a function of the form

$$f(t) = Q_j(t) + g(t),$$

where $Q_j(t)$ is the j-th Gegenbauer polynomial and where $g(t)$ has degree $2k-1$. To simplify the notation let us denote $\varphi(t) = f_{2k-1}^{(s)}(t)$. We like to specify $g(t)$ so that $f(t)$ has the form $f(t) = \varphi(t)\,h(t)$.

Dividing $Q_j(t)$ with $t\,\varphi(t)$ we get

$$Q_j(t) = t\,\varphi(t) + r(t), \quad \deg r(t) \leq 2k-1.$$

Choosing $g(t) = c\,\varphi(t) - r(t)$ gives us

$$Q_j(t) + g(t) = \varphi(t)\,(\eta(t)\,t + c).$$

It remains to choose the constant c. We proceed to show that it is possible to do that in such a way that $f(t) = Q_j(t)+g(t) \in D(n,s)$. In fact, it turns out that the only requirement is that c is large enough.

Recall that we have to establish the following conditions:

$$\begin{aligned} &(i) \quad f(t) = \varphi(t)\,(\eta(t)\,t + c) \leq 0, \quad -1 \leq t \leq s, \\ &(ii) \quad f(t) = Q_j(t) + g(t) = f_0 + f_1 Q_1(t) + \cdots + f_j Q_j(t), \end{aligned}$$

where $f_0 > 0$, $f_1, f_2, \ldots, f_j \geq 0$. Consider the first condition. We know that $\varphi(t)$ is a valid test function. It follows that it satisfies the condition $\varphi(t) \leq 0$, $-1 \leq t \leq s$. The function $f(t) = \varphi(t)\,(\eta(t)\,t + c)$ will satisfy the same condition if c is chosen such that $h(t) = \eta(t)\,t + c$ is nonnegative in the interval $-1 \leq t \leq s$. This will obviously be the case if c is large enough.

We now turn to the second condition. Consider the following Gegenbauer expansion

$$a(t) = \varphi(t)\,\eta(t)\,t = a_0 + a_1 Q_1(t) + \cdots + a_j Q_j(t).$$

We clearly have $a_{2k} = a_{2k+1} = \ldots = a_{j-1} = 0$, $a_j = 1$. For the remaining coefficients we get $g_i = a_i + c\,\varphi_i$, $i = 0, 1, \ldots, 2k-1$ where g_i and φ_i are the coefficients in the Gegenbauer expansions of the polynomials $g(t)$ and $\varphi(t)$ respectively. We see that with c large enough all of the coefficients g_i will be positive. With such a c the Gegenbauer expansion of $f(t) = Q_j(t)+g(t)$ has also only nonnegative coefficients, and so also condition (ii) will be satisfied.

In conclusion both conditions will be satisfied if c is large enough. Assume such a c has been chosen. We get

$$\begin{aligned} f_0 = g_0 &= \sum_{i=0}^{k} \rho_i \, g(\alpha_i) \\ &= \sum_{i=0}^{k} \rho_i \left(f(\alpha_i) \ - \ Q_j(\alpha_i) \right) = \\ &= \rho_k \, f(1) + \sum_{i=0}^{k-1} \rho_i \, f(\alpha_i) - P_j(s). \end{aligned} \tag{2.6.4}$$

As $\varphi(t) = f_{2k-1}^{(s)}(t)$ satisfies the bound $f_0 \leq \rho_k \, f(1)$ with equality it follows that we must have $\varphi(\alpha_i) = 0, \; i = 0, 1, \ldots, k-1$. As a consequence we also have $f(\alpha_i) = \varphi(a_i) \, h(\alpha_i) = 0, \; i = 0, 1, \ldots, k-1$. By assumption we have $P_j(s) < 0$. We get the following inequality:

$$\frac{f(1)}{f_0} < \frac{1}{\rho_k} = L_{2k-1}(s).$$

The conclusion is that the polynomial $f(t)$ gives a better bound than does the Levenshtein test function. This means that the Levenshtein bound can be improved whenever the condition $P_j(s) < 0$ holds for some $j > 2k$. □

The practical use of the Boyvalenkov–Danev–Bumova criterion is complicated by the fact that an infinite number of test functions are involved. On the other hand, it appears that in practice usually only a few cases actually need to be tested.

It has been proved that for fixed n improvements upon the Levenshtein bound $L^{(n)}(s)$ are always possible for large enough s. On the other hand, it appears that for fixed s the Levenshtein bound might be the best possible linear programming bound for all n large enough.

2.7 Properties of the Levenshtein bound

Although—as we have seen—the Levenshtein bound is not universally the best possible bound of the linear programming type, it is in most cases a very good bound—sometimes even tight. It has a number of very interesting properties. We like to mention a few of them. We start with the following basic fact.

Theorem 2.7.1 *The Levenshtein bound $L(s)$ is a continuous, monotonically increasing function on the interval $s \in (-1, 1)$.*

Proof: Recall that the function $L(s)$ is defined by a sequence of rational functions $L_m(s)$, where each of the functions $L_m(s)$ is used over a half-open interval $\mathcal{I}_m = [\tau_{m-1}, \tau_m)$. In order to establish continuity it is enough to establish the equalities $L_m(\tau_m) = L_{m+1}(\tau_m)$. Again we consider only the case $m = 2k - 1$ and refer to Levenshtein [158] for the other (slightly more difficult) case. Recall that we have $\tau_{2k-1} = t_k^{1,0}$, so we must show the following identity:

$$L_{2k-1}(t_k^{1,0}) = L_{2k}(t_k^{1,0})\,. \tag{2.7.1}$$

By (2.5.7) we have

$$L_{2k-1}(t_k^{1,0}) = K_{k-1}(1,1)\left(1 - \frac{Q_{k-1}^{1,0}(t_k^{1,0})}{Q_k(t_k^{1,0})}\right)$$

and by (2.5.8) we have

$$L_{2k}(t_k^{1,0}) = K_k(1,1) = K_{k-1}(1,1) + \gamma_k\,.$$

Using (2.5.6) we obtain

$$K_{k-1}(1,1)\,Q_{k-1}^{1,0}(t_k^{1,0}) = K_{k-1}(t_k^{1,0},1)$$

and by definition of $K_i(t,s)$ we get

$$\begin{aligned} K_{k-1}(t_k^{1,0},1) + Q_k(t_k^{1,0})\,r_k &= \\ = \sum_{i=0}^{k} r_i\,Q_i(t_k^{1,0}) = K_k(1,1)\,Q_k^{1,0}(t_k^{1,0}) &= 0\,. \end{aligned}$$

Combining these identities we obtain (2.7.1), which establishes the continuity of the Levenshtein function $L(s)$ at the points $s = \tau_{2k-1} = t_k^{1,0}$. It is straightforward to check—by differentiation and using the

Christoffel–Darboux identities—that all of the functions $L_m(s)$ are increasing in their respective intervals $\mathcal{I}_m$. We omit the details, referring to Levenshtein [158]. □

As a by-product of the proof we have also established the following identities:

$$\begin{aligned} L_{2k-1}(t_k^{1,0}) &= L_{2k}(t_k^{1,0}) = K_k(1,1) = \\ &= \sum_{i=0}^{k} r_i = dim\, Pol(n,k) = \\ &= \binom{n-1+k}{k} + \binom{n-2+k}{k-1}. \end{aligned}$$

In fact, extending the argument slightly we can actually prove the following identities:

$$L(\tau_m) = \begin{cases} \binom{n-1+k}{k} + \binom{n-2+k}{k-1}, & m = 2k-1, \\ 2\binom{n-1+k}{k}, & m = 2k. \end{cases}$$

These expressions are formally identical with a lower bound for *spherical designs*. Designs meeting this bound are said to be *tight*. A few such designs are known. We conclude that tight designs are also optimal spherical codes. Spherical designs will be further discussed in Chapter 9.

It turns out that several of the codes discussed in our examples in Chapter 1 satisfy the Levenshtein bound with equality. That means that those codes are best possible. The bound is said to be *tight* in those cases.

There are three infinite families of codes satisfying the Levenshtein bound with equality. They are the *Simplex* codes $\mathcal{S}_n$, with parameters $(n, \rho, M) = (n, 2+2/n, n+1)$, the *Biorthogonal* codes $\mathcal{B}_n$, with parameters $(n, \rho, M) = (n, 2, 2n)$, and a special family of codes with parameters (n, ρ, M) of the form

$$n = q\,(q^2 - q + 1), \quad \rho = 2 - 2/q^2, \quad M = (q+1)(q^3+1), \quad (2.7.2)$$

n	s	M	*Reference*
3	0.21014	7	Schütte–Van der Waerden [193]
3	$(2\sqrt{2}-1)/7$	8	Schütte–Van der Waerden [193]
3	1/3	9	Schütte–Van der Waerden [193]
3	0.40439	10	Danzer [77]
3	$1/\sqrt{5}$	12	Fejes Tóth [106]
3	0.72308	24	Robinson [185]
4	$(\sqrt{5}+1)/4$	120	Böröczky [53]
5	1/5	16	Delsarte–Goethals–Seidel [82]
6	1/4	27	Delsarte–Goethals–Seidel [82]
7	1/3	56	Delsarte–Goethals–Seidel [82]
8	1/2	240	Levenshtein [155], Odlyzko–Sloane [176]
21	1/9	112	Delsarte–Goethals–Seidel [82]
21	1/7	162	Delsarte–Goethals–Seidel [82]
22	1/11	100	Delsarte–Goethals–Seidel [82]
22	1/6	275	Delsarte–Goethals–Seidel [82]
22	1/4	891	Delsarte–Goethals–Seidel [82]
23	1/5	552	Delsarte–Goethals–Seidel [82]
23	1/3	4600	Delsarte–Goethals–Seidel [82]
24	1/2	196 560	Levenshtein [155], Odlyzko–Sloane [176]

Table 2.1: All isolated cases of spherical codes for which optimality has been proven. Except for the cases $M = 7, 10, 24$ in dimension $n = 3$ the exact value of the maximal correlation s has been given. For the three cases mentioned we give just a numerical approximation. We refer to Chapter 3 for a detailed discussion of these cases.

where $q \geq 2$ is a power of a prime. These codes were first constructed by Cameron, Goethals and Seidel [60] as spherical 3-designs obtained from generalized quadrangles. Levenshtein [158] noticed that viewed as spherical codes they are optimal. It is straightforward to check that the cardinalities of these codes meet the Levenshtein bound $L_3(s)$. The codes will be further discussed in Chapter 9.

In addition to these infinite families—and except for the trivial case $n = 2$—there are only 12 known isolated cases of codes satisfying the bound with equality. They are all listed in Table 2.1. The first 5 of them were used as examples in Chapter 1, the remaining ones will appear in later chapters.

In Table 2.1 we have actually included all non-trivial codes for which optimality has been established. With the exception for a few of the codes in dimension $n = 3$ optimality can be proved by the linear programming bound. The Levenshtein bound can be used in all dimensions $n \geq 5$, while a different form of the bound must be used to establish optimality for the code of size $M = 120$ in dimension $n = 4$, [4]. Most of the cases were actually observed already in the pioneering paper by Delsarte–Goethals–Seidel [82].

Chapter 3

Codes in dimension n=3

3.1 Introduction

The case $n = 3$ deserves special attention. It is the first case where interesting problems arise and it is the last case where a complete visualization is possible. The problems here belong mainly to classical geometry and most results have been obtained using geometrical arguments, quite different from those normally encountered in coding theory. Our presentation is tutorial, emphasizing only those aspects which are of interest from a coding theory point of view.

Point-sets in Ω_3 have interested mathematicians since long and is a topic with applications in many different parts of science. It should therefore be no surprise that quite a lot more is known about spherical codes in dimension $n = 3$ than in any higher dimension. More surprising is perhaps the fact that the knowledge available nevertheless is quite incomplete and that much of what *is* known was obtained fairly recently. For instance, optimal (in the sense of maximizing the squared minimum distance) configurations of M points in Ω_3 is known only for $M \leq 12$ and $M = 24$. The cases $M = 3, 4, 6, 12$ were proved by Fejes Tóth [105] in 1943, the cases $M = 5, 7, 8, 9$ were proved by Schütte–van der Waerden [193] in 1951, the cases M=10, 11 were proved by Danzer [77] in 1965 and the case $M = 24$ was proved by Robinson [185] in 1961. The case $M = 2$ is of course trivial.

We will discuss each one of the cases for which optimal codes are known. For $M = 2, 3, 4, 6, 7, 12$ we will prove optimality. In the remaining cases we will just describe the codes.

3.2 The optimal codes

M=2

For $M = 2$ there is essentially no choice: two points can always be connected by a line and all we need to establish is that for the distance to be maximal the line must also contain the origin, i.e. the points must be antipodal. This is trivial. The case is of course degenerated because the code is actually located in Ω_1. The parameters are $(n, \rho, M) = (3, 4, 2)$.

M=3

The optimal configuration of three points on the surface of a 3-dimensional sphere is, of course, an *equilateral triangle* centered at the origin. The coordinates can be chosen so as to locate one vertex to the point $(1, 0, 0)$ and the other two vertices to the points $(-1/2, \pm\sqrt{3}/2, 0)$. Alternatively—if we want to avoid radicals—we have the description $(1, -1, 0)^C$, where the superscript C indicates that all cyclic shifts should be taken.

We note in passing that if instead we take $(1, -1, 0)^S$, where S indicates the full symmetric group, i.e. if we take *all* possible permutations, we obtain the regular *hexagon*, as shown in Figure 3.1. The hexagon is an optimal spherical code in Ω_2, but not in Ω_3.

Both the triangle and the hexagon are located in the subspace Ω_2. Notice, however, that in order to avoid radicals in the characterizations we need to use *three* coordinates in order to describe them.

Optimality of the triangle can easily be established by a direct argument. Alternatively it follows from Fejes Tóth's bound [105] or from Rankin's second bound [184]. The parameters are $(n, \rho, M) = (3, 3, 3)$.

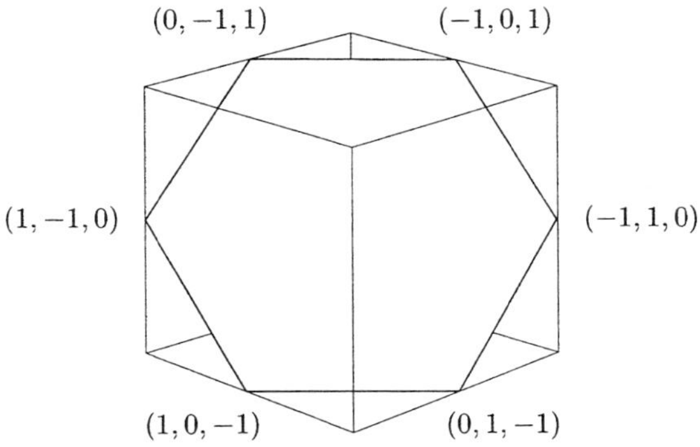

Figure 3.1: Inscribing a regular hexagon in a cube.

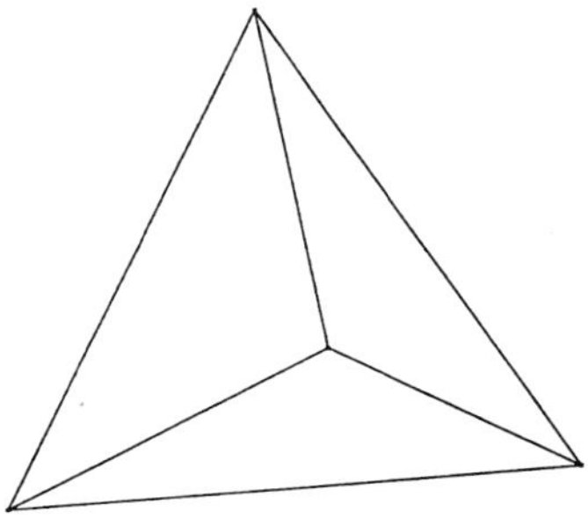

Figure 3.2: The tetrahedron.

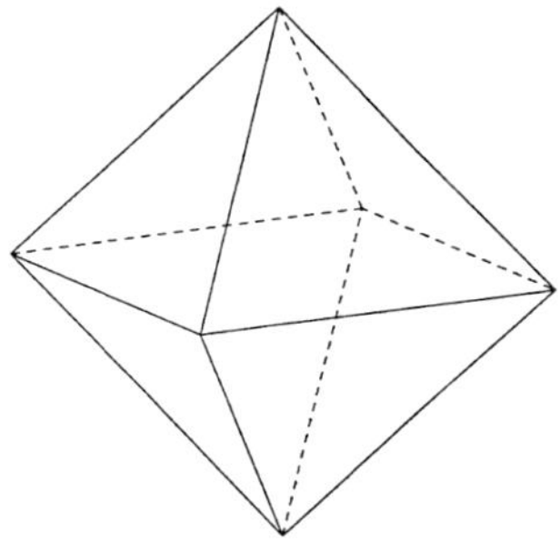

Figure 3.3: The regular octahedron.

M=4
This is the first case where the optimal code is not located in any smaller space, so in a sense this is the first non-degenerated case. The optimal configuration is of course the *Simplex* in Ω_3 i.e. the *tetrahedron*, see Figure 3.2. The standard coordinate representation is $(1, 1, 1, -3)^C$. Alternatively the tetrahedron can be specified by the coordinates $(1, 1, 1)$, $(1, -1, -1)^C$. The parameters are $(n, \rho, M) = (3, 8/3, 4)$.

M=5
It is an interesting fact that there is no optimal configuration of 5 points in Ω_3. More precisely: to any configuration of 5 points we can add a sixth point without decreasing the minimum distance. A

very nice explanation to this phenomenon is given by Coxeter [73], p. 323. Essentially the argument is this. The largest number of equal circles on Ω_3 with the property that they all touch each other is 4. Therefore, if in a configuration of 5 points we surround each point with a circle of maximum size there must be two circles that do not touch each other. In an optimal configuration the remaining three must touch both of the two. If we place the two at the poles the three will be located along the equator and it is fairly easy to see that there will be a continuous range of possibilities. If we put them as close to each other as possible there will actually be space enough for a sixth circle of the same size.

M=6

The optimal configuration of 6 points in Ω_3 is the *Biorthogonal* code. The codewords are located at the vertices of a regular *octahedron*, see Figure 3.3. The standard coordinate representation is $(0,0,\pm 1)^C$, where the notation means that we should take all cyclic shifts and all possible sign-changes. The parameters are $(n, \rho, M) = (3, 2, 6)$.

M=7

This is the first non-obvious case. The optimal configuration can be described as the union of two *equilateral triangles* of different sizes and a single point. The two triangles are placed in parallel above each other and with shifted orientations. If we put the smaller of them at the bottom the single point will be located at the top, above the larger triangle. A possible coordinate description is as follows:

$$\begin{array}{ll} \textit{Large triangle}: & ((a,-a,0)^C, x) \\ \textit{Small triangle}: & ((-b,b,0)^C, -y) \\ \textit{Single point}: & (0,0,0,1) \end{array}$$

with $0 < x < y < 1$, $0 < b < a$. The parameters a, b, x, y should be adjusted so that all points get unit norm and so that the minimum distance occurs at the appropriate places. The unit norm condition gives us

$$2a^2 + x^2 = 1, \quad 2b^2 + y^2 = 1,$$

while the conditions on the minimum distances—which alternatively can be formulated in terms of maximal correlations—is captured by the following equations:

$$-b^2 + y^2 = ab - xy = x.$$

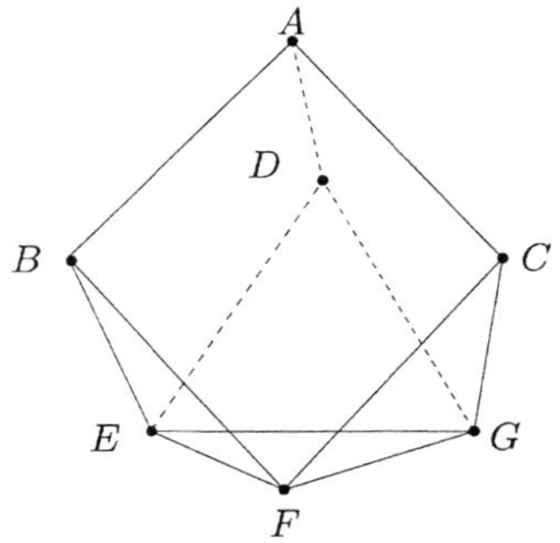

Figure 3.4: The optimal configuration of 7 points in Ω_3.

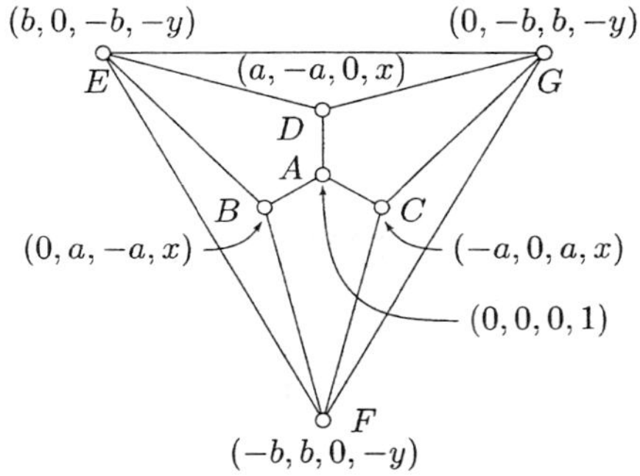

Figure 3.5: A minimum distance graph for the optimal configuration of 7 points in Ω_3.

A two-dimensional projection of the configuration is shown in Figure 3.4. The vertices of the smaller triangle at the bottom are indicated by E, F and G, while the vertices of the larger triangle above are indicated by B, C and D. The single point at the top is indicated by A. Notice that the triangles EFG and BCD are parallel.

The *graph* obtained by taking the codewords in a spherical code $\mathcal{X} \subseteq \Omega_3$ as the *vertices* and letting the *edges* be defined by all pairs of codewords which are located on *minimum distance* is called a *minimum distance graph*. It is easy to show that such a graph is always a *plane graph*, i.e. a graph which can be drawn on a piece of paper in such a way that no edges are crossing each other.

In Figure 3.5 such a graph is shown for the optimal code of size $M = 7$. Notice that the minimum distance graph is a topological device: it shows in a nice way all the connections between neighbors, but it usually gives entirely wrong impressions of geometric dimensions. For instance, in the present case the triangle with parameters a and x is really the bigger one although the minimum distance graph in Figure 3.5 gives the opposite impression.

We have four equations and four unknowns, so the 7-point code is fully characterized by the coordinates and the equations above. Straightforward computations reveal that the solution can be expressed as $\rho = 3x$, where x is a solution to the equation

$$729x^4 - 648x^3 - 129x^2 + 1728x - 512 = 0.$$

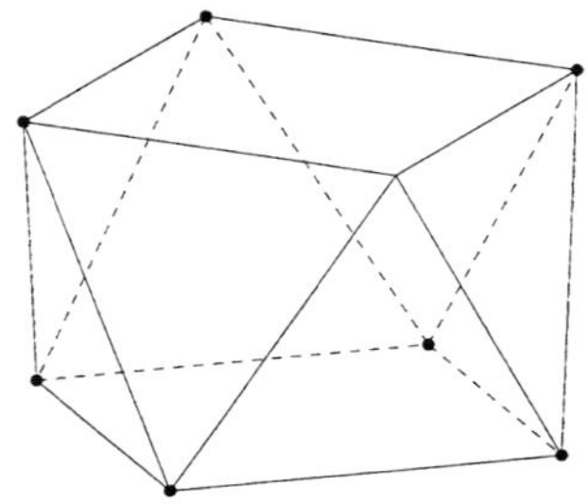

Figure 3.6: The square antiprism.

A numerical evaluation gives us the following values for the parameters: $(n, \rho, M) \approx (3, 1.57972, 7)$.

A more extensive discussion will be provided in Section 3.5, where a proof of optimality will be given. We also remark that this code is the first member of an infinite family of codes having size $M = 2n+1$ in dimension $n = 3, 4, \ldots$, see Chapter 5, Example 5.10.4.

M=8

The first configuration that comes to mind for the case $M = 8$ is maybe the cube, with parameters $(n, \rho, M) = (3, 4/3, 8)$. A moment of thought reveals, however, that the cube cannot be optimal. Think of the cube as a pair of squares placed one on top of the other. It is clear that we can increase the distance between any pair of vertical points by simply turning the squares 45 degrees out of phase from each. By then slightly increasing both squares while keeping all vertices on the unit sphere we can obviously generate a code with larger distance.

The above argument suggests the following coordinates,

$$\begin{array}{ll} \textit{Bottom square}: & (\pm(1/\sqrt{2})\sin\theta,\ \pm(1/\sqrt{2})\sin\theta,\ -\cos\theta), \\ \textit{Top square}: & (\pm\sin\theta,\ 0,\ \cos\theta),\ (0, \pm\sin\theta,\ \cos\theta)\,, \end{array}$$

where the parameter θ should be chosen such that the following equality holds:

$$\cos^2\theta = \frac{1}{\sqrt{2}}\sin^2\theta - \cos^2\theta\,.$$

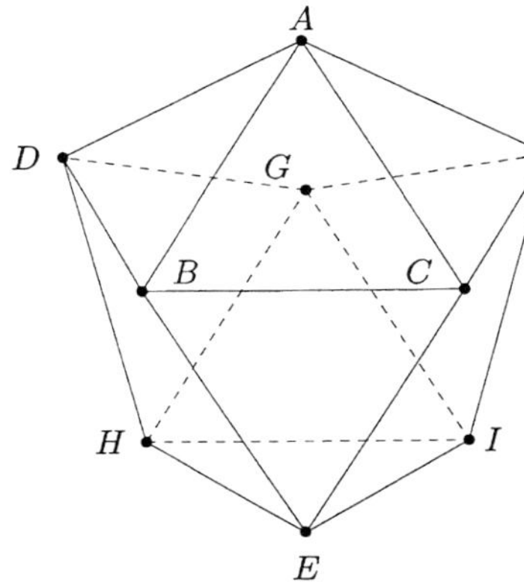

Figure 3.7: The optimal configuration of 9 points in Ω_3.

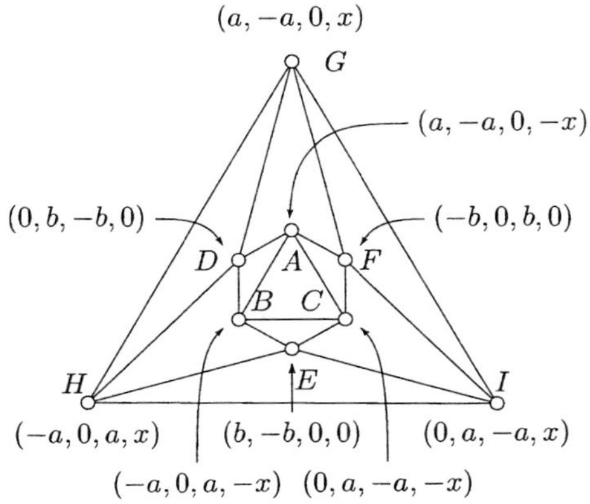

Figure 3.8: A minimum distance graph for the optimal configuration of 9 points in Ω_3.

Solving the equation we get

$$\cos^2\theta = \frac{1}{1+2\sqrt{2}},$$

which gives us the following squared minimum distance: $\rho = (4/7)(4-\sqrt{2}) \approx 1.47759$. It turns out that this is really an optimal configuration. The corresponding polyhedron is usually called a *square antiprism*. It is illustrated in Figure 3.6. The parameters are $(n, \rho, M) = (3, (4/7)(4-\sqrt{2}), 8)$. In Chapter 5, Example 5.10.1 this construction will be generalized so as to apply to an infinite sequence of codes of increasing dimensions.

M=9

The case $M = 9$ is very similar to the case $M = 7$, but actually somewhat simpler in the sense that here an explicit solution can be obtained. The optimal configuration consists of three parallel triangles as indicated in Figure 3.7. The three triangles are placed above each other. The two outer triangles ABC and GHI are smaller and both have the same orientation. The triangle DEF in the middle is bigger and has an opposite orientation. The coordinates can be given as follows:

$$\begin{array}{rl} \textit{Outer triangles}: & ((a,-a,0)^C, \pm x) \\ \textit{Middle triangle}: & ((-b,b,0)^C, 0)\,, \end{array}$$

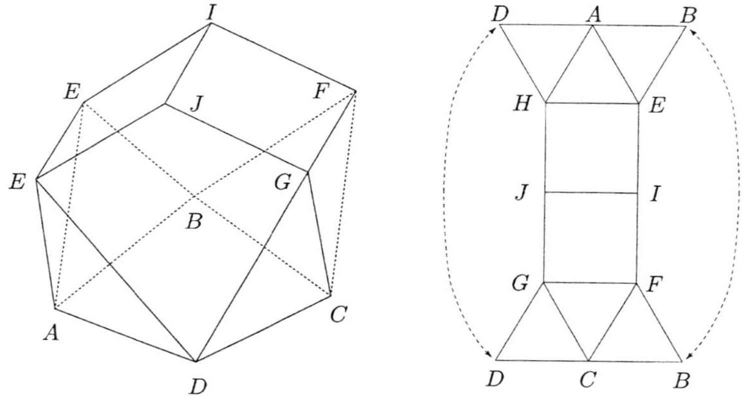

$$\begin{aligned} A, C: &\quad (\pm \sin\alpha, 0, \cos\alpha) \\ B, D: &\quad (0, \pm \sin\beta, \cos\beta) \\ E, F, G, H: &\quad (\pm \cos\theta \sin\gamma, \pm \sin\theta \sin\gamma, \cos\gamma) \\ I, J: &\quad (0, \pm \sin\delta, \cos\delta) \end{aligned}$$

Figure 3.9: The optimal configuration of 10 points in Ω_3.

where again the parameters should be chosen such that all vectors get unit norm and such that the minimum distance is established between appropriate points. The following equations guarantee all these conditions:

$$2a^2 + x^2 = 2b^2 = 1, \quad -a^2 + x^2 = -ab.$$

A minimum distance graph is shown in Figure 3.8. An easy calculation reveals that the minimum squared distance is $\rho = 4/3$. It is interesting to notice that this is exactly the same minimum distance as for the cube. The parameters for the optimal 9-point configuration are $(n, \rho, M) = (3, 4/3, 9)$.

M=10

The optimal configuration of 10 points in Ω_3 is rather complicated. In Figure 3.9 we offer three different characterizations: a projection on $\mathbb{R}^2$, an unfolding of the plane faces, and finally a set of possible Cartesian coordinates for the vertices. The unfolding of the plane faces can be used for constructing a model if the faces are cut out and

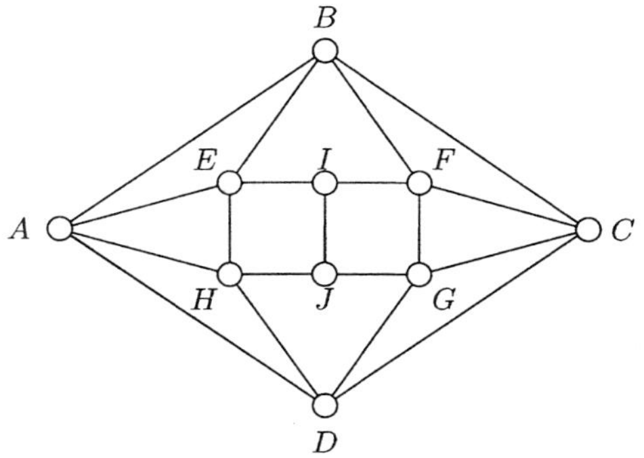

Figure 3.10: A minimum distance graph for the optimal configuration of 10 points in Ω_3.

folded back appropriately. Notice that $ABCD$, $BEIF$, and $DGJH$ are *skew* rhombs. A minimum distance graph is shown in Figure 3.10. The coordinates as given in Figure 3.9 are

$$A, C : (\pm \sin\alpha, 0, \cos\alpha)$$

$$B, D : (0, \pm \sin\beta, \cos\beta)$$

$$E, F, G, H : (\pm \cos\theta \sin\gamma, \pm \sin\theta \sin\gamma, \cos\gamma)$$

$$I, J : (0, \pm \sin\delta, \cos\delta)$$

The five parameters $\alpha, \beta, \gamma, \delta, \theta$ should be chosen so that the minimum distance occurs at all the edges in the triangles and in the squares. Taking inner products we obtain

$$AB : \cos\alpha \cos\beta$$

$$AE : \sin\alpha \cos\theta \sin\gamma + \cos\alpha \cos\gamma$$

$$BE : \sin\beta \sin\theta \sin\gamma + \cos\beta \cos\gamma$$

$$EH : 1 - 2\sin^2\theta \sin^2\gamma$$

$$EI : \sin\theta \sin\gamma \sin\delta + \cos\gamma \cos\delta$$

$$IJ : 1 - 2\sin^2\delta$$

It is clear that the five equations obtained by setting all these expressions equal are exactly what is needed in order to determine all five

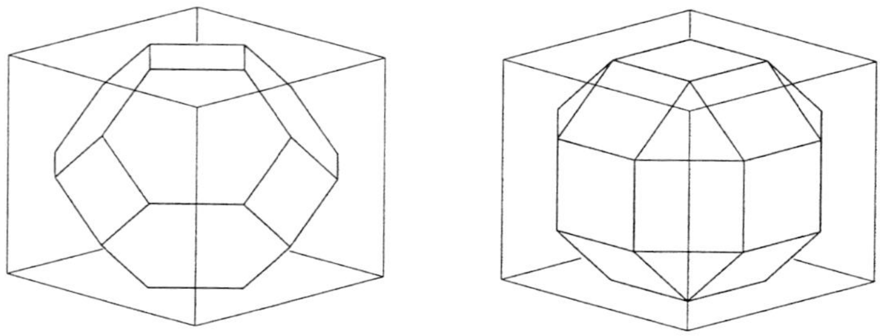

$X_1 = (0, \pm 1, \pm 2)^S$ $\quad$ $X_2 = (\pm 1, \pm 1, \pm(\sqrt{2}+1))^C$

$\rho_1 = \frac{2}{5}$ $\quad$ $\rho_2 = \frac{4}{5+2\sqrt{2}}$

Figure 3.11: Two suboptimal spherical codes of size $M = 24$ in Ω_3.

parameters. Thus the configuration is fully characterized by these conditions. Unfortunately the equations are somewhat complicated. A numerical evaluation gives $\rho \approx 1.19121$.

M=11

Here a phenomenon somewhat similar to what we saw for $M = 5$ occurs: to any optimal configuration of eleven points a twelfth point can always be added without decreasing the minimum distance. There is, however, an important difference. For the 5 points we had a certain degree of freedom in putting them on a largest possible distance. That is not the case for $M = 11$: whenever 11 points in Ω_3 are at a maximum distance from each other they are necessarily located at the vertices of a regular icosahedron. This remarkable fact was established by Danzer [77], see also Böröczky [54].

M=12

The solution for 12 points is the expected one: the optimal code is given by the vertices of the regular *icosahedron*. (This follows already from our discussion of the case $M = 11$). A possible parametric characterization of the icosahedron is given by $(0, \pm\tau, \pm 1)^C$, where $\tau = (1+\sqrt{5})/2$ is the *golden section*. The squared minimum distance is $\rho = 2 - 2/\sqrt{5} \approx 1.10557$. For an exciting an detailed discussion we refer to Coxeter ([75] p. 52). The parameters of the code are $(n, \rho, M) = (3, 2 - 2/\sqrt{5}, 12)$.

M=24

It is tempting to try to construct a 24-point code in Ω_3 by putting four points on each face of a cube. In Figure 3.11 we have indicated the two most obvious ways of doing that. The resulting codes are (in non-normalized notation) $\mathcal{X}_1 = (0, \pm 1, \pm 2)^S$ and $\mathcal{X}_2 = (\pm 1, \pm 1, \pm(\sqrt{2}+1))^C$. The squared minimum distances are $\rho_1 = 2/5$ and $\rho_2 = 4(5 - 2\sqrt{2})/17 \approx 0.51096$, respectively. Unfortunately, however, none of them is optimal. This is easy to see: in both cases we can increase the minimum distance between squares on different faces by slightly twisting all the squares. If we then let all the squares also increase we obtain—in both cases—codes with larger minimum distances. Thus, both constructions are not even locally optimal.

However, by continuing the turn of the squares we eventually will —after 45 degrees—transform one code to the other. Thus, it is clear that during this process the minimum distance—which initially increases—reaches a maximum and then decreases again. It is natural to ask: what are the parameters of the best code generated during this process?

If we start the process with $\mathcal{X}_1$ and denote the angle of turn with β we can compute—including the normalization to unit norm—the following sets of coordinates:

$$\begin{aligned} &\textit{Even}: \quad (\pm\cos\alpha, \pm\sin\alpha\cos\beta, \pm\sin\alpha\sin\beta)^C \\ &\textit{Odd}: \quad (\pm\cos\alpha, \pm\sin\alpha\sin\beta, \pm\sin\alpha\cos\beta)^C \end{aligned}$$

where "*Even*" and "*Odd*" indicate that the selection of signs is restricted so as always to have an even—respectively an odd—number of negative terms. The parameter α should be adjusted so that the minimum distance between points from two adjacent faces always equals the minimum distance between points on one and the same face. The inner product between points on one and the same face is $\cos^2\alpha$. For small angles β we have the condition

$$\cos^2\alpha = \sin\alpha\cos\alpha\sin\beta + \sin\alpha\cos\alpha\cos\beta + \sin^2\alpha\sin\beta\cos\beta,$$

while for large β the condition is

$$\cos^2\alpha = 2\sin\alpha\cos\alpha\cos\beta - \sin^2\alpha\sin^2\beta.$$

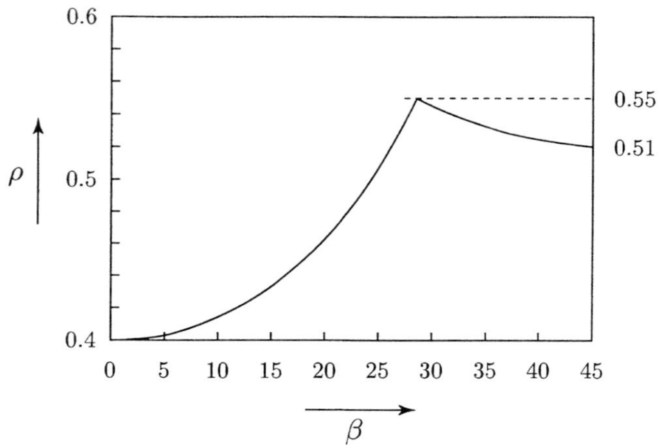

Figure 3.12: Minimum squared distance ρ as a function of the twist angle β for the 24-points codes generated by twisted squares on the cube.

Clearly the best code is the one obtained when both of these conditions are satisfied simultaneously.

In Figure 3.12 we show the squared minimum distance as a function of β in the range $0 \leq \beta \leq \pi/4$. As seen there is a sharp maximum. By straightforward computations we find that this maximum is a solution to the equation

$$7\rho^3 - 44\rho^2 + 80\rho - 23 = 0.$$

A numerical evaluation gives $\rho \approx 0.55384$.

It turns out that the corresponding code is optimal. Proving this fact is, however, quite an intriguing task. The result was conjectured by Schütte-van der Waerden [193] in 1951 and proved by Robinson [185] ten years later. The optimal code is shown in Figure 3.13 along with a plane map of half the surface. The code corresponds to a polyhedron known as "the snub cube" (Coxeter [75], p. 326). It is bounded by 32 equilateral triangles and 6 squares. One square and four triangles meet at each vertex. The triangles are actually of two kinds: 24 of them share an edge with one of the squares while the remaining 8 triangles are entirely surrounded by other triangles.

The parameters of the optimal code are $n = 3$, $\rho \approx 0.55384$, $M = 24$.

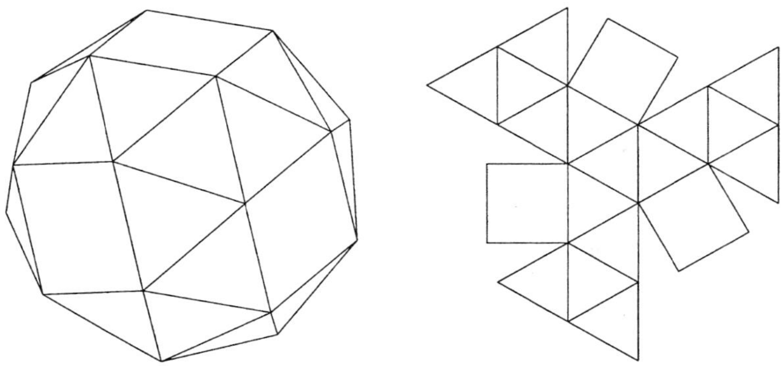

Figure 3.13: The snub cube along with a plane map of half its surface. The vertices correspond to an optimal spherical code with parameters $n = 3$, $\rho \approx 0.55384$, $M = 24$.

3.3 Additional comments

The problem of placing M points on the unit sphere such that the minimum distance between any two points is maximal—i.e. the problem of finding an optimal configuration for M codewords in Ω_3—was studied in the 1930-th by the Dutch botanist Tammes [220]. It is sometimes referred to as Tammes' problem. A very nice account of this problem including a lot of interesting historical remarks can be found in Coxeter [75].

The cases described in Section 3.2 are the only ones for which optimality has been proved. However, conjectures exist for many other cases, see for instance Coxeter [75]. Clare–Kepert [68] used numerical methods to derive a number of configurations in the range $20 \leq M \leq 40$. They also gave several best known codes in the range $M > 40$, along with a lot of detailed information about many of the codes. An interesting observation of theirs is that the density of the packings is close to 0.82 in all of the range $20 \leq M \leq 40$. As they pointed out this is surprisingly far away from the ultimate limit, which equals the density $\pi/(2\sqrt{3}) \approx 0.9069$ of the hexagonal packing in the plane.

An extensive numerical search for spherical codes in the range $3 \leq n \leq 5$, $M \leq 120$ was performed by Sloane [214]. The results are available by automatic e-mail.

Some of the best spherical codes in Ω_3 are closely connected to well known polyhedra. More precisely: for many polyhedra the set of vertices form a spherical code. The regular tetrahedron, the regular octahedron and the regular icosahedron generate in this way optimal codes of sizes $M = 4, 6, 12$. The optimal code of size $M = 8$ corresponds to a square antiprism (Coxeter [75], p. 4) and the codewords of the optimal 24-point configuration are the vertices of a "snub cube". The cases $M = 7, 9, 10$ are more complicated and the cases $M = 2, 3, 5, 11$ are degenerated in various senses.

It is worthwhile noting that among the five Platonic solids only those with triangular faces—i.e. the tetrahedron, the octahedron and the icosahedron—generate optimal codes in this direct way. As a spherical code the vertex set of the cube is dominated both by the square antiprism (same M and larger ρ) and the optimal 9-point code (same ρ and larger M). By instead placing a point at the center of each *face* of the cube we obtain again the *Biorthogonal* code, and by placing a *pair* of points or a *quadruple* of points on each face we might generate the icosahedron or the snub cube. The icosahedron can also be generated by placing a point on each one of the faces of the dodecahedron.

For applications there are, of course, many other aspects beside size and distance that are of importance. One such is the size of the *alphabet* used to generate the code. The representation we suggested for the vertex set of the icosahedron, namely $(0, \pm\tau, \pm 1)^C$, requires a *non-uniform* alphabet of size $q = 5$ for generating a 12-point spherical code with squared distance $\rho = 2 - 2/\sqrt{5} \approx 1.10557$. In contrast the standard representation $(0, \pm 1, \pm 1)^C$ for the vertex set of the cubeoctahedron uses a *uniform* alphabet of size $q = 3$ for generating a 12-point code with only slightly worse performance ($\rho = 1$). In some applications the advantage of the smaller alphabet might well compensate for the degradation in performance.

Both the icosahedron and the cubeoctahedron provide possible solutions to the so called *kissing number problem*, i.e. the problem of arranging a largest possible number of equal spheres in such a way that they all touch another equal sphere in the middle. Coxeter ([75], p. 321) mentions that this problem was discussed by Newton and Gregory in 1694. At that occasion Newton claimed that 12 is the maximal number while Gregory insisted that 13 might be possible.

It is easy to see that 12 is a *possible* number: a natural configuration is that suggested by the regular icosahedron. However, the squared distances between the outer spheres in an icosahedral configuration is $\rho = 2 - 2/\sqrt{5} > 1$, which means that the *outer* spheres do not touch *each other*. That means that there are a large number of alternative possibilities. The cubeoctahedron represents just one of them. The question is: if the outer spheres are pushed together as much as possible, will there be space enough for an additional thirteenth sphere?

The problem is certainly not trivial. Finding the solution is equivalent to proving that ρ is strictly less than 1 for any spherical code in Ω_3 of size $M = 13$. This is true, but far from easy to prove. According to Coxeter ([75], p. 321), the first proof was not given until 1874, by Hoppe [131], 180 years after the discussion between Newton and Gregory. However, Zong ([242], p. 21) claims that this proof is not complete. As we will see shortly none of the standard general bounds is strong enough to prove this fact. In Section 7 of the present chapter we offer a proof based on Leech, [150].

The cubeoctahedron has the interesting property that it can be generated by taking the midpoints of all *edges* from either the cube or the octahedron. This is a consequence of the fact that the cube and the octahedron are *reciprocal polyhedra* (Coxeter [75], p. 17). In this context we like to mention that interesting spherical codes in Ω_3 can also be generated by simply taking the union of the cube and the octahedron (gives a code of size $M = 14$ and squared minimum distance $\rho = 2 - 2/\sqrt{3} \approx 0.8453$) or by taking the union of the cubeoctahedron and the octahedron (gives a code of size $M = 18$ and squared minimum distance $\rho = 2 - \sqrt{2} \approx 0.5858$). Both are rather good but none of them is optimal.

3.4 The Fejes Tóth bound

Fejes Tóth [105] in 1943 derived a general bound for spherical codes in Ω_3. This appears to be the first systematic approach to the problem of finding best spherical codes. A generalization to arbitrary dimensions n was suggested by Coxeter [74] in 1963 and proved by K. Böröczky [53] in 1978. Unfortunately the generalization is considerably more complicated and usually very difficult to compute. We discuss in detail only the simpler bound by Fejes Tóth.

The proof of the Fejes Tóth bound is based on *spherical geometry*. In Appendix C we offer a brief introduction and derive some elementary results needed in the proof. A crucial role is played by the function

$$T^*(\alpha, b) \triangleq \alpha - \arcsin(\cos b \sin \alpha)\,.$$

In Appendix C we show that this function gives the area of a *spherical triangle* with angles (α, β, γ), where $\gamma = \pi/2$, and with side-lengths (a, b, c). We notice that this function is convex in α and increasing in b.

We recall that the *Voronoi region* to a point x in a code $\mathcal{X} \subseteq \Omega_3$ is the set of points in Ω_3 which are closer to x than to any other point in the code. We also notice that the borderlines between the Voronoi regions are parts of great-circles, i.e. *lines* in the spherical geometry. Two codewords corresponding to Voronoi regions with a common borderline will be called *neighbors*. By definition a line connecting two neighboring codewords cross the borderline between their Voronoi regions perpendicularly.

Consider a given code $\mathcal{X} \subseteq \Omega_3$. Let $\{V_i : i = 1, 2, \ldots, M\}$ be the set of Voronoi regions, let $E = \{e_k : k = 1, 2, \ldots, K\}$ be the set of all borderlines between the Voronoi regions and let $V = \{v_j : j = 1, 2, \ldots, N\}$ be the set of points where the borderlines meet. The Voronoi regions form a *partition* of the sphere Ω_3. A partition of this kind is usually called a *map* ([74], p. 6, see also Appendix C). We will refer to this particular map as the *Voronoi map* of the code $\mathcal{X} \subseteq \Omega_3$.

Theorem 3.4.1 (Fejes Tóth, [105], 1943) *Let $\mathcal{X}$ be a spherical code in Ω_3 of size $M \geq 3$. Then the squared minimum distance ρ satisfies the following inequality:*

$$\rho \leq \rho_{FT}(M) \triangleq 4 - \frac{1}{\sin^2 \omega_M}\,,$$

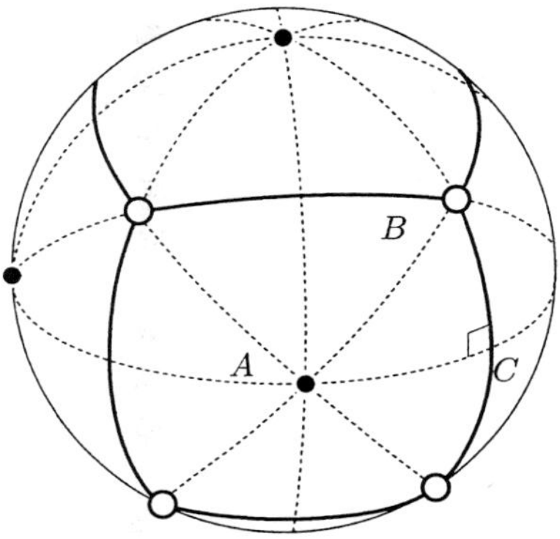

Figure 3.14: The Voronoi map and its triangulation for the *Biorthogonal* code (the octahedron). The codewords are indicated by black dots while the points in the vertex set V—the crossing points for the borderlines to the Voronoi regions—are indicated with white dots. The borders of the Voronoi regions are indicated by solid lines, while those of the triangles are indicated with dotted lines.

where

$$\omega_M \triangleq \frac{\pi}{6} \cdot \frac{M}{M-2}.$$

Proof: Consider the Voronoi map of the code $\mathcal{X} \subseteq \Omega_3$. From each codeword x_i we draw a line to each one of its neighboring codewords and to each one of the vertices v_j in its Voronoi region. In this way each Voronoi region is divided into *triangles*, with two triangles for each one of its borderlines, see Figure 3.14. As each edge in E is the borderline of *two* Voronoi regions it follows that there are *four* triangles associated with each edge. Thus, altogether there are $4K$ triangles, where $K = |E|$ denotes the number of edges in the Voronoi map.

Consider one of these triangles. It has one corner A in the code $\mathcal{X}$, one corner B in the vertex set V and one corner C located on one of the edges in E, halfway between two codewords, see Figure 3.14. The angle at C is $\gamma = \pi/2$. Let the angle at A be α and let the angular length of the side AC be b. By Appendix C—and as remarked above—the area of this triangle is $T^*(\alpha, b)$.

Let the minimum squared *Euclidean* distance be $\rho = 2 - 2\cos\theta$, where θ is the minimum *angular* distance. We enumerate the triangles by indices i, $i = 1, 2, \ldots, 4K$. For the i-th triangle, let α_i and b_i be defined as α and b above. Notice that the sum of the angles α_i is $2\pi M$ and that the inequality $b_i \geq \theta/2$ holds for all indices i.

As the triangles form a partition of the sphere the sum of their areas equals that of the sphere. Thus we have

$$4\pi = \sum_{i=1}^{4K} T^*(\alpha_i, b_i) \geq \sum_{i=1}^{4K} T^*(\alpha_i, \theta/2)\,,$$

where the inequality follows because T^* is *increasing* as a function of its second argument. As it is *convex* as a function of its first argument we can apply Jensen's inequality. The result is

$$\begin{aligned} 4\pi &\geq \sum_{i=1}^{4K} T^*(\alpha_i\ ,\ \theta/2) \geq 4K \cdot T^*(\frac{1}{4K}\sum_{i=1}^{4K} \alpha_i\ ,\ \theta/2) = \\ &= 4K \cdot T^*((\pi M)/(2K)\ ,\ \theta/2) = \\ &= 4K \cdot \left[\frac{\pi M}{2K} \quad - \quad \arcsin\left(\cos(\theta/2)\sin(\frac{\pi M}{2K})\right)\right], \end{aligned}$$

where in the second step we have utilized the fact mentioned above that the sum of the angles α_i equals $2\pi M$. Solving with respect to $\cos(\theta/2)$ we get

$$\cos(\theta/2) \geq \frac{\sin(\pi(M-2)/2K)}{\sin(\pi M/2K)}\,. \tag{3.4.1}$$

At this stage we need an estimate on the quantity K. Recall that N denotes the number of vertices in the set V. By Euler's formula (Appendix C) we have

$$M + N - K = 2\,.$$

Let k_i denote the number of edges connected to vertex v_i. As each edge is connected to precisely two vertices, and as the number of edges connected to each vertex is at least three, we get

$$2K = \sum_{i=1}^{N} k_i \geq 3N = 3(K - M + 2).$$

Solving with respect to K gives us

$$K \leq 3(M-2).$$

It is easy to establish that the right hand member of equation (3.4.1) is a decreasing function of K. Thus, the inequality remains valid if we replace K with the number $3(M-2)$. We notice that by this operation the nominator is replaced by $\sin(\pi/6) = 1/2$, while the denominator is replaced by $\sin \omega_M$, where $\omega_M \triangleq (\pi/6) \cdot M/(M-2)$. From the new inequality we obtain

$$\frac{1}{\sin^2 \omega_M} \leq 4\cos^2(\theta/2) = 2\cos\theta + 2 = 4 - \rho.$$

□

The quantity ω_M has an interesting interpretation. Consider an equilateral spherical triangle with angle $2\omega_M$. By the *Excess Theorem* (Appendix C) the area of such a triangle is

$$T = 6\omega_M - \pi = \frac{4\pi}{2M-4}.$$

A *triangulation* of a spherical code $\mathcal{X}$ is a map on Ω_3 with the code $\mathcal{X}$ itself as the vertex set and with edges selected so that the map has triangular faces only. Notice that the triangles formed in this way are *not* the same as those employed in the proof of the Fejes Tóth theorem. It is easy to see that such a map always exists for any point set $\mathcal{X}$. Actually, in general there are many different triangulations. However, the number F of faces is always given by $F = 2M - 4$. This is a simple consequence of *Euler's formula* (Appendix C). It follows that the area T of the equilateral spherical triangle with angle $2\omega_M$ equals the average size of the triangles in any triangulation of a code $\mathcal{X}$ of size M.

One consequence of this observation is that any code which admits a triangulation such that all faces are equilateral triangles of the same size satisfies the Fejes Tóth bound with equality.

There are four such cases, namely the equilateral *plane* triangle and the three regular polyhedra with triangular faces: the tetrahedron, the octahedron, and the icosahedron. The parameters are as follows:

M	3	4	6	12
ρ	3	8/3	2	$2-2/\sqrt{5}$

The theorem implies that all these codes are optimal. The first three cases follow also by the Rankin bounds (which, however, were proved only 12 years later—in 1955). We saw in Chapter 2 that the fourth case (optimality of the icosahedron) also follows from the Levenshtein bound. For $M = 13$ we have $\rho_{FT} \approx 1.0279 > 1$, which means that Fejes Tóth's bound is *not* strong enough to resolve the controversy between Newton and Gregory.

In Table 3.1 we have listed the parameters of all best known codes in the range $3 \leq M \leq 24$ as given by Clare–Kepert [68], along with the bounds by Levenshtein [156], [157], Fejes Tóth [105], and Boyvalenkov–Danev–Bumova [45]. We see that none of these bounds is strong enough to establish optimality except in the four cases mentioned above (and the trivial case $M = 2$ in case of the Levenshtein bound).

There is a strong agreement between the Levenshtein and the Fejes Tóth bounds, although they really coincide only occasionally. Usually the Levenshtein bound gives slightly better results for large distances, while the Fejes Tóth bound tends to be better for small distances. For $M = 13$ the Levenshtein bound gives $\rho \leq \rho_L(3, 13) \approx 1.0213$. This is slightly better than the Fejes Tóth bound, but still not enough to prove that Gregory was wrong.

The close agreement between the two bounds is rather remarkable, taking into account the fact that they are derived by entirely different techniques.

In all cases where the exact value is not known the Boyvalenkov–Danev–Bumova improvement of the Levenshtein bound is strictly better than both the Fejes Tóth bound and the Levenshtein bound. Still, not even this improvement is strong enough to establish the result proved by Hoppe, namely that $\rho^{(n)}(13)$ is strictly less than one.

3.5 Optimality in the case M=7

In addition to the cases covered by the Fejes Tóth and the Levenshtein bounds there are six other cases where optimality has been proved

M	ρ_{CK}	ρ_L	ρ_{FT}	ρ_{BDB}	
24	0.5538	0.5795	0.5788	0.5760	*
23	0.5538	0.6026	0.6028	0.5996	
22	0.5862	0.6286	0.6288	0.6254	
21	0.5996	0.6580	0.6572	0.6550	*
20	0.6470	0.6907	0.6883	0.6880	*
19	0.6470	0.7216	0.7225	0.7181	
18	0.7025	0.7577	0.7602	0.7529	
17	0.7421	0.8000	0.8020	0.7950	
16	0.7754	0.8494	0.8487	0.8397	*
15	0.8148	0.8956	0.9011	0.8866	
14	0.8772	0.9520	0.9604	0.9447	
13	0.9147	1.0213	1.0279	1.0136	
12	1.1056	1.1056	1.1056	1.1056	
11	1.1056	1.1821	1.1957	1.1744	
10	1.1912	1.2835	1.3016	1.2711	
9	1.3333	1.4203	1.4276	1.3926	
8	1.4776	1.5426	1.5797	1.5223	
7	1.5797	1.7239	1.7665	1.7121	
6	2.0000	2.0000	2.0000	2.0000	
5	2.0000	2.2223	2.2959	2.2223	
4	2.6667	2.6667	2.6667	2.6667	
3	3.0000	3.0000	3.0000	3.0000	

Table 3.1: Best known codes in the range $n = 3$, $3 \leq M \leq 24$, according to Clare–Kepert [68] (CK), along with evaluations of the Fejes Tóth (FT), the Levenshtein (L), and the Boyvalenkov–Danev–Bumova (BDB), bounds. The star indicates where the Fejes Tóth bound is slightly better than the Levenshtein bound. In all cases where the exact value is not known the Boyvalenkov–Danev–Bumova bound outperforms both of the other two upper bounds.

for codes in dimension $n = 3$, namely $M = 7, 8, 9, 10, 11$, and 24. As we have already remarked, the techniques employed for proving optimality in those cases are geometric and very different from the ones otherwise employed in connection with spherical codes. It is not our ambition to discuss these matters in detail. However, in order to illustrate the ideas involved we present a proof for the simplest of these cases, namely $M = 7$.

As for the Fejes Tóth bound the proof relies on several results from spherical geometry. Again we refer to Appendix C.

In spherical geometry the natural distance is the *angular* distance θ. We recall that this distance is related to the squared *Euclidean* distance ρ by the formula $\rho = 2 - 2\cos\theta$.

Let $\mathcal{X}$ be an arbitrary spherical code in Ω_3 and let its minimum angular distance be θ. Recall that the minimum distance graph is obtained by taking all codewords $x \in \mathcal{X}$ as vertices and letting the set E of edges be those lines connecting pairs (x, y) of codewords in $\mathcal{X}$ which are located on minimum distance. We denote this graph with $\Gamma(\mathcal{X}, E)$. As usual the number of edges connected to a given vertex will be called the *valency* of the vertex. A point x corresponding to a vertex of valency 0 will be called an *isolated point.*

A point x which by a small displacement not affecting any other word in the code can be transformed to an isolated point is said to be *loose*. This, of course, is the case with all already isolated points, but also with those points for which there are only one or two other points on minimum distance. In fact, in order for a point P *not* to be loose it must be surrounded by at least *three* other points. In addition the surrounding points must be located so that all the angles between edges connected to P are strictly less than π.

A code $\mathcal{X} \in \Omega_3$ containing non-isolated loose points is said to be *reducible*; if *all* its loose points are isolated it will be called *irreducible*, ([106], p. 159).

It is clear that any reducible code can be transformed to an irreducible code without decreasing the minimum distance. Therefore, in the sequel we restrict ourselves—without loss of generality—to irreducible codes.

The edges in the graph $\Gamma(\mathcal{X}, E)$ do not intersect. This is a simple consequence of the triangle inequality and the fact that angular distance is a *metric*. It follows that $\Gamma(\mathcal{X}, E)$ is a *plane* graph. A *cycle* of length m in this graph is a sequence $x_1x_2, x_2x_3, \ldots, x_mx_1$ of m connected edges such that all the m vertices $x_1, x_2, \ldots, x_m$ are different. Clearly $m \geq 3$. We notice that any irreducible code contains at least one cycle and that any isolated point must be located inside a cycle.

A cycle within which no other point is contained will be called a *face*. The length m of a cycle defining a face is called the *degree* of the face. If the code is irreducible none of the vertices defining a face is loose. It follows that all the inner angles at the various vertices are strictly less than π. In particular this implies that any face in an irreducible code is a convex region.

We need a few lemmas. The first one applies to *any* spherical code in Ω_3, while the remaining ones are special properties of the case $M = 7$.

Lemma 3.5.1 *Let $\mathcal{X}$ be a spherical code in Ω_3 and let E be the set of all pairs of codewords on minimum distance. The minimum distance graph $\Gamma(\mathcal{X}, E)$ has the property that any cycle of size* 3 *or* 4 *corresponds to a face.*

Proof: A cycle of length 3 generates an equilateral triangle. A circle of radius θ with center at one of the vertices of the triangle covers all of the triangle. It follows that the triangle cannot contain any further point from the code.

A cycle of length 4 generates a quadrangle with all edges of equal length θ. The two circles of radius θ with centers in two opposite vertices of the quadrangle cover all of the quadrangle. It follows that the quadrangle cannot contain any further points from the code. □

Lemma 3.5.2 *An* optimal irreducible *code of size $M = 7$ in Ω_3 has the property that* any *face has degree either* 3 *or* 4.

Proof: Let $\mathcal{X}$ be an optimal irreducible code of size $M = 7$. We denote the minimum angular distance—which is still to be determined—by θ. Consider any one of the faces in the graph $\Gamma(\mathcal{X}, E)$. Let the degree be m. We like to show that m is less than 5. As all the inner angles are less than π it is clear that the face is entirely located inside a half-sphere. The total length of the edges defining the face is $m\theta$ and the length of any great circle is 2π. Both the face and the half-sphere are convex regions. It follows that the inequality $m\theta \leq 2\pi$ must hold.

With this inequality established it is clear that any lower bound on θ produces an upper bound on m. Employing any of the constructions for $M = 7$ or $M = 8$ described in Section 3.2 we easily establish the desired inequality $m < 5$. □

One consequence of the above lemmas is that there is no isolated point in an irreducible optimal code of size $M = 7$. Indeed, in an optimal code $\mathcal{X} \subseteq \Omega_3$ there must be *some* non-isolated points. Any such point must be in a cycle. By Lemma 3.5.1 the degree must be $m \geq 5$. In the case $M = 7$ there are just two possibilities. The case $m = 6$ is excluded by Lemma 3.5.2. The only remaining possibility is a cycle of degree $m = 5$, with one isolated point on each side of its line of edges. This possibility, however, is excluded by the assumption that the code $\mathcal{X}$ is *irreducible.*

It is convenient to introduce the quantity

$$\alpha = \alpha(\theta) \triangleq \arccos\left(\frac{\cos\theta}{1+\cos\theta}\right), \qquad 0 \leq \theta \leq 2\pi/3. \qquad (3.5.1)$$

This is simply an expression for the angle α of an equilateral spherical triangle with edge length θ. It is easy to show that $\alpha(\theta)$ is a monotonically increasing function in the interval $0 < \theta \leq 2\pi/3$.

Lemma 3.5.3 *Any vertex in the minimum distance graph $\Gamma(\mathcal{X}, E)$ of an optimal irreducible code $\mathcal{X} \subseteq \Omega_3$ of size $M = 7$ has valency either 3 or 4.*

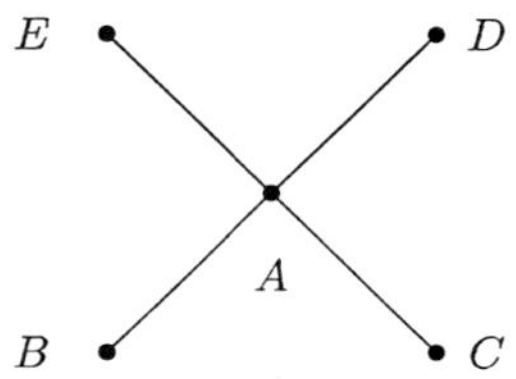

Figure 3.15: The first step in the construction of the graph Γ: the vertex A and its four neighbors.

Proof: Let $\mathcal{X}$ be an irreducible optimal code of size $M = 7$ and let its minimum distance be θ. Let A be an arbitrary vertex in the graph $\Gamma(\mathcal{X}, E)$ and let AB, AC be two edges connected to A. As the distance between the points B and C is at least θ the angel BAC must be at least α.

It is clear that any lower bound on θ produces a lower bound on α. From the codes we considered in Section 3.2 (either one of the codes for $M = 7$ or $M = 8$ will do) we easily establish the inequality $\alpha > 2\pi/5$. It follows that the number of edges is at most 4. □

Lemma 3.5.4 *The minimum distance graph of an irreducible optimal code of size $M = 7$ has at least one point of valency 4.*

Proof: The result is a direct consequence of Lemma 3.5.3 in combination with the well known fact that in any graph the number of vertices of odd valency is always even. In the particular case of interest here the argument is as follows. Let K denote the number of vertices of valency 4. It follows that the number of vertices of valency 3 is $7 - K$. As each edge is connected to precisely two vertices the following equality holds:

$$2|E| = 3(7 - K) + 4K = 21 + K.$$

It follows that K is an odd number, so zero is excluded. □

We are now ready to establish optimality for the code of size $M = 7$ as described in Section 3.2. It is clearly enough to show that the optimal

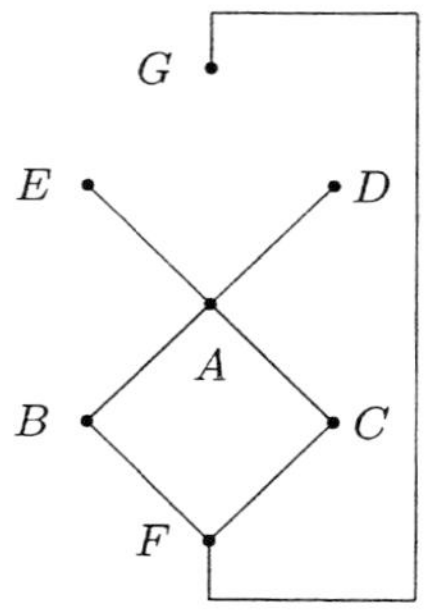

Figure 3.16: Assuming F connected to B and C.

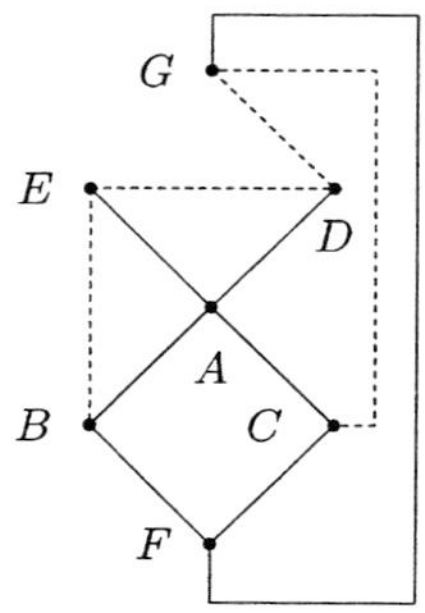

Figure 3.17: The point$G$ cannot be connected to C and D.

code of size $M = 7$ has a minimum distance graph coinciding with that of the code we described in Section 3.2. We will proceed by first demonstrating that the above lemmas impose restrictions which essentially determine the graph of the code. More precisely, they lead to the conclusion that the code must belong to a certain two-parameter family of codes. In the last step we will then perform an optimization within this family.

The first part of the argument is topological in nature. We will actually construct the graph $\Gamma(\mathcal{X}, E)$ from the constraints imposed by the above lemmas. We proceed in three steps.

(i) By Lemma 3.5.4 there is at least *one* vertex of degree 4. So let A be one such vertex and let it be connected to B, C, D, E, see Figure 3.15.

(ii) There are two more points in the code, say F and G. For both of them holds that they must be connected to at least 3 other points. By assumption they are not connected to A. Thus they must both be connected to at least two of the points B, C, D, E. None of them could be connected to two non-consecutive ones of these points, because if for instance F would be connected to B and D the point C would be inside in a loop of size 4, which is prohibited by Lemma 3.5.1. It follows that each one of F and G are connected to precisely two consecutive points from the set B, C, D, E. As they must have degree at least 3 it follows that they are also connected to each other.

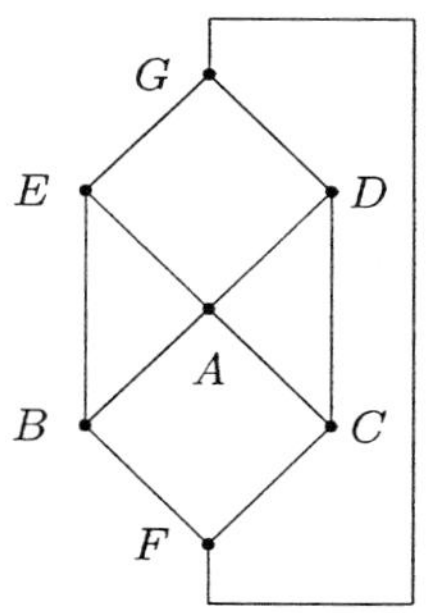

Figure 3.18: A subgraph in agreement with Lemmas 3.5.1–3.5.4

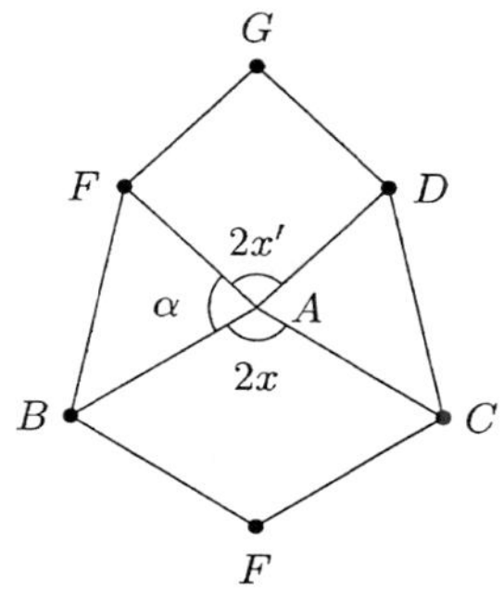

Figure 3.19: The remaining possibilities.

(*iii*) Let F be connected to B and C, (see Figure 3.16). What are the possibilities for G? It certainly cannot be connected to B and C, because this again would violate Lemma 3.7.1: $ABGC$ would form a cycle of length 4 with F on one side and D, E on the other. Neither could it be connected to the pair $\{C, D\}$ nor to the pair $\{B, E\}$. For suppose it were connected to $\{C, D\}$, (Figure 3.17). That would force the point E to be connected to B and D, because by construction all vertices must have degree at least 3. But then B, F, G, D, E would form a face of degree 5, contradictory to Lemma 3.5.2. It follows that G must be connected to D and E. Lemma 3.5.2 then forces BE and CD to be edges.

At this stage we have arrived at a graph which is in agreement with all the restrictions expressed in Lemmas 3.5.1–3.5.4, (Figure 3.18). However, the graph we have derived is not the only one possible: there might also be connections between one or two of the pairs $\{B, C\}, \{D, E\}$.

Actually, all the possibilities left can be captured by introducing a parameter x indicating half the angle BAC and the angle α as defined in equation (3.5.1), see Figure 3.19. Notice that the angle DAE equals $2x' = 2\pi - 2\alpha - 2x$. If $2x = \alpha$ the distance between the vertices B and C is θ and BC is an edge; if $2x' = \alpha$ the distance between the vertices D and E is θ and DE is an edge. If θ is such that both of these equalities are satisfied simultaneously both BC

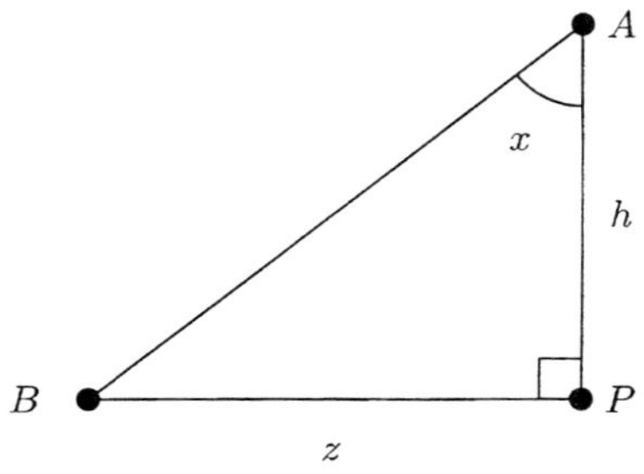

Figure 3.20: The triangle ABP.

and DE are edges. We proceed to determine which one of these cases prevails in an optimal irreducible code.

Consider the graph indicated in Figure 3.19. Except for the edge FG—which deliberately has been omitted in the figure—it contains all the edges that must appear in an irreducible optimal configuration with 7 points. Suppose that this graph is drawn on the sphere, with a fixed distance θ between all those points which are connected by edges and in such a way that the distance between any other pair of points is at least θ. What are the possible values for θ?

In order for the distances between the points B and C and between the points D and E to be at least θ both of the angles $2x$ and $2x'$ must be at least α. As $2x' = 2\pi - 2\alpha - 2x$ we conclude that x is at least α and at most $\pi - 3\alpha/2$. These two bounds coincide for $a = \pi/2$, in which case the code degenerates to an octahedron with one vertex missing. We conclude that α must be less than $\pi/2$.

In order for the lengths of the edges AF and AG to be at least θ we must have $2x \leq 2\alpha$ and $2x' = 2\pi - 2\alpha - 2x \leq 2\alpha$. That means that x is at least $\pi - 2\alpha$ and at most α. These two bounds coincide if $\alpha = \pi/3$, which leads to $2x = 2x' = 2\pi/3$. This would imply that we would have six equilateral triangles meeting in one point. But that is impossible on a sphere. We conclude that α must be larger than $\pi/3$.

In summary we conclude that the possibility that both BC and DE are edges in the graph for an irreducible optimal code of size $M = 7$

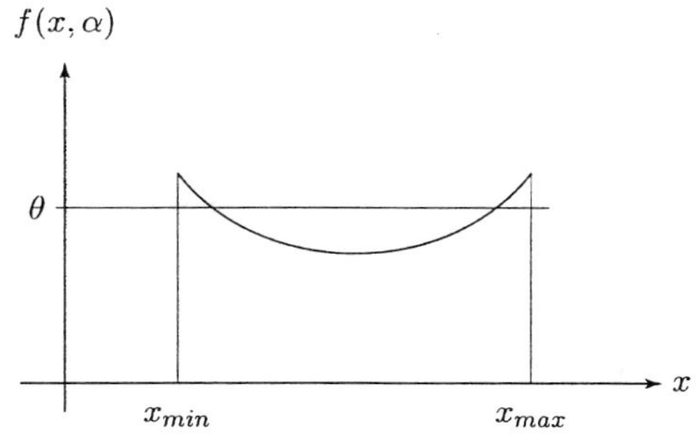

Figure 3.21: The function $f(x,\alpha)$.

must be excluded and that all the still possible cases are captured by letting the variables x and α vary in the range $\pi/3 < \alpha < \pi/2$, where $\max\{\alpha, 2\pi - 4\alpha\} \le 2x \le \min\{2\pi - 3\alpha, 2\alpha\}$. We proceed to reduce further the set of possibilities.

Let P be the point on the line AF, on equal distance from A and F. Consider the triangle ABP, with a straight angle at P, (Figure 3.20). Denote by h and z the lengths of the sides AP and BP respectively. By the *Sinus Theorem* and the Pythagorean theorem for spherical triangles (Appendix C) we have

$$\sin z = \sin\theta \sin x,$$

$$\cos z = \frac{\cos\theta}{\cos h}.$$

Squaring and adding gives us

$$1 = \sin^2\theta \sin^2 x + \frac{\cos^2\theta}{\cos^2 h}.$$

This equation expresses a relation between the parameters h, x, θ. Solving for h we get

$$h = \arctan(\tan\theta \cdot \cos x). \qquad (3.5.2)$$

The distance AF is $2h$. By a similar argument the distance AG is $2h'$, where h' is determined by replacing x with x' in the right hand member of (3.5.2).

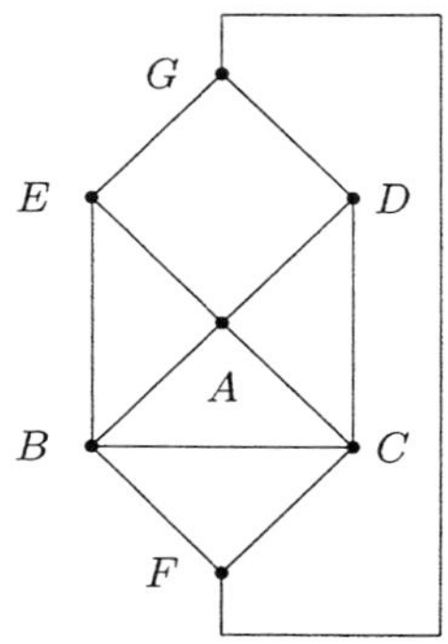

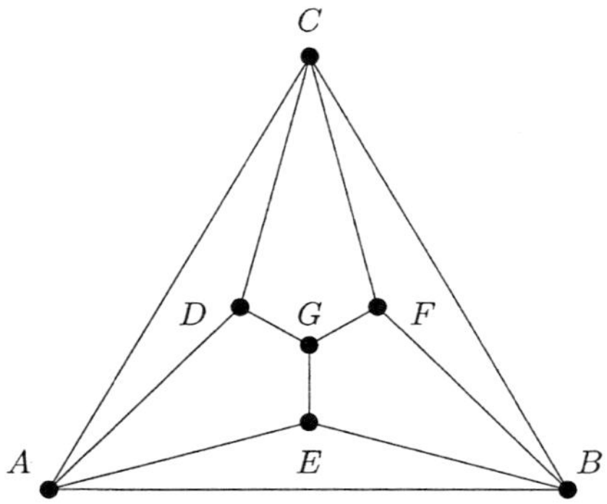

Figure 3.22: Two equivalent pictures of the minimum distance graph for the optimal code of size $M = 7$.

In order to finally determine the optimal configuration we consider the function

$$f(x, \alpha) = 2\pi - 2\arctan(\tan\theta \cdot \cos x) - $$
$$- 2\arctan(\tan\theta \cdot \cos(\pi - \alpha - x)),$$

(recall that θ and α are related by Equation (3.5.1)). It gives the distance between the points F and G. It is easy to establish that for fixed α the function $f(x, \alpha)$ is convex and symmetric around the center of the interval over which x varies. It follows that a maximum with respect to x is always—for any possible value of θ—attained at the two endpoints of the interval. By a geometrical argument it is also easy to establish that for fixed x the function $f(x, \alpha)$ is decreasing in the variable α (recall that θ is a function of α). It follows that as θ increases the two possible solutions to the equation $f(x, \alpha) = \theta$ gets successively closer to the endpoints of the allowed interval for the variable x, (Figure 3.21). For the largest possible value of θ the solutions to the equation occur at the endpoints. In particular this means that either BC or DE—but not both of them—is an edge in the graph of the optimal code. If we choose BC to be an edge and DE not to be an edge we get the following equations for determining the angle θ:

$$f(\alpha/2) = 2\pi - 2\arctan(\tan\theta\cos(\alpha/2)) - \\ -2\arctan(\tan\theta\cos(\pi - 3\alpha/2)),$$

$$\cos\alpha = \frac{\cos\theta}{1+\cos\theta}.$$

The optimum code we have derived is equivalent to the code we described in Section 3.1. The topological equivalence is indicated in Figure 3.22.

3.6 The Coxeter–Böröczky extension

Although we do not intend to provide a full discussion we like at least to mention the Coxeter–Böröczky extension of the Fejes Tóth bound. We quote the result from Coxeter [74], with just some minor adjustments of the notation.

The bound is expressed in terms of a function $F_n(\alpha)$, originally introduced by Schläfli [190]. An iterative definition is as follows (see Coxeter [74]):

$$F_0(\alpha) = F_1(\alpha) = 1,$$

$$F_{n+1}(\alpha) = \frac{2}{\pi}\int_{(1/2)\arccos(1/n)}^{a} F_{n-1}(\beta(t))dt,$$

$$n = 1, 2, \ldots\ ,$$

where

$$\beta(t) = \frac{1}{2}\arccos\left(\frac{\cos 2t}{1-2\cos 2t}\right).$$

The Coxeter–Böröcsky bound reads as follows.

Theorem 3.6.1 (Coxeter–Böröczky) *Let $\mathcal{X}$ be a spherical code with parameters (n, ρ, M). Then the following inequality holds:*

$$M \leq A_{CB}^{(n)}(\rho) \triangleq \frac{2F_{n-1}(\alpha)}{F_n(\alpha)},$$

where

$$\alpha \triangleq \frac{1}{2}\arccos\left(\frac{2-\rho}{2+(n-2)(2-\rho)}\right).$$

The proof is complicated and will not be considered here (see Böröczky, [53]).

Not only the proof is difficult: it is usually also quite difficult to *use* the bound. This is so because evaluating the Schläfli function $F_n(\alpha)$ is quite complicated in most cases. The first few cases are easy:

$$F_2(\alpha) = \frac{2\alpha}{\pi}, \qquad F_3(\alpha) = \frac{2\alpha}{\pi} - \frac{1}{3},$$

but already $F_4(\alpha)$ cannot be explicitly evaluated in general. Using some elaborate arguments it can be shown, however, that $F_4(\pi/5)$ equals $1/900$. Recalling the definition of σ as occurring in Examples 1.2.7 and 1.2.8 of Chapter 1 ($\sigma = 1/2 - \sqrt{5}/2$) and utilizing the standard identity $\cos(2\pi/5) = \left(\sqrt{5}-1\right)/4 = -\sigma/2$ we obtain by inserting $\alpha = \pi/5$ and after some straightforward manipulations

$$M \leq \frac{2\,F_3(\pi/5)}{F_4(\pi/5)} = \frac{2\left(\,(2/\pi)\cdot(\pi/5) - 1/3\right)}{1/900} = 120\,,$$

where

$$\rho = \sigma^2 = (3-\sqrt{5})/2,$$

showing—as we pointed out already in Chapter 2—that the code with parameters $(n, \rho, M) = (4, \sigma^2, 120)$ is optimal.

It has recently been shown that optimality for this case can also be established by the linear programming bound, [4].

3.7 Thirteen spheres

As we have already mentioned the dispute on the thirteen spheres was finally decided in favor of Newton. We offer here a proof based on Leech [150].

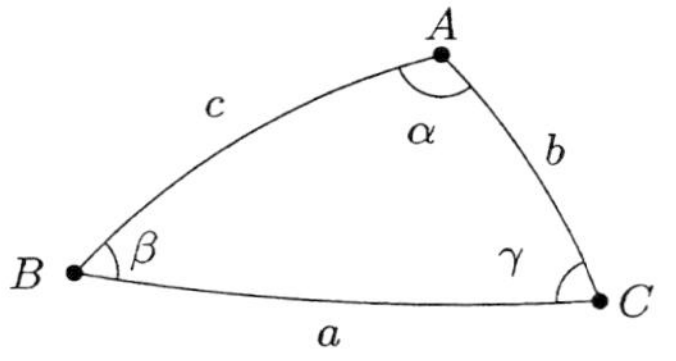

Figure 3.23: Standard notations for triangles in spherical geometry.

The proof is by contradiction. It is entirely elementary and relies only upon some simple facts from spherical geometry and graph theory. The idea is to show that if a spherical code with squared minimum distance $\rho = 1$ and cardinality $M = 13$ did exist, then there would also exist a graph with certain properties. The non-existence of the spherical code is then proved by showing that a graph with these properties cannot exist.

Let ABC be a triangle on the unit sphere in Ω_3. Let the edges have lengths a, b, c and let the opposite angles be α, β, γ, see Figure 3.23. Recall the Cosine Theorem (Appendix C):

$$\cos c = \cos a \cos b + \sin a \sin b \cos \gamma. \tag{3.7.1}$$

Also recall the formula for the triangle area:

$$T = \alpha + \beta + \gamma - \pi.$$

We will need three types of triangles on the sphere, defined by the following table:

	$\cos a$	$\cos b$	$\cos c$	$\cos \alpha$	$\cos \beta$	$\cos \gamma$
1	1/2	1/2	1/2	1/3	1/3	1/3
2	1/2	1/2	1/7	1/2	1/2	−1/7
3	1/2	1/7	1/7	47/96	1/12	1/12

The validity of this table follows from equation (3.7.1). The areas are as follows:

$$T_1 = 3 \arccos(1/3) - \pi \approx 0.5513,$$

$$T_2 = 2 \arccos(1/2) + \arccos(-1/7) - \pi \approx 0.6669,$$

$$T_3 = \arccos(47/96) + 2 \arccos(1/12) - \pi \approx 0.8923\,.$$

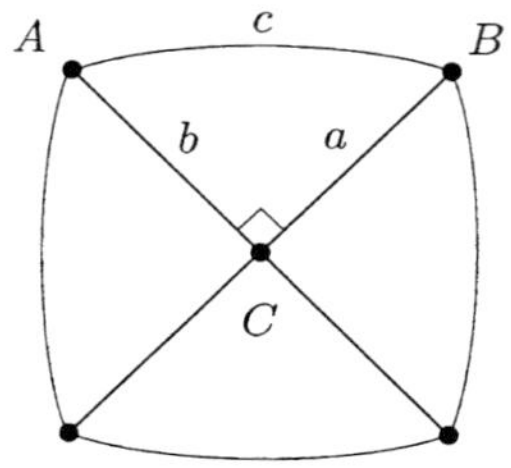

Figure 3.24: A square on the sphere.

We will need two lemmas.

Lemma 3.7.1 *Any spherical quadrangle such that all interior angles are less than π and all sides are larger than or equal to $\pi/3$ has at least one diagonal of length at least $\pi/2$.*

Proof: Consider a square, i.e. a quadrangle such that all four edges and all four angles are the same. Clearly then also the diagonals are the same. Let c be the length of the edges and let the length of the diagonals be $2a$, see Figure 3.24. Let A, B be two adjacent vertices and let C be the point where the diagonals cross. Clearly the angle at C is $\gamma = \pi/2$. The Cosine Theorem reduces to the following simple formula:

$$\cos c = (\cos a)^2.$$

Now suppose $c = \pi/3$. We get $a = \pi/4$, and so both diagonals have length $2a = \pi/2$. Deforming the square while keeping the lengths of the edges fixed will decrease one of the diagonals but increase the other. Increasing the length of any of the edges can only increase the diagonals. The lemma follows. □

A graph on the sphere Ω_3 such that no edges cross is said to be a *plane graph*. The edges of such a graph partitions the sphere in certain regions called *faces*. A face which is bounded by q edges will be called a *q-face*. The number of edges connected to a given vertex is called the *valency* of the vertex. Recall that a *cycle* is a sequence of connected edges such that all the connecting vertices are different. In a cycle the number of edges equals the number of vertices; this common number is usually called the *length* of the cycle. A *loop* is an edge both ends of which are connected to the same vertex. Two edges are said to be *parallel* if they are connecting the same pair of vertices. A graph is *simple* if it has no loops and no parallel edges. A graph is 2-*connected* if at least 2 edges must be removed in order to separate the graph in two disconnected parts. For any plane graph there is another plane graph called the *dual* graph which is obtained by defining one vertex for each face in the original graph, and one edge for each edge. Clearly the dual graph will have one face for each vertex in the original graph. The dual of the dual graph is the original graph.

Lemma 3.7.2 *There is no plane graph on Ω_3 with* 13 *faces and satisfying all of the following properties:*

(*i*) *All faces are pentagons except one, which is a quadrangle.*

(*ii*) *All vertices have valency* 3 *except one, which has valency* 4.

(*iii*) *The dual graph is simple and* 2-*connected.*

Proof: We first notice that the conditions required in the lemma imply that any cycle—except for the unique quadrangle—must have length at least 5. To see this first notice that any cycle divides the sphere into two parts. With one possible exception each vertex has valency 3. Therefore for each vertex in the cycle there must be one edge (or possibly two edges in at most one case) that doesn't belong to the cycle, and so must be in one of the two parts. For a cycle of length 3 or 4 the number of such extra edges is at most 5, and so in at least one of the two parts there are at most 2 such edges. If the number is 1 the dual graph has a loop and if the number is 2 the dual graph has a double edge. Both cases are excluded by assumption (*iii*), according to which the dual graph is simple. If the number is 0 the cycle defines a face. By assumption there is no 3-face and just one unique 4-face. The assertion follows.

We now prove the lemma by demonstrating that it is impossible to construct a graph with all the properties listed. Without loss of

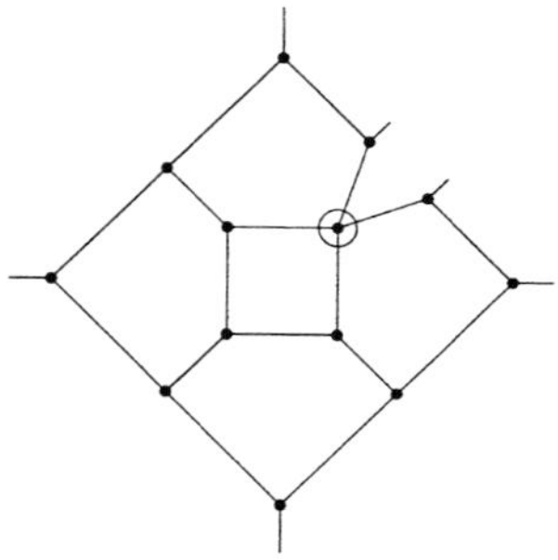

Figure 3.25: Subgraph implied by Lemma 3.7.2, supposing the single vertex of valency 4 is on the quadrangle.

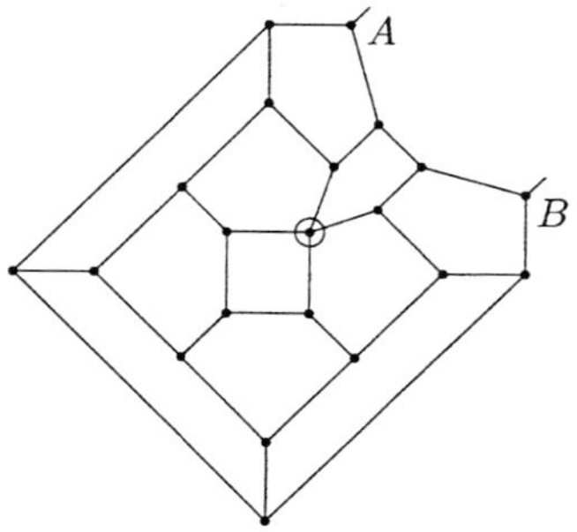

Figure 3.26: The points A and B cannot be neither merged nor connected.

generality we start the construction with the unique quadrangle. The construction proceeds by successively adjoining adjacent faces, all of which must be pentagons according to property (*i*). Moreover, all vertices—with one exception—must have valency 3. Clearly these severe restrictions leave very little freedom. In addition we must observe the restriction we just proved, namely that any cycle—except the unique quadrangle—must have length at least 5. The argument is slightly different depending on the location of the unique vertex of valency 4. Three cases need to be considered.

Case 1

Suppose the vertex with valency 4 is on the quadrangle. We necessarily arrive at the subgraph shown in Figure 3.25. Notice that this subgraph has diameter 4. It follows that all the vertices are really distinct, because otherwise there would be a cycle of length shorter than 5. Outside this first layer of pentagons the next layer also consists of pentagons. Observing the requirement that all vertices must have valency 3 we obtain the subgraph indicated in Figure 3.26. Here we run into difficulties. Notice that except for the two vertices A and B all vertices already have valency 3, which means that no more edges can be connected. For the vertices A and B there are two possibilities. Either they could be merged, but that would create a triangle, which is not permitted, or else they could be connected to a new vertex C. But than there would be no further vertex available which could be

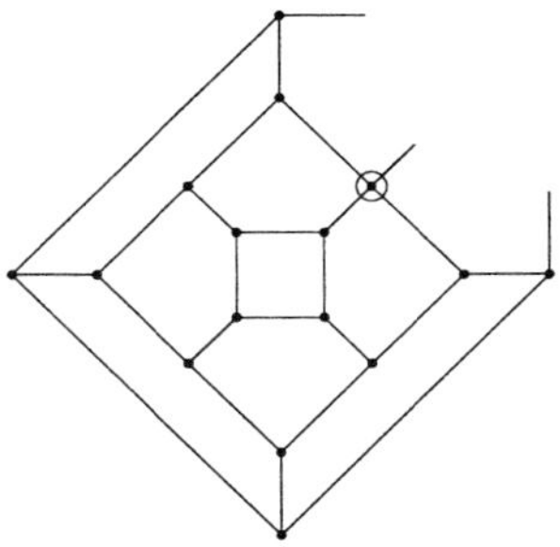

Figure 3.27: Subgraph implied by Lemma 3.7.2, supposing the single vertex of valency 4 is on distance 1 from the quadrangle.

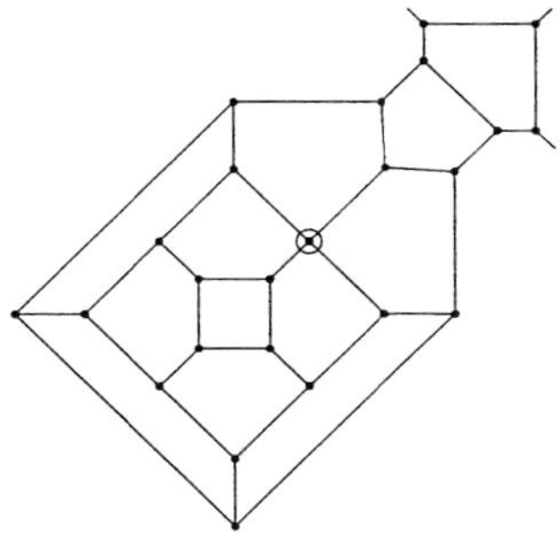

Figure 3.28: Too many faces.

connected to C, and so either this vertex would get valency 2 or else the graph would be only 1-connected. Both cases are excluded. The conclusion is that the graph cannot be completed.

Case 2
Now suppose the vertex with valency 4 is on graph distance 1 from the quadrangle. By the same kind of argument as in *Case 1* we see that the graph shown in Figure 3.27 must be contained in the desired graph. Continuing from this point we could try a construction as shown in Figure 3.28. However, the incomplete graph indicated in Figure 3.28 already contains 13 faces, and in order to respect the requirement that each vertex should have valency 3 the construction must obviously be continued. The conclusion is that the possibility indicated in Figure 3.28 must be abandoned. The other possibility is, of course, the one indicated in Figure 3.29. Then, however, the dual graph would contain two parallel edges, which again is not permitted. Again we must conclude that the graph cannot be completed.

Case 3
Finally we consider the case when the vertex having valency 4 is on graph distance 2 or more from the quadrangle. In this case we arrive at the subgraph shown in Figure 3.30, which is already a contradiction, because it contains a cycle of length 4. □

We are now ready for the main result. Recall that we use τ_n as a

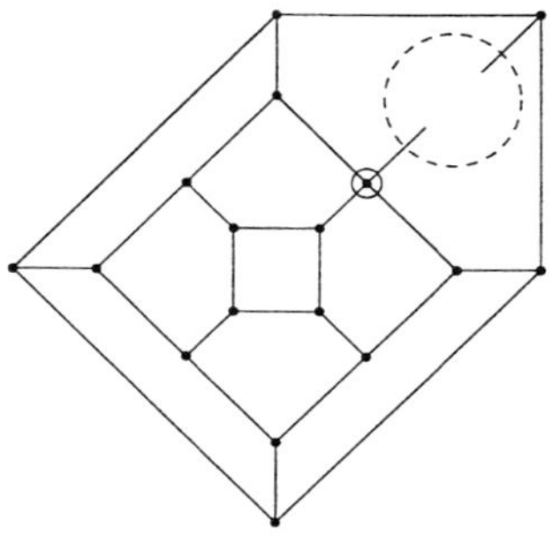

Figure 3.29: The dual graph has parallel edges.

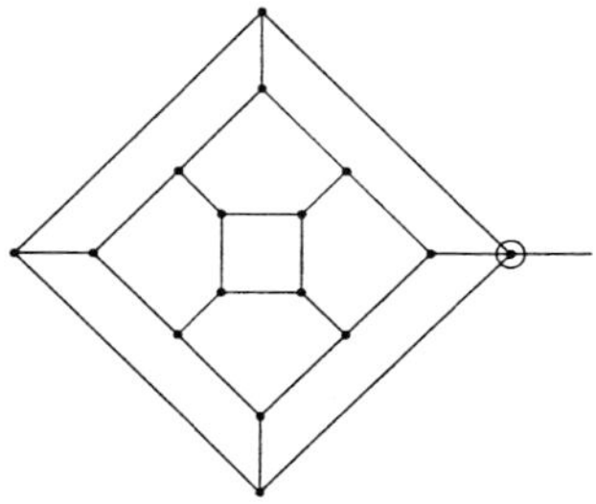

Figure 3.30: The graph has a cycle of length 4.

notation for the contact number in dimension n. Using this notation our main result can be formulated as follows.

Theorem 3.7.1 *In dimension $n = 3$ the kissing number is $\tau_3 = 12$.*

Proof: The cubeoctahedron induces a spherical code with parameters $(n, \rho, M) = (3, 1, 12)$. Therefore $\tau_3 \geq 12$. It remains to demonstrate that $\rho < 1$ for any code on Ω_3 of size $M = 13$.

With the purpose of establishing a contradiction, let $\mathcal{X}$ be a code with parameters $(n, \rho, M) = (3, 1, 13)$. Define a graph $\Gamma = (\mathcal{X}, E)$ with the codewords of $\mathcal{X}$ as the set of vertices and with an edge for each pair $\{x, y\}$ of codewords such that $1/7 < \langle x, y \rangle \leq 1/2$. Notice that this means that we connect with an edge all pairs of codewords x, y such that $1 \leq \|x - y\|^2 < 12/7$.

The graph is necessarily planar. This follows by Lemma 3.7.1, because suppose the opposite. Then there would be a quadrangle with both diagonals less than $\arccos(1/7) < \pi/2$, which contradicts the lemma. Without loss of generality we may also assume it to be 2-connected: if this is not the case with the graph induced by the original code we can easily move a few of the codewords so as to establish this property. Finally, Γ is simple by construction.

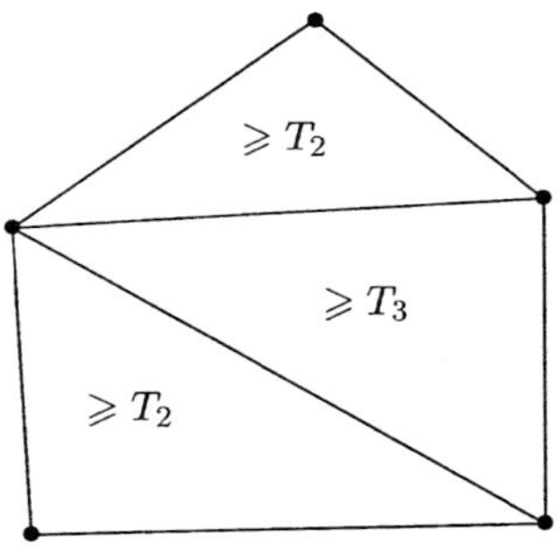

Figure 3.31: Triangulating a pentagon.

Now consider the graph Γ. Let us extend it by selecting for each q-face with $q \geq 4$ one fixed vertex and connecting that vertex with all the non-adjacent vertices in the q-face, see Figure 3.31.

Denote by Γ' the so extended graph. In the extended graph all the faces are triangles. These triangles can be partitioned in three classes according to size. Each triangular face that belongs already to the original graph Γ has area at least T_1. Suppose we have F_1 of those. For each q-face, $q \geq 4$ in Γ there will be two triangular faces in Γ', defined by the two edges closest to the fixed vertex used in the construction of Γ'. Each of these has an area at least equal to T_2 . Let the total number of those faces be F_2. In addition there will be $q-2$ further triangular faces, each with area at least T_3. Let the number of such faces be F_3. As the total area of the sphere is 4π we obtain the following inequality:

$$F_1T_1 + F_2T_2 + F_3T_3 \leq 4\pi .$$

We notice that F_2 is even.

Let $F = F_1 + F_2 + F_3$ be the total number of faces in Γ' and let the number of edges be E. By Euler's formula we have $M + F = E + 2$, and by counting edges in two ways we get $3F = 2E$. Combining these two identities we get $F = 2M - 4$. Inserting $M = 13$ we find $F = 22$, which gives us

$$(T_2 - T_1)F_2 + (T_3 - T_1)F_3 \leq 4\pi - 22 \cdot T_1 . \qquad (3.7.2)$$

From this we conclude

$$F_2 \leq \frac{4\pi - 22T_1}{T_2 - T_1} \leq 3.8.$$

As F_2 is even it follows that we must have $F_2 = 0$ or 2. In the first case we have triangles only. Then Euler's formula applied to Γ gives us $3 \cdot 22/2 = 33$ edges, and so the average vertex degree is $66/13 > 5$. However, each edge has length a such that $1/7 < \cos a \leq 1/2$. By the cosine theorem we easily establish that this implies that all angles α satisfy $\alpha > \pi/3$, and so all vertices have valency at most 5. Thus the case $F_2 = 0$ is excluded.

Now suppose $F_2 = 2$. Inserting numerical values for the quantities T_1, T_2, T_3 in the inequality (3.7.2) we find $F_3 = 0$. It follows that all faces in Γ are triangles, except one which is a quadrangle. The number f of faces in Γ will be one less that the corresponding number in Γ', and so we have $f = 22 - 1 = 21$. The number of edges is $e = 21 + 13 - 2 = 32$. Each vertex has valency 3, 4, or 5. Let v_3, v_4, and v_5 denote the number of vertices of each kind. We get the following conditions:

$$3v_3 + 4v_4 + 5v_5 = 2e = 64,$$

$$v_3 + v_4 + v_5 = M = 13,$$

$$v_3, v_4, v_5 \geq 0.$$

An easy evaluation reveals that $(v_3, v_4, v_5) = (0, 1, 12)$ is the only integer solution. It follows that the dual graph satisfies all of the properties listed in Lemma 3.7.2, which we have just shown is impossible. It follows that the graph Γ doesn't exist, and so neither does a code with parameters $(n, \rho, M) = (3, 1, 13)$. □

Chapter 4

Permutation codes

4.1 Introduction

Permutation modulation is one of the oldest principles for generating spherical codes. We will refer to the resulting codes as *permutation codes.* They were presented by Slepian in 1965, [211], and have since then been studied by many authors. The basic idea is very simple and comes in two versions. Take an arbitrary real-valued unit norm vector $x_0 = (x_{01}, x_{02}, \ldots, x_{0N})$ and form all possible permutations of it. The set of vectors so generated is obviously a spherical code. The construction is known as permutation modulation *Variant 1.* If in addition all possible sign changes are also allowed the construction is referred to as permutation modulation *Variant 2.* Both versions produce codes with a number of attractive properties: they are easy to characterize, they are easy to analyze and they can easily be implemented. Most remarkable of all: maximum likelihood decoding can be performed with the aid of a very simple algorithm.

For small dimensions the class of permutation codes contains some very good codes. Unfortunately, however, performance generally deteriorates quickly with increasing dimension. For dimensions larger than 6 no interesting codes are produced, with the exception of the *Simplex* and *Biorthogonal* codes, which are always generated as special cases.

We give a detailed discussion of the various versions, paying special attention to the problem of optimizing the initial vector. We also present tables of the best codes generated by this principle.

4.2 Variant 1

Let the initial vector x_0 be of the form

$$x_0 = (\mu_1, \mu_1, \ldots, \mu_1, \mu_2, \ldots, \mu_2, \mu_3, \ldots, \mu_K)$$

where $\mu_1 < \mu_2 < \ldots < \mu_K$ are given real numbers and where the number μ_1 occurs m_1 times, μ_2 occurs m_2 times etc. We collect the integers m_i in a vector $m = (m_1, m_2, \ldots, m_K)$, which we will refer to as the *distribution*. The code consists of all permutations of x_0.

We notice that the codewords are described as vectors with $N = m_1 + m_2 + \cdots + m_K$ components. We point out right away that this does not mean that the dimension is equal to N: as will be seen shortly *Variant 1* codes always have dimension $n = N - 1$.

We also notice that all codewords have the same energy, namely

$$E = \sum_{i=1}^{K} m_i \, \mu_i^2 .$$

The number of codewords is

$$M = \frac{(m_1 + m_2 + \cdots + m_K)!}{m_1! m_2! \ldots m_K!},$$

and the minimum squared distance is

$$d_E^2 = \min_{i \neq j} \{2(\mu_i - \mu_j)^2\}, \qquad i, j = 1, 2, \ldots, K .$$

For a given initial vector x_0 these formulas are sufficient for deriving the main parameters (n, ρ, M) of the code (recall that the normalized squared distance ρ is defined as $\rho = d_E^2/E$). However, without a proper choice of the initial vector x_0 we cannot expect to obtain any good codes. Therefore, let us see how x_0 should be chosen in order to generate codes with favorable parameters.

Without loss of generality we prescribe that the minimum distance between different levels μ_j should be 1. That fixes the value of d_E^2 to 2. Under this assumption the problem is to minimize the energy E. It is clear that the signal energy can only increase if we allow the distance between any pair of adjacent parameters m_j to exceed the minimum. Therefore without loss of generality we may assume

$$\mu_j = j - a, \quad j = 1, 2, \ldots, K. \tag{4.2.1}$$

It remains to determine the constant a and to choose the best distribution m. We consider first the selection of a and we begin with a few definitions.

Definition 4.2.1 *For any permutation code of Variant 1 we define*

$$A(m) \triangleq \sum_{i=1}^{K} m_i \cdot i^2, \qquad B(m) \triangleq \sum_{i=1}^{K} m_i \cdot i,$$

$$a_m \triangleq B(m)/N, \qquad V(m) \triangleq A(m) - B(m)^2/N\,,$$

$$\Delta(m, a) \triangleq \sum_{i=1}^{K} m_i (i - a)^2.$$

We get the following expression for the energy E for a scheme with distribution m and mean a:

$$E = \Delta(m, a) = V(m) + N(a - a_m)^2. \tag{4.2.2}$$

Choosing $a = a_m$ we obtain

$$E = \Delta(m, a_m) = V(m).$$

This is clearly the choice of a that minimizes E. The remaining problem is how to select the distribution $m = (m_1, m_2, \ldots, m_K)$.

We notice that any permutation of the components m_i in the vector $m = (m_1, m_2, \ldots, m_K)$ only affects the energy E but leaves the cardinality M unchanged. As a first step in our optimization with respect to m it is therefore natural to consider optimization over the set $\mathcal{T}$ of all permutations of a given distribution $m = (m_1, m_2, \ldots, m_K)$. A set of that kind is usually referred to as a *type*.

Let a type $\mathcal{T}$ be given and let us consider the problem of finding the best vector m in $\mathcal{T}$. A key to the solution is given by the following lemma.

Lemma 4.2.1 *Let a type $\mathcal{T}$ be given and suppose that $m \in \mathcal{T}$ minimizes $V(m)$, i.e. suppose*

$$V(m) = \min\{V(m') : m' \in \mathcal{T}\}. \tag{4.2.3}$$

Then the following implication holds:

$$(i - a_m)^2 > (j - a_m)^2 \;\Rightarrow\; m_i \le m_j\,, \quad 1 \le i, j \le K, \quad i \ne k. \tag{4.2.4}$$

Proof: Suppose $m \in \mathcal{T}$ does not satisfy (4.2.4), i.e. suppose there exists a pair (i, j) of indices such that

$$\begin{aligned} \varepsilon &\triangleq (i - a_m)^2 - (j - a_m)^2 > 0, \\ \delta &\triangleq m_i - m_j > 0\,. \end{aligned}$$

Define $m^* \in \mathcal{T}$ as follows:

$$m_k^* = \begin{cases} m_k, & if \quad k \ne i, j, \\ m_i, & if \quad k = j, \\ m_j, & if \quad k = i. \end{cases}$$

We get

$$\begin{aligned} \Delta(m^*, a_m) &= \sum_{k=1}^{K} m_k^*(k - a_m)^2 = \\ &= \sum_{k=1}^{K} m_k(k - a_m)^2 + (m_i^* - m_i)(i - a_m)^2 + \\ &+ (m_j^* - m_j)(j - a_m)^2 = V(m) - \delta\,\varepsilon. \end{aligned}$$

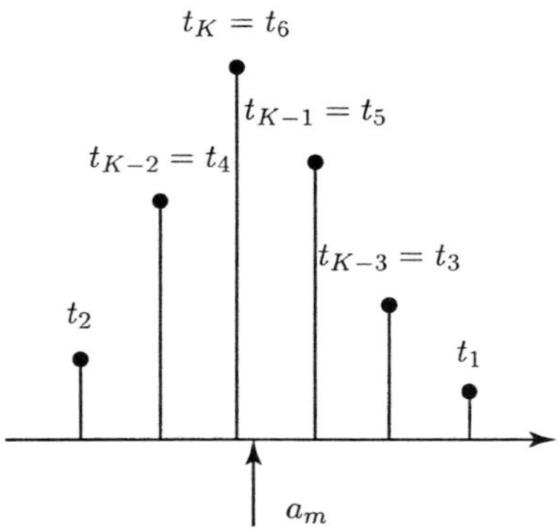

Figure 4.1: Permuting $t = (1, 2, 4, 7, 8, 10)$ in order to obtain the optimal distribution $m = (2, 7, 10, 8, 4, 1)$.

But by equation (4.2.2) we also have

$$\Delta(m^*, a_m) = V(m^*) + N(a_m - a_{m^*})^2 .$$

If follows that we must have $V(m^*) < V(m)$, so (4.2.3) cannot be satisfied. □

We notice next that the condition (4.2.4) essentially determines the $m \in \mathcal{T}$ minimizing V. Let $\mathcal{T}$ and $t = (t_1, t_2, \ldots, t_K) \in \mathcal{T}$ with $t_1 < t_2 < \ldots < t_K$ be given and let m be a distribution minimizing V over $\mathcal{T}$. The corresponding mean value is denoted a_m. Let us consider how t should be permuted in order to generate m.

The obvious procedure is illustrated in Figure 4.1. It is clear that t_K should be closest to a_m. Without loss of generality we assume that it is located to the left of a_m. The next symbol t_{K-1} is then immediately to the right of a_m. Continuing the argument we find that t_{K-2} must be just to the left of t_K, t_{K-3} must be just to the right of t_{K-1} etc. In summary we have that

$$m \triangleq \begin{cases} (t_1, t_3, \ldots, t_K, t_{K-1}, \ldots, t_4, t_2), & \mathit{if} \quad K \ \mathit{odd}, \\ (t_2, t_4, \ldots, t_K, t_{K-1}, \ldots, t_3, t_1), & \mathit{if} \quad K \ \mathit{even}, \end{cases} \tag{4.2.5}$$

is a possible choice for an $m \in \mathcal{T}$ minimizing V.

The choice in not unique. We could equally well have put t_K to the *right* of a_m. Further, whenever an optimal distribution happens to be such that a_m is just halfway between t_K and t_{K-1} there are still further alternative choices. However, it is easy to see that they all lead to the same value of V.

It is worth emphasizing that for a given type $\mathcal{T}$ the above argument leads to an optimal distribution m without prior knowledge of the mean a_m. This is important, because in order to compute a_m we need of course to know the distribution m, i.e. the quantity we are looking for.

It turns out that for *small* dimensions n different types $\mathcal{T}$ produce different spherical codes and that they are all incompatible, i.e. none of them dominates any other. Therefore, for *small* dimensions we have actually carried the optimization as far as it is possible to go. In other words: in order to produce all dominating codes we need to consider *all* different types $\mathcal{T}$.

For *large* dimensions this is no longer true: in those cases only *some* of the types produce dominating codes. The optimal distributions are those approximating Gaussian distributions, see Slepian [211] and Ingemarsson [137]. However, as we have already pointed out, for large dimensions the permutation codes unfortunately do not offer good performance.

The first example of a type *not* producing a dominating code is the one containing $m = (1, 4, 1, 1)$, which is dominated by the one containing $m = (2, 3, 2)$. This occurs in dimension $n = 6$.

All optimal codes satisfy (4.2.1) with $a = a_m$, i.e. each codeword x satisfies

$$\sum_{i=1}^{N} x_i = \sum_{k=1}^{K} m_k(k - a_m) = B(m) - N a_m = 0.$$

It follows that although specified by N coordinates the optimal codes are actually located in an $(N-1)$-dimensional subspace. It turns out that the dimension is always *exactly* $N - 1$, i.e. there is no even smaller subspace that contains any of these codes. For individual codes this is usually easy to establish. A general proof will be given in Section 4.6.

We summarize the properties of permutation codes, *Variant 1*:

$$Dimension: \ n = m_1 + m_2 + \cdots + m_K - 1,$$

$$Normalized\ distance: \ \rho = \frac{2N}{NA(m) - B(m)^2},$$

$$Cardinality: \ M = \frac{(m_1 + m_2 + \cdots + m_K)!}{m_1!\, m_2! \ldots m_K!}.$$

The parameters $A(m)$, $B(m)$ and N are obtained as

$$A(m) = \sum_{i=1}^{K} m_i \cdot i^2,$$

$$B(m) = \sum_{i=1}^{K} m_i \cdot i,$$

$$N = m_1 + m_2 + \cdots + m_K.$$

Finally, the alphabet is

$$\mathcal{L}_q = \{i - a_m : i = 1, 2, \ldots, q\}, \qquad a_m = B(m)/N,$$

with $q = K$. We notice that in general the alphabet is non-symmetric.

4.3 Best Variant 1 codes

In Table 4.1 we have listed all the dominating permutation codes of *Variant 1* for dimensions $1 \leq n \leq 6$. That actually covers all cases of interest: for $n > 6$ we generally obtain better codes from other constructions.

The table gives dimension n, normalized squared distance ρ, and cardinality M. For each code we have also indicated the vector m generating the code. As can be seen, for each n in the range considered all partitions of $N = n + 1$ produce dominating codes, with the distribution $m = (1, 4, 1, 1)$ mentioned above as the only exception. On the other hand, the discussion in Section 4.2 assures that

n	ρ	M	m	
1	4	2	1 1	$*$
2	1	6	1 1 1	$*$
2	3	3	2 1	$*$
3	2/5	24	1 1 1 1	
3	1	12	1 2 1	
3	2	6	22	$*$
3	8/3	4	3 1	$*$
4	1/5	120	1 1 1 1 1	
4	5/13	60	1 2 1 1	
4	5/7	30	1 2 2	
4	1	20	1 3 1	
4	5/3	10	3 2	B
4	5/2	5	4 1	$*$
5	4/35	720	1 1 1 1 1 1	
5	1/5	360	1 1 2 1 1	
5	4/11	180	1 2 2 1	
5	3/8	120	1 3 1 1	
5	1/2	90	2 2 2	
5	12/17	60	1 3 2	
5	1	30	1 4 1	
5	4/3	20	3 3	
5	3/2	15	4 2	
5	12/5	6	5 1	$*$
6	1/14	5040	1 1 1 1 1 1 1	
6	7/62	2520	1 1 2 1 1 1	
6	7/38	1260	1 1 2 2 1	
6	1/5	840	1 1 3 1 1	
6	7/26	630	2 2 2 1	
6	7/20	420	1 3 2 1	
6	1/2	210	2 3 2	
6	7/12	140	1 3 3	
6	7/10	105	1 4 2	
6	1	42	1 5 1	
6	7/6	35	4 3	
6	7/5	21	5 2	
6	7/3	7	6 1	$*$

Table 4.1: All dominating permutation codes of *Variant 1*, $1 \leq n \leq 6$. Codes known to be optimal have been indicated by $*$, while codes with best parameters known have been indicated with B.

by considering all possible partitions of $N = n+1$ we can be sure of actually obtaining also all dominating codes.

We have indicated with a star $(*)$ all cases where the codes are known to be optimal and with (B) all other cases where the codes are the best ones known.

For $n = 1$ we have just a fancy representation of the binary antipodal code. In dimension $n = 2$ we have the equilateral triangle and the regular hexagon. Both are optimal, although they represent rather trivial codes. We notice that both the triangle and the hexagon are described without using radicals.

In tree dimensions $m = (2,2)$ gives a very nice description of the *Biorthogonal* code (which in general is not included in this construction) and $m = (1,2,1)$ offers a description of the cubeoctahedron. The distribution $m = (1,1,1,1)$ generates the semi-regular polyhedron with 6 squares and 8 hexagons as faces. It is not in itself a particularly good spherical code. However, adding one point just outside the center of each hexagon produces a spherical code with parameters $(n, \rho, M) = (3, 2/5, 32)$, which although not optimum is reasonably good (for $n = 3$, $M = 32$ the best known code has $\rho \approx 0.4125$).

The distribution $m = (n, 1)$ provides the standard representation of the *Simplex* code. Thus, all *Simplex* codes are included in the construction. These codes, of course, are optimal for any value of n.

The code corresponding to $m = (3,2)$ has been marked with a B in the table as an indication that the parameters are best known, although no proof of optimality is available. The parameters are $(n, \rho, M) = (4, 5/3, 10)$. We will encounter this code again in Chapter 6 as a representation in Euclidean space of the Johnson space $J(5,2)$ and in Chapter 9 as a representation of the Petersen graph. It possesses a number of interesting properties. It is a reasonable guess that it is actually an optimal spherical code.

Apart from the cases mentioned there seems to be very few good ones among the *Variant 1* codes. We have performed a complete search in the range $n \leq 24$. In that range there is none. It is very unlikely that any further good codes at all should exist for larger n (except the *Simplex* codes).

The problem of finding the best vector x_0 is usually referred to as the *initial vector* problem (Slepian [212], p. 582, see also Ingemarsson [137], Karlof [141] and Chapter 8.1). The first few observations in Section 4.2 reduce this problem to that of finding the best distribution m. Lemma 4.2.1 addresses this latter part of the problem. It is worthwhile emphasizing that although the lemma does not fully solve the problem it quite substantially reduces the number of distributions needed to consider. Suppose we want to find all dominating codes in dimension $n = N - 1$. The total number of *distributions* $m = (m_1, m_2, \ldots, m_K)$ with $m_1 + m_2 + \cdots + m_K = N$ is 2^{N-1}, while the number of *types* equals the number $p(N)$ of partitions of $N = n + 1$. That number is much smaller. Hall [122] gives a table (p. 38) containing the first 100 values of $p(N)$ and also offers the following asymptotic formula (p. 44):

$$p(N) \approx \frac{1}{4N\sqrt{3}}\, e^{A\sqrt{N}},$$

where A is a constant. As an example, for $n = 24$ we have $p(N) = p(25) = 1958$, which should be compared with $2^{24} = 16777216$.

For n larger than 24 the above method becomes rather time consuming and more efficient methods are desirable. Karlof [141] has developed an algorithm for directly determining an optimal initial vector x_0. According to Ingemarsson [137] the best distributions m are approximately Gaussian, a result anticipated already by Slepian [211]. However, as we have already mentioned—and as was observed already by Slepian ([211], p. 233)—the resulting codes usually have poor performance.

4.4 Variant 2a

In *Variant 2* we allow, in addition to all permutations, also all possible sign changes in the components of the initial vector

$$x_0 = (\mu_1, \mu_1, \ldots, \mu_1, \mu_2, \ldots, \mu_2, \mu_3, \ldots, \mu_K, \ldots, \mu_K).$$

Clearly we may assume, without loss of generality, that all components μ_i are non-negative. It turns out to be convenient to distinguish between two cases: the case $\mu_1 = 0$, which we will refer to as *Variant 2a*, and the case $\mu_1 \neq 0$, which we will call *Variant 2b*. We consider *Variant 2a* first.

It is convenient to introduce an additional notation so as to distinguish between the zero symbols and the other symbols. Let the distribution m be of the form $m = (r, p)$, where $r = m_1$ and $p = (p_1, p_2, \ldots, p_J) = (m_2, m_3, \ldots, m_K)$. The size of a *Variant 2a* code with distribution $m = (r, p)$ is (with $J = K - 1$)

$$M = \frac{(r + p_1 + p_2 + \cdots + p_J)!}{r!\, p_1!\, p_2!\, \ldots\, p_J!} \cdot 2^{(p_1 + p_2 + \cdots + p_J)}. \tag{4.4.1}$$

Again without loss of generality we prescribe that the minimum distance between any pair of adjacent letters μ_j and μ_{j+1} should be 1. As before this fixes the minimum squared distance to 2. As we have already prescribed the condition $\mu_1 = 0$ we obviously have in this case

$$\mu_j = j - 1, \qquad j = 1, 2, \ldots, K,$$

corresponding to an alphabet $\mathcal{L}_q$ of the form

$$\mathcal{L}_q = \{0, \pm 1, \pm 2, \ldots, \pm (q-1)/2\}.$$

The size q of the alphabet is always odd: $q = 2J + 1 = 2K - 1$. In Section 4.6 we show that the dimension of the code is

$$n = r + p_1 + p_2 + \cdots + p_J.$$

The energy E is given by the formula

$$E = \sum_{j=1}^{J} p_j \cdot j^2. \tag{4.4.2}$$

All that remains is to determine the distributions $m = (r, p)$ that lead to dominating codes.

Denote by $\mathcal{T}_r$ the set of all distributions of the form $m = (r, p)$. From (4.4.1) and (4.4.2) we see that for fixed r the size M is constant

n	ρ	M	r	p	
2	2	4	1	1	*
3	2/5	24	1	11	
3	1	12	1	2	
3	2	6	2	1	*
4	1/7	192	1	1 1 1	
4	1/3	96	1	2 1	
4	2/5	48	2	1 1	
4	2/3	32	1	3	
4	1	24	2	2	B
4	2	8	3	1	*
5	1/15	1920	1	1 1 1 1	
5	2/15	960	1	2 1 1	
5	1/5	480	1	2 2	
5	2/13	320	1	3 1	
5	1/3	240	2	2 1	
5	2/3	80	2	3	
5	1	40	3	2	B
5	2	10	4	1	*
6	2/55	23040	1	1 1 1 1 1	
6	2/31	11520	1	2 1 1 1	
6	2/19	5760	1	2 2 1	
6	1/8	3840	1	3 1 1	
6	2/15	2880	2	2 1 1	
6	2/11	1920	1	3 2	
6	1/5	1440	2	2 2	
6	2/7	960	2	3 1	
6	1/3	480	3	2 1	
6	1/2	240	2	4	
6	2/3	160	3	3	
6	1	60	4	2	
6	2	12	5	1	*

Table 4.2: All dominating permutation codes of *Variant 2a*, $2 \leq n \leq 6$. Codes known to be optimal have been indicated by $*$, while codes with best parameters known have been indicated with B.

over the set $\mathcal{T}_r$ while the energy is not. It follows immediately that $p = (p_1, p_2, \ldots, p_J)$ should be chosen such that $p_1 \geq p_2 \geq \ldots \geq p_J$.

The selection of r is less obvious. It is easy to see that for fixed $p = (p_1, p_2, \ldots, p_J)$ the size M is an increasing function of r. However, the dimension $n = r + p_1 + p_2 + \cdots + p_J$ increases also with r, so different values of r give rise to codes of different dimensions.

In Table 4.2 we have listed all the dominating codes from *Variant 2a* in the range $2 \leq n \leq 6$. Again this range contains practically all codes of good quality. The case $n = 1$ does not occur in *Variant 2a.* In dimension $n = 2$ we have only the square. In dimension $n = 3$ we have alternative representations of the cubeoctahedron, with parameters $(n, \rho, M) = (3, 1, 12)$, and a semi-regular polyhedron with parameters $(n, \rho, M) = (3, 2/5, 24)$. Both were generated also in *Variant 1.* For larger n we obtain a number of new codes. Especially in dimension $n = 5$ we have several good codes.

The distributions $m = (r, p) = (n-1, 1)$ give the standard representation of the *Biorthogonal* codes; they are all included in *Variant 2a.* However, apart from this class of codes there does not seem to be any further optimal codes in *Variant 2a.*

We notice that in the range $n \leq 6$ all possible values of r and most of the possible partitions p are represented. As for *Variant 1* there are very few good codes from *Variant 2a* for $n \geq 6$.

4.5 Variant 2b

In *Variant 2b* we have $0 < \mu_1 < \mu_2 < \ldots < \mu_K$. Here we have to be a little bit more careful when determining the signal levels μ_i. Because of the sign changes there are two ways of generating nearest neighbors to a given word: either by permuting a pair of coordinates (as before) or (and this is new) by changing sign in one of the positions containing μ_1. A moment of thought reveals the following formula:

$$d_E^2 = \min\{4\mu_1^2, \min_{i \neq j} 2(\mu_i - \mu_j)^2\}.$$

n	ρ	M	m	
1	4	2	1	$*$
2	$2-\sqrt{2}$	8	1 1	$*$
2	2	4	2	$*$
3	$(52-2\sqrt{2})/97$	48	1 1 1	
3	$(20-8\sqrt{2})/17$	24	2 1	
3	$4/3$	8	3	
4	$(8-3\sqrt{2})/46$	384	1 1 1 1	
4	$(14-6\sqrt{2})/31$	192	2 1 1	
4	$(2-\sqrt{2})/2$	96	2 2	
4	$(6-\sqrt{2})/7$	64	3 1	
4	1	16	4	
5	$(52-16\sqrt{2})/685$	3840	1 1 1 1 1	
5	$(44-16\sqrt{2})/267$	1920	2 1 1 1	
5	$(68-32\sqrt{2})/161$	960	2 2 1	
5	$(20-8\sqrt{2})/51$	640	3 1 1	
5	$(36-16\sqrt{2})/49$	320	3 2	
5	$(28-8\sqrt{2})/41$	160	4 1	
5	$4/5$	32	5	
6	$(58-15\sqrt{2})/1457$	46080	1 1 1 1 1 1	
6	$(66-20\sqrt{2})/889$	23040	2 1 1 1 1	
6	$((18-7\sqrt{2})/113$	11520	2 2 1 1	
6	$(34-12\sqrt{2})/217$	7680	3 1 1 1	
6	$((26-12\sqrt{2})/97$	5760	2 2 2	
6	$(18-8\sqrt{2})/49$	3840	3 2 1	
6	$(8-3\sqrt{2})/23$	1920	4 1 1	
6	$(2-\sqrt{2})/3$	1280	3 3	
6	$(10-4\sqrt{2})/17$	960	4 2	
6	$(4-\sqrt{2})/7$	384	5 1	
6	$2/3$	64	6	

Table 4.3: All dominating permutation codes of *Variant 2b*, $1 \leq n \leq 6$. Codes known to be optimal have been indicated by $*$.

By a simple calculation we find that under the constraint $d_E^2 = 2$ the energy is minimized by

$$\mu_i = 1/\sqrt{2} + i - 1; \qquad i = 1, 2 \ldots, K.$$

The corresponding alphabet is

$$\mathcal{L}_q = \{\pm 1/\sqrt{2}, \pm(1/\sqrt{2}+1), \ldots, \pm(1/\sqrt{2}+K-1)\}.$$

The size of the alphabet is even: $q = 2K$.

For codes in *Variant 2b* the size of the code is

$$M = \frac{(m_1 + m_2 + \cdots + m_K)!}{m_1!\; m_2!\; \ldots\; m_K!} \cdot 2^{(m_1 + m_2 + \cdots + m_K)}$$

and the dimension n is (again, see Section 4.6)

$$n = m_1 + m_2 + \cdots + m_K.$$

Finding the best distributions is simple: clearly we should have $m_1 \geq m_2 \geq \ldots \geq m_K$.

In Table 4.3 we display the dominating codes for $n \leq 6$. As before this list contains practically all interesting codes. It contains several quite good codes, but none for which we are able to prove optimality (except for the trivial codes in dimensions 1 and 2).

In dimension $n = 3$ we notice that we have a better code of size 24 than we had in the two previous versions. The new code is the semi-regular polyhedron with 8 triangles and 18 squares as its faces. The minimum squared distance is $\rho = (20 - 8\sqrt{2})/17 \approx 0.51096$, which is considerably better than we had before and reasonably close to the optimal value, which approximately equals 0.55384.

4.6 Dimensionality

We like to establish the dimensionality of the various forms of permutation codes. The crucial result is the following lemma.

Lemma 4.6.1 *Let* $x = (x_1, x_2, \ldots, x_N)$ *be an element in* $\mathbb{R}^N$ *and let* $\mathcal{P}(x)$ *be the linear space spanned by all permutations of* x. *Let* $\mathbf{j} = (1, 1, \ldots, 1)$ *denote the vector with ones in all components and let* λ *be any non-zero real number. The dimensionality of* $\mathcal{P}(x)$ *depends on* x *in the following way:*

$$\begin{array}{l} A:\ x = 0 \quad \Rightarrow \quad dim\mathcal{P}(x) = 0, \\ B:\ x = \lambda\mathbf{j} \quad \Rightarrow \quad dim\mathcal{P}(x) = 1, \\ C:\ x \neq \mathbf{j},\ \ \sum_{i=1}^{N} x_i = 0 \quad \Rightarrow \quad dim\mathcal{P}(x) = N - 1, \\ D:\ x \neq \mathbf{j},\ \ \sum_{i=1}^{N} x_i \neq 0 \quad \Rightarrow \quad dim\mathcal{P}(x) = N. \end{array}$$

Proof: The statement is trivially true when $N = 1$. By induction, suppose it is true for N and consider $x = (x_1, x_2, \ldots, x_{N+1})$. We will prove the statement by considering each one of the above cases separately. Cases A and B are trivial, so we really only have to consider cases C and D.

Case C: Without loss of generality we may assume $x_{N+1} \neq 0$. Consider $x' = (x_1, x_2, \ldots, x_N)$. It is clear that x' cannot belong to case A or C. If x' belongs to case B we must have that x is proportional to $(1^N, -N)$. But then $\mathcal{P}(x)$ is spanned by a *Simplex*, and the dimension is obviously N, as it should.

If x' belongs to case D then $dim\mathcal{P}(x') = N$ by the induction hypothesis. Clearly we have $dim\mathcal{P}(x) \geq dim\mathcal{P}(x') = N$. But we also have $dim\mathcal{P}(x) \leq N$ because $\sum x_i = 0$. It follows that $dim\mathcal{P}(x) = N$ holds in this case too.

Case D: Again without loss of generality we assume $x_{N+1} \neq 0$ and consider $x' = (x_1, x_2, \ldots, x_N)$. If $x' = 0$ (belongs to case A) it is obvious that $dim\mathcal{P}(x) = N + 1$. If $x' = \lambda\mathbf{j}$ (belongs to case B) we have x of the form $x = (a^N, b)$, with $a \neq b$; a,$b \neq 0$; $b \neq -Na$. Alternatively we might write x in the form $x = c^{N+1} + ((a-c)^N, b-c)$ and choose c such that $N(a-c) + b - c = Na + b - (N+1)c = 0$, i.e. we may choose

$$c = \frac{Na + b}{N + 1}.$$

Then the vector $((a-c)^N, b-c)$ is proportional to $(1^N, -N)$ and $c \neq 0$. Upon permuting x the vector $((a-c)^N, b-c)$ will generate a *Simplex* spanning a space of dimension N while the vector c^{N+1} stays constant. By observing that c^{N+1} is orthogonal to all of the vectors in the space spanned by the *Simplex* we conclude that the permutations of x span a space of dimension $N+1$.

If x' belongs to case C we simply delete some other component than x_{N+1} in order to generate x'. There must be some component in x such that $\sum x'_i \neq 0$, because if all the different sums of this form are zero we might easily conclude $x_1 = x_2 = \cdots = x_{N+1}$, which would contradict the assumption. Thus, there remains only the case when x' belongs to case D.

If x' belongs to case D we may write $x = (0^N, x_{N+1}) + (x', 0)$. Consider the subset of all permutations of x that leave the last component unchanged. By the induction hypothesis those permutations generate a space of dimension N when operating on $(x', 0)$. Moreover, the vector $(0^N, x_{N+1})$ is unchanged by this subset of permutations and orthogonal to the space generated by $(x', 0)$. It follows that the space generated in this way has dimension $N+1$. Clearly the dimension does not decrease if we let all permutations operate on x. This observation concludes our argument. □

It is clear that optimized permutation codes of *Variant 1* always belong to case C while codes of *Variant 2* always belong to case D.

4.7 Decoding

One of the most fascinating properties of permutation codes is the fact that they can be optimally decoded with a very simple algorithm. Let a permutation code $\mathcal{X}$ be given. By decoding we understand in this context the following problem.

Problem: For any given vector $r = (r_1, r_2, \ldots, r_N) \in \mathbb{R}^N$ find a vector $x = (x_1, x_2, \ldots, x_N) \in \mathcal{X}$ with *minimal distance* to r.

Alternatively we could, of course, try to find a vector x having *maximal correlation* with r. We will use this second formulation.

An algorithm solving the problem is often referred to as a *maximum likelihood* decoding algorithm. This refers to decoding in the presence of additive white Gaussian noise.

We consider first codes from *Variant 1*. The crucial observation is the following lemma.

Lemma 4.7.1 *Let $\mathcal{X}$ be a permutation code of Variant 1 and, for a given vector $r = (r_1, r_2, \ldots, r_n) \in \mathbb{R}^n$, let $x = (x_1, x_2, \ldots, x_n) \in \mathcal{X}$ satisfy*

$$\langle x, r\rangle = \max\{\langle x', r\rangle : x' \in \mathcal{X}\}. \tag{4.7.1}$$

Then the following implication holds:

$$r_i < r_j \;\Rightarrow\; x_i \leq x_j. \tag{4.7.2}$$

Proof: The form of the proof is the same as in Lemma 4.2.1. Suppose $x \in \mathcal{X}$ violates (4.7.2), i.e. suppose that for some pair i, j we have both $r_i < r_j$ and $x_i > x_j$. Define

$$y_k = \begin{cases} x_k, & k \neq i, j, \\ x_i, & k = j, \\ x_j, & k = i. \end{cases}$$

Clearly $y = (y_1, y_2, \ldots, y_N) \in \mathcal{X}$. We have

$$\begin{aligned} \langle y, r\rangle - \langle x, r\rangle &= (y_i - x_i)r_i + (y_j - x_j)r_j \\ &= (x_i - x_j)(r_j - r_i) > 0. \end{aligned}$$

It follows that x cannot maximize $\langle x, r\rangle$. □

The lemma immediately leads to the following decoding algorithm. Let the initial vector be $x_0 = (x_{01}, x_{02}, \ldots, x_{0N})$ with $x_{01} \leq x_{02} \leq \ldots \leq x_{0N}$. Decoding can be thought of as assigning a permutation of x_0 to each vector $r = (r_1, r_2, \ldots, r_N) \in \mathbb{R}^N$. From the lemma follows that by assigning x_{01} to the smallest component in r, by assigning x_{02} to the second smallest and so on we can be sure to obtain a vector $x \in \mathcal{X}$ with a minimal distance to the given vector r.

A slight modification of the algorithm allows optimal decoding also of codes from *Variant 2*. In this case we need the following lemma.

Lemma 4.7.2 *Let $\mathcal{X}$ be a permutation code of Variant 2 and let $x \in \mathcal{X}$ satisfy*

$$\langle x, r\rangle = \max\{\langle x', r\rangle : x' \in \mathcal{X}\}. \tag{4.7.3}$$

Then the following implications hold:

$$(i) \qquad |r_i| < |r_j| \Rightarrow |x_i| \le |x_j|, \tag{4.7.4}$$

$$(ii) \qquad x_i, r_i \ne 0 \Rightarrow sign(r_i) = sign(x_i). \tag{4.7.5}$$

Proof: First suppose $x \in \mathcal{X}$ satisfies (4.7.4) and violates (4.7.5), i.e. assume that for some i we have $x_i, r_i \ne 0$ with $sign(r_i)sign(x_i) = -1$. Define

$$y_k = \begin{cases} x_k, & k \ne i, \\ -x_i, & k = i. \end{cases}$$

We obtain for $y = (y_1, y_2, \ldots, y_N) \in \mathcal{X}$

$$\langle y, r\rangle - \langle x, r\rangle = y_i r_i - x_i r_i = -2x_i r_i > 0.$$

Thus (ii) is established. Next suppose x satisfies (ii) but violates (i). That means that we assume that for some pair i, j we have $|r_i| < |r_j|$ and $|x_i| > |x_j|$. First assume $r_i \ne 0$ and define

$$y_k = \begin{cases} x_k, & k \ne i, j, \\ |x_i| sign(r_j), & k = j, \\ |x_j| sign(r_i), & k = i. \end{cases}$$

Clearly $y \in \mathcal{X}$. We obtain

$$\begin{aligned} \langle y, r\rangle - \langle x, r\rangle &= (y_i - x_i)r_i + (y_j - x_j)r_j = \\ &= |x_j|\, sign(r_i) r_i - |x_i|\, sign(x_i) r_i + \\ &+ |x_i|\, sign(r_j)\, r_j - |x_j|\, sign(x_j)\, r_j \end{aligned} \tag{4.7.6}$$

By (ii) we have $sign(x_i) = sign(r_i)$, which inserted in (4.7.6) gives us

$$\langle y, r\rangle - \langle x, r\rangle = |x_j||r_i| - |x_i||r_i| + |x_i||r_j| - |x_j||r_j|$$

$$= (|x_i| - |x_j|)(|r_j| - |r_i|) > 0.$$

If $r_i = 0$ the same inequality is easily established by a slight modification of the argument. Thus (i) is established and the lemma is proved. □

Consider a code of either *Variant 2a* or *Variant 2b*. Let the initial vector be $x_0 = (x_{01}, x_{02}, \ldots, x_{0N})$, with $0 \leq x_{01} \leq x_{02} \leq \ldots \leq x_{0N}$. By Lemma 4.7.2 follows that we obtain optimal decoding by first assigning the values a_i according to increasing values of $|r_i|$ and then choosing $sign(x_i) = sign(r_i)$.

We see that in both variants the decoding essentially consists of sorting the components of the vector $r = (r_1, r_2, \ldots, r_N)$ according to magnitude, an operation that requires $N(N-1)/2$ comparisons. It is worthwhile noticing that the number of comparisons needed depends only on the dimension n ($N-1$ in *Variant 1* and N in *Variant 2*) and *not* on the size M of the code.

4.8 General comments

It is an ironic fact that one of the nicest properties displayed by the permutation codes—namely the availability of a very simple optimal decoding algorithm—actually is also a property that excludes good asymptotic performance. This follows from an observation by Landau [148]. The observation is the following.

Decoding, as defined in Section 4.7, of a code $\mathcal{X}$ with parameters (n, ρ, M) corresponds to partitioning the Euclidean space $\mathbb{R}^n$ in M disjoint subregions, one for each codeword. Maximum likelihood decoding corresponds to assigning to each codeword its associated Voronoi region. Clearly the boundaries between the various Voronoi regions are defined by hyperplanes in $\mathbb{R}^n$. Denote by K the number of hyperplanes needed to specify all the Voronoi regions for the code $\mathcal{X}$ and define $t = \lfloor 4/\rho \rfloor$. Landau ([148], Theorem 2, p. 160) showed that for $M > 4$ the following inequality holds:

$$K \geq (M/2)^{1/t}\,.$$

The simplicity of the decoding algorithms we derived in Section 4.7 essentially means that the number of hyperplanes involved is small.

For codes of *Variant 1* we have to perform $N(N-1)/2$ comparisons. Each determines the position of the given vector r with respect to some hyperplane. For codes of *Variant 2* we also have to determine the sign of each one of the N coordinates in $r = (r_1, r_2, \ldots, r_N)$. Those decisions also determine the position of r with respect to certain hyperplanes, so for codes of *Variant 2* the number of hyperplanes can be estimated as $K \leq N(N-1)/2 + N$, ([148], p. 161). In both cases we have that for fixed ρ—and consequently also fixed t—the size M of the code $\mathcal{X}$ is bounded by $M \leq 2N^{2t}$. Thus, the exponential growth we know is possible for spherical codes in general is excluded for the class of permutation codes.

The above result is well in line with the observations mentioned above regarding the performance for large dimensions. It strongly supports our conjecture that essentially no good codes are produced for dimensions larger than 6. The *Simplex* and the *Biorthogonal* codes do not provide any counterexamples: although optimal they do not exhibit exponentially increasing size.

It is interesting to note that all three of those regular polytopes that generalize to arbitrary dimensions are found among the permutation codes: the *Simplex* codes are all generated by *Variant 1*, the *Biorthogonal* codes are generated by *Variant 2a*, and the hypercubes are generated by *Variant 2b*.

Chapter 5

Symmetric alphabets

5.1 Introduction

An extensive literature is available describing codes in various discrete spaces. In particular there is quite a vast literature on binary codes, usually characterized in terms of Hamming metric. It is natural to try to adapt those results to the case of spherical codes and Euclidean metric. This leads naturally to the concept of *code concatenation.*

Concatenation of codes is a way of building new codes from old ones. This is done by taking the codewords of one code as the letters in one or several other codes. Usually the code forming the letters is referred to as the *inner code.* Anyone of the codes using the codewords from the inner code as its letters is referred to as an *outer code.* When more than one outer code is employed the inner code is subject to so called *set partitioning.* The basic idea—using just one single outer code—belongs to Forney, [108]. The generalization—using several outer codes and code partitioning—was originally developed by Blokh–Zyablov [35], [36] (the case of linear codes) and by Zinoviev [236], [237] (the general nonlinear case). It was rediscovered independently by Ungerboeck [225] and Imai–Hirakawa [134] and has since then been the subject of interest of many authors. For spherical codes it was first studied by Ginzburg [115], Portnoi [182] and by Sayegh [189]. Bos [38] used the same idea for general metrics.

The idea of code concatenation can be instrumented in many different ways. In the present chapter we shall discuss in detail how it can be used in connection with alphabets constituting symmetric, equally spaced pointset on the real line. The alphabet is interpreted as an inner code. As outer codes we will use binary codes only. Non-symmetric alphabets will be considered in Chapter 6. The present chapter is a slightly modified version of [104].

5.2 An introductory example

Before discussing the general construction we like to illustrate a few of the basic ideas upon which it relies. We do this by considering first a simple construction by which a spherical code is generated from a *triple of constant weight* binary codes. We then demonstrate how this simple construction can be essentially improved under the conditions of interest in this chapter, namely for codes built up from *symmetric equally spaced* alphabets on the real line. The main theme of the present chapter is a generalization of the ideas illustrated in this introductory example.

Let $\mathcal{A}$, $\mathcal{C}_0$, and $\mathcal{C}_1$ be three constant weight binary codes with lengths, weights, distances and sizes given by $(n, w_{\mathcal{A}}, d_{\mathcal{A}}, M_{\mathcal{A}})$, $(n - w_{\mathcal{A}}, w_0, d_0, M_0)$, $(w_{\mathcal{A}}, w_1, d_1, M_1)$ respectively. Notice that the length of $\mathcal{C}_1$ equals the weight of $\mathcal{A}$ while the length of $\mathcal{C}_0$ equals the number of zeroes in each one of the codewords in $\mathcal{A}$. Let L_{00}, L_{01}, L_{10}, and L_{11} be four arbitrary real numbers. We will construct a spherical code $\mathcal{X} \subseteq \Omega_n$ such that each component in each codeword in $\mathcal{X}$ equals one of these four numbers. We say that the code $\mathcal{X}$ uses the *alphabet* $\mathcal{L} = \{L_{00}, L_{01}, L_{10}, L_{11}\}$.

For any $a = (a_1, a_2, \ldots, a_n) \subseteq \mathcal{A}$ define the integer functions φ and $\bar{\varphi}$ by the formulas

$$\varphi(a,i) \triangleq \sum_{j=1}^{i} a_j\,, \quad \bar{\varphi}(a,i) \triangleq \sum_{j=1}^{i} \bar{a}_j\,,$$

where $\bar{a}_i \triangleq a_i \oplus 1$ is the binary complement to a_i, interpreted as a real number. For each triple (a, c_0, c_1) of binary codewords $a =$

$(a_1, a_2, \ldots, a_n) \in \mathcal{A}$, $c_0 = (c_{0,1}, c_{0,2}, \ldots, c_{0,n-w_A}) \in \mathcal{C}_0$, and $c_1 = (c_{11}, c_{12}, \ldots, c_{1,w_A}) \in \mathcal{C}_1$ we define a codeword $x = (x_1, x_2, \ldots, x_n)$ in the constant energy code $\mathcal{X}$ by

$$x_i \triangleq \begin{cases} L_{00}, & if \quad a_i = 0, \; c_{0,\bar{\varphi}(a,i)} = 0, \\ L_{01}, & if \quad a_i = 0, \; c_{0,\bar{\varphi}(a,i)} = 1, \\ L_{10}, & if \quad a_i = 1, \; c_{1,\varphi(a,i)} = 0, \\ L_{11}, & if \quad a_i = 1, \; c_{1,\varphi(a,i)} = 1, \end{cases}$$

$$i = 1, 2, \ldots, n.$$

It is clear that we obtain a constant energy code where each codeword has energy

$$\begin{aligned} E = \|x\|^2 &= (n - w_A - w_0) L_{00}^2 + w_0\, L_{01}^2 + \\ &+ (w_A - w_1)\, L_{10}^2 + w_1\, L_{11}^2 = \\ &= n\, L_{00}^2 + w_A\, (L_{10}^2 - L_{00}^2) + \\ &+ w_0\, (L_{01}^2 - L_{00}^2) + w_1\, (L_{11}^2 - L_{10}^2). \end{aligned}$$

In general the energy depends on all three weights $w_{\mathcal{A}}$, w_0, w_1. However, under the special condition $L_{10} = -\, L_{00}$, and $L_{01} = -\, L_{11}$ the expression simplifies to

$$E = n\, L_{00}^2 + (w_0 + w_1)\, (L_{11}^2 - L_{00}^2).$$

We notice two things:

(*i*) *The weight $w_{\mathcal{A}}$ does not enter this new expression.*

(*ii*) *The weights w_0 and w_1 occur only as a sum.*

Our interest in this chapter is in spherical codes generated from equally spaced symmetric alphabets. Let $\mathbb{Z}$ be the set of all integers, let $\mathcal{L}$ denote the alphabet and let K denote its size. For K odd we assume $\mathcal{L} \subseteq \mathbb{Z}$ while for K even we assume $\mathcal{L} \subseteq \mathbb{Z} + 1/2$. For $K = 4$ we choose

$$L_{10} = -L_{00} = -1/2, \quad L_{01} = -L_{11} = -3/2.$$

For the construction above a simple argument leads to the following estimate of the minimum distance

$$d_E^2 \geq \min\{d_{\mathcal{A}}, 4d_0, 4d_1\}.$$

At this point it is easy to see, however, that for the special alphabet considered here, $\mathcal{L} = \{\pm 1/2, \pm 3/2\}$, the construction can be essentially improved. By observation (i) the weight of the code $\mathcal{A}$ does not affect the energy E. Therefore the code $\mathcal{A}$ need not be a constant weight code, but could be any binary code of length n. By observation (ii) the two codes $\mathcal{C}_0$ and $\mathcal{C}_1$ of lengths $n - w_{\mathcal{A}}$ and $w_{\mathcal{A}}$ and with weights w_0 and w_1 affect the energy E only through the sum $w_0 + w_1$. Therefore they can be replaced with one single binary code $\mathcal{C}$ of length n and weight $w = w_0 + w_1$. Thus we are lead to a new construction by which a constant energy code $\mathcal{X} \subset \mathbb{R}^n$ is determined by a *pair* of binary codes, one unrestricted code $\mathcal{A}$ and one constant weight code $\mathcal{C}$. In this new construction each codeword $x = (x_1, x_2, \ldots, x_n) \in \mathcal{X}$ is determined by a pair (a, c) of codewords

$$\begin{aligned} a &= (a_1, a_2, \ldots, a_n) \in \mathcal{A} \\ c &= (c_1, c_2, \ldots, c_n) \in \mathcal{C} \end{aligned}$$

through a component-wise mapping according to the rule

$$x_i \triangleq \begin{cases} L_{00}, & if \quad a_i = 0,\ c_i = 0, \\ L_{01}, & if \quad a_i = 0,\ c_i = 1, \\ L_{10}, & if \quad a_i = 1,\ c_i = 0, \\ L_{11}, & if \quad a_i = 1,\ c_i = 1, \end{cases}$$

$$i = 1, 2, \ldots, n.$$

Each codeword $x \in \mathcal{X}$ has energy

$$E = \|y\|^2 = (n + 8w)/4.$$

The minimum distance is estimated as before:

$$d_E^2 \geq \min\{d_{\mathcal{A}}, 4d_{\mathcal{C}}\},$$

where $d_{\mathcal{C}}$ is the minimum Hamming weight in the code $\mathcal{C}$.

As seen, the new construction is very similar to the old one. However, in the new construction we have much larger freedom in choosing the binary codes $\mathcal{A}$ and $\mathcal{C}$. As a special case we could still let $\mathcal{A}$ be a constant weight code and we could let $\mathcal{C}$ be a direct sum $\mathcal{C} = \mathcal{C}_0 \oplus \mathcal{C}_1$. That, of course, would generate exactly the same code as we had before. Our new construction is therefore a true generalization: any code that can be generated by the old construction can also be generated by the new one (recall that we assume the condition $L_{10} = -L_{00}$, $L_{01} = -L_{11}$). However, the new freedom of choice opens new possibilities and we usually obtain much better spherical codes if we choose $\mathcal{A}$ *not* to be a constant weight code and if we choose $\mathcal{C}$ *not* to be a direct sum. This, in a nutshell, is the idea upon which we build our constructions. As we shall see it can be extended to much more general situations.

In our example with size $K = 4$ our alphabet was $\mathcal{L} = \{\pm 1/2, \pm 3/2\}$. We utilized the fact that this alphabet can be partitioned according to

$$\mathcal{L} = \{-1/2, 3/2\} \cup \{1/2, -3/2\}.$$

The codeword $a \in \mathcal{A}$ determines the subset while the codeword $c \in \mathcal{C}$ determines the letter within the subset. Clearly the minimum distance in each of the two subsets is exactly twice that in the whole alphabet (a factor 4 in terms of squared distance). We like to retain this property also in the more general cases.

There is one more thing that must be observed: the construction should lead to a code where all codewords have the same energy. In the first construction this was easy because only constant weight binary codes were employed. In the second construction things are slightly more complicated. Here constant energy depends also on the odd symmetry of the partition:

$$\{-1/2, 3/2\} = -\{1/2, -3/2\}.$$

Odd symmetry is the reason why we obtain a constant energy spherical code even with an unrestricted binary code $\mathcal{A}$. We like also this property to be retained in general.

5.3 Binary labeling

Recall that we assume the alphabet $\mathcal{L} = \mathcal{L}_K$ to be a subset of $\mathbb{Z}$ for K odd and a subset of $\mathbb{Z} + 1/2$ for K even. In order to see how the above two properties—odd symmetry and increase of distance—can be retained in general we first investigate how the two infinite sets $\mathbb{Z}$ and $\mathbb{Z} + 1/2$ can be partitioned with these objectives in mind. We start by considering $\mathbb{Z} + 1/2$ and notice first that this set can be partitioned according to

$$\mathbb{Z} + 1/2 = (1/2) \cdot (2\mathbb{Z} + 1) = (1/2) \cdot (4\mathbb{Z} + 1) \cup (1/2) \cdot (4\mathbb{Z} - 1).$$

The distance in the subsets has increased by a factor of two in comparison with the distance in the original set. We have

$$4\mathbb{Z} - 1 = -(4\mathbb{Z} + 1),$$

so the odd symmetry is also retained in this first step. But more is true. In a next step we can partition according to

$$(1/2) \cdot (4\mathbb{Z} + 1) = (1/2) \cdot (8\mathbb{Z} + 1) \cup (1/2) \cdot (8\mathbb{Z} - 3),$$
$$(1/2) \cdot (4\mathbb{Z} - 1) = (1/2) \cdot (8\mathbb{Z} - 1) \cup (1/2) \cdot (8\mathbb{Z} + 3).$$

We see that not only do we have symmetry between the sets $4\mathbb{Z} - 1$ and $4\mathbb{Z} + 1$, but appropriately selected partitions are also symmetric. We also see that in the second partition the distance is increased by another factor of two. The process can be continued indefinitely. The general form is

$$(1/2) \cdot (2^k\mathbb{Z} + i) = (1/2) \cdot (2^{k+1}\mathbb{Z} + i) \cup (1/2) \cdot (2^{k+1}\mathbb{Z} + i - 2^k \, sgn(i)),$$

where i is odd and $0 < |i| < 2^{k-1}$, for $k = 2, 3, 4, \ldots$.

Clearly the distance increases by a factor of two in each step and the odd symmetry is retained through all the steps. The first few of them are illustrated in Figure 5.1 in the form of a tree diagram. Notice that the upper and the lower parts of the tree are the same apart from sign.

For the construction to be described we need an appropriate labeling. This can be achieved in many different ways. One simple possibility

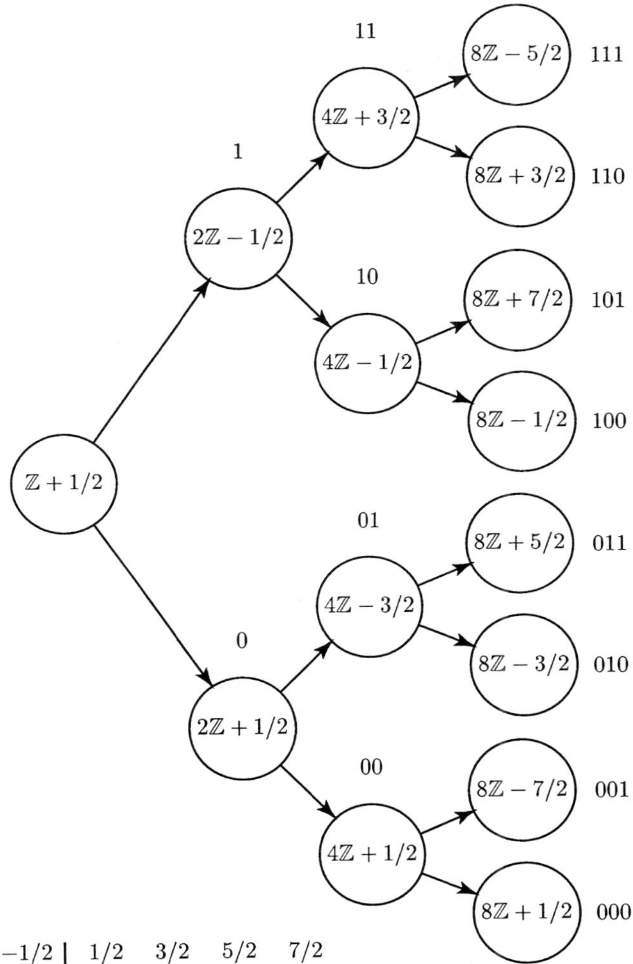

$-7/2$	$-5/2$	$-3/2$	$-1/2$	$1/2$	$3/2$	$5/2$	$7/2$
0	1	0	1	0	1	0	1
0	1	1	0	0	1	1	0
1	1	0	0	0	0	1	1

Figure 5.1: Binary partitionings of $\mathbb{Z} + 1/2$.

is to assign a binary number to each one of the vertices in the tree (except the root) in the way indicated in Figure 5.1. Notice that with this labeling the binary numbers describe the way to the vertex from the root according to the simple rule that a branching downwards is indicated by a zero while a branching upwards is indicated by a one. Clearly this labeling can also be seen as a labeling of the points in $\mathbb{Z}+1/2$, as also indicated in Figure 5.1. A general definition can be given as follows.

For any $i \geq 0$ let the binary expansion be

$$i = \sum_{j=0}^{\infty} b_j(i)\, 2^j\,.$$

The sequence $\{b_j(i) : 0 \leq j < \infty\}$ is a binary representation of the integer $i \geq 0$. For $i < 0$ we may use the representation

$$b_j(i) = b_j(-i-1) \oplus 1.$$

Unfortunately this well known binary labeling does not work directly for our purposes, but a simple modification does. Let $b_{-1}(i) = 0,\ i = 0, \pm 1, \pm 2, \ldots$ and define

$$\begin{aligned} &a_j(i) = b_{j-1}(i) \oplus b_j(i),\\ &j = 0, 1, 2, \ldots\ ,\\ &i = 0, \pm 1, \pm 2, \ldots\ . \end{aligned}$$

Further, define a binary labeling for $\mathbb{Z}+1/2$ by the rule

$$i + \frac{1}{2} \ \leftrightarrow\ a_0(i), a_1(i), a_2(i), \ldots\ . \tag{5.3.1}$$

It is easy to see that this labeling is the one used in Figure 5.1. We will see in the next section how it works in general. Before that, however, we like to consider the partitions and the appropriate binary labeling for $\mathbb{Z}$.

The first partitioning is simple:

$$\mathbb{Z} = 2\mathbb{Z} \cup 2(\mathbb{Z}+1/2).$$

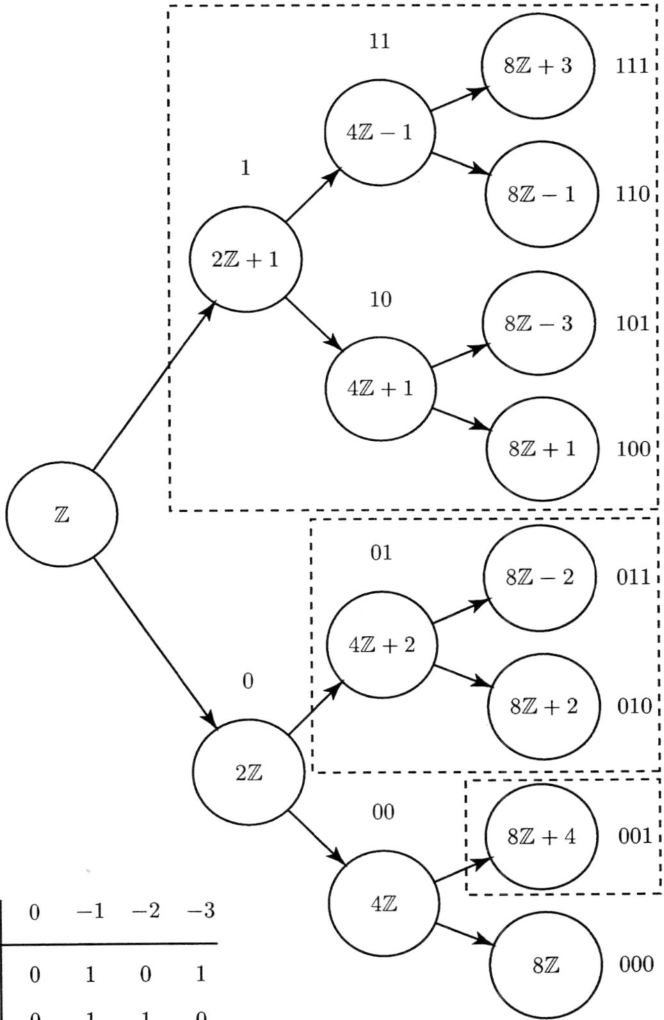

4	3	2	1	0	−1	−2	−3
0	1	0	1	0	1	0	1
0	1	1	0	0	1	1	0
1	1	0	0	0	0	1	1

Figure 5.2: Binary partitionings of $\mathbb{Z}$.

Again the minimum distance in the subsets is twice that in the original alphabet, but this time we do not have symmetry between the two subsets. We notice, however, that the first subset is of the same form as the set $\mathbb{Z}$ we started with while the second subset has the form $\mathbb{Z}+1/2$. It is clear that the set $2(\mathbb{Z}+1/2)$ can be further partitioned in exactly the same way as was $\mathbb{Z}+1/2$, while the set $2\mathbb{Z}$ can be again partitioned in the same way as was $\mathbb{Z}$:

$$2\mathbb{Z} = 4\mathbb{Z} \cup 4(\mathbb{Z}+1/2).$$

Clearly this process can also be continued indefinitely. The general form is

$$2^k\mathbb{Z} = 2^{k+1}\mathbb{Z} \cup 2^{k+1}(\mathbb{Z}+1/2),$$

where $k = 1, 2, 3, \ldots$. The appropriate labeling for the set $\mathbb{Z}$ is

$$i \leftrightarrow a_0(-i),\ a_1(-i),\ a_2(-i), \ldots \ . \tag{5.3.2}$$

In Figure 5.2 the partition and the labeling are illustrated by another tree diagram. It is easy to see that it is in agreement with the representation (5.3.2). We also see that by this labeling of $\mathbb{Z}$ we obtain exactly the same labeling for the symmetric subsets of the form $\mathbb{Z}+1/2$ as we had before. The symmetric subsets have been framed by dashed lines in Figure 5.2

5.4 The construction: $2 \le K \le 4$

We suggest a family of constructions by which spherical codes can be generated from binary codes. In each case the spherical code is generated from a set of binary codes. Both constant weight codes and unrestricted binary codes are used. All constructions use equally spaced symmetric alphabets from the real line $\mathbb{R}$. We normalize the internal distance in each alphabet to unit length, and define the alphabets as

$$\mathcal{L}_K = \{\pm 1/2, \pm 3/2, \ldots, \pm(K-1)/2\}, \quad K \ even,$$
$$\mathcal{L}_K = \{0, \pm 1, \pm 2, \ldots, \pm(K-1)/2\}, \quad K \ odd.$$

Each codeword $x = (x_1, x_2, \ldots, x_n)$ in the resulting code $\mathcal{X}$ will have a constant energy $E = x_1^2 + x_2^2 + \cdots + x_n^2$. The minimum Euclidean distance is

$$d_E^2 \triangleq \min_{x \neq y} \|x - y\|^2, \quad x,\ y \in \mathcal{X}.$$

The corresponding normalized code—which by definition is a subset of the unit sphere Ω_n—is obtained by a proper scaling. It is characterized by the normalized squared minimum distance $\rho = d_E^2/E$.

We give one construction for each size K of the alphabet $\mathcal{L}_K$. The constructions become increasingly more complex as the size K increases. We develop the general principles successively by first considering the first few simple cases in detail. We then formulate the general construction and derive its main properties.

K=2:
The case $K = 2$ is the conventional binary antipodal representation. In this case only one binary code is involved, and there is no need to impose any restriction regarding the weights of the codewords. Let $\mathcal{A} \subseteq \mathbb{F}_2^n$ have parameters (n, d, M), where n is the length, d is the minimum Hamming distance and M is the size. The spherical code $\mathcal{X} = Y_2(\mathcal{A})$ is generated by assigning a real-valued vector $x = (x_1, x_2, \ldots, x_n)$ to each codeword $a = (a_1, a_2, \ldots, a_n) \in \mathcal{A}$ according to the rule

$$x_i = \frac{1}{2}(-1)^{a_i}, \quad i = 1, 2, \ldots, n.$$

Clearly each codeword $x = (x_1, x_2, \ldots, x_n)$ in the code $\mathcal{X} = Y_2(\mathcal{A})$ has energy $E = n/4$ and the minimum squared Euclidean distance is $d_E^2 = d$. The normalized code has parameters (n, ρ, M), with $\rho = 4d/n$. The alphabet is $\mathcal{L}_2 = \{\pm 1/2\}$ (for the non-normalized code).

K=3:
For $K = 3$ we use the alphabet $\mathcal{L}_3 = \{0, \pm 1\}$ and two binary codes, one constant weight code $\mathcal{C} \subseteq \mathbb{F}_2^{n,w}$ and one unrestricted code $\mathcal{A} \subseteq \mathbb{F}_2^w$. Notice that the length w of the unrestricted code $\mathcal{A}$ equals the weight of the constant weight code $\mathcal{C}$. As in the preceding paragraph we introduce a function φ which for each codeword $c \in \mathcal{C}$ counts the

number of ones according to the rule

$$\varphi(c,i) = \sum_{j=1}^{i} c_j\,, \quad i = 1, 2, \ldots, n,$$

where the components c_j should be interpreted as real numbers. For each pair $(c, a) \in \mathcal{C} \times \mathcal{A}$ of binary codewords we define a codeword $x = (x_1, x_2, \ldots, x_n)$ in the spherical code $\mathcal{X} = Y_3(\mathcal{C}, \mathcal{A})$ according to the rule

$$x_i = \begin{cases} 0, & if \quad c_i = 0, \\ 1, & if \quad c_i = 1, \; a_{\varphi(c,i)} = 0, \\ -1, & if \quad c_i = 1, \; a_{\varphi(c,i)} = 1, \end{cases}$$

$$i = 1, 2, \ldots, n.$$

The idea is very simple: the codeword $c \in \mathcal{C}$ decides between 0 and ± 1 while the codeword $a \in \mathcal{A}$ decides between $+1$ and -1. Thus we use code concatenation with the partitioning $\{0\} \cup \{\pm 1\}$. Notice that the squared minimum distance in the original alphabet is 1 while in the subset $\{\pm 1\}$ it is $2^2 = 4$ (in the subset $\{0\}$ it is infinite). Clearly the energy in each codeword $x \in Y_3(\mathcal{C}, \mathcal{A})$ is $E = w$, and the squared minimum distance can be estimated by the inequality

$$d_E^2 \geq \min\{d_{\mathcal{C}}, 4d_{\mathcal{A}}\},$$

where $d_{\mathcal{C}}$ and $d_{\mathcal{A}}$ denote the minimum Hamming distance in $\mathcal{C}$ and $\mathcal{A}$ respectively. The normalized code has parameters (n, ρ, M), where

$$\rho \geq \frac{1}{w} \min\{d_{\mathcal{C}}, 4d_{\mathcal{A}}\}, \quad M = M_{\mathcal{A}} M_{\mathcal{C}}.$$

K=4:
The construction for $K = 4$ is the one we developed in Section 5.2. Recall that this construction also uses two binary codes, but unlike the case $K = 3$ the two codes have the *same* length. There is another difference worth noticing: the factor 4 in the estimation of the minimum distance occurs with the parameter $d_{\mathcal{A}}$ in the construction for $K = 3$ while it appears with the parameter $d_{\mathcal{C}}$ in the construction for $K = 4$. Thus in a certain sense the roles of the constant weight code and the unrestricted code are interchanged. We use the

notation $Y_4(\mathcal{A}, \mathcal{C})$ for the construction for $K = 4$ and notice that the normalized code has parameters (n, ρ, M), with

$$\rho \ge \frac{4}{n + 8w} \min\{d_{\mathcal{A}}, 4d_{\mathcal{C}}\}, \quad M = M_{\mathcal{A}} M_{\mathcal{C}}.$$

Although the above examples are not complex enough to exhibit all the properties of the general construction most of the basic ideas are already present. There are a few things worth mentioning.

- All three constructions can be viewed as examples of *code concatenation*, with the binary codes as outer codes and the alphabet $\mathcal{L}_K$ as the inner code. For $K > 2$ there are more than one outer code in play and the concatenation is of the *generalized* type.

- The binary labeling is in accordance with the a-labeling defined in the preceding paragraph. Thus, for K *even* letters of the same magnitude but different signs have the same binary labeling except for the first binary symbol. For K *odd* the binary labeling for such letters differs in the first binary symbol after an initial sequence of the form $0^{\nu-1}1$, $\nu = 1, 2, 3, \ldots$.

- In general codes of *different* lengths are employed.

- The binary partitions are such that the internal distances increase by a factor of 2 in each step (a factor of 4 in terms of *squared* Euclidean distance).

- Each component x_i in each codeword $x = (x_1, x_2, \ldots, x_n)$ in the code $\mathcal{X}$ is specified successively by the components in the binary codewords. Each step gives a further refinement until a unique letter from the alphabet $\mathcal{L}_K$ has been specified. *Binary* choices between non-symmetric alternatives are associated with constant weight codes while choices between symmetric alternatives are associated with unconstrained codes.

Now let us consider the general case.

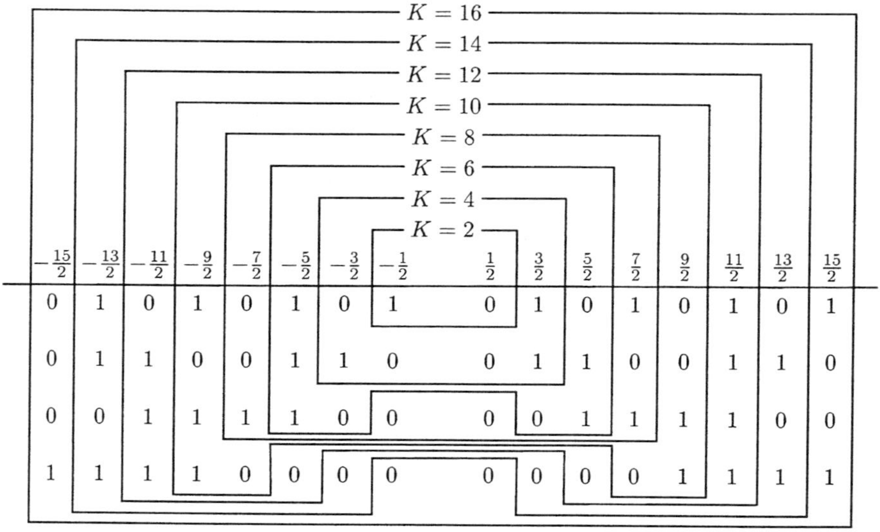

Figure 5.3: Binary labeling for alphabet $\mathcal{L}_K$ of even size K.

5.5 The construction: general case

We make a distinction between alphabets of odd or even size. In both cases we use the binary labeling as defined in Section 5.3, but with the sequences truncated to the shortest length needed in order to get a unique labeling for each letter in the alphabet $\mathcal{L}_K$. Of course, how to truncate depends on the size K. In Figures 5.3 and 5.4 we have indicated all cases of size $K \le 16$. The extension to larger K should be obvious.

Denote by $\mathcal{S}_K$ the set of all binary sequences $s = s_0 s_1 \dots s_{m-1}$ corresponding to the letters in the alphabet $\mathcal{L}_K$, and denote by $\mathcal{T}_K$ the set of all prefixes $s' = s_0 s_1 \dots s_{m'-1}$; $m' \le m$, including the empty sequence. For a given $s = s_0 s_1 \dots s_{m-1} \in \mathcal{S}_K$ denote by $L(s)$ the letter in $\mathcal{L}_K$ corresponding to s.

We will make use of certain directed graphs. We define one such graph Γ_K for each alphabet $\mathcal{L}_K$. In a sense the graph will contain all essential information about the corresponding spherical code.

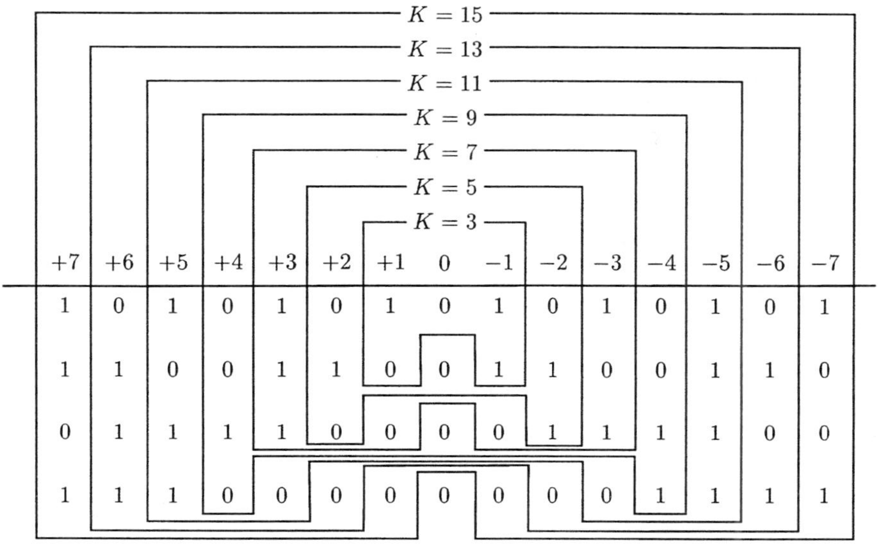

Figure 5.4: Binary labeling for alphabet $\mathcal{L}_K$ of odd size K.

In order to define the graph Γ_K corresponding to the alphabet $\mathcal{L}_K$ we need a slight modification of the sets $\mathcal{S}_K$ and $\mathcal{T}_K$. The modification is obtained by merging sequences which generate letters $L(s)$ of same magnitude but different signs. The modification is slightly different in the two cases K even and K odd.

Let us consider first the case K even. If the two sequences s and s' generate letters $L(s)$ and $L(s')$ of the same magnitude but with opposite signs, then s and s' differ only in the first symbol. Any such pair should be merged. More precisely, let $s = 0u$ and $s' = 1u$ be such a pair. We replace them with one single sequence $\tilde{s} = *u$, where the special symbol " $*$ " can be read as "either a zero or a one". We denote by $\tilde{\mathcal{S}}_K$ and $\tilde{\mathcal{T}}_K$ respectively the sets of ternary sequences (with symbols 0, 1, $*$) which are obtained in this way from the sets $\mathcal{S}_K$ and $\mathcal{T}_K$.

For K odd the definition is slightly more complicated. As can be seen from Figure 5.4 two sequences s and s' generate letters $L(s)$ and $L(s')$ of same magnitude but opposite signs if and only if they differ in the first binary symbol following an initial sequence of the form $0^{\nu-1}1$,

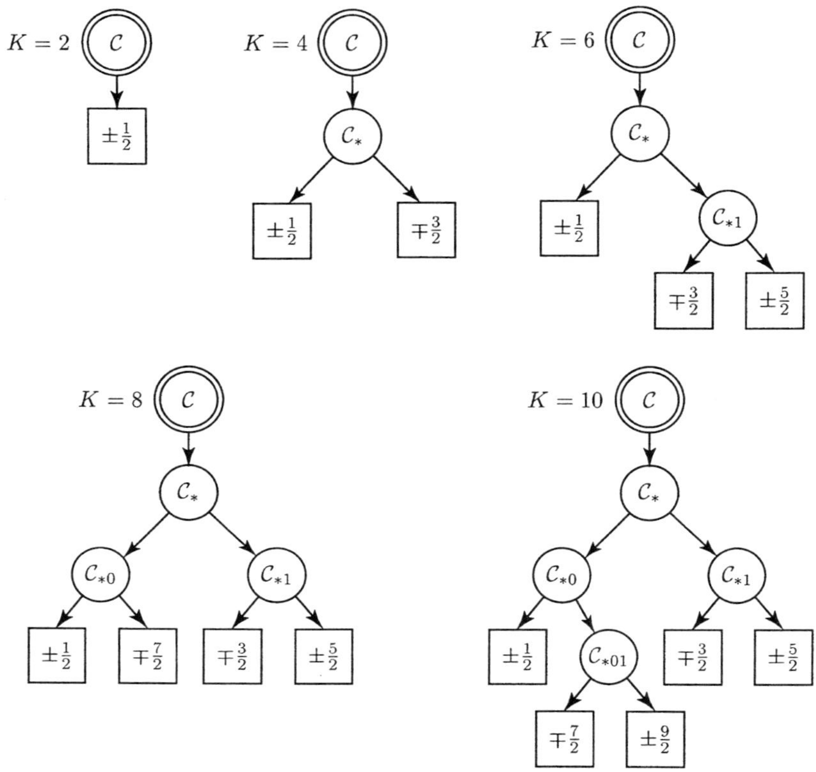

Figure 5.5: Code trees for K even.

where $\nu = 1, 2, \ldots$. With exception of the all-zero sequences all binary sequences s have an initial part of this form for some positive integer ν. Thus, for K odd we merge any sequences s and s' of the forms $s = 0^{\nu-1}10u$ and $s' = 0^{\nu-1}11u$ respectively into the single sequence $\tilde{s} = 0^{\nu-1}1 * u$. Again we use the notations $\tilde{\mathcal{S}}_K$ and $\tilde{\mathcal{T}}_K$ for the modified versions of the sets $\mathcal{S}_K$ and $\mathcal{T}_K$.

The graph Γ_K corresponding to the alphabet $\mathcal{L}_K$ is now defined by assigning a vertex to each sequence $\tilde{t} = \tilde{t}_0\tilde{t}_1 \ldots \tilde{t}_{m-1} \in \tilde{\mathcal{S}}_K \cup \tilde{\mathcal{T}}_K$ and by assigning an edge from the vertex $\tilde{u}$ to the vertex $\tilde{t}$ if and only if $\tilde{t}$ is an immediate successor to $\tilde{u}$, i.e. if and only if $\tilde{t}$ has the form $\tilde{t} = \tilde{u}x$, where $x \in \{0, 1, *\}$. Clearly the graph so generated is a tree, with $\tilde{\mathcal{T}}_K$ as the set of internal vertices and with $\tilde{\mathcal{S}}_K$ as the set

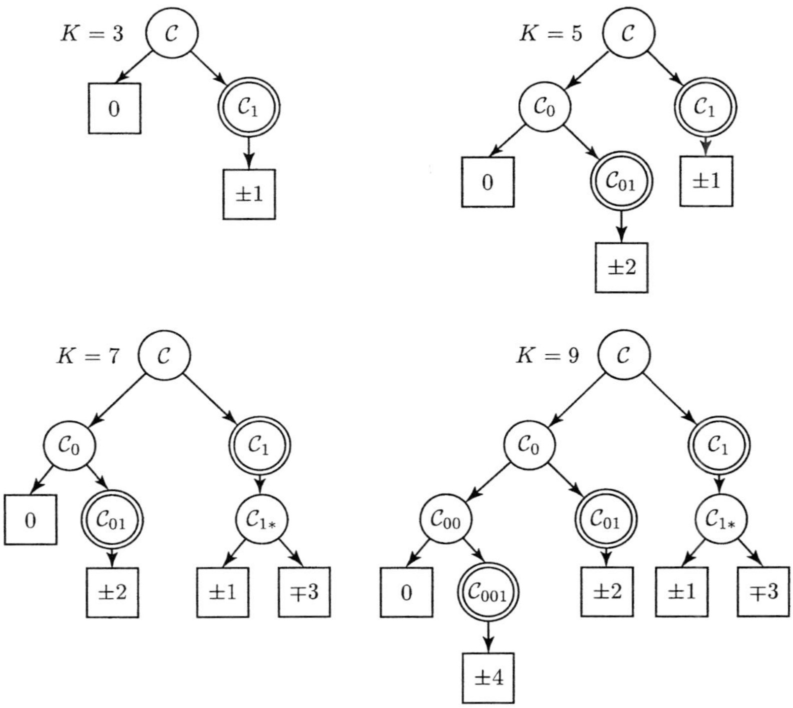

Figure 5.6: Codes trees for K odd.

of terminating vertices. We will refer to the terminating vertices as *leaves*. The empty sequence corresponds to the *root* of the tree.

In Figures 5.5 and 5.6 we have indicated all graphs for K in the range $2 \leq K \leq 10$. In order to simplify the interpretation we have indicated all leaves with squares and all internal vertices with circles. In addition certain vertices have been indicated by double circles. For K even it is only the root that has been so indicated. For K odd all nodes of the form $0^{\nu-1}1$, $\nu = 1, 2, \ldots$ have been specially indicated in this way. The reason for this special indication will soon be explained. For the time being we simply notice that each one of the doubly encircled vertices has exactly one edge leaving. All other vertices have either two leaving edges or none at all. With the exception of the doubly encircled vertices all internal vertices have exactly two edges leaving. The leaves, of course, have no leaving edges at all.

When drawing the graphs we have followed the convention of letting a zero correspond to a branching to the left while a one corresponds to a branching to the right. The single edges leaving the doubly encircled vertices are vertical in our diagrams.

In the leaves we have indicated the corresponding letters $L(s) \in \mathcal{L}_K$. The sign-convention for the leaves is the following. First notice that in each path from the root to a leave at most one doubly encircled vertex is passed: if the leave corresponds to the letter 0 no doubly encircled vertex is passed; in all other cases exactly one such vertex is passed. Recall that each letter is specified by a sequence $s = s_0 s_1 \ldots s_{m-1} \in \mathcal{S}_K$ where the magnitude is determined by the corresponding sequence $\tilde{s} = \tilde{s}_0 \tilde{s}_1 \ldots \tilde{s}_{m-1} \in \tilde{\mathcal{S}}_K$ and the sign depends on the symbol s_i in $s = s_0 s_1 \ldots s_{m-1}$ which has been replaced by the symbol " $*$ " in $\tilde{s}$. When indicating the two antipodal letters corresponding to the leave $\tilde{s}$ we have followed the convention that the upper sign indicates the letter generated if s_i is zero while the lower sign indicates the letter produced when s_i is one.

We are now ready to describe the general construction. Let the number $K \geq 2$ be given and consider the graph Γ_K corresponding to the alphabet $\mathcal{L}_K$. We assume that there is one binary code $C_{\tilde{t}}$ available for each vertex $\tilde{t} \in \tilde{\mathcal{T}}_K$. If K is even the code $\mathcal{C} = \mathcal{C}_\lambda$ corresponding to the root is an unrestricted code while all other codes are constant weight codes. If K is odd all codes corresponding to vertices of the form $\tilde{t} = 0^{\nu-1}1$ are unrestricted and all other codes are constant weight codes. In both cases the code parameters should satisfy the following constrains:

$$\forall \tilde{t} \in \tilde{\mathcal{T}}_K \setminus \{\lambda\} :$$

$$n_{\tilde{t}} = \begin{cases} n_{\tilde{u}} - w_{\tilde{u}}, & if \quad \tilde{t} = \tilde{u}0, \\ w_{\tilde{u}}, & if \quad \tilde{t} = \tilde{u}1, \\ n_{\tilde{u}}, & if \quad \tilde{t} = \tilde{u}*, \end{cases} \tag{5.5.1}$$

where the parameters indicate lengths and weights in an obvious way. Apart from these constrains, however, the binary codes can be chosen arbitrarily.

Now we see the significance of the doubly encircled vertices: they are exactly the vertices associated with unrestricted codes.

The encoding is a mapping

$$f : \bigotimes_{\tilde{t} \in \tilde{\mathcal{T}}_K} \mathcal{C}_{\tilde{t}} \to \mathbb{R}^n, \tag{5.5.2}$$

i.e. a mapping that associates a vector $x = (x_1, x_2, \dots, x_n) \in \mathbb{R}^n$ to each set $\{c_{\tilde{t}} \in \mathcal{C}_{\tilde{t}} : \tilde{t} \in \tilde{\mathcal{T}}_K\}$ of binary codewords. In order to describe this mapping a few additional definitions are needed.

For each $\tilde{t} \in \tilde{T}_K$ and for each codeword $c_{\tilde{t}} \in \mathcal{C}_{\tilde{t}}$ we define the integer functions

$$\varphi(c_{\tilde{t}}, i) \triangleq \sum_{j=1}^{i} c_{\tilde{t}}(j), \quad \bar{\varphi}(c_{\tilde{t}}, i) \triangleq \sum_{j=1}^{i} \bar{c}_{\tilde{t}}(j), \quad i = 1, 2, \dots, n_{\tilde{t}},$$

where $\bar{c}_{\tilde{t}}(j)$ denotes the binary complement to $c_{\tilde{t}}(j)$. As before the symbol "$\oplus$" denotes addition modulo 2. The summation signs indicate ordinary arithmetic sums. The functions simply denote the numbers of ones and zeroes respectively in the first i components of the codeword $c_{\tilde{t}} = (c_{\tilde{t}}(1), c_{\tilde{t}}(2), \dots, c_{\tilde{t}}(n_{\tilde{t}})) \in \mathcal{C}_{\tilde{t}}$.

Next we define for each set $\{c_{\tilde{t}} \in \mathcal{C}_{\tilde{t}} : \tilde{t} \in \tilde{\mathcal{T}}_K\}$ and for each $\tilde{t} \in \tilde{\mathcal{S}}_K \bigcup \tilde{\mathcal{T}}_K$ a function $\Psi_{\tilde{t}} : \{1, 2, \dots, n\} \to \{1, 2, \dots, n_{\tilde{t}}\}$. These functions are a form of book-keeping device by which we keep track of which components from the binary codewords should affect which component in the codeword $x = (x_1, x_2, \dots, x_n)$. In order not to overburden the notation we suppress the dependence on the set $\{c_{\tilde{t}} \in \mathcal{C}_{\tilde{t}} : \tilde{t} \in \tilde{\mathcal{T}}_K\}$.

Let $|\tilde{t}|$ denote the length of the sequence $\tilde{t}$. The definition is iterative and goes as follows:

$$\forall i \in \{1, 2, \ldots, n\}:$$

$$|\tilde{t}| = 0: \qquad \Psi(i) = i;$$

$$|\tilde{t}| > 0: \qquad \Psi_{\tilde{t}}(i) = \begin{cases} \bar{\varphi}(c_{\tilde{u}}, \Psi_{\tilde{u}}(i)), & \text{if } \tilde{t} = \tilde{u}0, \\ \varphi(c_{\tilde{u}}, \Psi_{\tilde{u}}(i)), & \text{if } \tilde{t} = \tilde{u}1, \\ \Psi_{\tilde{u}}(i), & \text{if } \tilde{t} = \tilde{u}*. \end{cases}$$

Finally we extend the definition to the domain $\mathcal{S}_K \bigcup \mathcal{T}_K$. This is done in the natural way: if $\tilde{t} \in \tilde{\mathcal{T}}_K$ corresponds to $t \in \mathcal{T}_K$ we simply define

$$\forall t \in \mathcal{S}_K \bigcup \mathcal{T}_K, \quad \forall i \in \{1, 2, \ldots, n\}:$$

$$\Psi_t(i) \triangleq \Psi_{\tilde{t}}(i).$$

Notice that the definition of $\Psi_{\tilde{t}}(i)$ is such that the definition of $\Psi_t(i)$ is unambiguous.

We are now ready for the main definition. It is a component-wise mapping which defines each component x_i of the codeword $x = (x_1, x_2, \ldots, x_n) \in \mathcal{X}$ successively as a function of the components $c_{\tilde{t}}(j)$ of the binary codewords. The difficulty in the definition arises because the binary codes have different lengths. This is where the functions Ψ_t come into play.

Let K be fixed and recall that each sequence $s = s_0 s_1 \ldots s_{m-1} \in \mathcal{S}_K$ is associated with a letter $L(s)$ in the alphabet $\mathcal{L}_K$. For each index $i \in \{1, 2, \ldots, n\}$ we should assign one such letter to the i-th component x_i in the codeword $x = (x_1, x_2, \ldots, x_n) \in \mathcal{X}$. This amounts to assigning a sequence $s(i) \in \mathcal{S}_K$ to each $i \in \{1, 2, \ldots, n\}$. We do this by defining the elements of the sequence $s(i) = s_0(i) s_1(i) \ldots s_{m-1}(i) \in \mathcal{S}_K$ successively as functions of the components $c_{\tilde{t}}(j)$ in the given codewords $c_{\tilde{t}} = (c_{\tilde{t}}(1), c_{\tilde{t}}(2), \ldots, c_{\tilde{t}}(n_{\tilde{t}})) \in \mathcal{C}_{\tilde{t}}$ in the following way:

$$s_0(i) = c(\Psi(i)) = c(i),$$
$$s_1(i) = c_{s_0(i)}\left(\Psi_{s_0(i)}(i)\right),$$
$$s_2(i) = c_{s_0(i)s_1(i)}\left(\Psi_{s_0(i)s_1(i)}(i)\right),$$
$$\ldots\ldots$$
$$s_{m-1}(i) = c_{s_0(i)s_1(i)\ldots s_{m-2}(i)}\left(\Psi_{s_0(i)s_1(i)\ldots s_{m-2}(i)}(i)\right),$$
$$i = 1, 2, \ldots, n.$$

Each subsequence $\tilde{t} = \tilde{s}' = \tilde{s}_0\tilde{s}_1 \ldots \tilde{s}_{m'-1} \in \tilde{\mathcal{S}}_K$ of a given sequence $\tilde{s} = \tilde{s}_0\tilde{s}_1 \ldots \tilde{s}_{m-1} \in \tilde{\mathcal{S}}_K$ corresponds to a code $\mathcal{C}_{\tilde{t}}$. Thus if a codeword $c_{\tilde{t}} \in \mathcal{C}_{\tilde{t}}$ is specified for each code $\mathcal{C}_{\tilde{t}}$, $\tilde{t} \in \tilde{\mathcal{T}}_K$, the above rule clearly specifies for each index $i \in \{1, 2, \ldots, n\}$ a sequence $\tilde{s} = \tilde{s}_0\tilde{s}_1 \ldots \tilde{s}_{m-1} \in \tilde{\mathcal{S}}_K$. The sequence $s(i) = s_0(i)s_1(i) \ldots s_{m-1}(i) \in \mathcal{S}_K$ is obtained by replacing the single symbol " $*$ " with the appropriate symbol in the corresponding unrestricted code.

We finally define

$$x_i \triangleq L(s(i)), \quad i = 1, 2, \ldots, n.$$

This completes the definition of the function

$$f : \bigotimes_{\tilde{t} \in \tilde{\mathcal{T}}_K} \mathcal{C}_{\tilde{t}} \to \mathbb{R}^n.$$

5.6 A simple example

Although the formal description of the general case might look complicated the underlying ideas are quite simple and easy to apply. In order to illustrate this fact we describe in detail a very simple example using the alphabet $\mathcal{L}_5$.

Consider Figure 5.7. We have indicated the tree and the labeling suggested by our construction. Four codes are involved: $\mathcal{C}$, $\mathcal{C}_0$, $\mathcal{C}_1$,

$K = 5$

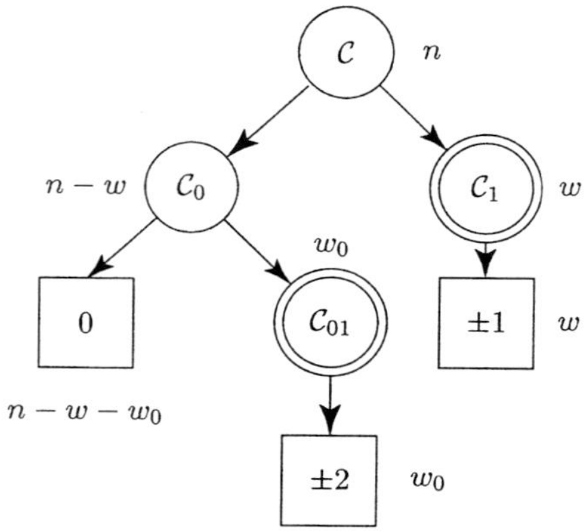

2	1	0	−1	−2
0	1	0	1	0
1	0	0	1	1
0				1

$\mathcal{C}$	$\mathcal{C}_0$	$\mathcal{C}_1$	$\mathcal{C}_{01}$	$\mathcal{X}$		
001	01	0	0	0	2	1
		0	1	0	−2	1
		1	0	0	2	−1
		1	1	0	−2	−1
001	10	0	0	2	0	1
		0	1	−2	0	1
		1	0	2	0	−1
		1	1	−2	0	−1
010	01	0	0	0	1	2
		0	1	0	1	−2
		1	0	0	−1	2
		1	1	0	−1	−2
010	10	0	0	2	1	0
		0	1	−2	1	0
		1	0	2	−1	0
		1	1	−2	−1	0
100	01	0	0	1	0	2
		0	1	1	0	−2
		1	0	−1	0	2
		1	1	−1	0	−2
100	10	0	0	1	2	0
		0	1	1	−2	0
		1	0	−1	2	0
		1	1	−1	−2	0

Figure 5.7: A simple example: $K = 5$.

and $\mathcal{C}_{01}$. The first two of them are constant weight codes while the two last ones are unrestricted codes. The parameters are:

$$\begin{array}{ll} \mathcal{C} : (n, w, d, M), & \mathcal{C}_0 : (n_0 = n - w, w_0, d_0, M_0), \\ \mathcal{C}_1 : (n_1 = w, d_1, M_1), & \mathcal{C}_{01} : (n_{01} = w_0, d_{01}, M_{01}). \end{array}$$

In each vertex of the tree we have indicated the corresponding code. As an explicit example we have indicated in Figure 5.7 also the detailed construction for the case

$$\begin{array}{l} (n, w, d, M) = (3, 1, 2, 3), \\ (n_0, w_0, d_0, M_0) = (2, 1, 2, 2), \\ (n_1, d_1, M_1) = (n_{01}, d_{01}, M_{01}) = (1, 1, 2). \end{array}$$

The resulting constant energy code $\mathcal{X}$ can be expressed in the following more compressed form:

$$\mathcal{X} = (0, \pm 1, \pm 2)^S,$$

where the notation as usual indicates that all possible combinations of signs and all permutation of coordinates should be taken. It is easy to see that the resulting spherical code has parameters

$$(n, \rho, M) = (3, 2/5, 24).$$

A graphic representation is indicated in Figure 5.8. The geometric constellation is sometimes referred to as a "truncated octahedron" ([71], p. 34).

It should be pointed out that the resulting code is not a particularly good one. The example serves only to illustrate the principle. The best possible constellation of 24 points on the three-dimensional sphere is the "snub cube" (see Chapter 3.2).

In the sequel we shall offer several examples where our constructions *do* provide good codes. First, however, we shall see how the main parameters can be estimated in general.

5.7 Analysis

In order to evaluate the constructions we need to determine the parameters (n, ρ, M). The difficult part of the problem is to determine

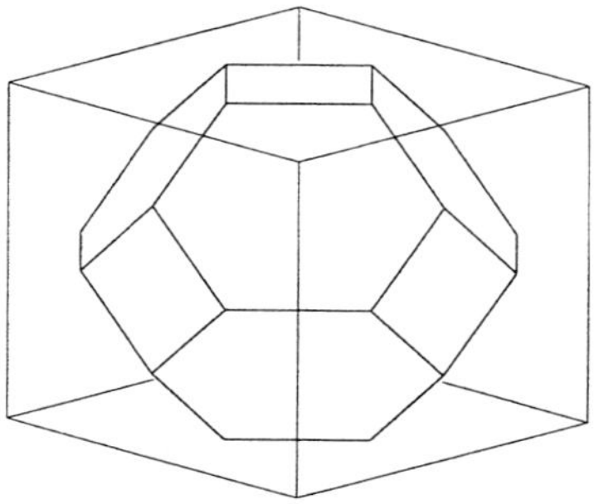

Figure 5.8: Graphical representation of the spherical code $\mathcal{X} = (0, \pm 1, \pm 2)^S$.

the normalized squared distance ρ. Actually we will derive only a lower bound to this parameter. We do this by first computing the constant squared norm $E = \|x\|^2 = x_1^2 + x_2^2 + \cdots + x_n^2$ and then deriving a lower bound for the squared distance d_E^2 of the non-normalized code. In both cases the analysis is based on the associated graph.

Recall that each sequence $\tilde{s} = \tilde{s}_0\tilde{s}_1 \ldots \tilde{s}_{m-1} \in \tilde{\mathcal{S}}_K$ is generated from a pair (s, s') of sequences $s, s' \in \mathcal{S}_K$ such that $L(s) = -L(s')$. It is therefore possible to define $\tilde{L}(\tilde{s}) = |L(s)|$, where $s = s_0 s_1 \ldots s_{m-1} \in \mathcal{S}_K$ is any one of the two sequences defining $\tilde{s}$. The common cost associated with the letters $L(s)$ and $L(s')$ can now be expressed as $\tilde{L}(\tilde{s})^2$. The significant thing is that this is a function of the *merged* sequence $\tilde{s} \in \tilde{\mathcal{S}}_K$.

For a given set $\{c_{\tilde{t}} \in \mathcal{C}_{\tilde{t}} : \tilde{t} \in \tilde{\mathcal{T}}_K\}$ of binary codewords denote by $n_{\tilde{s}}$ the number of occurrences of the sequence $\tilde{s} = \tilde{s}_0\tilde{s}_1 \ldots \tilde{s}_{m-1} \in \tilde{\mathcal{S}}_K$ when employing the mapping (5.5.2)

$$f : \bigotimes_{\tilde{t} \in \tilde{\mathcal{T}}_K} \mathcal{C}_{\tilde{t}} \to \mathbb{R}^n .$$

It is evident from the construction that this number is independent of the set $\{c_{\tilde{t}} \in \mathcal{C}_{\tilde{t}} : \tilde{t} \in \tilde{\mathcal{T}}_K\}$; for each such set it is given by the obvious extension of (5.5.1):

$$\forall \tilde{s} \in \tilde{\mathcal{S}}_K : \quad n_{\tilde{s}} = \begin{cases} n_{\tilde{u}} - w_{\tilde{u}}, & if \quad \tilde{s} = \tilde{u}0, \\ w_{\tilde{u}}, & if \quad \tilde{s} = \tilde{u}1, \\ n_{\tilde{u}}, & if \quad \tilde{s} = \tilde{u}*, \end{cases} \tag{5.7.1}$$

where of course $\tilde{u} \in \tilde{\mathcal{T}}_K$. We have the following theorem.

Theorem 5.7.1 *For given $K \geq 2$ and for a given set of binary codes satisfying the condition (5.7.1) let $\mathcal{X}$ be the constant energy code generated by construction Y_K. Any codeword $x = (x_1, x_2, \ldots, x_n) \in \mathcal{X}$ has a squared norm $\|x\|^2$ given by the expression*

$$\|x\|^2 = \sum_{\tilde{s} \in \tilde{\mathcal{S}}_K} n_{\tilde{s}}\, \tilde{L}(\tilde{s})^2.$$

Proof: By definition we have

$$\|x\|^2 = \sum_{i=1}^{n} x_i^2 = \sum_{i=1}^{n} L(s(i))^2.$$

By the above remarks the sequence $s(i) = s_0(i)s_1(i)\ldots s_{m-1}(i) \in \mathcal{S}_K$ has the same cost as its associated merged sequence $\tilde{s}(i) = \tilde{s}_0(i)\tilde{s}_1(i)\ldots\tilde{s}_{m-1}(i) \in \tilde{\mathcal{S}}_K$. Moreover, the frequency of occurrence of any sequence $\tilde{s} \in \tilde{\mathcal{S}}_K$ is given by the number $n_{\tilde{s}}$ as given by (5.7.1) and is the same for each codeword $x \in \mathcal{X}$. This proves the theorem. □

Remark 5.7.1: Notice that the frequency of occurrence of the sequences $s(i) \in \mathcal{S}_K$ is *not* the same for all codewords $x \in \mathcal{X}$, i.e. the various letters in the alphabet $\mathcal{L}_K$ do not have the same frequency of occurrence in each codeword $x \in \mathcal{X}$. This is actually a key point in the construction, as we tried to illustrate in our initial example in Section 5.2. However, the frequency of occurrence of letters from antipodal *pairs* in $\mathcal{L}_K$ is the same for each codeword $x \in \mathcal{X}$. This is the reason why we have to rely on the sequences $\tilde{s} \in \tilde{\mathcal{S}}_K$ in the proof.

Next we consider the squared distance d_E^2 of the code $\mathcal{X}$ generated by construction Y_K. We assume that for each node $\tilde{t} \in \tilde{\mathcal{T}}_K$ we have a binary code $C_{\tilde{t}}$ and that the lengths and weights of the codes satisfy the condition (5.5.1). Notice that the length $k = |\tilde{t}|$ of the sequence $\tilde{t} = \tilde{t}_0\tilde{t}_1\ldots\tilde{t}_{k-1} \in \tilde{\mathcal{T}}_K$ uniquely determines the level of the corresponding vertex in the graph Γ_K, where by "the level" we understand the number of steps that are needed in order to reach the corresponding vertex, starting from the root. Also notice the following basic feature of the partitions employed in the construction and of our binary a-labeling.

Lemma 5.7.1 *Let $s, s' \in \mathcal{S}_K$, $s \neq s'$, have a common prefix, i.e. suppose $s = uv$, $s' = uv'$, where u, v, v' are binary sequences and $v \neq v'$. Then the corresponding letters $L(s), L(s') \in \mathcal{L}_K$ satisfy the following inequality:*

$$|L(s) - L(s')| \geq 2^{|u|}.$$

Proof: The result is an obvious consequence of the binary labeling. □

Denote by $d(\mathcal{C}_{\tilde{t}})$ the minimum Hamming distance of the code $\mathcal{C}_{\tilde{t}}$, and define

$$d_k \triangleq \min\{d(\mathcal{C}_{\tilde{t}}) : |\tilde{t}| = k\}, \quad 0 \leq k \leq m-1. \tag{5.7.2}$$

In words: d_k is the smallest Hamming distance of any code on level k. The parameter m denotes the "height" of Γ_K, i.e. the deepest level occurring in the tree.

Theorem 5.7.2 *Let $\mathcal{X}$ be the constant energy code generated by construction Y_K and let the parameters d_k, $0 \leq k \leq m-1$, be defined as in (5.7.2). The minimum squared distance d_E^2 of the code $\mathcal{X}$ satisfies the inequality*

$$d_E^2 \geq \min_{0 \leq k \leq m-1} \{d_k \cdot 4^k\}.$$

Proof: Let x and x' be two different codewords in $\mathcal{X}$. Let the corresponding sets of binary codewords be $\{c_{\tilde{t}} \in \mathcal{C}_{\tilde{t}} : \tilde{t} \in \tilde{\mathcal{T}}_K\}$ and $\{c'_{\tilde{t}} \in \mathcal{C}_{\tilde{t}} : \tilde{t} \in \tilde{\mathcal{T}}_K\}$ respectively. As x and x' are different there must be some $\tilde{t} \in \tilde{\mathcal{T}}_K$ such that $c_{\tilde{t}} \neq c'_{\tilde{t}}$. Let $\tilde{u}$ be such a sequence of shortest length. This means that there must be at least $d(\mathcal{C}_{\tilde{u}})$ indices i such that the sequences $s(i) = s_0(i)s_1(i)\ldots s_{m-1}(i) \in \mathcal{S}_K$ and $s'(i) = s'_0(i)s'_1(i)\ldots s'_{m-1}(i) \in \mathcal{S}_K$ specifying x_i and x'_i respectively are of the forms $s(i) = u(i)v(i)$, $s'(i) = u'(i)v'(i)$ respectively, where both $u(i) \in \mathcal{T}_K$ and $u'(i) \in \mathcal{T}_K$ are such that they correspond to the fixed sequence $\tilde{u} \in \tilde{\mathcal{T}}_K$. By Lemma 5.7.1 we have

$$|L(s(i)) - L(s'(i))| \geq 2^{|u(i)|} = 2^{|\tilde{u}|}$$

in all those cases. It follows that for the squared distance between x and x' we have the following bound:

$$\|x - x'\|^2 \geq d(\mathcal{C}_{\tilde{u}}) \cdot (2^{|\tilde{u}|})^2 = d(\mathcal{C}_{\tilde{u}}) \cdot 4^{|\tilde{u}|}.$$

Thus, for any pair (x, x') of codewords in $\mathcal{X}$ the squared distance is lower bounded by an expression of the form $d_k \cdot 4^k$. This proves the theorem. □

Remark 5.7.2: Of course Theorem 5.7.2 is nothing else than the usual product bound for generalized concatenated codes.

We are now ready to formulate estimates for all of the basic parameters. Clearly we obtain a bound on the normalized squared distance $\rho = d_E^2/E$ by combining Theorems 5.7.1 and 5.7.2. As for the dimension we notice that all codewords $x \in \mathcal{X}$ are expressed as vectors $x = (x_1, x_2, \ldots, x_n)$ with n coordinates, where $n = n_\lambda$ is the length of the binary code $\mathcal{C}_\lambda$ at level zero. Thus $n = n_\lambda$ is an obvious upper bound for the dimension. The size M, finally, is clearly given by the product of the sizes $M(\mathcal{C}_{\tilde{t}})$ of all the binary codes $\mathcal{C}_{\tilde{t}}$ employed in the construction. Combining these observations we are lead to the following theorem.

Theorem 5.7.3 *For given $K \geq 2$ let Γ_K be the graph associated to the construction Y_K, with $\tilde{\mathcal{T}}_K$ as its set of internal nodes and with $\tilde{\mathcal{S}}_K$ as its set of leaves. Let a set $\{\mathcal{C}_{\tilde{t}} : \tilde{t} \in \tilde{\mathcal{T}}_K\}$ of binary codes satisfying (5.5.1) be given, and let the parameters $\{n_{\tilde{t}} : \tilde{t} \in \tilde{\mathcal{S}}_K\}$ be generated by (5.5.1) and (5.7.1). Finally, let m be the height of the tree Γ_K and let $\{d_k : 0 \leq k \leq m-1\}$ be given by (5.7.2). The spherical code $\mathcal{X} = Y_K(\mathcal{C}_{\tilde{t}} : \tilde{t} \in \tilde{\mathcal{T}}_K)$ generated by the binary codes $\{\mathcal{C}_{\tilde{t}} : \tilde{t} \in \tilde{\mathcal{T}}_K\}$ has parameters (n, ρ, M) satisfying the following conditions:*

$$n \leq n_\lambda, \quad \rho = d_E^2/E, \quad M = \prod_{\tilde{t} \in \tilde{\mathcal{T}}_K} M(\mathcal{C}_{\tilde{t}}),$$

where

$$E = \sum_{\tilde{s} \in \tilde{\mathcal{S}}_K} n_{\tilde{s}}\, \tilde{L}(\tilde{s})^2, \quad d_E^2 \geq \min_{0 \leq k \leq m-1} \{d_k \cdot 4^k\}.$$

Remark 5.7.3: Recall that the parameter n denotes the dimension of the space used for the description of the codes as given by the construction. In most cases of interest this is also the true dimension of the code, i.e. in most cases of interest it is not possible to embed the code in a space of smaller dimension. Still the parameter n_λ as given above is an *upper* bound of the dimension of the code.

5.8 Explicit constructions

In order to illustrate the constructions we consider several explicit examples.

$K = 2$:
Let H_k be a Hadamard matrix of order k (see Appendix A, or Beth–Jungnickel–Lenz [22], Section 1.9, or Hall [122], Chapter 14). Without loss of generality we assume that it is normalized so that the first row and the first column contain positive entities only. Three different spherical codes of the Y_2-type are generated by any such matrix.

(*i*) A code with parameters

$$(n, \rho, M) = (n, 2 + 2/n, n + 1); \quad n = k - 1.$$

This code is obtained by first deleting the first column in H_k. The code is given by normalized versions of the rows in the remaining matrix.

(*ii*) A code with parameters

$$(n, \rho, M) = (n, 2 - 2/n, 2n + 2); \quad n = k - 1.$$

This code is obtained by adjoining to each codeword x in the previous code also its antipodal inverse $-x$.

(*iii*) A code with parameters

$$(n, \rho, M) = (n, 2, 2n); \quad n = k.$$

This code is generated by taking as codewords normalized versions of all rows in H_k and $-H_k$, including the first column.

We recognize the first code as a version of the *Simplex* code while the third code is a version of the *Biorthogonal* code. We have already noticed that both of these codes exist for any value of the parameter n. We have also noted that both are optimal. The observation worth making here is simply that if the parameter k is such that a Hadamard matrix exists, then a *Simplex* code with dimension $n = k - 1$ and a *Biorthogonal* code with dimension $n = k$ can be implemented using a *binary symmetric* alphabet, and so by the Y_2-construction.

The second construction is obtained by taking the union of a *Simplex* code and its negative. Such a code also exists in any dimension n. It has a Y_2-realization if and only if the *Simplex* code has such a realization. Unlike the *Simplex* this construction is *not* optimal.

The Y_2-construction is widely used in applications. Whenever there is an estimate available for the parameters of the binary code the general formulas derived in Section 5.4 give us corresponding estimates for the parameters of the spherical code. It should be observed, however, that a good binary code does not necessarily generate a good spherical code. Generally speaking good spherical codes are obtained from the Y_2-construction only with binary codes of *low* rate. The reason is simple: for *high* rates the limitation to a binary alphabet becomes quite restrictive.

The *Kerdock* codes [143] generate interesting spherical codes under the Y_2-construction. The binary codes are non-linear, with parameters given by

$$n = 2^{2r}, \quad d = (1/2) \cdot (n - \sqrt{n}), \quad M = n^2; \quad r = 2, 3, \dots .$$

For $r = 2$ we have the Nordstrom–Robinson code (see MacWilliams–Sloane [166], p. 73 and Semakov–Zinoviev [198]) as the first member of this family. The corresponding spherical codes have parameters

$$n = 2^{2r}, \quad \rho = 2 - 2/\sqrt{n}, \quad M = n^2; \quad r = 2, 3 \dots .$$

This is quite close to the Levenshtein bound. Still, as we shall see in the next section, all of these codes can be improved.

K =3:
The *Biorthogonal* codes are most easily generated using construction Y_3: choosing $\mathcal{C}$ with parameters $(n, w, d, M) = (n, 1, 2, n)$ and choosing the code $\mathcal{A}$ with parameters $(n, d, M) = (1, 1, 2)$ we obtain a spherical code $\mathcal{X} = Y_3(C, A)$ with parameters $(n, \rho, M) = (n, 2, 2n)$, $n = 1, 2, \ldots$. Besides being quite important in themselves these codes are also extremely useful as building blocks in constructions of other codes.

For the Y_3-construction the following estimate applies:

$$\rho \geq \frac{1}{w} \min\{d_{\mathcal{C}}, 4d_{\mathcal{A}}\}.$$

This estimate suggests choosing the binary codes $\mathcal{C}$ and $\mathcal{A}$ so as to satisfy the relation $d_{\mathcal{C}} = 4d_{\mathcal{A}}$. The simplest possibility is to choose $d_{\mathcal{A}} = 1$. This calls for a constant weight code $\mathcal{C}$ with distance $d_{\mathcal{C}} = 4$. Let us choose $w = 3$. For this weight optimal (in terms of maximizing size) codes with distance $d_C = 4$ are known for all lengths n. The sizes are given by the following formula due to Schönheim, (see MacWilliams–Sloane [166], p. 529)

$$M_{\mathcal{C}} = \begin{cases} \left\lfloor \frac{n}{3} \left\lfloor \frac{n-1}{2} \right\rfloor \right\rfloor, & if\ n \not\equiv 5 \mod 6, \\ \left\lfloor \frac{n}{3} \left\lfloor \frac{n-1}{2} \right\rfloor \right\rfloor - 1, & if\ n \equiv 5 \mod 6, \end{cases}$$

where $\lfloor a \rfloor$ denotes integer part of a. The unrestricted code $\mathcal{A}$ matching a constant weight code $\mathcal{C}$ of this type is the trivial code of length $n_{\mathcal{A}} = w = 3$, distance $d_{\mathcal{A}} = 1$ and size $M_{\mathcal{A}} = 8$. The resulting spherical code has squared distance $\rho = 4/3$ and size $M = 8M_{\mathcal{C}}$. The first few codes of this kind have sizes as follows:

n	5	6	7	8	9	10	11	12	13	14	15
M	16	32	56	64	96	104	136	160	208	224	280

Some of these codes are quite good. In dimension $n = 7$ we even have an optimal code. For dimensions $n = 6, 8, 9, 10, 11, 13,$ and 14 these codes have best known cardinalities.

For $n \geq 15$ we generally get better spherical codes by using constant weight codes with larger weights and larger distances. Such a choice for the constant weight code allows us to use larger unrestricted codes. In the range $15 \leq n \leq 22$ quite good spherical codes are generated by constant weight codes $\mathcal{C}$ of weight 6 and distance 8 in combination with an unrestricted code $\mathcal{A}$ with parameters $(n, d, M) = (6, 2, 32)$ (i.e. a simple parity-check code). This produces spherical codes with squared minimum distance $\rho \geq 4/3$. Using the best known constant weight codes with the above parameters [51] we obtain spherical codes of the following sizes:

n	15	16	17	18	19	20	21	22
M	320	512	544	672	896	1280	1792	2464

For the lengths $n = 15, 16, 18, \ldots, 22$ these codes have the best known parameters.

The above considerations suggest that it might be a good idea to let w and d increase with the dimension n. There are many such possibilities. We indicate here just two constructions with interesting properties.

Let q be a power of a prime such that $q + 1$ is a multiple of 4. Then there is a maximal constant weight code $\mathcal{C}$ equivalent to the incidence matrix of a finite projective plane ([199], see also [22]) of order q and having the following parameters:

$$n_{\mathcal{C}} = q^2 + q + 1, \quad w = q + 1, \quad d_{\mathcal{C}} = 2q, \quad M_{\mathcal{C}} = n_{\mathcal{C}}.$$

As $q+1$ is a factor of 4 there exists a Hadamard matrix from the Paley construction (see Appendix A), which generates a binary *Biorthogonal* code $\mathcal{A}$ with parameters

$$n_{\mathcal{A}} = q + 1, \quad d_{\mathcal{A}} = n_{\mathcal{A}}/2, \quad M_{\mathcal{A}} = 2n_{\mathcal{A}}.$$

This is the same idea as used above for generating *Biorthogonal* codes from the Y_2-construction. Again the relation $d_{\mathcal{C}} = 4d_{\mathcal{A}}$ is satisfied. By our general estimates we get

$$\begin{aligned} n &= q^2 + q + 1, \\ \rho &\geq 2 - 2/(q + 1), \\ M &= 2(q + 1)(q^2 + q + 1). \end{aligned}$$

For a second example let q be a power of 2 and suppose there exists a conference matrix of order $q+2$ (see Appendix A). It is known that such a matrix exists when $q+1$ is a power of a prime (this is also a Paley construction, see Appendix A, or MacWilliams–Sloane [166], p. 57). Then there exists a conference matrix code $\mathcal{A}$ with parameters

$$n_{\mathcal{A}} = q+1, \quad d_{\mathcal{A}} = q/2, \quad M_{\mathcal{A}} = 2(q+2).$$

With a constant weight code $\mathcal{C}$ generated from a projective plane as above this choice of the unrestricted code $\mathcal{A}$ gives us another infinite family of spherical codes with the following parameters:

$$\begin{aligned} n &= q^2+q+1, \\ \rho &\geq 2 - 2/(q+1), \\ M &= 2(q+2)(q^2+q+1). \end{aligned}$$

K = 4:
For the Y_4-construction we have the following estimate for the minimum distance:

$$\rho \geq \frac{4}{n+8w} \min\{d_A, 4d_C\}.$$

This formula suggests that we should try to choose codes $\mathcal{A}$ and $\mathcal{C}$ such that $d_{\mathcal{A}} = 4d_{\mathcal{C}}$. As $d_{\mathcal{C}} \geq 2$ the smallest possible value for $d_{\mathcal{A}}$ is 8, which in turn also gives us a lower bound for the possible lengths of such codes. Of course it is possible to use the Y_4-construction also for smaller dimensions, but good spherical codes from the Y_4-construction occur mainly in dimensions n larger than or equal to 8.

K = 5:
Let $\mathcal{X} = Y_5(\mathcal{C}, \mathcal{C}_0, \mathcal{C}_1, \mathcal{C}_{01})$ be generated by codes with the following parameters:

$$\begin{aligned} &\mathcal{C} : (n, w, d, M) \quad &&\mathcal{C}_0 : (n_0, w_0, d_0, M_0), \\ &\mathcal{C}_1 : (n_1, d_1, M_1) \quad &&\mathcal{C}_{01} : (n_{01}, d_{01}, M_{01}). \end{aligned}$$

From Theorem 5.7.3 we have the following estimate for the energy and the minimum distance:

$$E = w + 4w_0, \quad d_E^2 \geq \min\{d, 4d_0, 4d_1, 16d_{01}\}.$$

Again it is tempting to match the distances, which in this case leads to the following relations:

$$d = 4d_0 = 4d_1 = 16d_{01}.$$

The first dimension in which a complete match is possible is $n = 16$, where the following choice suggests itself:

$$\begin{aligned} \mathcal{C} &: (n, w, d, M) = (16, 8, 16, 2), \\ \mathcal{C}_0 &: (n_0, w_0, d_0, M_0) = (8, 4, 4, 14), \\ \mathcal{C}_1 &: (n_1, d_1, M_1) = (8, 4, 16), \\ \mathcal{C}_{01} &: (n_{01}, d_{01}, M_{01}) = (4, 1, 16). \end{aligned}$$

At first glance this design appears quite attractive: all the involved binary codes meet well known optimality criteria. The resulting spherical code has parameters $(n, \rho, M) = (16, 2/3, 7168)$. This, however, is not at all impressive. By choosing instead the following binary codes

$$\begin{aligned} \mathcal{C} &: (n, w, d, M) = (16, 8, 8, 30), \\ \mathcal{C}_0 &: (n_0, w_0, d_0, M_0) = (8, 1, 2, 8), \\ \mathcal{C}_1 &: (n_1, d_1, M_1) = (8, 2, 128), \\ \mathcal{C}_{01} &: (n_{01}, d_{01}, M_{01}) = (1, 1, 2) \end{aligned}$$

we obtain a code of size $M = 61440$ having the same dimension and the same minimum distance. This number coincides with the number of points in the second shell of the *Barnes–Wall* lattice (see Chapter 10). However, even this is not the best code for given n and ρ. The example illustrates the fact that a match of distances does not always produce the best spherical codes.

5.9 Unions

Often quite good spherical codes can be obtained by forming the union of two or more codes. Sometimes good codes are generated in this way even by unions of codes which individually are not very good.

In order to estimate the parameters of a spherical code formed as the union of two or more known codes we need a useful estimate of the minimum distance between the constituent codes. We describe a few cases where such estimates can be found for unions of codes generated by the Y_K-constructions.

Given two spherical codes $\mathcal{X} \subseteq \Omega_n$ and $\mathcal{Y} \subseteq \Omega_n$ we define an *inner product* $s(\mathcal{X}, \mathcal{Y})$ as the *largest* inner product between two codewords $x \in \mathcal{X}$ and $y \in \mathcal{Y}$. We also define a squared *distance* $\rho(\mathcal{X}, \mathcal{Y})$ as the *smallest* squared Euclidean distance between the components of any pair $(x, y) \in \mathcal{X} \times \mathcal{Y}$. We clearly have

$$\rho(\mathcal{X}, \mathcal{Y}) = 2 - 2s(\mathcal{X}, \mathcal{Y}).$$

Usually we prefer to express our bounds in terms of the inner product.

For any binary words $a, b \in \mathbb{F}_2^n$ we denote by $\lambda(a, b)$ the number of coordinates k such that $a_k = b_k = 1$. This quantity is referred to as the *binary correlation* between a and b.

Let $\mathcal{X} = Y_3(\mathcal{C}_1, \mathcal{A}_1)$ and $\mathcal{Y} = Y_3(\mathcal{C}_2, \mathcal{A}_2)$ be two spherical codes generated by the Y_3-construction. Recall that the normalized alphabet used in the Y_3-construction is $\mathcal{L} = \{0, \pm 1/\sqrt{w}\}$, where w is the *weight* of the constant weight code. Let the weights in the present case be w_1 and w_2, and let $x \in \mathcal{X}$ and $y \in \mathcal{Y}$ be generated by $(c_1, a_1) \in \mathcal{C}_1 \times \mathcal{A}_1$ and $(c_2, a_2) \in \mathcal{C}_2 \times \mathcal{A}_2$ respectively. The following bound is evident:

$$\langle x, y \rangle \leq \frac{\lambda(c_1, c_2)}{\sqrt{w_1 w_2}}.$$

This bound is often useful in order to determine the parameters of a union between codes from the Y_3-construction.

Example 5.9.1: Let the binary codes have the following parameters:

$$\begin{aligned} &\mathcal{C}_1 : (8, 1, 2, 8), \quad && \mathcal{A}_1 : (1, 1, 2), \\ &\mathcal{C}_2 : (8, 4, 4, 14), \quad && \mathcal{A}_2 : (4, 1, 16). \end{aligned}$$

The maximal correlation between words from the codes $\mathcal{C}_1$ and $\mathcal{C}_2$ is $\lambda = 1$. It follows that the minimum distance between the spherical codes $\mathcal{X} = Y_3(\mathcal{C}_1, \mathcal{A}_1)$ and $\mathcal{Y} = Y_3(\mathcal{C}_2, \mathcal{A}_2)$ satisfies $\rho(\mathcal{X}, \mathcal{Y}) \geq 1$.

As the minimum distance in each one of the constituent codes $\mathcal{X}$ and $\mathcal{Y}$ also satisfies this bound it follows that the union $\mathcal{Z} = \mathcal{X} \cup \mathcal{Y}$ has parameters $(n, \rho, M) = (8, 1, 240)$. This code is known to be optimal and unique, (Bannai–Sloane [10], Chapter 14, Theorem 5). An alternative construction was suggested in Example 1.2.4 in Chapter 1 and will be reconsidered in the next example of the present chapter. Still another construction will be considered in Example 5.9.4. □

The construction Y_2 can be seen as a special case of the construction Y_3, using the degenerated code consisting of the all-one codeword only as the constant weight code. By this observation we get the bound

$$\langle x, y \rangle \leq \sqrt{w/n} \tag{5.9.1}$$

for the inner product between a codeword x from a Y_2-code and a codeword y from a Y_3-code, where n is the dimension and w is the weight of the constant weight code used in the Y_3-code. The following two examples illustrate the use of this bound.

Example 5.9.2: Let $\mathcal{X}$ be generated from the binary code $\mathcal{A}$ using the Y_2-construction and let $\mathcal{Y}$ be generated by the binary codes $\mathcal{B}$ and $\mathcal{C}$ using the Y_3-construction. The codes $\mathcal{A}$ and $\mathcal{B}$ are unrestricted codes while $\mathcal{C}$ is a constant weight code. Let the parameters be as follows:

$$\mathcal{A} : (8, 2, 128), \quad \mathcal{B} : (2, 1, 4), \quad \mathcal{C} : (8, 2, 2, 28).$$

The standard estimates for the constructions Y_2 and Y_3 reveal that $\mathcal{X}$ and $\mathcal{Y}$ are spherical codes with parameters $(n, \rho, M) = (8, 1, 128)$ and $(8, 1, 112)$ respectively. The bound (5.9.1) reveals that the distance between the codes satisfies $\rho(\mathcal{X}, \mathcal{Y}) \geq 1$. It follows that the union $\mathcal{Z} = \mathcal{X} \cup \mathcal{Y}$ has parameters $(n, \rho, M) = (8, 1, 240)$. This is the construction we encountered already in Chapter 1, Example 1.2.4. □

Example 5.9.3: Let $\mathcal{X}$ be the spherical code generated by the Kerdock codes from the Y_2-construction. Recall the $\mathcal{X}$ has parameters $(n, \rho, M) = (2^{2r}, 2(1-2^{-r}), 2^{4r})$, $r = 2, 3, \ldots$. Let $\mathcal{Y}$ be the *Biorthogonal* code of the same dimension, generated by the Y_3-constriction using a constant weight code with weight $w = 1$. The bound (5.9.1) gives us

$$s(\mathcal{X}, \mathcal{Y}) \le \sqrt{w/n} = 2^{-r}.$$

The *Biorthognal* code has squared distance $\rho = 2$. It follows that the union $\mathcal{Z} = \mathcal{X} \cup \mathcal{Y}$ has the same squared distance as the code $\mathcal{X}$, while the size is $M = n^2 + 2n = 2^{4r} + 2^{2r+1}$, $r = 2, 3, \dots$. The code is regular with the inner product distribution given by

s_i	1	$1/\sqrt{n}$	0	$-1/\sqrt{n}$	-1
k_i	1	$n^2/2$	$2n-2$	$n^2/2$	1

The first code in this sequence of codes is of particular interest. It has parameters $(n, \rho, M) = (16, 3/2, 288)$.

These codes are mentioned in Levenshtein [157], where it is also proved that they are asymptotically optimal within the class of codes containing antipodal vectors. Calderbank–Cameron–Kantor–Seidel [56], [57] considered this class of spherical codes in connection with extremal Euclidean line sets. □

The following theorem contains a useful estimate for the union of a code from the Y_2-construction with one from the Y_4-construction.

Theorem 5.9.1 *Let $\mathcal{X} \subseteq \Omega_n$ be generated by the Y_2-construction using the binary code $\mathcal{A} \subseteq \mathbb{F}_2^n$ and let $\mathcal{Y}$ be generated by the Y_4-construction using the unrestricted code $\mathcal{B} \subseteq \mathbb{F}_2^n$ and the constant weight code $\mathcal{C} \subseteq \mathbb{F}_2^{n,w}$. Let the minimal Hamming distance between the binary codes $\mathcal{A}$ and $\mathcal{B}$ be d. Then the following inequality holds:*

$$s(\mathcal{X}, \mathcal{Y}) \le \frac{n + 4w - 2d}{\sqrt{n(n+8w)}} . \tag{5.9.2}$$

Proof: The alphabets employed in the two constructions are $\mathcal{L}_2 = \{\pm 1/\sqrt{n}\}$ and $\mathcal{L}_4 = \{\pm 1/\sqrt{n+8w}, \pm 3/\sqrt{n+8w}\}$ respectively. It follows that for any codewords $x = (x_1, x_2, \dots, x_n) \in \mathcal{X}$ and $y = (y_1, y_2, \dots, y_n) \in \mathcal{Y}$ we have

$$\sqrt{n(n+8w)}\, x_i y_i \in \{\pm 1, \pm 3\}, \quad i = 1, 2, \dots, n.$$

Let $\alpha, \beta, \gamma, \delta$ denote the number of coordinates i for which each one of the cases $1, -1, 3, -3$ occurs. Then the inner product satisfies the following identity:

$$\sqrt{n(n+8w)}\langle x, y\rangle = \alpha - \beta + 3\gamma - 3\delta.$$

Let $x \in \mathcal{X}$ be generated by $a = (a_1, a_2, \ldots, a_n) \in \mathcal{A}$ and let $y \in \mathcal{Y}$ be generated by $b = (b_1, b_2, \ldots, b_n) \in \mathcal{B}$ and $c = (c_1, c_2, \ldots, c_n) \in \mathcal{C}$. The following table applies:

$+1$	$c_i = 0$	$a_i = b_i$	α
-1	$c_i = 0$	$a_i \neq b_i$	β
$+3$	$c_i = 1$	$a_i \neq b_i$	γ
-3	$c_i = 1$	$a_i = b_i$	δ

We conclude that the following constraints must be satisfied:

$$\alpha + \delta \leq n - d, \quad \beta + \gamma \geq d, \quad \alpha + \beta = n - w, \quad \gamma + \delta = w.$$

Eliminating α, β, and δ we obtain

$$\alpha - \beta + 3\gamma - 3\delta \leq n - 2d + 8\gamma - 4w.$$

Inserting the obvious inequality $\gamma \leq w$ we obtain

$$\sqrt{n(n+8w)}\langle x, y\rangle \leq n + 4w - 2d.$$

As this inequality holds for *any* pair $(x, y) \in \mathcal{X} \times \mathcal{Y}$ the theorem follows. □

There are also good codes produced from unions of codes from the Y_3 and the Y_4 constructions. The general result is as follows.

Theorem 5.9.2 *Let $\mathcal{X} \subseteq \Omega_n$ and $\mathcal{Y} \subseteq \Omega_n$ be generated by the constructions Y_3 and Y_4 respectively. Let the constant weight codes involved have weights w_3 and w_4 respectively. The following bound holds for the inner product:*

$$s(\mathcal{X}, \mathcal{Y}) \leq \frac{\min\{w_3 + 2w_4, 3w_3\}}{\sqrt{w_3(n + 8w_4)}}.$$

Proof: The proof is straightforward and follows directly from an investigation of worst cases. □

This theorem can sometimes be improved if certain relations hold between the binary codes involved. In particular the following corollary is often useful.

Corollary 5.9.1 *Let $\mathcal{X} \subseteq \Omega_n$ be generated by the Y_3 construction using the binary codes $\mathcal{A}_3$ and $\mathcal{C}_3$ respectively, where $\mathcal{C}_3$ is a constant weight code with weight w_3. Similarly, let $\mathcal{Y} \subseteq \Omega_n$ be generated by the Y_4 construction using the binary codes $\mathcal{A}_4$ and $\mathcal{C}_4$ respectively, where $\mathcal{C}_4$ is a constant weight code with weight w_4. Suppose we have $w_4 \leq w_3$ and let the following conditions be fulfilled:*

(i) For any codewords $c \in \mathcal{C}_3$ and $a \in \mathcal{A}_4$ the binary correlation satisfies

$$\lambda(c, a) \equiv \lambda \bmod 2, \quad (\lambda = constant).$$

(ii) For any codeword $a \in \mathcal{A}_3$ the Hamming weight satisfies

$$w_H(a) \not\equiv w_4 + \lambda \bmod 2.$$

Then the correlation between the codes $\mathcal{X}$ and $\mathcal{Y}$ is bounded as follows:

$$s(\mathcal{X}, \mathcal{Y}) \leq \frac{w_3 - 2 + 2w_4}{\sqrt{w_3(n + 8w_4)}} .$$

Proof: The result is easily obtained by an analysis of worst cases. □

Example 5.9.4: Let $\mathcal{X}$ and $\mathcal{Y}$ be generated by the Y_3 and the Y_4 constructions and let the binary codes involved be as follows:

	n	d	M
$\mathcal{A}_3$	4	2	8
$\mathcal{A}_4$	8	4	16

	n	w	d	M
$\mathcal{C}_3$	8	4	4	14
$\mathcal{C}_4$	8	1	2	8

Let $\mathcal{C}_3$ be a subcode in $\mathcal{A}_4$. Then we have $\lambda(c, a) \equiv 0 \bmod 2$ for any pair $(c, a) \in \mathcal{C}_3 \times \mathcal{A}_4$. We choose $\mathcal{A}_3$ so as to be an even weight code.

Then we have $w_H(a) \neq w_4 = 1$. The conditions of Corollary 5.9.1 are satisfied, and we conclude

$$s(\mathcal{X},\mathcal{Y}) \leq \frac{4-2+2}{\sqrt{4(8+8)}} = \frac{1}{2}.$$

We notice that $\mathcal{X}$ has parameters $(n,\rho,M) = (8,1,112)$ and that $\mathcal{Y}$ has parameters $(n,\rho,M) = (8,1,128)$. It follows that the union $\mathcal{Z} = \mathcal{X} \cup \mathcal{Y}$ has parameters $(n,\rho,M) = (8,1,240)$. This an alternative construction of the unique optimal code we encountered already in Chapter 1, Example 1.2.4. and which we considered also in Examples 5.9.1 and 5.9.2. above. □

We conclude by an example involving two codes from the construction Y_3 and one code from the construction Y_4.

Example 5.9.5: For $i = 1,2,3$ let $\mathcal{A}_i$ be unrestricted binary codes and let $\mathcal{C}_i$ be constant weight binary codes. Let the parameters be as follows:

	n	d	M
$\mathcal{A}_1$	2	1	4
$\mathcal{A}_2$	8	2	128
$\mathcal{A}_3$	24	8	2^{12}

	n	w	d	M
$\mathcal{C}_1$	24	2	2	276
$\mathcal{C}_2$	24	8	8	759
$\mathcal{C}_3$	24	1	2	24

Here $\mathcal{A}_3$ is the extended binary Golay code while $\mathcal{C}_2$ is the subcode formed by all codewords of weight 8. From the Y_3 and the Y_4 constructions we obtain spherical codes as follows:

$$\mathcal{X}_1 = Y_3(\mathcal{C}_1,\mathcal{A}_1), \quad \mathcal{X}_2 = Y_3(\mathcal{C}_2,\mathcal{A}_2), \quad \mathcal{X}_3 = Y_4(\mathcal{A}_3,\mathcal{C}_3).$$

By the standard estimates the parameters are as follows:

	n	ρ	M
$\mathcal{X}_1$	24	1	1104
$\mathcal{X}_2$	24	1	97152
$\mathcal{X}_3$	24	1	98304

We notice that all codes have squared minimum distance $\rho = 1$. We claim that also the union

$$\mathcal{X} = \mathcal{X}_1 \cup \mathcal{X}_2 \cup \mathcal{X}_3$$

has the same squared minimum distance. In order to see that we first notice that the alphabets are as follows:

$$\mathcal{L}_1 = \{0, \pm 1/\sqrt{2}\}, \quad \mathcal{L}_2 = \{0, \pm 1/(2\sqrt{2})\},$$
$$\mathcal{L}_3 = \{\pm 1/(4\sqrt{2}), \pm 3/(4\sqrt{2})\}.$$

Let L_i, $i = 1, 2, \ldots 9$ be the letters in the union

$$\mathcal{L} = \mathcal{L}_1 \cup \mathcal{L}_2 \cup \mathcal{L}_3,$$

taken in increasing order. We notice that the various levels are attained according to the following scheme:

	L_1	L_2	L_3	L_4	L_5	L_6	L_7	L_8	L_9
c_1	1				0				1
a_1	0								1
c_2			1		0		1		
a_2			0				1		
a_3		0		1		0		1	
c_3		1		0		0		1	

The fact that the union has the same minimum distance as each of the constituent subcodes is established by checking the various cases. The least obvious case is that involving $\mathcal{X}_2$ and $\mathcal{X}_3$. The smallest distance between the alphabets $\mathcal{L}_2$ and $\mathcal{L}_3$ is $1/(4\sqrt{2})$. Therefore, if there would be a pair $(x, y) \in \mathcal{X}_2 \times \mathcal{X}_3$ such that all components (x_i, y_i) were on minimum distance, those codewords would be on distance $24/(4\sqrt{2})^2 = 3/4$. However, if at least one pair of components is located on second smallest distance the codewords will satisfy $\|x - y\|^2 \geq 23/(4\sqrt{2})^2 + 3^2/(4\sqrt{2})^2 = 1$. Thus all that we need to establish is the fact that there is no pair of codewords $(x, y) \in \mathcal{X}_2 \times \mathcal{X}_3$ for which *all* pairs of components (x_i, y_i) are on minimum distance.

Let us define events α, β, ... as follows:

	c_2	a_2	a_3	c_3
α	1	0	0	1
β	1	0	1	0
γ	0		1	0
δ	0		0	0
ϵ	1	1	0	0
φ	1	1	1	1

and let N_α, N_β, ... denote the number of occurrences of these events in a given word $x = (x_1, x_2, \ldots, x_{24}) \in \mathcal{X}$, generated by the binary codewords $(a_i, c_i) \in \mathcal{A}_i \times \mathcal{C}_i$, $i = 2, 3$. The following constraints apply:

$$
\begin{aligned}
w_H(c_2) &= 8, && \Rightarrow && N_\alpha + N_\beta + N_\epsilon + N_\varphi = 8,\\
w_H(a_2) &\equiv 0 \bmod 2, && \Rightarrow && N_\epsilon + N_\varphi \equiv 0 \bmod 2,\\
w_H(a_3) &\equiv 0 \bmod 4, && \Rightarrow && N_\beta + N_\gamma + N_\varphi \equiv 0 \bmod 4,\\
w_H(c_3) &= 1, && \Rightarrow && N_\alpha + N_\varphi = 1,\\
d_H(c_2, a_3) &\equiv 0 \bmod 4, && \Rightarrow && N_\alpha + N_\gamma + N_\epsilon \equiv 0 \bmod 4,\\
n &= 24, && \Rightarrow && N_\alpha + N_\beta + N_\gamma + N_\delta + N_\epsilon + N_\varphi = 24.
\end{aligned}
$$

The fact that these equations cannot be simultaneously satisfied is easily established by considering the two cases $N_\alpha = 1$, $N_\varphi = 0$ and $N_\alpha = 0$, $N_\varphi = 1$. It follows that codewords from $\mathcal{X}_2$ and $\mathcal{X}_3$ are always separated by squared Euclidean distance at least 1. Similar—but simpler—arguments show that the same holds also for pairs of codewords involving $\mathcal{X}_1$. Therefore the minimum squared distance in the union $\mathcal{X}$ is $\rho = 1$. The number of codewords is $M = 1104 + 97152 + 98304 = 196560$. It follows that $\mathcal{X}$ is a realization of the unique ([10]) optimal code with parameters $(n, \rho, M) = (24, 1, 196560)$. This code also appears as the inner shell in the Leech lattice. □

5.10 Extensions

There are many kinds of minor modifications and tricks that can be used in order to produce new spherical codes out of old ones. At this point we like to mention just one additional idea that quite often produces good codes. We call it "extension", or "extended union".

Let $\mathcal{Y} = \{y_1, y_2, \ldots, y_m\}$ be a spherical code with dimension n_2 and assume that for each codeword $y_i \in \mathcal{Y}$ we have a spherical code $\mathcal{X}_i$, each with dimension n_1. The *extension with tail* $\mathcal{Y}$ of the codes $\mathcal{X}_i$, $i = 1, 2, \ldots, m$ is the following code:

$$\mathcal{Z} = \bigcup_{i=1}^{m} (\mathcal{X}_i \sin\theta_i, y_i \cos\theta_i), \tag{5.10.1}$$

where θ_i are parameters to be optimized in a moment. The name "extension" is motivated by the fact that this is a simple modification of the union we considered in the previous section. We can think of the code $\mathcal{Y}$ as a means of increasing the distances between the subcodes $\mathcal{X}_i$ by adding an identification tail for each subcode. Of course there is a price for this increase in distance: the dimension of the resulting code increases from n_1 to $n_1 + n_2$ and the energy of the codes $\mathcal{X}_i$ must be decreased, with a corresponding decrease of the minimum distance as an unavoidable consequence. Nevertheless, this construction often produces very good codes.

Consider first the special case $\theta_i = \theta$, $i = 0, 1, \ldots, m$. Define

$$\begin{aligned}
\rho_1 &\triangleq \min\{\|x - x'\|^2 : x \in \mathcal{X}_i,\ x' \in \mathcal{X}_j, 1 \le i < j \le m\},\\
\rho_0 &\triangleq \min\{\|x - x'\|^2 : x, x' \in \mathcal{X}_i,\ x \ne x',\ 1 \le i \le m\},\\
\rho_2 &\triangleq \min\{\|y - y'\|^2 : y, y' \in \mathcal{Y},\ y \ne y'\}.
\end{aligned}$$

It is easy to see that the minimum distance in $\mathcal{Z}$ is

$$\rho_{\mathcal{Z}} = \min\{\rho_0 \sin^2\theta,\ \rho_1 \sin^2\theta + \rho_2 \cos^2\theta\}. \tag{5.10.2}$$

In order to maximize this quantity we choose

$$\cot^2\theta = \frac{1}{\sin^2\theta} - 1 = \frac{\rho_0 - \rho_1}{\rho_2}. \tag{5.10.3}$$

With this choice the code $\mathcal{Z}$ gets the following parameters:

$$n = n_1 + n_2\,, \quad \rho = \frac{\rho_0 \rho_2}{\rho_0 + \rho_2 - \rho_1}\,, \quad M = \sum_{i=1}^{m} M_i\,, \tag{5.10.4}$$

where M_i is the size of the code $\mathcal{X}_i$.

As an example we consider explicitly the case when the suffix code $\mathcal{Y}$ is the *Simplex* code $\mathcal{Y} = \mathcal{S}_{m-1}$. For that case we get

$$\rho = \frac{2m\rho_0}{2m + (m-1)(\rho_0 - \rho_1)}\,.$$

Extended unions using a *Simplex* code in the suffix produce many of the best known spherical codes. The case $m = 2$ is of special interest. In that case the code $\mathcal{Y}$ is the binary antipodal code $\mathcal{Y} = \{1, -1\}$ with parameters $(n, \rho, M) = (1, 4, 2)$, and the code $\mathcal{X}$ is subdivided into just two subcodes: $\mathcal{X} = \mathcal{X}_1 \bigcup \mathcal{X}_2$. The general formulas reduce to

$$n = n_1 + 1, \quad \rho = \frac{4\rho_0}{4 + \rho_0 - \rho_1}, \quad M = M_1 + M_2. \tag{5.10.5}$$

Choosing all angles θ_i equal is natural when all the constituent codes $\mathcal{X}_i$ have the same minimum distance. When that is *not* the case better results are often obtained by choosing different angles for the different codes. We consider the case $m = 2$. Let $\mathcal{X}_1$ and $\mathcal{X}_2$ be spherical codes such that the maximal inner products are s_1 and s_2 respectively and let the maximal inner product between the codes be s_{12}. Consider the extended union

$$\mathcal{Z} = (\mathcal{X}_1 \sin\theta_1, \cos\theta_1) \cup (\mathcal{X}_2 \sin\theta_2, -\cos\theta_2).$$

The internal inner products are bounded by $s_1 \sin^2\theta_1 + \cos^2\theta_1$ and $s_2 \sin^2\theta_2 + \cos^2\theta_2$ respectively, while the cross product is bounded by $s_{12} \sin\theta_1 \sin\theta_2 - \cos\theta_1 \cos\theta_2$. Let s denote the maximal inner product in the code $\mathcal{Z}$. Choosing angles θ_1, θ_2 so as to make both of the internal inner products equal to s we arrive at the conditions

$$\sin^2\theta_1 = \frac{1-s}{1-s_1}, \quad \cos^2\theta_1 = \frac{s-s_1}{1-s_1},$$

$$\sin^2\theta_2 = \frac{1-s}{1-s_2}, \quad \cos^2\theta_2 = \frac{s-s_2}{1-s_2}.$$

Inserting these expressions in the formula for the cross product we obtain the following equation for determining s:

$$s = s_{12}\sqrt{\frac{1-s}{1-s_1}}\sqrt{\frac{1-s}{1-s_2}} - \sqrt{\frac{s-s_1}{1-s_1}}\sqrt{\frac{s-s_2}{1-s_2}}.$$

A simple reordering of terms gives us

$$\sqrt{(s-s_1)(s-s_2)} = s_{12} - (s_{12} + \sqrt{(1-s_1)(1-s_2)}\,)s.$$

We notice that this is a quadratic equation from which s can be easily solved. The result is

$$s = \frac{s_{12}^2 - s_1 s_2}{(s_{12} + \sqrt{(1-s_1)(1-s_2)}\ )^2 - 1} .$$

The general case can, of course, be analyzed in a similar way, but the formulas get more complicated and the analysis is best performed case by case. We illustrate the ideas by several examples.

Example 5.10.1: Let n be an integer such that a Hadamard matrix of order $n-1$ exists and let $\mathcal{B}_{n-1}^*$ be the *Biorthogonal* code in dimension $n-1$ using a binary representation according to the Y_2-construction as outlined in Chapter 5.8. Let $\mathcal{B}_{n-1}$ be the same code in the usual Y_3-construction:

$$\mathcal{B}_{n-1}^* = \frac{1}{\sqrt{n-1}}(\pm 1^{n-1}), \quad \mathcal{B}_{n-1} = (\pm 1, 0^{n-2})^C.$$

Define

$$\mathcal{X} = (\mathcal{B}_{n-1}\sin\theta,\ \cos\theta), \quad \mathcal{Y} = (\mathcal{B}_{n-1}^*\sin\theta, -\cos\theta).$$

The maximal inner products are as follows:

$$s_{\mathcal{X}} = s_{\mathcal{Y}} = \cos^2\theta; \quad s_{\mathcal{XY}} = \frac{\sin^2\theta}{\sqrt{n-1}} - \cos^2\theta.$$

Choosing θ such that $s_{\mathcal{X}} = s_{\mathcal{Y}} = s_{\mathcal{XY}} = s$ we obtain

$$s = \frac{2\sqrt{n-1} - 1}{4n-5},$$

which corresponds to the following squared minimum distance:

$$\rho = 2 - 2s = 4 \cdot \frac{2(n-1) - \sqrt{n-1}}{4n-5} .$$

We get codes of size $M = 4(n-1)$. The first few cases are as follows:

n	ρ	M
3	$(4/7)(4-\sqrt{2})$	8
5	$8/5$	16
9	$(8/31)(8-\sqrt{2})$	32
13	$(8/47)(12-\sqrt{3})$	48
17	$16/9$	64
21	$(8/79)(20-\sqrt{5})$	80

The first two cases are known to give the parameters of optimal codes. All cases give parameters of best known codes. □

Example 5.10.2: Consider a code generated by the construction Y_3. For the constant weight code we choose $(N, w, d, T) = (12, 4, 4, 51)$. For the unrestricted code we take the trivial code with parameters $(n, d, M) = (4, 1, 16)$. The resulting spherical code has parameters $(n, \rho, M) = (12, 1, 816)$. Adding a dummy zero to each codeword we get a code $\mathcal{X}_1$ with parameters $(n, \rho, M) = (13, 1, 816)$.

Consider the Y_2-construction using a binary code with parameters $(n, d, M) = (12, 4, 144)$. For the resulting spherical code we get $(n, \rho, M) = (12, 4/3, 144)$. Taking the extended union of this code with itself we get a spherical code $\mathcal{X}_2$, which by formula (5.10.5) has parameters $(n, \rho, M) = (13, 1, 288)$. We use the optimal weighting factor, which according to (5.10.3) is $\sin^2\theta = 3/4$.

Consider the *Biorthogonal* code with dimension $n = 12$ generated by the Y_3-construction. We take again the extended union of this code by itself, and again we use weight factors such that $\sin^2\theta = 3/4$, which however in this case is *not* optimal. By (5.10.2) we obtain a code $\mathcal{X}_3$ with parameters $(n, \rho, M) = (13, 1, 48)$.

Let $\mathcal{X}_4 = (0^{12}, \pm 1)$ be a very redundant version of the binary antipodal code. The parameters are $(n, \rho, M) = (13, 4, 2)$

Finally, consider the union $\mathcal{Z} = \mathcal{X}_1 \cup \mathcal{X}_2 \cup \mathcal{X}_3 \cup \mathcal{X}_4$. We have

$$s(\mathcal{X}_1, \mathcal{X}_2) = s(\mathcal{X}_2, \mathcal{X}_4) = s(\mathcal{X}_3, \mathcal{X}_4) = 1/2,$$

while

$$s(\mathcal{X}_1, \mathcal{X}_3) = \sqrt{3}/4 < 1/2, \quad s(\mathcal{X}_1, \mathcal{X}_4) = 0,$$
$$s(\mathcal{X}_2, \mathcal{X}_3) = (\sqrt{3} + 2)/8 < 1/2.$$

It follows that the mutual distance between any pair of codes is unity. As the internal distance in each one of the constituent codes is also at least one we conclude that the union $\mathcal{Z}$ is a spherical code with parameters $(n, \rho, M) = (13, 1, 1154)$. This construction provides the best known lower bound for contact number in dimension $n = 13$, ([239]). □

Example 5.10.3: Consider the Y_3-construction, using a constant weight code with parameters $(N, w, d, T) = (13, 4, 4, 65)$ and a trivial unrestricted code with parameters $(n, d, M) = (4, 1, 16)$. Adding a redundant zero to each codeword we obtain a spherical code $\mathcal{Y}_1$ with parameters $(n, \rho, M) = (14, 1, 1040)$.

By shortening three times the extended binary Hamming code of length $n = 16$ we obtain a linear binary code $\mathcal{A}$ with parameters $(n, d, M) = (13, 4, 256)$. From this code we can generate a non-linear code $\mathcal{B}$ with the same parameters by simply adding a fixed word of weight one to each codeword in the code $\mathcal{A}$. By the Y_2-construction each one of these binary codes generates a spherical code with parameters $(n, \rho, M) = (13, 16/13, 256)$. Moreover, any word $a \in \mathcal{A}$ differs in at least one position from any word $b \in \mathcal{B}$. It follows that the squared distance between the corresponding codewords x and y in the spherical codes satisfies $\|x - y\|^2 \geq 4/13$. By (5.10.5) the optimized extended union is a spherical code $\mathcal{Y}_2$ with parameters $(n, \rho, M) = (14, 1, 512)$.

By the same construction as in Example 5.10.2 and by using the same non-optimized weights we obtain from the *Biorthogonal* code of length $n = 13$ a spherical code $\mathcal{Y}_3$ with parameters $(n, \rho, M) = (14, 1, 52)$. We also have a redundant version $\mathcal{Y}_4$ of the binary antipodal code, with parameters $(n, \rho, M) = (14, 4, 2)$. By arguments similar to those in Example 5.10.2 one easily establishes that the union $\mathcal{Z} = \mathcal{Y}_1 \cup \mathcal{Y}_2 \cup \mathcal{Y}_3 \cup \mathcal{Y}_4$ is a spherical code with parameters $(n, \rho, M) = (14, 1, 1606)$. This code provides a best known estimate for the contact number in dimension $n = 14$, [239]. □

Example 5.10.4: Let $\mathcal{S}_{n-1}$ be the *Simplex* code in Ω_{n-1} and consider the following three codes:

$$\mathcal{X}_1 = \{(\mathcal{S}_{n-1} \sin \alpha, -\cos \alpha)\},$$

$$\mathcal{X}_2 = \{(-\mathcal{S}_{n-1} \sin \beta, \cos \beta)\},$$

$$\mathcal{X}_3 = \{(0^{n-1}, 1)\}.$$

Here α and β are two constants that should be appropriately chosen, and $\mathcal{X}_3$ is a code consisting of one single codeword only. Clearly the resulting code has size $M = 2n + 1$. Define

$$s_1 \triangleq -\frac{1}{n-1}\sin^2\alpha + \cos^2\alpha,$$

$$s_{12} \triangleq \frac{1}{n-1}\sin\alpha\sin\beta - \cos\alpha\cos\beta,$$

$$s_{23} \triangleq \cos\beta.$$

The quantity s_1 denotes the largest inner product within the code $\mathcal{X}_1$, while s_{12} and s_{23} denote the largest inner products between the codes $\mathcal{X}_1$ and $\mathcal{X}_2$ and between $\mathcal{X}_2$ and $\mathcal{X}_3$ respectively. Assuming $0 < \alpha < \beta < \pi/2$ it is easy to show that the largest minimum distance is obtained by choosing α and β so that the three quantities s_1, s_{12}, and s_{23} are equal. By this construction we get a sequence of best known spherical codes of size $M = 2n + 1$ in dimensions $n = 3, 4, \ldots$. The optimal code of size $M = 7$ in dimension $n = 3$ is the first member of this infinite family. □

5.11 Concluding remarks

Sometimes one and the same code can be obtained in several different ways. For instance, a code with parameters $(4, 2/3, 32)$ is generated by Y_3 (cyclic permutations of $(0, \pm 1, \pm 1, \pm 1)$), by Y_4 (cyclic permutations of $(\pm 1/2, \pm 1/2, \pm 1/2, \pm 3/2)$, where signs should always be chosen so as to maintain even parity), and by Y_5 (union of $(0, \pm 2, \pm 1, \pm 1)$, $(\pm 2, 0, \pm 1, \pm 1)$, $(\pm 1, \pm 1, 0, \pm 2)$, $(\pm 1, \pm 1, \pm 2, 0)$). It is interesting to note that this code can be implemented using phase shift keying with two different amplitudes and two different phase values, see Figure 5.9. The same code without different amplitudes (but keeping the phase shifts) is an example of the well known principle of set partitioning as suggested by Ungerboeck [225]). In the present example Ungerboeck's scheme gives $\rho = 2 - \sqrt{2} \simeq 0.58578$ while our scheme is about 0.56 dB better. The Y_5-construction differs from the Y_3-construction only by a change of the relation between amplitude and phase reference.

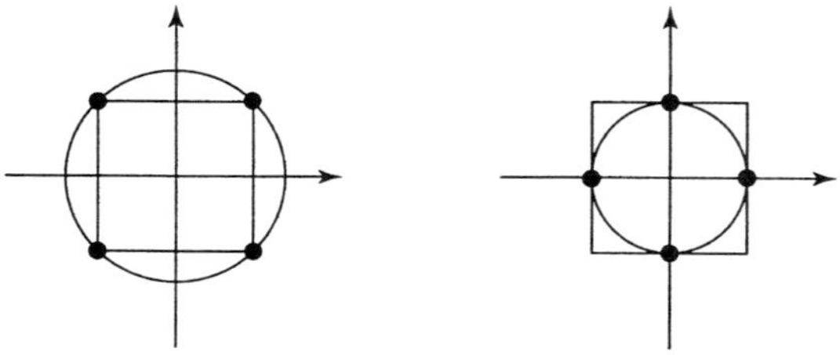

Figure 5.9: Part of the code generating $(n, \rho, M) = (4, 2/3, 32)$ by construction Y_3.

Figure 5.10 offers a graphical presentation of the performance obtained by all the best codes from the constructions Y_2–Y_8 in dimensions $n = 8, 16, 24$. We have plotted the rate

$$R = \frac{1}{n} \log M$$

versus normalized distance ρ. The staircase functions indicate the parameters of all dominating codes. Only codes obtained directly from the Y_K-constructions are included in the diagram. Unions and extended unions often produce much better codes.

We have also plotted the upper bound by Levenshtein [157], (a),(c),(e) and the results obtained from the shells in the Gosset lattice (b), the Barnes–Wall lattice (d) and the Leech lattice (f). Only parameters of spherical codes obtained from the point-centered lattices are displayed. Spherical codes from lattices will be extensively discussed in Chapter 10.

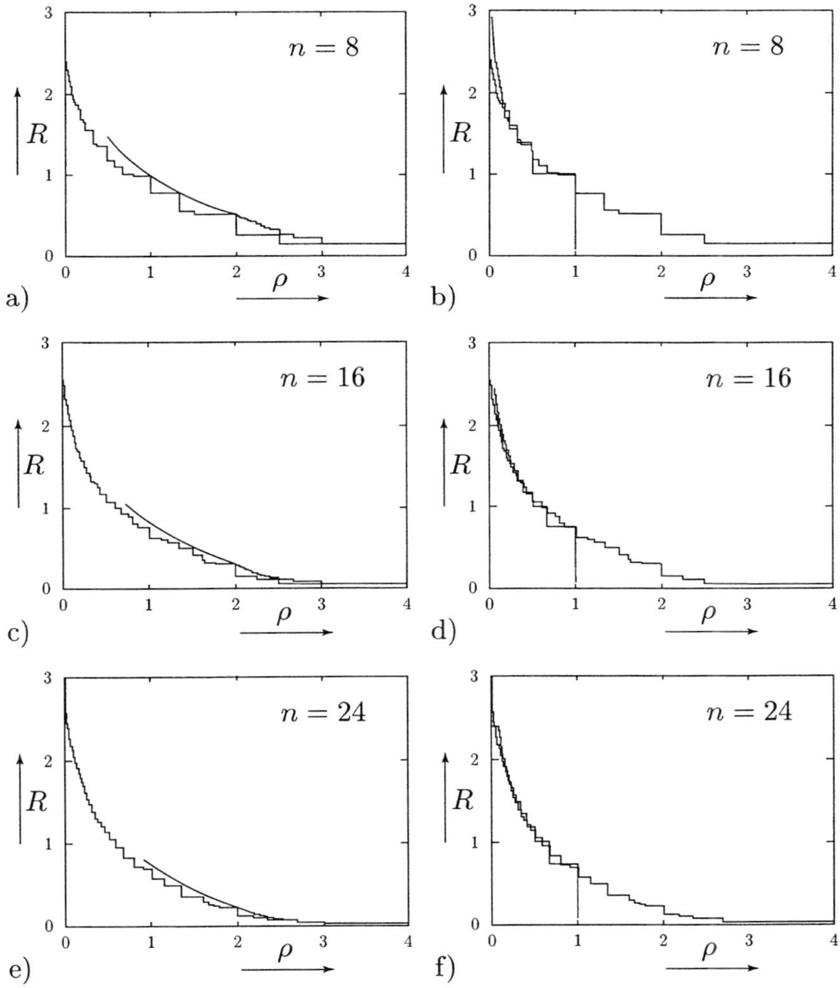

Figure 5.10: Dominating codes from the constructions Y_2–Y_8 compared with Levenshtein bound (a),(c),(d), and with shells in the Gosset lattice (b), the Barnes–Wall lattice (d) and the Leech lattice, $n = 24$, (f).

Chapter 6

Non-symmetric alphabets

6.1 Introduction

Spherical codes in Euclidean space resemble in an obvious way constant weight codes in Hamming space. In both cases the characteristic feature of the code is that all codewords are restricted to have a fixed norm. It is therefore a very natural idea to try to construct spherical codes from binary constant weight codes. In this chapter we shall study several such constructions.

6.2 The binary balanced mapping

Let $\mathcal{C} \subseteq \mathbb{F}_2^{N,w}$ be a constant weight binary code of length N and weight w; let the minimum Hamming distance be d and let the size be T. We say that $\mathcal{C}$ is an (N, w, d, T)-code. We can map this code into a spherical code by using for each codeword the following componentwise mapping, which we will refer to as the *binary balanced mapping*:

$$0 \mapsto \sqrt{\frac{w}{N(N-w)}}, \quad 1 \mapsto -\sqrt{\frac{N-w}{Nw}}.$$

It is clear that all codewords satisfy the following equation:

$$x_1^2 + x_2^2 + \cdots + x_N^2 = 1,$$

so the code is indeed a spherical code. It is also clear that each codeword satisfies the equation

$$x_1 + x_2 + \cdots + x_N = 0.$$

The latter fact implies that all codewords are located in the hyperplane defined by the vector $(1, 1, \ldots, 1)$, and that the dimension n of the code is upper-bounded by $N - 1$. Usually this bound is tight.

The construction will be referred to as the CW_2-construction and the spherical code $\mathcal{X}$ obtained in this way by the CW_2-mapping will be denoted $\mathcal{X} = CW_2(\mathcal{C})$. The parameters are:

CW_2:

$$n \leq N - 1, \quad \rho = \frac{Nd}{w(N-w)}, \quad M = T.$$

We observe in passing that the CW_2-construction can be used to produce bounds on the size of binary constant weight codes in the same way as the Y_2-construction can be used to produce bounds for unrestricted binary codes. Inserting the expression for ρ into the first Rankin bound (Theorem 1.3 in Chapter 1.5) we get the following bound for the size T of a code with weight w, minimum distance d and length N:

$$T \leq \left\lfloor \frac{\rho}{\rho - 2} \right\rfloor = \left\lfloor \frac{Nd}{2w^2 - 2Nw + Nd} \right\rfloor,$$

$$2 < \rho = \frac{Nd}{w(N-w)} \leq 4.$$

This bound was first discovered by Johnson [138], and is usually referred to as the *Second Johnson bound* (see MacWilliams-Sloane [167], p. 525). Recently Agrell et. al. [3] derived several new bounds for binary constant weight codes by using this same idea together with the *Levenshtein* bound and other bounds for spherical codes.

To any constant weight code $\mathcal{C}$ with parameters (N, w, d, T) there is another constant weight code $\mathcal{C}^*$ called the *complement* to $\mathcal{C}$ and obtained by interchanging zeros and ones. The complement $\mathcal{C}^*$ has parameters $(N, N-w, d, T)$ and gives rise to the spherical code $\mathcal{Y} = CW_2(\mathcal{C}^*) = -CW_2(\mathcal{C})$. By this little observation it is clear that we impose no loss of generality by restricting the weight w to the range $0 < w \leq N/2$. For the rest of this chapter this restriction on w will always be assumed.

For a given constant weight code $\mathcal{C}$ with parameters (N, w, d, T) let the spherical codes $\mathcal{X}$ and $\mathcal{Y}$ be defined as $\mathcal{X} = CW_2(\mathcal{C})$ and $\mathcal{Y} = -CW_2(\mathcal{C})$ respectively. The union $\mathcal{Z} = \mathcal{X} \cup \mathcal{Y}$ is another spherical code also generated by the binary code $\mathcal{C}$; it will be denoted $UCW_2(\mathcal{C})$. In order to determine the parameters of this code a crucial step is to estimate the minimum distance or—what is the same—the maximum correlation between codewords $x \in \mathcal{X}$ and $y \in \mathcal{Y}$. Let a and b be the codewords in $\mathcal{C}$ generating x and y respectively. Denote the binary correlation between a and b by $\lambda(a,b)$, and recall that this quantity simply denotes the number of coordinates k where $a_k = b_k = 1$. It follows that a and b differ in $2w - 2\lambda(a,b)$ positions and that they are simultaneously 0 in $N - 2w + \lambda(a,b)$ positions. Using these observations one easily computes the following expression for the *Euclidean* correlation between x and y:

$$\langle x, y \rangle = \frac{w^2 - N\lambda(a,b)}{w(N-w)}.$$

Let λ_{min} be the *minimal binary* correlation between any pair of codewords in $\mathcal{C}$. It is now a simple task to show that the parameters of the union $\mathcal{Z} = \mathcal{X} \cup \mathcal{Y}$ can be estimated as follows:

UCW_2:

$$n \leq N-1, \quad M = 2T,$$

$$\rho \geq \min\{\frac{Nd}{w(N-w)}, \frac{2(N-2w)}{N-w} + \frac{2\lambda_{min}N}{w(N-w)}\}.$$

Quite good codes are sometimes obtained by this construction. As an example we notice that the simple constant weight code $\mathcal{C}$ with parameters $(N, w, d, T) = (8, 2, 2, 28)$ generates the optimal spherical code with parameters $(n, \rho, M) = (7, 4/3, 56)$, ([10]). It is known that this code is unique, [10]. We recall that the same code can also be generated by the Y_3-construction.

In cases where the internal distance $\rho(\mathcal{X})$ exceeds the distance $\rho(\mathcal{X}, \mathcal{Y})$ between the code $\mathcal{X}$ and the antipodal code $\mathcal{Y}$ the *extended union*—as defined in Chapter 5—often produces good results. The simplest case, using the binary antipodal code $\{\pm 1\}$ as the suffix code, will be denoted ECW_2. An easy calculation reveals the following estimates for the parameters:

ECW_2:

$$n \leq N, \quad \rho = \frac{4d}{d + 2w - 2\lambda_{min}}, \qquad M = 2T.$$

6.3 Comments

In general the three constructions for non-symmetric alphabets are incompatible. This means that each one of them has its own region of applicability. In a few cases, however, a comparison in terms of dominance can be made. The construction UCW_2 is most interesting under the special condition $d = 2(w + \lambda_{min}) - 4w^2/N$. When this relation is satisfied all three constructions produce spherical codes with the same squared distance:

$$\rho_{CW_2} = \rho_{UCW_2} = \rho_{ECW_2} = \frac{4d}{2w + d}.$$

Dimension and size are then as follows:

	n	M
CW_2	$N-1$	T
UCW_2	$N-1$	$2T$
ECW_2	N	$2T$

The trivial code $\mathcal{C}$: $(N, w, d, T) = (N, 1, 2, N)$ generates the following spherical codes:

Code	n	ρ	M	
CW_2	$N-1$	$2+2/n$	$n+1$	*Simplex*
UCW_2	$N-1$	$2-2/n$	$2n+2$	
ECW_2	N	2	$2n$	*Biorthogonal*

We recognize precisely the same three classes of codes that we observed in Section 5.8 in our discussion of the Y_2-construction. The difference is that the above constructions exist in *all* dimensions n while the Y_2-constructions were depending on the existence of certain Hadamard matrices.

The simple binary code C with parameters $(N, w, d, T) = (N, 2, 2, \binom{N}{2})$ generates spherical codes with the following parameters:

Code	Domain	ρ	M
CW_2	$n \geq 2$	$(n+1)/(n-1)$	$n(n+1)/2$
UCW_2	$4 \leq n \leq 6$	$(n-3)/(n-1)$	$n(n+1)$
UCW_2	$n \geq 7$	$(n+1)/(n-1)$	$n(n+1)$
ECW_2	$n \geq 4$	$4/3$	$n(n-1)$

We notice that the construction CW_2 produces a code with parameters $(n, \rho, M) = (4, 5/3, 10)$, while the construction UCW_2 produces codes with parameters $(n, \rho, M) = (7, 4/3, 56)$ and $(n, \rho, M) = (8, 9/7, 72)$. All of these codes are best known. We have already noticed that the code in dimension $n = 7$ is even optimal. From the construction ECW_2 we obtain a code with parameters $(n, \rho, M) = (5, 4/3, 20)$, which is also best known. In general, however, we need more elaborate binary codes or more sophisticated constructions in order to obtain interesting spherical codes. We proceed to describe a few of those.

6.4 Unions from the CW_2-construction

It is also natural to take unions of spherical codes generated from two *different* constant weight codes. Let $\mathcal{A} : (N, w_{\mathcal{A}}, d_{\mathcal{A}}, T_{\mathcal{A}})$ and $\mathcal{B} : (N, w_{\mathcal{B}}, d_{\mathcal{B}}, T_{\mathcal{B}})$ be two constant weight codes and let $\mathcal{X} = CW_2(\mathcal{A})$ and $\mathcal{Y} = CW_2(\mathcal{B})$ be the spherical codes generated from them by the CW_2-construction. Let $x = (x_1, x_2, \ldots, x_N) \in \mathcal{X}$ and $y = (y_1, y_2, \ldots, y_N) \in \mathcal{Y}$ be generated by $a = (a_1, a_2, \ldots, a_N) \in \mathcal{A}$ and

$b = (b_1, b_2, \ldots, b_N) \in \mathcal{B}$ respectively. The following table applies:

a_k	b_k	$x_i\ y_i$
0	0	$\frac{1}{N}\sqrt{\frac{w_{\mathcal{A}}w_{\mathcal{B}}}{(N-w_{\mathcal{A}})(N-w_{\mathcal{B}})}}$
0	1	$-\frac{1}{N}\sqrt{\frac{w_{\mathcal{A}}(N-w_{\mathcal{B}})}{(N-w_{\mathcal{A}})w_{\mathcal{B}}}}$
1	0	$-\frac{1}{N}\sqrt{\frac{(N-w_{\mathcal{A}})w_{\mathcal{B}}}{w_{\mathcal{A}}(N-w_{\mathcal{B}})}}$
1	1	$\frac{1}{N}\sqrt{\frac{(N-w_{\mathcal{A}})(N-w_{\mathcal{B}})}{w_{\mathcal{A}}w_{\mathcal{B}}}}$

For $i, j \in \{0, 1\}$ let λ_{ij} denote the number of coordinates k where $a_k = i$, $b_k = j$. Let us use λ as a simplified notation for λ_{11}. We get

$$\begin{aligned}
\lambda_{11} &= \lambda, \\
\lambda_{01} &= w_{\mathcal{B}} - \lambda, \\
\lambda_{10} &= w_{\mathcal{A}} - \lambda, \\
\lambda_{00} &= N - w_{\mathcal{A}} - w_{\mathcal{B}} + \lambda\,.
\end{aligned}$$

Straightforward calculations now reveal the following expression:

$$\langle x, y \rangle = \frac{N\lambda - w_{\mathcal{A}}w_{\mathcal{B}}}{\sqrt{w_{\mathcal{A}}w_{\mathcal{B}}(N-w_{\mathcal{A}})(N-w_{\mathcal{B}})}}\,.$$

Let λ_{max} denote the *largest* binary correlation between $\mathcal{A}$ and $\mathcal{B}$. We have the following result.

Theorem 6.4.1 *Let $\mathcal{A}$ and $\mathcal{B}$ be binary constant weight codes with parameters $(N, w_{\mathcal{A}}, d_{\mathcal{A}}, T_{\mathcal{A}})$ and $(N, w_{\mathcal{B}}, d_{\mathcal{B}}, T_{\mathcal{B}})$ respectively and let*

λ_{max} denote the maximal binary correlation between two codewords $a \in \mathcal{A}$ and $b \in \mathcal{B}$. The union $\mathcal{Z} = \mathcal{X} \cup \mathcal{Y}$ between the spherical codes $\mathcal{X} = CW_2(\mathcal{A})$ and $\mathcal{Y} = CW_2(\mathcal{B})$ generated from $\mathcal{A}$ and $\mathcal{B}$ by the CW_2-construction is a spherical code with parameters (n, ρ, M) satisfying the following bounds:

$$n \leq N - 1, \qquad M = T_{\mathcal{A}} + T_{\mathcal{B}},$$

$$\rho \geq min\{\frac{Nd_{\mathcal{A}}}{w_{\mathcal{A}}(N - w_{\mathcal{A}})}, \frac{Nd_{\mathcal{B}}}{w_{\mathcal{B}}(N - w_{\mathcal{B}})},$$

$$2 - 2\frac{N\lambda_{max} - w_{\mathcal{A}}w_{\mathcal{B}}}{\sqrt{w_{\mathcal{A}}w_{\mathcal{B}}(N - w_{\mathcal{A}})(N - w_{\mathcal{B}})}}\}.$$

Remark 6.4.1: Replacing $\mathcal{Y} = CW_2(\mathcal{B})$ with $\mathcal{Y}^* = CW_2(\mathcal{B}^*) = -CW_2(\mathcal{B})$ the same bounds apply, with the only change that the last component in the bound for ρ should be replaced by the expression

$$2 - 2\frac{w_{\mathcal{A}}w_{\mathcal{B}} - N\lambda_{min}}{\sqrt{w_{\mathcal{A}}w_{\mathcal{B}}(N - w_{\mathcal{A}})(N - w_{\mathcal{B}})}},$$

where λ_{min} denotes the *minimum* binary correlation between two codewords $a \in \mathcal{A}$ and $b \in \mathcal{B}$.

Example 6.4.1: The binary code $\mathcal{C}$ corresponding to the Steiner system $S(24, 8, 5)$ has parameters $(N, w, d, T) = (24, 8, 8, 759)$. From this code it is possible to derive two codes $\mathcal{A}$ and $\mathcal{B}$, both with parameters $(N, w, d, T) = (21, 6, 8, 56)$ and such that the mutual binary correlation between any pair $(a, b) \in \mathcal{A} \times \mathcal{B}$ is either 1 or 3. Define $\mathcal{X} = CW_2(\mathcal{A})$, $\mathcal{Y} = CW_2(\mathcal{B})$. The union $\mathcal{Z} = \mathcal{X} \cup (-\mathcal{Y})$ has parameters $(n, \rho, M) = (20, 5/3, 112)$. The corresponding *extended* union has parameters $(n, \rho, M) = (21, 16/9, 112)$, which is one of the known examples of *optimal* code parameters, see Chapter 2.7. □

6.5 Non-symmetric ternary alphabet

Binary constant weight codes can also be used in connection with ternary alphabets. Again the code and the alphabet should be related

The functions $f_1(\beta)$, $f_2(\beta)$, $f_3(\beta)$ are displayed in Figure 6.1. A straightforward calculation shows that the bound on the normalized distance $\rho = d_E^2/E$ obtained from (6.5.2) and (6.5.3) is maximized by

$$\beta = \begin{cases} \dfrac{w - 2w_1}{N - w + 2w_1}, & d \le 4d_1, \\ \\ \dfrac{w - w_1\sqrt{d/d_1}}{N - w + w_1\sqrt{d/d_1}}, & d > 4d_1. \end{cases}$$

We refer to this construction as the CW_3-construction. The spherical code $\mathcal{X}$ generated in this way from a pair $\{\mathcal{C}, \mathcal{C}_1\}$ of constant weight codes with parameters (N, w, d, T) and (N_1, w_1, d_1, T_1), where $N_1 = w$, will be denoted $\mathcal{X} = CW_3(\mathcal{C}, \mathcal{C}_1)$. By lengthy but straightforward computations the following explicit expression can be computed for the parameters:

CW_3 :

$$n = N - 1, \quad M = T \cdot T_1, \quad \rho =$$

$$= \begin{cases} \dfrac{Nd}{Nw - (w - 2w_1)^2}, & if \quad d \le 4d_1, \\ \dfrac{Nd}{w(N - w) + k^2 w_1(N - w_1) - 2kw_1(N - w)}, & if \quad d > 4d_1, \end{cases}$$

where $k = \sqrt{d/d_1}$.

Remark 6.5.1: Unions and extended unions of codes from the CW_2- and the CW_3-constructions produce almost all optimal spherical codes in dimensions $n \le 24$. Including also unions and extended unions of codes from the Y_2-, Y_3-, and Y_4-constructions we actually obtain *all* known optimal codes in dimensions $n \le 24$, see Appendix D.

6.6 The general balanced construction

It is clear that the above ideas can be generalized. We can use constructions very similar to the Y_K-constructions discussed in Chapter 5 in order to produce balanced non-symmetric codes over an alphabet of size K for any integer $K \geq 2$. We refer to this construction as *general balanced binary concatenation.* We proceed to outline these ideas.

Let the alphabet be

$$\mathcal{L}_K = \{\xi, \xi + 1, \dots, \xi + K - 1\},$$

where the parameter ξ should be chosen so that the balance condition is met: $x_1 + x_2 + \cdots + x_N = 0$. We use the same mappings as for the Y_K-constructions, and again we associate a graph with each construction. This time, however, we need not worry about symmetries. Therefore we prefer to use a more direct labeling.

For any binary sequence $s = s_0 s_1 \dots s_m$ let $f(s)$ denote the corresponding integer according to the usual binary representation:

$$f(s) = \sum_{i=0}^{m} s_i 2^i. \tag{6.6.1}$$

The set $\mathcal{S}$ of leaves is labeled by the set of all binary sequences s such that $f(s) \leq K - 1$. The labeling we suggest is characterized by the following relation:

$$\forall\ s \in \mathcal{S}: \quad L(s) = \xi + f(s). \tag{6.6.2}$$

The principle is illustrated in Figure 6.2.

The corresponding trees are shown in Figures 6.3 and 6.4 for the cases $2 \leq K \leq 6$. It is easy to see that exactly as for the labeling employed in Chapter 5 the distance increases with a factor of two with each level in the tree. Let $\mathcal{T}$ denote the set of all internal nodes. The squared minimum distance d_E^2 can be estimated as follows:

$$d_E^2 \geq \min\{d_t 4^{|t|} : t \in \mathcal{T}\}.$$

Notice that in the present case there are no merges. Therefore there is a direct correspondence between the alphabet $\mathcal{L}_K$ and the set $\mathcal{S}$ of leaves in the tree.

	ξ	$\xi+1$	$\xi+2$	$\xi+3$	$\xi+4$	$\xi+5$	$\xi+6$	$\xi+7$
s_0	0	1	0	1	0	1	0	1
s_1	0	0	1	1	0	0	1	1
s_2	0	0	0	0	1	1	1	1

$\underline{K=2}$

ξ	$\xi+1$
0	1

$\underline{K=3}$

ξ	$\xi+1$	$\xi+2$
0	1	0
0		1

$\underline{K=4}$

ξ	$\xi+1$	$\xi+2$	$\xi+3$
0	1	0	1
0	0	1	1

$\underline{K=5}$

ξ	$\xi+1$	$\xi+2$	$\xi+3$	$\xi+4$
0	1	0	1	0
0	0	1	1	0
0				1

$\underline{K=6}$

ξ	$\xi+1$	$\xi+2$	$\xi+3$	$\xi+4$	$\xi+5$
0	1	0	1	0	1
0	0	1	1	0	0
0	0			1	1

Figure 6.2: Binary labeling for the balanced binary concatenation.

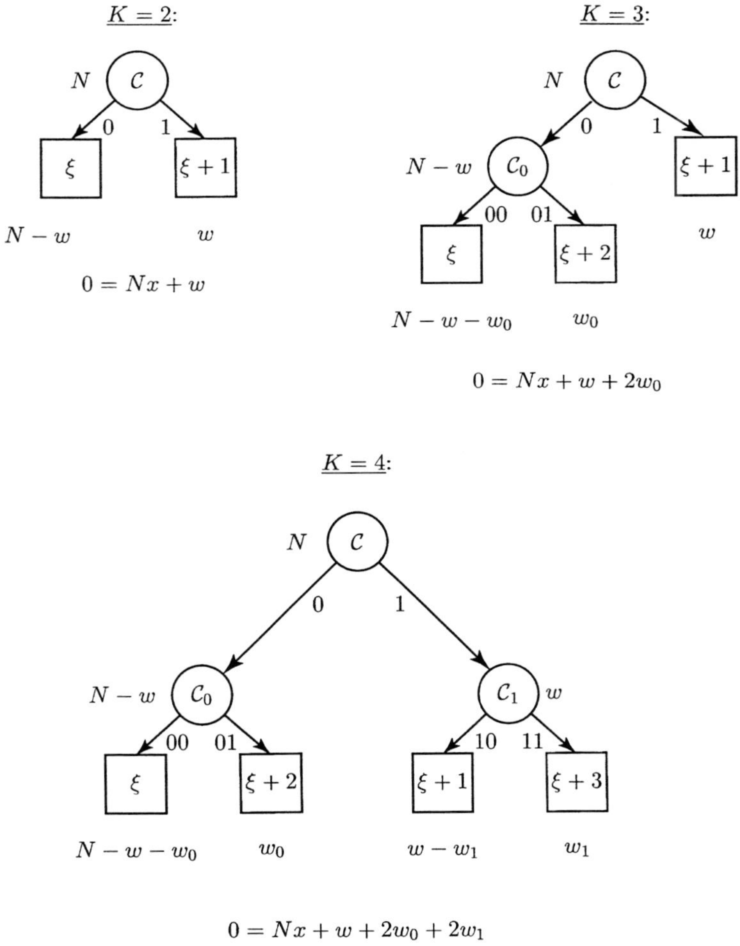

Figure 6.3: Graphs for the balanced binary concatenation; $K = 2, 3, 4$.

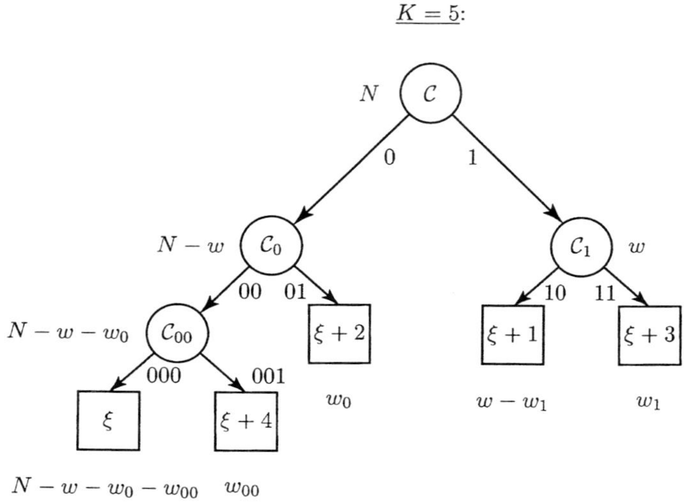

$$0 = Nx + w + 2(w_0 + w_1) + 4w_{00}$$

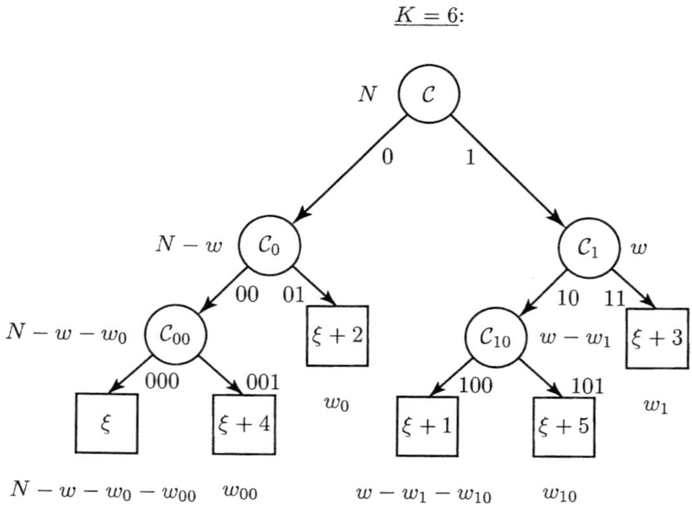

$$0 = Nx + w + 2(w_0 + w_1) + 4(w_{00} + w_{10})$$

Figure 6.4: Graphs for the balanced binary concatenation; $K = 5, 6$.

Each branch is assigned a weight. All left branches have weight 0 while a right branch on level i has weight 2^i. The letter $L(s) = \xi + f(s)$ corresponding to the leaf $s = s_0 s_1 \ldots s_m$ can be computed by adding all weights of the branches connecting the leaf with the root of the tree. This is a direct consequence of the formula (6.6.2).

In Figures 6.3 and 6.4 we have also assigned numbers to all nodes. For the internal nodes these numbers indicate the lengths of the corresponding binary codes. For the leaves they indicate the number of occurrences of the corresponding letters. The following relation holds:

$$\forall\, t \in \mathcal{S} \cup \mathcal{T} : \; n_t = \begin{cases} n_u - w_u, & if \quad t = u0, \\ w_u, & if \quad t = u1, \end{cases}$$

where $\mathcal{S}$ denotes the set of leaves and $\mathcal{T}$ denotes the set of internal nodes.

Finally we have in Figures 6.3 and 6.4 also indicated the equations which determine the parameter ξ so that the balance condition is met. Notice that all weights occur with a factor corresponding to the level at which the corresponding code is located. A quick look at the graphs explains this phenomenon. It follows that the parameter ξ in general is given by the following expression:

$$\xi = -\frac{1}{N} \sum_{t \in \mathcal{T}} |t| \cdot w_t,$$

where w_t denotes the weight of the code $\mathcal{C}_t$ located at node t.

The energy is given by the formula

$$E = \sum_{s \in \mathcal{S}} n_s L_s^2.$$

We denote by CW_K the general balanced binary concatenation as we have just described it. In summary we have the following characterization:

$$n \leq N - 1, \quad \rho = d_E^2 / E, \quad M = \prod_{t \in \mathcal{T}} M_t,$$

where

$$d_E^2 \geq \min\{d_t 4^{|t|} : t \in T\}, \quad E = \sum_{s \in \mathcal{S}} n_s L_s^2.$$

Example 6.6.1: As an example, consider the construction CW_4 with $\mathcal{C}_\lambda$ having parameters $(N, w, d, T) = (16, 8, 8, 30)$ and with both $\mathcal{C}_0$ and $\mathcal{C}_1$ having parameters $(N, w, d, T) = (8, 2, 2, 28)$. The resulting spherical code $\mathcal{X} = CW_4(\mathcal{C}, \mathcal{C}_0, \mathcal{C}_1)$ has parameters $(n, \rho, M) = (15, 1/2, 23520)$. This code can be improved by taking the union with another balanced code. Consider the construction $UCW_2(\mathcal{C})$ and let the binary code $\mathcal{C}$ have parameters $(N, w, d, T) = (16, 4, 2, 1820)$. The resulting spherical code $\mathcal{Y} = UCW_2(\mathcal{C})$ has parameters $(n, \rho, M) = (15, 2/3, 3640)$. It is easy to show that the union $\mathcal{Z} = \mathcal{X} \cup \mathcal{Y}$ is a spherical code with parameters $(n, \rho, M) = (15, 1/2, 27160)$, (this is *not* a best possible set of parameters). □

Chapter 7

Polyphase codes

7.1 Introduction

A code generated from an alphabet $\mathcal{L}_q$ consisting of q equally spaced points on the unit circle in $\mathbb{R}^2$ is obviously a constant energy code. Codes of this form are known as *polyphase* codes. They are of great importance in applications because of their natural implementation in terms of phase shift keying. The normalized codes are of course spherical codes.

Polyphase codes have been studied by a large number of authors, including Zetterberg [233], [234], Einarsson [89], [90], Ottosson [178], [179], Zetterberg–Brändström [235], Nilsson [174], Piret [181], Ginzburg [115], Sayegh [189], Portnoi [182], Zyablov–Portnoi [244], Zinoviev– Litsyn–Portnoi [239]. Poly-phase codes are also employed in so-called trellis modulation (see Ungerböck [225], Imai–Hirakawa [134], Forney–Gallager–Lang–Longstaff–Quershi [109]). This latter topic has recently attracted a lot of interest, but falls outside the scope of this book.

7.2 General properties

It is quite natural to describe polyphase codes using a complex notation. Therefore, in this chapter we interpret a spherical code $\mathcal{X}$ as a subset in the complex space $\mathbb{C}^N$. We associate the complex space $\mathbb{C}$ with the two-dimensional real space $\mathbb{R}^2$ in the natural way using the mapping $z = x + iy \mapsto (x, y) \in \mathbb{R}^2$. Thus a code $\mathcal{X} \subseteq \mathbb{C}^N$ will also be regarded as a code in $\mathbb{R}^n$, with $n = 2N$. In the latter interpretation it is of course always a code of even length. The characterization will be in terms of the usual parameters (n, ρ, M), interpreting $\mathcal{X}$ as a code in $\mathbb{R}^n$. We notice that the squared distance $\|x - y\|^2$ between two unit vectors in complex notation is given by

$$\|x - y\|^2 = 2 - 2Re\{\langle x, y\rangle\}, \tag{7.2.1}$$

where "Re" denotes the real part of the complex inner product

$$\langle x, y\rangle \triangleq \sum_{i=1}^{N} x_i \bar{y}_i.$$

For fixed $q \geq 3$ let θ be a q-th root of unity in $\mathbb{C}$. Define the alphabet $\mathcal{L}_q$ as follows:

$$\mathcal{L}_q \triangleq \{\theta^i : 0 \leq i \leq q - 1\}.$$

Contrary to the cases studied earlier in this book the present alphabet is itself a spherical code. As such it has parameters $(n, \rho, M) = (2, 4\sin^2(\pi/q), q)$. Any polyphase code can be interpreted as a concatenated code having this trivial spherical code as inner code. If the outer code is characterized in terms of Hamming distance the usual product bound can be employed for estimating the Euclidean distance of the polyphase code. This leads often—but not always—to a useful estimate.

As a spherical code the alphabet $\mathcal{L}_q$ has the property that each element θ^i has the same distance distribution to all the other elements. A code with this property is said to be *regular*. The possible squared distances are $\{4\sin^2(\pi i/q) : 0 \leq i \leq q - 1\}$. If q is *even* the maximal squared distance is 4. In this case the squared distances 0 and 4 occur only once while all other distances occur twice. If q is *odd* the maximum squared distance is less than 4 and all non-zero distances occur twice.

Let k_i denote the number of occurrences of the squared distance ρ_i. We notice the following special cases:

$q = 6$:

ρ_i	0	1	3	4
k_i	1	2	2	1

$q = 8$:

ρ_i	0	$2-\sqrt{2}$	2	$2+\sqrt{2}$	4
k_i	1	2	2	2	1

Notice that the cases $q = 3$ and $q = 4$ occur as subsets of the above cases.

7.3 The case $q = 3$

In the case $q = 3$ the alphabet consists of the vertices of an equilateral triangle. Let $\mathcal{C} \subseteq \mathbb{F}_3^N$ be a ternary code with parameters $(N, d, T)_3$, where d denotes minimum Hamming distance. We map $\mathbb{F}_3 = \{0, 1, 2\}$ into $\mathbb{C}$ using the natural map $a \mapsto \theta^a$, where θ is a third root of unity, and we map $\mathcal{C} \subseteq \mathbb{F}_3^N$ into $\mathbb{C}^N$ by applying this natural map component-wise on all codewords in $\mathcal{C}$. The resulting spherical code has parameters $(n, \rho, M) = (2N, 3d/N, T)$. This is a natural map to use in any Euclidean implementation of a ternary code. For instance, from the extended ternary Golay code with parameters $(n, d, M)_q = (24, 6, 729)_3$ we obtain a polyphase code with parameters $(n, \rho, M) = (24, 3/2, 729)$. However, from the construction ECW_2 we can obtain another spherical code with the same length and minimum distance, but with size $M = 807$. This code, of course, is not a polyphase code, but as a spherical codes it is certainly better.

The example is typical: we are not aware of one single example where a ternary polyphase code is not outperformed by a code from some other construction. Sometimes, however, good spherical codes can be generated using various modifications. In particular, some good spherical codes are obtained by using an *hexagonal* alphabet, which can be viewed as an union of two different ternary alphabets.

+	0	1	2	3
0	0	1	2	1
1	1	0	1	2
2	2	1	0	1
3	1	2	1	0

$\mathbb{Z}_4$, Lee metric

+	00	01	11	10
00	0	1	2	1
01	1	0	1	2
11	2	1	0	1
10	1	2	1	0

$\mathbb{F}_2^2$, Hamming metric

$\cdot$	1	θ	θ^2	θ^3
1	0	2	4	2
θ	2	0	2	4
θ^2	4	2	0	2
θ^3	2	4	2	0

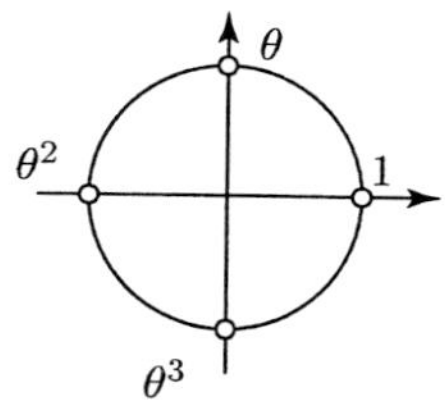

$\mathcal{L}_4$, Squared Euclidean metric

Figure 7.1: Three isometric spaces.

7.4 The case $q = 4$

Consider the following three metric spaces, see Figure 7.1:

1. The ring $\mathbb{Z}_4 = \{0, 1, 2, 3\}$ under Lee metric.
2. The linear space $\mathbb{F}_2^2 = \{00, 01, 10, 11\}$ under Hamming metric.
3. The space $\mathcal{L}_4 = \{0, 1, \theta, \theta^2\}$ under squared Euclidean metric.

We notice that although the *squared* Euclidean metric in general is *not* a metric—it fails to satisfy the triangle inequality—its *restriction* to the *four* points in $\mathcal{L}_4$ indeed *is* a metric.

It is clear that all three spaces are mutually isometric. This fact has interesting implications with respect to coding. The isometry between $\mathbb{F}_2^2$ under the Hamming metric and $\mathcal{L}_4$ under the squared Euclidean metric implies that any binary code of even length can be implemented as a polyphase code. The isometry between $\mathbb{Z}_4$ under the Lee metric and $\mathbb{F}_2^2$ under the Hamming metric implies that codes designed over the ring $\mathbb{Z}_4$ and with respect to the Lee metric can also be interpreted as binary codes with respect to Hamming metric. This observation has recently led to new characterizations of several codes (Hammons–Kumar–Calderbank–Sloane–Sole [123]) like the Nordstrom–Robinson code, the Preparata codes, Kerdock codes and the Goethals codes. If viewed as subsets in a *binary* vector space all of these codes are *nonlinear*. In the new interpretation they are viewed as *linear* codes over the ring $\mathbb{Z}_4$. The isometry with $\mathcal{L}_4$ indicates an implementation as *polyphase* codes.

The isometries between the three spaces imply that problems which have been formulated in one of the spaces can often be equivalently formulated in any of the other. When this is the case we use the formulation which appears most convenient. Often—but not always—this means that a polyphase code over $\mathcal{L}_4$ is most conveniently regarded as a binary code of even length.

7.5 The case $q = 6$

The case $q = 6$ corresponds to an alphabet $\mathcal{L}_6$ in the form of a regular hexagon. A natural idea for constructing codes over this alphabet is to use *set partitioning* and *generalized concatenation*. There are two possibilities: either we partition into three binary subsets or else we partition into two ternary subsets. Let $\mathcal{A} \subseteq \mathbb{F}_2^N$ be a binary code with parameters $(N, d_\mathcal{A}, M_\mathcal{A})_2$ and let $\mathcal{B} \subseteq \mathbb{F}_3^N$ be a ternary code with parameters $(N, d_\mathcal{B}, M_\mathcal{B})_3$. Using $\mathcal{A}$ and $\mathcal{B}$ as the outer codes the two different possibilities lead to polyphase codes with parameters (n, ρ, M), where $n = 2N$, $M = M_\mathcal{A} \cdot M_\mathcal{B}$, and where the normalized squared distance can be estimated by either of the following two formulas,

$$\rho \geq \frac{1}{N} \min\{d_\mathcal{A}, 3d_\mathcal{B}\} \quad \text{or} \quad \rho \geq \frac{1}{N} \min\{4d_\mathcal{A}, d_\mathcal{B}\},$$

depending on which set partitioning is employed. As an example, take

$$(N, d_\mathcal{A}, M_\mathcal{A})_2 = (3, 3, 2)_2, \quad (N, d_\mathcal{B}, M_\mathcal{B})_3 = (3, 1, 27)_3.$$

Partitioning into three binary subsets we obtain a spherical code $\mathcal{X}$ with parameters $(n, \rho, M) = (6, 1, 54)$. This is not in itself very impressive, but a simple improvement is possible, leading to a code with best known parameters. Let us join the point $\{0\}$ to the alphabet $\mathcal{L}_6$ and consider the new alphabet $\mathcal{L}^* = \mathcal{L}_6 \cup \{0\}$. Over this alphabet we can construct a code of length $N = 3$ with $M = 3 \times 6 = 18$ codewords. We do this by choosing a point from $\mathcal{L}_6$ in one position and the letter 0 in the two remaining positions. By proper normalization this leads to a spherical code $\mathcal{Y}$ with parameters $(n, \rho, M) = (6, 1, 18)$. The union $\mathcal{Z} = \mathcal{X} \cup \mathcal{Y}$ has parameters $(n, \rho, M) = (6, 1, 72)$. This is a best known construction for contact number in dimension $n = 6$. Of course neither the code $\mathcal{Y}$ nor the code $\mathcal{Z}$ is a polyphase code in a strict sense.

7.6 The case $q = 8$

Polyphase codes over the alphabet $\mathcal{L}_8$ are of particular practical interest. One form of such codes (using so called *trellis codes*, which are outside the scope of our treatment) was suggested by Ungerböck [225]. The basic ideas employed are binary *set partitioning* and *generalized concatenation.* The natural binary partitioning gives subcodes with increasing Euclidean distances as indicated in the table given in Section 7.2. Using binary codes with parameters (N, d_0, M_0), (N, d_1, M_1) and (N, d_2, M_2) respectively at the successive levels gives us a polyphase code with parameters (n, ρ, M), where $n = 2N$, $M = M_0 M_1 M_2$, and where the squared normalized Euclidean distance can be estimated as

$$\rho \geq \frac{1}{N} \min\{(2-\sqrt{2})d_0, 2d_1, 4d_2\}.$$

It is common to compare various polyphase coding schemes with "uncoded 4PSK " or—what is the same—with a binary antipodal representation of a binary code with distance $d = 1$. As we know such a scheme is a spherical code with parameters $(n, \rho, M) = (n, 4/n, 2n)$, provided n is even. From the formulas above we see that an 8-PSK coded scheme with $(N, d_0, M_0) = (N, 4, M_0)$ and $(N, d_1, M_1) = (N, d_2, M_2) = (N, 1, 2^N)$ will always dominate an uncoded 4-PSK scheme by having the same normalized distance but a size exceeding that for 4-PSK by the factor M_0.

7.7 Two special constructions

For any odd prime p there is a polyphase realization of the *Simplex* code with parameters $(n, \rho, M) = (p-1, 2p/(p-1), p)$. The construction is simple. Let p be an odd prime and define $N = (p-1)/2$. Let θ be a complex p-root of unity. We notice the following identity:

$$0 = \sum_{i=0}^{p-1} \theta^i = 1 + 2 \sum_{i=1}^{(p-1)/2} Re\{\theta^i\}. \tag{7.7.2}$$

Let the spherical code $\mathcal{X}$ consist of all complex vectors $x \in \mathbb{C}^N$ of the form

$$x = \frac{1}{\sqrt{N}}(\theta^u, \theta^{2u}, \ldots, \theta^{Nu}), \quad u \in \mathbb{Z}_p.$$

Clearly this is a spherical code with dimension $n = 2N = p - 1$ and size $M = p = n + 1$. In order to show that this code is a form of the *Simplex* code all that remains to establish is that any pair of distinct codewords have the same squared distance $\rho = 2p/(p-1)$.

Let x and y be two distinct words in $\mathcal{X}$, generated by $u, v \in \mathbb{Z}_p$, $u \neq v$. Remember that we use complex notation. The complex inner product is

$$\langle x, y \rangle = \frac{1}{N} \sum_{i=1}^{N} \theta^{(u-v)i} . \tag{7.7.3}$$

As p is prime and u and v are distinct we have

$$Re\{\sum_{i=1}^{N} \theta^{(u-v)i}\} = \sum_{i=1}^{(p-1)/2} Re\{\theta^i\} = -1/2, \tag{7.7.4}$$

where in the last step we have also utilized the identity (7.7.2). Now inserting (7.7.4) in (7.7.3) and observing (7.2.1) the result follows immediately.

Actually, for any odd prime p we also have a polyphase representation of the *Biorthogonal* code with parameters $(n, \rho, M) = (p, 2, 2p)$. To see this we first notice that the complex vectors x of the form $x = (1, \theta^u, \theta^{2u}, \ldots, \theta^{(p-1)u})$, $u \in \mathbb{Z}_p$, form an orthogonal basis in the complex space $\mathbb{C}^p$. Using the standard mapping $z = x + iy \mapsto (x, y) \in \mathbb{R}^2$ these vectors also generate an orthogonal basis in the real space $\mathbb{R}^{2p}$. Notice that the number of vectors is p, so the space spanned in $\mathbb{R}^{2p}$ actually has dimension p, although all vectors are expressed in a representation using $2p$ real components. The *Biorthogonal* code is obtained by taking all vectors in this orthogonal basis along with all the antipodal vectors $-x$ and then normalizing to unit length.

7.8 A general comment

Although of great *practical* interest, in terms of the parameters used for evaluation of *spherical* codes the polyphase codes are not very impressive. Except for the *Simplex* codes and the *Biorthogonal* codes we have not been able to find any polyphases codes with best known parameters. Generally they are outperformed by the Y_2- and the Y_3-constructions.

Chapter 8

Group codes

8.1 Introduction

An orthogonal transformation T that transforms each codeword x—represented as a *row vector* here and throughout all of the book—in a spherical code $\mathcal{X} \subseteq \Omega_n$ into another codeword $x\,T$ in the same code is said to be a *symmetry* of the code. The product TS of two symmetries is of course another symmetry, the identity transformation I is a trivial symmetry of any code, and if T is a symmetry the inverse T^{-1} is clearly also a symmetry. It follows that the set of all symmetries of a given spherical code $\mathcal{X}$ forms a *group*; it will be referred to as the *symmetry group* and it will be denoted $Symm(\mathcal{X})$. By definition it is a subgroup of the orthogonal group $\mathcal{O}_n$.

Let $\mathcal{G} = Symm(\mathcal{X})$ be the symmetry group to the spherical code $\mathcal{X}$ and let x be an arbitrary fixed codeword in $\mathcal{X}$. We denote by $x\mathcal{G}$ the set of all codewords $y = x\,T$ generated by transformations of this fixed codeword $x \in \mathcal{X}$ by elements T from the symmetry group $\mathcal{G}$:

$$x\mathcal{G} = \{y = x\,T : T \in \mathcal{G}\}\,.$$

The set $x\,\mathcal{G}$ is of course a subset in $\mathcal{X}$. It is called the *orbit* of x under the *action* of $\mathcal{G}$.

In general a spherical code consists of many orbits generated in this way by its symmetry group. As a special case, however, the code might consist of *one single orbit.* Such codes are called *group codes.* The concept was introduced by Slepian [212]. It is a natural generalization of his earlier idea of permutation codes, as discussed in Chapter 4. Recently the same term has been used in a slightly different meaning by Forney–Trott [110]. We remain by the original definition suggested by Slepian. Group codes in this sense are the topic of the present chapter.

8.2 Basic properties

Let $\mathcal{G}$ be a group in $\mathcal{O}_n$. It is clear that each $x \in \Omega_n$ generates an orbit $x\,\mathcal{G}$ in Ω_n and that the orbits are disjoint. More precisely: for any vectors $x,\ y \in \Omega_n$ we have either $x\,\mathcal{G} = y\,\mathcal{G}$ or $x\,\mathcal{G} \bigcap y\,\mathcal{G} = \emptyset$. Thus to each group $\mathcal{G} \subseteq \mathcal{O}_n$ corresponds a *partitioning* of Ω_n into disjoint orbits. By definition each orbit $x\,\mathcal{G}$ is a *group code.* We refer to $\mathcal{G}$ as the *generating group* and to x as the *initial vector.* The initial vector will usually be denoted x_0 . We say that the group code $\mathcal{X} = x_0\mathcal{G}$ is *generated* by the initial vector x_0 and the group $\mathcal{G}$. We notice several things.

Firstly, the initial vector is not unique: any vector x in a given group code $\mathcal{X} = x_0\mathcal{G}$ generates the same code. On the other hand, any vector $y \notin x_0\,\mathcal{G}$ generates a different code. The codes $x\,\mathcal{G}$ and $y\,\mathcal{G}$ might have very different parameters.

Secondly, also the generating group $\mathcal{G}$ is usually *not* unique. For a given group code $\mathcal{X} \subseteq \Omega_n$ the group $Symm\,(\mathcal{X})$ is by definition the *largest* group generating $\mathcal{X}$. In general, however, there might be several *subgroups* in $Symm\,(\mathcal{X})$ which also generate $\mathcal{X}$. In order to demonstrate that a given code $\mathcal{X} \subseteq \Omega_n$ is a group code it is enough to find *one* group generating $\mathcal{X}$. Finding $Symm\,(\mathcal{X})$ might be a much harder task.

A third observation is that the size $M = |\mathcal{X}|$ of a group code $\mathcal{X} = x_0\,\mathcal{G}$ in general is *not* the same as the size $|\mathcal{G}|$ of the group $\mathcal{G}$. The reason is that there might be many different elements $T \in \mathcal{G}$ mapping the initial vector x_0 into one and the same codeword $x \in \mathcal{X}$. More precisely, let $\mathcal{H}$ be the set of all elements T in $\mathcal{G}$ mapping the initial vector x_0 into itself. It is easy to see that $\mathcal{H}$ is a subgroup in $\mathcal{G}$, and that all the elements of the right coset $\mathcal{H}T$ generate the same codeword while different right cosets $\mathcal{H}T$ and $\mathcal{H}T'$, where $T \neq T'$, generate different codewords. It follows that there is precisely one codeword x for each right coset $\mathcal{H}T$ and that the size M of the code $\mathcal{X}$ is given by

$$M = \frac{|\mathcal{G}|}{|\mathcal{H}|}.$$

The group $\mathcal{H}$ is called the *stabilizer.*

All the above observations are illustrated in the following simple example.

Example 8.2.1: Consider the space $\mathbb{R}^2$ and let $\mathcal{G}$ be the set of all transformations of the form $T : (x, y) \mapsto (\pm x, \pm y)^C$, that is: all possible combinations of sign changes and permutations of coordinates. It is clear that $\mathcal{G}$ is a group of orthogonal transformations and that the order of the group is 8. (This is the dihedral group $\mathcal{D}_4$, a simple example of a *reflection group*; a more extensive treatment of such groups and their corresponding codes is given in Sections 8.7 and 8.8 of the present chapter). Let the initial vector be $x_0 = (\cos\alpha, \sin\alpha)$, where $0 \leq \alpha \leq \pi/4$. The code is given by $\mathcal{X}(\alpha) = (\pm\cos\alpha, \pm\sin\alpha)^C$. A simple consideration reveals that for $0 < \alpha < \pi/4$ we obtain 8 codewords with squared minimum distance $\rho = \min\{2 - 2\sin 2\alpha, 2 - 2\cos 2\alpha\}$, while for $\alpha = 0$ and $\alpha = \pi/4$ we obtain 4 codewords with squared minimum distance $\rho = 2$. The cases $\alpha = 0$ and $\alpha = \pi/4$ can be seen as degenerated forms of the first case, with pair-wise coinciding codewords. In the case $\alpha = 0$ the initial vector $x_0 = (1, 0)$ is invariant under the subgroup corresponding to the transformations $T : (x, y) \mapsto (x, \pm y)$ while in the case $\alpha = \pi/4$ the initial vector $x_0 = (1/\sqrt{2})(1, 1)$ is invariant under the subgroup corresponding to the transformations $T : (x, y) \mapsto (y, x)$. It is clear that $\mathcal{X}(\alpha) \neq \mathcal{X}(\beta)$ for $0 \leq \alpha < \beta \leq \pi/4$, and that the union of all codes $\mathcal{X}(\alpha)$ covers the whole space: $\Omega_2 = \cup_{0 \leq \alpha \leq \pi/4} \mathcal{X}(\alpha)$. For $\alpha = 0$, $\pi/4$ the same code is generated also by the cyclic group C_4, while for $\alpha = \pi/8$ the same code is generated by the cyclic group C_8. □

As seen by the example codes generated by one and the same group but with different initial vectors might have very different properties. This fact gives rise to the so called *initial vector problem.* Informally this problem is that of finding—for a given group $\mathcal{G}$—all the best initial vectors x_0. A more precise formulation is as follows.

Recall that a code $\mathcal{X}$ with parameters (n, ρ, M) *dominates* another code $\mathcal{X}'$ with parameters (n, ρ', M') if both of the inequalities $\rho \geq \rho'$ and $M \geq M'$ are satisfied. If the reversed inequalities hold we say that $\mathcal{X}'$ dominates $\mathcal{X}$ while in all other cases we say that $\mathcal{X}$ and $\mathcal{X}'$ are *incompatible* (cf. Section 6.3). In a given set of codes we can always find a subset of incompatible codes such that each code in the given set is dominated by one in the subset. We say that such a subset is a *dominating set.*

Let $\mathcal{G}$ be a fixed group in $\mathcal{O}_n$ and denote by $[\mathcal{G}]$ the set of all group codes corresponding to $\mathcal{G}$:

$$[\mathcal{G}] = \{x\,\mathcal{G} : x \in \Omega_n\}\,.$$

The initial vector problem is that of finding for the given group $\mathcal{G}$ a dominating set of codes from the set $[\mathcal{G}]$ of all group codes generated by $\mathcal{G}$.

The problem is complicated. It was stated by Slepian in his pioneering work [212]; until today solutions are known only in a few special cases. We shall discuss some of those in detail.

For the above simple example, let the codes $\mathcal{X}$ and $\mathcal{Y}$ be given by $\mathcal{X} = (\pm\cos(\pi/8), \pm\sin(\pi/8))^{C}$ and $\mathcal{Y} = (0, \pm 1)^{\mathcal{C}}$. It is clear that $\{\mathcal{X}, \mathcal{Y}\}$ is a dominating set, and that the parameters of the dominating codes are $(n, \rho, M) = (2, 2-\sqrt{2}, 8)$ and $(n, \rho, M) = (2, 2, 4)$ respectively. Notice that we obtain another equally valid dominating set by replacing $\mathcal{Y}$ with $\mathcal{Y}' = (\pm 1, \pm 1)/\sqrt{2}$.

As for spherical codes in general our interest concerning group codes is focused on *nonplanar* codes, (cf. Section 1.1).

Nonplanar group codes do not exist for all values of n and M. For instance, it is fairly easy to show that there exists no such group code with parameters $(n, M) = (3, 5)$. The question "For which parameters (n, M) do nonplanar group codes exist?" was raised already by Slepian ([212], p. 579). A complete answer is still lacking, although essential progress was made by Biglieri–Elia [23]. Their results can be summarized as follows:

- If n is *even* a nonplanar group code exists for *any* $M = 2, 3, 4, \ldots$.
- If n is *odd* a nonplanar group code exists for *any even* M.
- If n is *odd* and M is an *odd prime* no nonplanar group code exists.

The case $(n, M) = (3, 5)$ clearly belongs to the third category.

The only case left open by Biglieri–Elia [23] is that when n is odd and M is odd and composite. A partial answer for this case was given by Downey–Karlof [86] and Karlof–Downey [142]:

- For $n = 3$ there is no nonplanar group code with *odd* size M.
- For *any odd* $n \neq 3$ there is always a nonplanar group code for *some* odd M.

The subsequent discussion is concerned solely with nonplanar codes.

Let $\mathcal{X} \subseteq \Omega_n$ be a given spherical code. For any codeword $x \in \mathcal{X}$ let $D_{\mathcal{X}}(x) = \{\rho_0(x), \rho_1(x), \ldots, \rho_{d(x)}(x)\}$ be the set of all possible values of squared Euclidean distances from x to some other codeword in $\mathcal{X}$. For any $u \in D_{\mathcal{X}}(x)$ denote by $k_u(x)$ the number of codewords in $\mathcal{X}$ on squared distance u from x. We define the *distance distribution* at the point $x \in \mathcal{X}$ as the set

$$\Delta_{\mathcal{X}}(x) = \{(u,\ k_u(x)) : u \in D_{\mathcal{X}}(x)\}.$$

The distance distribution for a given codeword $x \in \mathcal{X}$ indicates all possible distances to other codewords as well as the number of codewords on each given distance.

In general different codewords may have different distance distributions. A code having the property that *all* codeword have the *same* distance distribution is said to be *regular*. For such codes we write k_u, $\Delta_{\mathcal{X}}$, $D_{\mathcal{X}}$ rather than $k_u(x)$, $\Delta_{\mathcal{X}}(x)$, $D_{\mathcal{X}}(x)$.

Any group code is regular. This follows from the simple observation that the distance $\|x - y\|$ equals the distance $\|xT - yT\|$. As $\mathcal{G}$ is a group it is clear that if we let y run through all of $\mathcal{X}$ so does yT. Therefore, if we keep x and T fixed the sets of distances generated by $\|x - y\|$ and by $\|xT - yT\|$ are the same.

It should be pointed out that while any group code is regular the converse is not true: *a regular code need not necessarily be a group code.* This fact was established by Slepian [213] by means of an explicit example.

Let $\mathcal{X} = x_0\mathcal{G}$ be a group code in Ω_n. For any $x \in \mathcal{X}$ we define the *Voronoi region* $V(x) \subseteq \mathbb{R}^n$ as the set of all points in $\mathbb{R}^n$ that are closer to x than to any other codeword in $\mathcal{X}$:

$$V(x) \triangleq \{y \in \mathbb{R}^n : \|y - x\| < \|y - x'\|,\ x' \in \mathcal{X},\ x' \neq x\}.$$

By definition the Voronoi regions are disjoint open sets. The closure of their union equals $\mathbb{R}^n$.

For any set $A \subseteq \mathbb{R}^n$ and for any transformation $T \in \mathcal{O}_n$ denote by AT^{-1} the set of points $x \in \mathbb{R}^n$ such that $xT \in A$. We say that the sets A and AT^{-1} are *congruent*. It is easy to see that for the Voronoi regions we have $V(x)T^{-1} = V(xT^{-1})$, so all the Voronoi regions are congruent. This property is of great importance in connection with decoding.

8.3 Groups represented by matrices

The study of group codes is closely related to the theory of groups represented by orthogonal matrices. This is an old topic in mathematics known as *representation theory*. We recall a few fundamental facts.

Let $\mathcal{G}$ be a subgroup in $\mathcal{O}_n$ and let $\mathcal{G}^*$ be an arbitrary abstract group. If there is a mapping $\varphi : \mathcal{G}^* \to \mathcal{G}$ such that $\varphi(ab) = \varphi(a)\varphi(b)$ holds for any $a, b \in \mathcal{G}^*$, then we say that $\mathcal{G}$ is a *representation* of $\mathcal{G}^*$. The mapping φ is said to be an *homomorphism* from $\mathcal{G}^*$ to $\mathcal{G}$ and the subset $\mathcal{K}^* = \{a \in \mathcal{G}^* : \varphi(a) = I\}$ is called the *kernel.* If the mapping is onto and if the kernel contains the identity element only—i.e. if φ is an *isomorphism*—then the representation is said to be *faithful.* The space on which it operates is referred to as a *representation space.*

It is well known (see, for example, Serre [201]) that any finite group has a representation by matrices. As a matter of fact, there always exists an infinity of such representations. A central issue in *representation theory* is the *classification* of such representations and the *identification* of possible structures.

Let W be a linear subspace in $\mathbb{R}^n$. The space W is said to be *stable* under $\mathcal{G}$ if for *each* element $T \in \mathcal{G}$ it is true that any vector $x \in W$ is mapped into another vector $y = xT$ in the same space W. For any subspace W which is stable under $\mathcal{G}$ there is another subspace W^c also stable under $\mathcal{G}$ and with the property that any vector $x \in \mathbb{R}^n$ has a unique representation on the form $x = x_1 + x_2$, where $x_1 \in W$ and $x_2 \in W^c$. The space W^c is said to be a *complement* to W and we say that $\mathbb{R}^n$ is a *direct sum* of the subspaces W and W^c. Among other things this means that with respect to some appropriate basis *all* the matrices $T \in \mathcal{G}$ have the following form

$$T = \begin{bmatrix} T_1 & 0 \\ 0 & T_2 \end{bmatrix},$$

where T_1 and T_2 belong to *subrepresentations,* say $\mathcal{G}_1$ and $\mathcal{G}_2$, of the representation $\mathcal{G}$. The group $\mathcal{G}$ is a *direct sum* of the subgroups $\mathcal{G}_1$ and $\mathcal{G}_2$.

The subspaces $W = \{0\}$ and $W = \mathbb{R}^n$ are said to be *trivial.* Clearly these two subspaces are stable with respect to *any* representation. A representation having no other stable subspaces is said to be *irreducible.* Every representation is a direct sum of irreducible representations (Serre [201], p. 7).

In the study of group codes the irreducible representations can naturally be associated with *alphabets.* To see this suppose first that

$n = mN$ is composite and let $\mathcal{G}$ be a group of matrices acting on $\mathbb{R}^n$. Suppose that $\mathcal{G}$ is a direct sum of N identical groups $\mathcal{G}_0$, each acting on a subspace W_i of dimension $m = n/N$. This means that each matrix $T \in \mathcal{G}$ has the form

$$T = diag\,(T_1, T_2, \ldots, T_N) = \begin{bmatrix} T_1 & 0 & 0 & \cdots & 0 \\ 0 & T_2 & 0 & \cdots & 0 \\ \vdots & \vdots & \vdots & & \vdots \\ 0 & 0 & 0 & \cdots & T_N \end{bmatrix},$$

where all the submatrices T_i are elements from the same group $\mathcal{G}_0$. Suppose in addition that the initial vector $x_0 \in \mathbb{R}^n$ has the form

$$x_0 = (x_{00}, x_{00}, \ldots, x_{00})/\sqrt{N},$$

where x_{00} is a unit length element in any of the spaces W_i upon which $\mathcal{G}_0$ acts. Then clearly the code $\mathcal{X} = x_0\,\mathcal{G}$ can be interpreted as a code of length N over the alphabet $\mathcal{L} = x_{00}\,\mathcal{G}_0/\sqrt{N}$. By a proper normalization the alphabet is in itself a group code with elements in $\mathbb{R}^m$, where $m = n/N$. The elements of $\mathcal{L} = x_{00}\,\mathcal{G}_0$ are referred to as *letters.*

The simplest example is provided by the binary antipodal representation of a binary code. In this case we have $N = n$, $m = n/N = 1$, $W = \mathbb{R}$ and $\mathcal{L} = \{\pm 1\}$. We will discuss this case in detail in the next section.

Another simple example is provided by polyphase codes. As we have already seen for those codes a complex notation is natural. We may take $N = n/2$, $m = 2$, $W = \mathbb{C}$, and $\mathcal{L} = \{1, \theta, \theta^2, \ldots, \theta^{q-1}\}$, where q is the size of the alphabet and θ is a complex q-th root of unity. This case will be further discussed in Section 8.6.

In general the various submatrices T_i are elements of different groups $\mathcal{G}_i$ acting on different subspaces W_i and the initial vector x_0 has the form

$$x_0 = (\theta_1\,x_{01}, \theta_2\,x_{02}, \ldots, \theta_N\,x_{0N}),$$

where $\theta_i \geq 0$, $\theta_1^2 + \theta_2^2 + \cdots + \theta_N^2 = 1$, and where x_{0i} is a unit norm vector in the space W_i upon which T_i acts, $i = 1, 2, \ldots, N$. Defining

$\mathcal{L}_i = \theta_i x_{0i} \mathcal{G}_i$, for $i = 1, 2, \ldots, N$, we can regard the spherical code $\mathcal{X} = x_0 \mathcal{G}$ as a subset of the product $\mathcal{L}_1 \times \mathcal{L}_2 \times \cdots \times \mathcal{L}_N$. The sets $\mathcal{L}_i$ can still be regarded as alphabets. We refer to this situation as a representation with *mixed alphabets*.

Of course there are also many other ways in which alphabets can be associated with group codes. The Y_K-constructions, for instance, are usually most conveniently described by alphabets which are not easily related to the decomposition of the generating group $\mathcal{G}$ into irreducible representations.

We shall give simple criteria under which several of the constructions we have described in earlier chapters can actually be interpreted as group codes. Sometimes interesting group codes are generated from irreducible groups directly. The class of permutation codes considered in Chapter 4 is one example of this. Later in this chapter we shall describe a generalization based on Coxeter's reflection groups.

8.4 Group codes in binary Hamming spaces

The concept of group code is not at all restricted to Euclidean spaces. In fact it occurs naturally in any metric space. In any such space the distance preserving transformations form a group. Any orbit generated by a subgroup in this group is a group code.

Many constructions for spherical codes are based on binary codes. With the purpose of arriving at conditions under which these constructions lead to group codes in Euclidean space we consider first a few basic facts regarding group codes in binary Hamming space.

The binary Hamming space is usually represented as the linear space $\mathbb{F}_2^n$ of all n-tuples $a = (a_1, a_2, \ldots, a_n)$ of elements a_i from the binary field $\mathbb{F}_2$. We adhere to this convention, but we emphasize right away that in the present context it is only the *additive group* structure in $\mathbb{F}_2^n$ that will be of concern.

The metric associated with this space is the *Hamming metric* d_H, by which the distance $d_H(a, b)$ between two vectors a and b in $\mathbb{F}_2^n$ equals the number of coordinates in which they differ. A code $\mathcal{C}$ in $\mathbb{F}_2^n$ is called a *binary code*. The *Hamming space* is the linear space $\mathbb{F}_2^n$ along with the Hamming metric d_H. It will be denoted $(\mathbb{F}_2^n, d_H)$.

In the space $(\mathbb{F}_2^n, d_H)$ there are two types of distance preserving transformations: *translations* and *permutations of coordinates*. They are defined as follows.

For any $a \in \mathbb{F}_2^n$ we define $T_a : \mathbb{F}_2^n \to \mathbb{F}_2^n$ by $x \mapsto x + a$. A transformation of this type is said to be a *translation* in $\mathbb{F}_2^n$. The set of all such translations will be denoted $\mathcal{G}_{\mathbb{F}_2^n}$. Together with the natural composition of transformations it forms a group isomorphic to $\mathbb{F}_2^n$. The natural isomorphism is given by the correspondence $a \in \mathbb{F}_2^n \leftrightarrow T_a \in \mathcal{G}_{\mathbb{F}_2^n}$. We clearly have $T_a T_b = T_{a+b}$. Any translation $T_a \in \mathcal{G}_{\mathbb{F}_2^n}$ is distance preserving. Recall that in the present context a *linear* code $\mathcal{A} \subseteq \mathbb{F}_2^n$ is regarded as an additive *group* . The corresponding subgroup in $\mathcal{G}_{\mathbb{F}_2^n}$ will be denoted $\mathcal{G}_{\mathcal{A}}$.

Let s be a permutation acting on the index set $I_n = \{1, 2, \ldots, n\}$; denote by $s(i)$ the image of i under s and denote by $\mathcal{S}_n$ the set of all permutations of I_n. The set $\mathcal{S}_n$ is usually referred to as the *symmetric group*. With any $s \in \mathcal{S}_n$ we associate a transformation $\mathbb{F}_2^n \to \mathbb{F}_2^n$, sending $a = (a_1, a_2, \ldots, a_n) \in \mathbb{F}_2^n$ to $as = (a_{s(1)}, a_{s(2)}, \ldots, a_{s(n)})$. The latter transformation will be referred to as a *permutation among coordinates.* By a slight abuse of notation we use the same letter s to denote both the permutation on the index set $I_n = \{1, 2, \ldots, n\}$ and the transformation $\mathbb{F}_2^n \to \mathbb{F}_2^n$. It is clear that a permutation among coordinates preserves distance. Translations and permutations among coordinates are related as follows

$$\forall a \in \mathbb{F}_2^n,\ \forall s \in \mathcal{S}_n : \quad T_{as} = s^{-1} T_a\, s. \tag{8.4.1}$$

It is easy to see that all translations and all permutations among coordinates as well as all combinations of such transformations preserve Hamming distance. Less obvious is perhaps the fact that *any* distance preserving transformation must be of this form. We have the following theorem (mentioned in [140] without proof).

Theorem 8.4.1 *Let Ψ be a distance preserving transformation in $(\mathbb{F}_2^n, d_H)$. Then there is some $a \in \mathbb{F}_2^n$ and some $s \in \mathcal{S}_n$ such that $\Psi = sT_a$.*

Proof: Any distance preserving transformation Ψ is bijective. Choose $y \in \mathbb{F}_2^n$ such that $\Psi(y) = 0$. Denote by e_i the binary vector with zeroes in all but the i-th position. For any $i \in \{1, 2, \ldots, n\}$ we obtain

$$1 = d_H(y, y + e_i) = d_H(\Psi(y), \Psi(y + e_i)) = w_H(\Psi(y + e_i)),$$

where $w_H(x)$ denotes the Hamming weight of $x \in \mathbb{F}_2^n$. The identity implies that Ψ transforms each vector $y + e_i$ into some vector $e_{s(i)}$ of weight one, i.e. that we have $\Psi(y + e_i) = e_i s$ for some $s \in S_n$. Now let $z \in \mathbb{F}_2^n$ be arbitrary. We have just noticed that the identity $\Psi(y + z) = zs$ holds for any z of weight one. By a simple induction argument one easily shows that it holds for any $z \in \mathbb{F}_2^n$. With $x = y + z$, $a = ys$ we obtain

$$\Psi\,(x) = \Psi\,(y + z) = zs = (x + y)s = xs + a = xsT_a.$$

As $x \in \mathbb{F}_2^n$ is arbitrary the theorem follows. □

Corollary 8.4.1 *If* $\Psi : \mathbb{F}_2^n \to \mathbb{F}_2^n$ *is a* weight preserving *transformation, i.e. if* $w_H(x) = w_H(\Psi(x))$ *for all* $x \in \mathbb{F}_2^n$*, then* $\Psi(x) = xs$ *for some* $s \in \mathcal{S}_n$*.*

Proof: A simple modification of the proof of Theorem 8.4.1 □

The following result is immediate.

Theorem 8.4.2 *Any linear code* $\mathcal{A} \in \mathbb{F}_2^n$ *is a group code. Moreover, any coset* $\mathcal{A} + u$ *is also a group code, with the same generating group but with a different initial vector.*

Proof: For the linear code $\mathcal{A}$ take $\mathcal{G} = \mathcal{G}_\mathcal{A}$ and let the initial vector be $a_0 = 0^n = (0, \ldots, 0)$. For the coset $\mathcal{A} + u$ let the initial vector be $a_0 = u$. □

We notice several things.

Remark 8.4.1: As with spherical group codes the initial vector for binary group codes is not unique. Actually any codeword a in $\mathcal{A}$ could be used.

The group $aut(\mathcal{C})$ consisting of all permutations leaving the code $\mathcal{C}$ invariant is called the *automorphism group* of $\mathcal{C}$ (see [166], p. 229).

Remark 8.4.2: The group $\mathcal{G}_{\mathcal{A}}$ is not the only possible generating group. Let $aut(\mathcal{A})$ be the automorphism group of $\mathcal{A}$. It is easy to see—and follows from Lemma 8.4.1 below—that $\mathcal{G}_{\mathcal{A}} \times aut(\mathcal{A})$ is also a group. It follows that $\mathcal{G}_{\mathcal{A}} \times aut(\mathcal{A})$ is another group generating $\mathcal{A}$.

Remark 8.4.3: The linear codes and their cosets are certainly not the only group codes in $\mathbb{F}_2^m$. For instance, let $\mathcal{G}$ be a group containing permutations of coordinates only. Then all codewords in the code $\mathcal{C} = a\mathcal{G}$ have the same weight. In this way many constant weight codes are also group codes. They are obviously not linear. For another example, let $\mathcal{G}$ be the group generated by $T : (x, y, z) \mapsto (z, x+1, y+1)$ and let the initial vector be $a_0 = (0, 0, 0)$. We obtain the group code

$$\mathcal{C} = a_0\mathcal{G} = \{(0,0,0), (0,1,1), (1,1,0)\}.$$

Further examples are easily constructed.

The observation made in Remark 8.4.2 above is a special case of the following result, which for future purposes we state in its general form.

Lemma 8.4.1 *Let $\mathcal{H}$ and $\mathcal{K}$ be subgroups in the group $\mathcal{G}$ and let $\mathcal{H} \times \mathcal{K}$ denote the set of all products hk formed by one element $h \in \mathcal{H}$ and one element $k \in \mathcal{K}$. The set $\mathcal{H} \times \mathcal{K}$ is a subgroup in $\mathcal{G}$ if and only if*

$$\mathcal{H} \times \mathcal{K} = \mathcal{K} \times \mathcal{H}.$$

Proof: See for instance Herstein [126], p. 44, Lemma 2.5.1. □

Remark 8.4.4: Notice that the requirement is *not* that the elements in $\mathcal{H}$ and $\mathcal{K}$ should commute: we might well have $hk \neq kh$ for certain individuals $h \in \mathcal{H}$ and $k \in \mathcal{K}$. What is important is that the *set*

$\mathcal{H} \times \mathcal{K}$ generated by all possible products hk equals the *set* $\mathcal{K} \times \mathcal{H}$ generated by all possible products kh. In other words: even if $\mathcal{H} \times \mathcal{K}$ is a group it need not necessarily be commutative.

The following relation holds whether $\mathcal{H}\times\mathcal{K}$ is a group or not (Herstein [126], p. 45, Theorem 2.5.1):

$$|\mathcal{H} \times \mathcal{K}| = \frac{|\mathcal{H}| \times |\mathcal{K}|}{|\mathcal{H} \cap \mathcal{K}|} \; .$$

As we saw above there is a bijective correspondence in $\mathbb{F}_2^n$ between linear codes and groups of translations. The following theorem characterizes those groups which generate linear codes.

Theorem 8.4.3 *Let $\mathcal{G}_\mathcal{A}$ be the group of translations in $\mathbb{F}_2^n$ corresponding to the linear code $\mathcal{A}$ and let $\mathcal{S}$ be a group of permutations of coordinates in $\mathbb{F}_2^n$. Then $\mathcal{G}_\mathcal{A}\times\mathcal{S}$ is a group if and only if $\mathcal{S} \subseteq aut\,(\mathcal{A})$.*

Proof: By Lemma 8.4.1 $\mathcal{G}_\mathcal{A} \times \mathcal{S}$ is a group if and only if for any pair $(a, s) \in \mathcal{A} \times \mathcal{S}$ there exists another pair $(b, t) \in \mathcal{A} \times \mathcal{S}$ such that $s\,T_a = T_b\,t$ and conversely for any pair $(b, t) \in \mathcal{A}\times\mathcal{S}$ there exists a pair $(a, s) \in \mathcal{A} \times \mathcal{S}$ such that the same relation holds. By symmetry it is enough to show just one of the conditions. As sT_at^{-1} is a translation if and only if $t = s$ if follows by (8.4.1) that b must be of the form $b = as$. Thus $\mathcal{G}_\mathcal{A} \times \mathcal{S}$ is a group if and only if $b = as$ belongs to $\in \mathcal{A}$ for any $s \in \mathcal{S}$ and for any $a \in \mathcal{A}$. But this is precisely the condition characterizing $aut(\mathcal{A})$. □

Corollary 8.4.2 *The set $\mathcal{G}_{\mathbb{F}_2^n} \times \mathcal{S}_n$ forms a group: it is the group of all distance preserving transformations in $\mathbb{F}_2^n$.*

Remark 8.4.5: Notice that for $n \geq 2$ the group $\mathcal{G}_{\mathbb{F}_2^n} \times \mathcal{S}_n$ is *not abelian.*

8.5 Group codes from binary codes

The groups of translations and permutations of coordinates in $\mathbb{F}_2^n$ have natural isomorphic counterparts in $\mathbb{R}^n$. Extending the convention from the previous paragraph we denote by s the orthogonal transformation that sends $x = (x_1, x_2, \ldots, x_n) \in \mathbb{R}^n$ into $xs = (x_{s(1)}, x_{s(2)}, \ldots, x_{s(n)}) \in \mathbb{R}^n$. It is clear that any subgroup S in $\mathcal{S}_n$ defines in this way also a subgroup in $\mathcal{O}_n$.

For any $a = (a_1, a_2, \ldots, a_n) \in \mathbb{F}_2^n$ we define a diagonal matrix $D(a)$ as follows:

$$D(a) \triangleq \begin{bmatrix} (-1)^{a_1} & 0 & 0 & \cdots & 0 \\ 0 & (-1)^{a_2} & 0 & \cdots & 0 \\ \vdots & \vdots & \vdots & & \vdots \\ 0 & 0 & 0 & \cdots & (-1)^{a_n} \end{bmatrix},$$

where we employ the natural convention

$$(-1)^{a_i} \triangleq \begin{cases} 1, & if \quad a_i = 0, \\ -1, & if \quad a_i = 1. \end{cases}$$

For any subset $\mathcal{A} \subseteq \mathbb{F}_2^n$ we denote by $\mathcal{O}_\mathcal{A}$ the set of all such matrices corresponding to elements $a \in \mathcal{A}$:

$$\mathcal{O}_\mathcal{A} \triangleq \{D(a) \in \mathcal{O}_n : a \in \mathcal{A}\}.$$

If $\mathcal{A}$ is a subgroup of $\mathbb{F}_2^n$—i.e. if it is a linear code in $\mathbb{F}_2^n$—it is clear that $\mathcal{O}_\mathcal{A}$ is a group in $\mathbb{R}^n$ and that the groups $\mathcal{A}$, $\mathcal{G}_\mathcal{A}$, and $\mathcal{O}_\mathcal{A}$ are isomorphic.

As an immediate consequence we have the following counterpart in $\mathbb{R}^n$ to Theorem 8.4.3:

Theorem 8.5.1 *Let $\mathcal{O}_\mathcal{A}$ be a group of diagonal matrices in $\mathbb{R}^n$ corresponding to the linear code $\mathcal{A}$ in $\mathbb{F}_2^n$ and let $\mathcal{S}$ be a subgroup in $\mathcal{S}_n$. Then $\mathcal{O}_\mathcal{A} \times \mathcal{S}$ is a group if and only if $\mathcal{S} \subseteq aut(\mathcal{A})$.*

Proof: Indeed, $\mathcal{O}_{\mathcal{A}} \times \mathcal{S} \subseteq \mathcal{O}_n$ is isomorphic to $\mathcal{G}_{\mathcal{A}} \times \mathcal{S} \subseteq \mathcal{G}_{\mathbb{F}_2^n} \times \mathcal{S}_n$. □

One thing should be pointed out immediately: while in $\mathbb{F}_2^n$ all distance preserving transformations belong to the group $\mathcal{G}_{\mathbb{F}_2^n} \times \mathcal{S}_n$ generated by translations and permutations it is by no means true that the corresponding group $\mathcal{O}_{\mathbb{F}_2^n} \times \mathcal{S}_n$ exhausts the family of distance preserving transformations in $\mathbb{R}^n$. Still, for the study of group properties of spherical codes generated from binary codes the group $\mathcal{O}_{\mathbb{F}_2^n} \times \mathcal{S}_n$ is often sufficient.

Another thing worth noticing is the dependence on the coordinate system: while the groups $\mathcal{O}_{\mathcal{A}}$ and $\mathcal{S}_n$ remain geometrically the same in any coordinate system (the geometric properties by definition are precisely those that are independent of the basis) their representations as diagonal and permutation matrices are valid with respect to a fixed appropriate basis only.

By the isomorphism between the groups $\mathcal{G}_{\mathcal{A}} \subseteq \mathcal{G}_{\mathbb{F}_2^n}$ and $\mathcal{O}_{\mathcal{A}} \subseteq \mathcal{O}_n$ we obtain the following immediate counterpart to Theorem 8.4.2.

Theorem 8.5.2 *Any* linear *code $\mathcal{A} \subseteq \mathbb{F}_2^n$ generates a group code in $\mathcal{O}_n$ under the binary antipodal mapping $Y_2(\mathcal{A})$.*

Proof: Let $\mathcal{X}$ be the spherical code generated by $\mathcal{A} \in \mathbb{F}_2^n$ under the Y_2-mapping. It is clear that we have $\mathcal{X} = x_0\,\mathcal{G}$ with $\mathcal{G} = \{D(a) \in \mathcal{O}_n : a \in \mathcal{A}\}$ and $x_0 = (1, 1, \ldots, 1)/\sqrt{n}$. □

The natural mapping into Euclidean space of a binary constant weight code is offered by the CW_2-mapping described in Chapter 6. We recall that under this mapping a constant weight code $\mathcal{C} \subseteq \mathbb{F}_2^{N,w}$ is mapped component-wise into Ω_N according to the rule

$$0 \mapsto \sqrt{\frac{w}{N(N-w)}}, \quad 1 \mapsto -\sqrt{\frac{N-w}{Nw}}. \tag{8.5.1}$$

Suppose $\mathcal{C} \subseteq \mathbb{F}_2^{N,w}$ is a group code in $\mathbb{F}_2^n$, i.e. suppose that $\mathcal{C}$ has the form $\mathcal{C} = a_0\,\mathcal{S}$ for some permutation group $\mathcal{S} \subseteq \mathcal{S}_n$ and for some $a_0 \in \mathbb{F}_2^{N,w}$. Then we clearly have $CW_2(\mathcal{C}) = x_0\,\mathcal{S}$, where x_0 is obtained from a_0 by employing the component-wise mapping (8.5.1) defining CW_2. It follows that the spherical code $CW_2(\mathcal{C})$ is a group code.

If $2w \neq N$ the alphabet employed in the CW_2-construction is non-symmetric. In those cases it is clear that if $CW_2(\mathcal{C})$ is a group code the generating group cannot contain any diagonal matrices except the identity matrix. However, if $2w = N$ more general groups might sometimes be employed. As an example let $\mathcal{C}$ be the constant weight code corresponding to the Steiner system $S(8,4,3)$. It is obtained by taking all cyclic shifts of the first seven components in the vector $(1,1,0,1,0,0,0,1)$ and in the complementary vector $(0,0,1,0,1,1,1,0)$. It has parameters $(N,w,d,M_{\mathcal{C}}) = (8,4,4,14)$. We have

$$\mathcal{X} = CW_2(\mathcal{C}) = x_0\,\mathcal{H} \times \mathcal{O}_{\mathcal{A}},$$

where $\mathcal{H} \subset \mathcal{S}_8$ is the group of cyclic shifts among the seven first components of the vectors in $\mathbb{R}^8$, and where $\mathcal{O}_{\mathcal{A}} = \{I, -I\}$ is the group in $\mathcal{O}_8$ corresponding to $\mathcal{A} = \{0^8, 1^8\}$ in $\mathbb{F}_2^8$. The initial vector x_0 can be taken as the CW_2-map of $(1,1,0,1,0,0,0,1)$. The resulting spherical code has parameters $(n,\rho,M) = (7,2,14)$ and is a realization of the biorthogonal code. A few comments are in order.

Remark 8.5.1: The binary code $\mathcal{C}$ corresponding to $S(8,4,3)$ is a group code in $\mathbb{F}_2^8$ with $\mathcal{G}_{\mathcal{A}} \times \mathcal{S}$ as the generating group. Alternatively it can be interpreted as two orbits from the group $\mathcal{S}$ (as we did in our definition above).

Remark 8.5.2: For $(N,w) = (8,4)$ the normalized alphabet is $\mathcal{L} = \{\pm 1/(2\sqrt{2})\}$. It follows that the mapping of $\mathcal{C}$ into $\mathbb{R}^8$ can also be interpreted as a mapping under Y_2. Notice, however, that our general estimate for the dimension under the Y_2-mapping fails to recognize that the dimension in this particular case is actually 7 rather than 8.

Remark 8.5.3: The union $\mathcal{A} \cup \mathcal{C}$ is the binary extended Hamming code with parameters $(n,d,M) = (8,4,16)$. Under the Y_2-mapping this code generates the biorthogonal code with parameters $(n,\rho,M) = (8,2,16)$.

All of the constructions considered in Chapters 5 and 6 can be seen as variations of the two ideas considered above. Each of them generates a spherical code by generalized concatenation using binary outer codes, and in each of them the mapping from the set of binary codes into the Euclidean space can be defined by a set of matrices from the group $\mathcal{O}_{\mathbb{F}_2^n} \times \mathcal{S}_n$ operating on some fixed vector x_0. In general there are many different sets of such matrices generating the same spherical code $\mathcal{X}$. The code is a group code if and only if among all the possible sets of matrices generating $\mathcal{X}$ there is at least one set which forms a group. Theorem 8.5.1 is often quite useful for determining that a given code $\mathcal{X}$ of this kind is a group code. We illustrate these statements by considering the case Y_3.

Consider the code $\mathcal{X} = Y_3(\mathcal{C}, \mathcal{A}_1)$, where $\mathcal{C}$ and $\mathcal{A}_1$ have parameters $(n, w, d_\mathcal{C}, M_\mathcal{C})$ and $(n_\mathcal{A} = w, d_\mathcal{A}, M_\mathcal{A})$ respectively. Choose $c_0 \in \mathbb{F}_2^{n,w}$ and let $\mathcal{S}$ be a set of permutations generating $\mathcal{C}$ by operating on c_0. Let x_0 be the vector in Ω_n obtained from c_0 by the map $0 \mapsto 0$, $1 \mapsto 1/\sqrt{w}$ and let $\mathcal{O}_\mathcal{A}$ be a set of diagonal matrices of the form $D(a)$, with $a = (a_0, a_1)$, where a_0 belongs to some set $\mathcal{A}_0 \subseteq \mathbb{F}_2^{n-w}$ and where a_1 belongs to the code $\mathcal{A}_1$. It is clear that we have $\mathcal{X} = x_0 \, \mathcal{O}_\mathcal{A} \times \mathcal{S}$. It is also clear that in general this is *not* the same as $x_0 \, \mathcal{S} \times \mathcal{O}_\mathcal{A}$. In two statements below we give sufficient conditions under which equality prevails, i.e. conditions under which the resulting spherical code $\mathcal{X}$ is a *group code.*

Notice that we have a considerable degree of freedom in choosing the sets $\mathcal{S}$ and $\mathcal{O}_\mathcal{A}$. Any given set $\mathcal{S}$ can be augmented with any set of matrices leaving c_0 invariant. All matrices in the set $\mathcal{A}_0$ can be altered in those positions corresponding to zero components in the vector x_0.

As in Chapter 5 let us employ the coordinate function

$$\varphi(c, i) = \sum_{j=1}^{i} c_j,$$

where in the summation the components c_j are interpreted as real numbers. For any $c \in \mathcal{C}$ denote by π_c the projection $\mathbb{F}_2^n \to \mathbb{F}_2^{n,w}$ which sends the vector $a = (a_1, a_2, \ldots, a_n) \in \mathbb{F}_2^n$ to the vector $a\,\pi_c \in \mathbb{F}_2^{n,w}$:

$$a \mapsto a\pi_c = \big(c_1 a_{\varphi(c,1)}, c_2 a_{\varphi(c,2)}, \ldots, c_n a_{\varphi(c,n)}\big).$$

The simple observations above are summarized in the following theorem.

Theorem 8.5.3 *Let $\mathcal{X} \subseteq \Omega_n$ be generated by $\mathcal{C} \subseteq \mathbb{F}_2^{n,w}$ and $\mathcal{A} \subseteq \mathbb{F}_2^w$ under the Y_3-construction and suppose the following conditions are satisfied:*

1. *The constant weight code $\mathcal{C}$ is a group code of the form $\mathcal{C} = c_0\, \mathcal{S}$ for some $\mathcal{S} \subseteq \mathcal{S}_n$ and some $c_0 \in \mathbb{F}_2^{n,w}$.*
2. *There is a linear code $\mathcal{A}^* \subseteq \mathbb{F}_2^n$ such that $\mathcal{A} = \pi_c\, \mathcal{A}^*$ for any $c \in \mathcal{C}$.*
3. *The permutation group $\mathcal{S}$ is contained in the automorphism group of the linear code $\mathcal{A}^* : \mathcal{S} \subseteq aut(\mathcal{A}^*)$.*

Then the code $\mathcal{X} \subseteq \Omega_n$ is a group code with $\mathcal{O}_{\mathcal{A}^*} \times \mathcal{S}$ as a generating group and with the vector $x_0 \in \Omega_n$ obtained from the vector $c_0 \in \mathbb{F}_2^{n,w}$ by the mapping $0 \mapsto 0,\ 1 \mapsto 1/\sqrt{w}$ as a possible initial vector.

Proof: From Theorem 8.5.1 we have that $\mathcal{O}_{\mathcal{A}^*} \times \mathcal{S}$ is a group in $\mathcal{O}_n$. Therefore for any initial vector $x_0 \in \Omega_n$ the code $\mathcal{X} = x_0\mathcal{O}_{\mathcal{A}^*} \times \mathcal{S}$ is a group code. Now take as x_0 the result of the mapping above upon the vector c_0. The corresponding group code $\mathcal{X}$ coincides with the spherical code obtained by construction Y_3 from the given codes $\mathcal{C}$ and $\mathcal{A}$. □

When the code $\mathcal{A}$ is the whole space $\mathcal{A} = \mathbb{F}_2^w$ (which is quite often the case) we have the following special case of the theorem above.

Corollary 8.5.1 *If the code $\mathcal{A}$ is the whole space $\mathcal{A} = \mathbb{F}_2^w$ the Y_3-construction leads to a group code in Ω_n whenever the constant weight code $\mathcal{C}$ is a group code in $\mathbb{F}_2^{n,w}$.*

Similar results—of various degree of complexity—can be derived for all of the constructions in Chapters 5 and 6. The following result for the Y_4-construction is particularly simple.

Theorem 8.5.4 *Let $\mathcal{A} \subseteq \mathbb{F}_2^n$ be a linear code and let $\mathcal{C} \subseteq \mathbb{F}_2^{n,w}$ be a group code generated by a permutation group $\mathcal{S} \subseteq aut(\mathcal{A})$. Then $\mathcal{X} = Y_4(\mathcal{A}, \mathcal{C}) \in \Omega_n$ is a group code with $\mathcal{O}_{\mathcal{A}} \times \mathcal{S}$ as a possible generating group.*

Proof: The proof is straightforward. □

8.6 Dual codes and MacWilliams' identity

Group codes generated from abelian groups have a number of special properties. In particular we have the concepts of duality and MacWilliams identity. We devote the present section to a discussion of those matters.

By a well known theorem in group theory any finite abelian group is a direct product of cyclic components (Herstein [126], Theorem 2.14.1, p. 109). This implies that group codes generated from abelian groups are naturally interpreted as polyphase codes. More precisely, with respect to an appropriate basis each matrix T in an abelian group $\mathcal{G}$ can be represented in the form

$$T = \begin{bmatrix} \theta_1^{a_1} & 0 & 0 & \dots & 0 \\ 0 & \theta_2^{a_2} & 0 & \dots & 0 \\ \vdots & \vdots & \vdots & & \vdots \\ 0 & 0 & 0 & \dots & \theta_N^{a_N} \end{bmatrix},$$

where each θ_i is an appropriate complex root of unity and where as in Chapter 7 we use a complex notation. For simplicity we restrict ourselves to the case where all the cyclic components have the same order $q \geq 2$. For $q = 2$ this corresponds to the binary antipodal mapping Y_2 where the dimension of the spherical code is $n = N$. For $q > 2$ we have a polyphase representation as discussed in Chapter 7 and the dimension of the spherical code is $n = 2N$. In either case we assume an initial vector of the form $x_0 = (1, 1, \dots, 1)/\sqrt{N}$ and employ an alphabet of the form $\mathcal{L} = \{1, \theta, \theta^2, \dots, \theta^{q-1}\}$, where θ is a q-th root of unity in $\mathbb{C}$.

Although it is still the group structure that is at the focus it turns out to be convenient to introduce the additional structure of a *ring*. Thus let us consider the isomorphism $a \leftrightarrow \theta^a$ between the additive group in the ring $\mathbb{Z}_q$ and the multiplicative group $\mathcal{L}_q$ in $\mathbb{C}$. To each $a \in \mathbb{Z}_q$ we associate a weight

$$w(a) \triangleq 4\sin^2(a\pi/q)$$

and to each pair $a, b \in \mathbb{Z}_q$ we associate a distance $d(a,b) = w(a-b)$. We extend both the weight and the distance additively in the natural way to the product space $\mathbb{Z}_q^N$ according to the following formulas: for any $a = (a_1, a_2, \ldots, a_N)$ and $b = (b_1, b_2, \ldots, b_N) \in \mathbb{Z}_q^N$:

$$w(a) \triangleq \sum_{i=1}^{N} w(a_i), \quad d(a,b) \triangleq \sum_{i=1}^{N} d(a_i, b_i).$$

For any $a = (a_1, a_2, \ldots, a_N) \in \mathbb{Z}_q^N$ we define a diagonal matrix over $\mathbb{C}$:

$$\Lambda(a) \triangleq \begin{bmatrix} \theta^{a_1} & 0 & 0 & \ldots & 0 \\ 0 & \theta^{a_2} & 0 & \ldots & 0 \\ . & . & . & \ldots & . \\ 0 & 0 & 0 & \ldots & \theta^{a_N} \end{bmatrix}.$$

For any vectors $a, b \in \mathbb{Z}_q^N$ we have $\Lambda(a+b) = \Lambda(a)\Lambda(b)$. Thus the mapping Λ is an homomorphism from the additive group in $\mathbb{Z}_q^N$ into the multiplicative group of diagonal matrices over the complex field $\mathbb{C}$. We have

$$\|x_0\Lambda(a) - x_0\Lambda(b)\|^2 = \frac{1}{N}\sum_{i=1}^{N} |\theta^{a_i} - \theta^{b_i}|^2$$

$$= \frac{4}{N}\sum_{i=1}^{N} \sin^2 \frac{\pi}{q}(a_i - b_i) = \frac{1}{N}\, d(a,b).$$

Thus there is also an obvious isometry from the space $(\mathbb{Z}_q^N, d)$ into the space (Ω_n, d_E^2). It is clear that the spherical code $\mathcal{X} = \{x_0\Lambda(a) : a \in \mathcal{C}\}$ is a group code if and only if $\mathcal{C}$ is an additive group in $\mathbb{Z}_q^N$. For the study of *abelian* group codes in Ω_n we prefer the representation in $\mathbb{Z}_q^N$.

The multiplication in $\mathbb{Z}_q$ is useful for the definition of a *dual* code. We define in $\mathbb{Z}_q^N$ the following inner product :

$$\langle a, b\rangle \triangleq a_1b_1 + a_2b_2 + \cdots + a_Nb_N.$$

Let $\mathcal{C}$ be an additive subgroup in $\mathbb{Z}_q^N$. Using the inner product we next define the dual code $\mathcal{C}^\perp$ as follows:

$$\mathcal{C}^\perp = \{a \in \mathbb{Z}_q^N : \langle a, b\rangle = 0,\ b \in \mathcal{C}\}.$$

It is clear that $\mathcal{C}^\perp$ is a subgroup in $\mathbb{Z}_q^N$. The key relation between the codes $\mathcal{C}$ and $\mathcal{C}^\perp$ is the following.

Theorem 8.6.1 *Let $\mathcal{C}$ be a group code in $\mathbb{Z}_q^N$ and let $\mathcal{C}^\perp$ be the corresponding dual code. Then*

$$\sum_{b\in\mathcal{C}} \theta^{\langle a,b\rangle} = \begin{cases} |\mathcal{C}|, & \text{if } a \in \mathcal{C}^\perp, \\ 0, & \text{if } a \notin \mathcal{C}^\perp. \end{cases}$$

where θ is an arbitrary complex q-th root of unity.

Proof: For any $a \in \mathbb{Z}_q^N$ and for any $x \in \mathbb{Z}_q$ let us define

$$\mathcal{C}_a(x) \triangleq \{b \in \mathcal{C} : \langle a, b\rangle = x\}.$$

It is clear that $\mathcal{C}_a(0)$ is always a subgroup in $\mathcal{C}$. By definition, if $a \in \mathcal{C}^\perp$ we have $\mathcal{C}_a(0) = \mathcal{C}$ and the equality

$$\sum_{b\in\mathcal{C}} \theta^{\langle a,b\rangle} = |\mathcal{C}|$$

follows immediately. If $a \notin \mathcal{C}^\perp$ we have $\langle a, b\rangle \neq 0$ for at least one $b \in \mathcal{C}$. In this case $\mathcal{C}_a(0)$ is a subgroup in $\mathcal{C}$ which does *not* coincide with $\mathcal{C}$. It is easy to see that the sets $\mathcal{C}_a(x)$ constitute the various cosets, $x \in \mathbb{Z}_q$ of $\mathcal{C}_a(0)$. They are all of the same size as $\mathcal{C}_a(0)$. We obtain

$$\sum_{b\in\mathcal{C}} \theta^{\langle a,b\rangle} = |\mathcal{C}_a(0)| \cdot (1 + \theta^u + \theta^{2u} + \cdots + \theta^{(q-1)u}) = 0,$$

where u is a factor in q. □

Let us define the *indicator function* for the code $\mathcal{C} \subseteq \mathbb{Z}_q^N$:

$$\chi_{\mathcal{C}}(a) \triangleq \begin{cases} 1, & if\ a \in \mathcal{C}, \\ 0, & if\ a \notin \mathcal{C}. \end{cases}$$

By Theorem 8.6.1 we have the following remarkable formula:

$$\chi_{\mathcal{C}^\perp}(a) = \frac{1}{|\mathcal{C}|} \sum_{b \in \mathbb{Z}_q^N} \chi_{\mathcal{C}}(b)\, \theta^{\langle a,b \rangle}. \tag{8.6.1}$$

It says that the Fourier transform of the indicator function of the group code $\mathcal{C}$ equals $|\mathcal{C}|$ times the indicator function of the dual code.

Although the above relation is quite interesting in its own right the main interest for us is one of its implications, namely the so called *MacWilliams identities.* By these identities the distance distributions of an abelian group code $\mathcal{C}$ and its dual $\mathcal{C}^\perp$ are related by a simple formula. While the formula (8.6.1) is a pure group property, independent of any metric, the MacWilliams identity presupposes a metric appropriately matched to the group structure. We begin by noticing the following generalization of Theorem 8.4.1.

Theorem 8.6.2 *Let Ψ be a distance preserving transformation in the space $(\mathbb{Z}_q^N, d)$. Then for any $a = (a_1, \ldots, a_N) \in \mathbb{Z}_q^N$ the mapping Ψ has the form:*

$$\Psi\,(a_1, \ldots, a_N) = (\pm a_1, \ldots, \pm a_N)s + (b_1, \ldots, b_N),$$

where $s \in \mathcal{S}_N$ is a permutation of coordinates. The vector $b = (b_1, b_2, \ldots, b_N)$ is an element in $\mathbb{Z}_q^N$ and "$\pm$" indicates that arbitrary sign changes are permitted in all coordinates.

Proof: The result follows by a minor modification of the proof of Theorem 8.4.1. □

Corollary 8.6.1 *Any weight preserving transformation* $\Psi : \mathbb{Z}_q^N \to \mathbb{Z}_q^N$ *has the form*

$$\Psi(a_1, \ldots, a_N) = (\pm a_1, \ldots, \pm a_N)s$$

for some permutation of coordinates $s \in \mathcal{S}_N$.

Remark 8.6.1: For $N = 1$ we have $\Psi(a) = \pm a + b$. The constant $b \in \mathbb{Z}_q^N$ performs a rotation while $a \mapsto -a$ is a reflection. It follows that the group of distance preserving transformations in $(\mathbb{Z}_q, d)$ is the *dihedral group* $\mathcal{D}_q$. In the Cartesian product space $(\mathbb{Z}_q^N, d)$ the distance preserving transformations are given by the dihedral transformations in the individual spaces $(\mathbb{Z}_q, d)$ along with all possible permutations among coordinates in these spaces.

Next we notice that the inner product $\langle a, b\rangle$ is invariant under weight preserving transformations: $\langle a\Psi, b\Psi\rangle = \langle a, b\rangle$ for any pair $a, b \in \mathbb{Z}_q^N$ and for any weight preserving transformation Ψ. This simple property is a direct consequence of the above corollary. It leads in turn to the following consequence. We define

$$j_u(a) \triangleq \begin{cases} 1, & if \quad w(a) = u, \\ 0, & if \quad w(a) \neq u. \end{cases}$$

The function $j_u(a)$ is a *characteristic function* of the weight u of the vector $a \in \mathbb{Z}_q^N$. Consider the Fourier transform of this function $j_u(a)$ for some element $b \in \mathbb{Z}_q^N$,

$$J_u(b) \triangleq \sum_{a \in \mathbb{Z}_q^N} j_u(a)\theta^{\langle a,b\rangle}.$$

The somewhat surprising fact is that this Fourier transform $J_u(b)$ depends only on the *weight* $w(b)$ of the argument $b \in \mathbb{Z}_q^N$. For any $N \geq 1$ and any $q \geq 2$ denote by $\mathcal{W}(N, q)$ the set of all possible weights in $(\mathbb{Z}_q^N, d)$. The following result is true.

Lemma 8.6.1 *For any* $u \in \mathcal{W}(N, q)$, *for any* $b \in \mathbb{Z}_q^N$, *and for any weight preserving transformation* Ψ *the following identity is valid:*

$$J_u(b\Psi) = J_u(b).$$

Proof: As Ψ is bijective we have

$$J_u(b\,\Psi) = \sum_{a\in\mathbb{Z}_q^N} j_u(a)\theta^{\langle a,\, b\Psi\rangle} = \sum_{a\in\mathbb{Z}_q^N} j_u(a\Psi)\theta^{\langle a\Psi,\, b\Psi\rangle}.$$

However, $j_u(a\Psi) = j_u(a)$ because $w(a\,\Psi) = w(a)$ and $\langle a\Psi, b\Psi\rangle = \langle a, b\rangle$ by the observation above. Thus we have

$$J_u(b\Psi) = \sum_{a\in\mathbb{Z}_q^N} j_u(a)\theta^{\langle a,b\rangle} = J_u(b).$$

□

The property guaranteed by this Lemma is precisely the link needed between the metric and the group structure. For any $b \in \mathbb{Z}_q^N$ and any $u, v \in \mathcal{W}(N, q)$ such that $w(b) = v$ we define

$$K(u,v) \triangleq J_u(b) = \sum_{a\in\mathbb{Z}_q^N} j_u(a)\theta^{\langle a,b\rangle}. \tag{8.6.2}$$

Lemma 8.6.1 guarantees that the definition is unambiguous. For $q = 2$ the function $K(u, v)$ is given by the so called *Krawtchouk polynomials* (see MacWilliams-Sloane [166], pp. 150-153). It is clear that weight and distance distributions coincide for group codes. For a given group code $\mathcal{C} \in \mathbb{Z}_q^N$ denote by k_u the number of codewords with weight u and let $k_u^\perp$ denote the corresponding number in the dual code $\mathcal{C}^\perp$. The MacWilliams identity is established by the following theorem.

Theorem 8.6.3 (MacWilliams identity) *Let $\mathcal{C}$ be a group code in $\mathbb{Z}_q^N$ with distance distribution $k_v : v \in \mathcal{W}(N, q)$, and let $k_u^\perp : u \in \mathcal{W}(N, q)$ be the distance distribution of the dual code $\mathcal{C}^\perp$. The following relation holds:*

$$k_u^\perp = \frac{1}{|\mathcal{C}|} \sum_{v\in\mathcal{W}(N,q)} K(u,v)\,k_v\,, \qquad u \in \mathcal{W}(N,q).$$

Proof: By definition and by (8.6.1) we have

$$k_u^\perp = \sum_a j_u(a)\chi_{\mathcal{C}^\perp}(a) = \sum_a j_u(a)\frac{1}{|\mathcal{C}|}\sum_b \chi_{\mathcal{C}}(b)\theta^{\langle a,b\rangle},$$

where the summations are over all $a, b \in \mathbb{Z}_q^N$. Multiplying with the identity

$$\sum_{v \in \mathcal{W}(N,q)} j_v(b) = 1$$

and rearranging terms we obtain

$$k_u^\perp = \frac{1}{|\mathcal{C}|} \sum_{v \in \mathcal{W}(N,q)} \sum_b j_v(b)\chi_\mathcal{C}(b) \sum_a j_u(a)\theta^{\langle a,b \rangle}.$$

The theorem follows upon observing that we have

$$K(u,v) = \sum_a j_u(a)\theta^{\langle a,b \rangle}$$

by (8.6.2) and

$$k_v = \sum_b j_v(b)\chi_\mathcal{C}(b)$$

by definition. □

Remark 8.6.2: Notice that it is an essential requirement in the proof that the quantity

$$\sum_a j_u(a)\theta^{\langle a,b \rangle}$$

depends only on the weight $w(b)$.

Remark 8.6.3: It is worthwhile pointing out that the concept of duality is a *relative* one: the dual $\mathcal{C}^\perp$ of a given code $\mathcal{C}$ is always defined with respect to a given abelian group $\mathcal{A}$ in which both $\mathcal{C}$ and $\mathcal{C}^\perp$ are subgroups. If $\mathcal{C}$ is regarded as a subgroup in another group $\mathcal{A}'$ the dual might be different. For instance, in the group $\mathbb{Z}_6$ the triangle $\mathcal{C} = 2\mathbb{Z}_3$ has the dual $\mathcal{C}^\perp = 3\mathbb{Z}_2$. If instead the same code is regarded as a subgroup in $\mathbb{Z}_9$ it gets the description $\mathcal{C} = 3\mathbb{Z}_3$ and is selfdual.

Example 8.6.1: We illustrate MacWilliams identity by considering the triangle $\mathcal{C} = \mathbb{Z}_3$ in $\mathbb{Z}_6$. The possible distances in $\mathbb{Z}_6$ are $\mathcal{W}(6,1) = \{0, 1, 3, 4\}$ and the distance distribution of $\mathcal{C} = 2\mathbb{Z}_3$ is $k = [1, 0, 2, 0]^T$. As we saw above the dual code with respect to $\mathbb{Z}_6$ is $\mathcal{C}^\perp = 3\mathbb{Z}_2$ with distance distribution $k^\perp = [1,0,0,1]^T$. The function $K(u,v)$ in $\mathbb{Z}_6$ is given by the matrix

$$K = \begin{bmatrix} 1 & 1 & 1 & 1 \\ 2 & 1 & -1 & -2 \\ 2 & -1 & -1 & 2 \\ 1 & -1 & 1 & -1 \end{bmatrix}.$$

One readily verifies the following identities: $k^\perp = (1/3)K \cdot k$ and $k = (1/2)K \cdot k^\perp$.

8.7 Finite reflection groups

Finite groups generated by reflections constitute an interesting subclass in $\mathcal{O}_n$. Coxeter [72] gave a complete classification and introduced an efficient graphic notation. Slepian [212] employed some special reflection groups in his work on permutation codes [211] (see also Chapter 4), and indicated—without working out the details—the possibility of further generalizations. The general case was recently considered by Lahtonen–Mittelholzer [171]. It turns out that several extremely good spherical codes are generated by reflection groups. Unfortunately the good codes occur only in small dimensions.

Before discussing the codes we need to recall a few basic facts about the groups. The results will be useful also in Chapter 10, where we discuss spherical codes from lattices. For details and for most proofs we refer to Humphreys [133], Grove–Benson [121] or Coxeter [72], [75].

A *reflection* is an orthogonal transformation of the form

$$R_a : x \mapsto x - 2a\langle a, x\rangle,$$

where a is a fixed vector of unit norm (i.e. a vector in Ω_n) and where $\langle a, x\rangle$ denotes the usual inner product. We say that R_a is a reflection in the hyperplane

$$a^\perp \triangleq \{x \in \mathbb{R}^n : \langle a, x\rangle = 0\},$$

and we say that a is a *root* corresponding to the reflection R_a. It is clear that any vector $a \in \Omega_n$ uniquely determines a reflection $R_a \in \mathcal{O}_n$. The vector $-a$ determines the same reflection, but any $b \neq \pm a$ determines a different reflection. In other words: R_a equals R_b if and only if $a = \pm b$.

It should be pointed out that it might sometimes be convenient not to restrict the roots to have unit length but allow vectors of arbitrary lengths. The modifications in the above definitions are obvious. Some authors prefer this more flexible approach. For our present purposes we find it more convenient to insist on the unit norm assumption.

In matrix notation we have $R_a = I - 2a^T a$, where a^T denotes the transpose of $a = (a_1, a_2, \ldots, a_n) \in \Omega_n$. It is clear that with respect to some appropriate basis the matrix R_a is diagonal with $n-1$ of the diagonal elements equal to 1 and the single remaining one equal to -1. It follows that $\det(R_a) = -1$. If R_a and R_b are two reflections the product $R_a R_b$ has determinant $\det(R_a R_b) = \det(R_a)\det(R_b) = 1$, and so is *not* a reflection. It follows that a group cannot contain reflections *only*: at least half of the group elements (and usually many more) must be transformations of other kinds.

A word on terminology is in order. In accordance with the established convention we use the term "reflection" only in the meaning "*reflection in a hyperplane*". There are, of course, many other orthogonal transformations that one intuitively might want to refer to as "reflections". For instance, in $\mathbb{R}^3$ we have the following transformations:

(*i*) $R_1 : (x, y, z) \mapsto (-x, y, z)$ (reflection in the yz-plane),
(*ii*) $R_2 : (x, y, z) \mapsto (-x, -y, z)$ (reflection in the z-axis),
(*iii*) $R_3 : (x, y, z) \mapsto (-x, -y, -z)$ (reflection in the origin),

with the following matrix representations:

$$R_1 = \begin{bmatrix} -1 & 0 & 0 \\ 0 & 1 & 0 \\ 0 & 0 & 1 \end{bmatrix}, \quad R_2 = \begin{bmatrix} -1 & 0 & 0 \\ 0 & -1 & 0 \\ 0 & 0 & 1 \end{bmatrix},$$

$$R_3 = \begin{bmatrix} -1 & 0 & 0 \\ 0 & -1 & 0 \\ 0 & 0 & -1 \end{bmatrix}.$$

In the strict terminology only R_1 qualifies as a reflection. Notice, however, that R_2 and R_3 in a simple and obvious way are generated by matrices of the same type as R_1, i.e. by reflections. Also notice that $\det R_2 = 1$, while $\det R_3 = -1$. Thus: while a reflection always has negative determinant a non-reflection can have determinant of either sign.

Let R_a, R_b be two different reflections. Their product is

$$\begin{aligned} R_a R_b &= (I - 2a^T a)(I - 2b^T b) \\ &= I - 2a^T a - 2b^T b + 4a^T ab^T b\,. \end{aligned}$$

Expanding $R_b R_a$ in the same way and comparing the results we see that we have $R_a R_b = R_b R_a$ if and only if $ab^T = \langle a, b\rangle = 0$, i.e. if and only if the vectors a and b are *orthogonal.* Thus, two different reflections commute if and only if they correspond to *orthogonal hyperplanes* .

Let $\{R_1, R_2, \ldots, R_k\}$ be a set of reflections. The group $\mathcal{G}$ in $\mathcal{O}_n$ *generated* by $\{R_1, R_2, \ldots, R_k\}$ will be denoted $< R_1, R_2, \ldots, R_k >$. By definition it is the smallest group in $\mathcal{O}_n$ containing all of these transformations. Let $\{a_1, a_2, \ldots, a_k\}$ be a corresponding set of roots. We say that $a \in \Omega_n$ *belongs* to $\mathcal{G}$ if the reflection R_a is an element in $\mathcal{G}$. Clearly all of the roots $\{a_1, a_2, \ldots, a_k\}$ belong to $\mathcal{G}$, but they are certainly not the only roots belonging to $\mathcal{G}$: for instance the negative $-a$ of any root $a \in \{a_1, a_2, \ldots, a_k\}$ also belongs to $\mathcal{G}$. In addition there might be many further reflections in $\mathcal{G}$ except the generating ones. Still, for any finite reflection group $\mathcal{G}$ it is clear that the set Φ of all roots belonging to $\mathcal{G}$ is a finite subset in Ω_n. It is also clear that Φ has the following properties:

$$\begin{aligned} &(i) \quad a \in \Phi \Rightarrow -a \in \Phi, \\ &(ii) \quad a, b \in \Phi \Rightarrow bR_a \in \Phi\,. \end{aligned}$$

The first property is a direct consequence of the simple fact we observed above, namely that a and $-a$ generate the same reflection. The second property is an immediate consequence of the general formula

$$S^{-1} R_b S = S^T (I - 2b^T b) S = I - 2S^T b^T bS = R_{bS}\,,$$

which holds for any reflection R_b and any $S \in \mathcal{O}_n$: choosing $S = R_a$ and $R_b \in \mathcal{G}$ we obtain $R_{bS} = S^{-1}R_bS \in \mathcal{G}$, implying $bR_a \in \Phi$.

We call any set $\Phi \subseteq \Omega_n$ satisfying the properties (*i*) and (*ii*) above a *root system*. From the remarks motivating the definition follows that the set of roots belonging to any reflection group is a root system. It is also clear that any reflection group is fully determined by its root system. Actually, the root system provides a very convenient characterization of the reflection group whenever one exists. However, a given root system Φ—i.e. a subset in Ω_n satisfying properties (i) and (ii) above—need not necessarily generate a finite reflection group.

The geometric description is not accidental: the concept of *reflection group* is really a *geometric* one; the group structure itself is *not* a full characterization of a reflection group. For an easy illustration let us consider again the three transformations R_1, R_2, R_3 considered above. It is clear that $\{I, R_1\}$, $\{I, R_2\}$, $\{I, R_3\}$ are three geometrically different representations of one and the same abstract group, namely the unique group of order 2. Only the first of them is a reflection group.

It is worthwhile to consider the case $n = 2$ in some detail. There are only two kinds of orthogonal transformations in $\mathbb{R}^2$: rotations and reflections. For $\alpha \in \mathbb{R}$ we define

$$T_\alpha \triangleq \begin{bmatrix} \cos\alpha & -\sin\alpha \\ \sin\alpha & \cos\alpha \end{bmatrix}, \quad S_\alpha \triangleq \begin{bmatrix} \cos\alpha & \sin\alpha \\ \sin\alpha & -\cos\alpha \end{bmatrix}.$$

It is easy to see that we have $\det T_\alpha = 1$ and $\det S_\alpha = -1$ and that any transformation in $\mathcal{O}_2$ must be of one of these forms. The transformation T_α performs a rotation by the angle α in positive direction; S_α performs a reflection in the line $a^\perp = \{x \in \mathbb{R}^2 : \langle x, a\rangle = 0\}$, where $a = \pm(-\sin\alpha/2, \cos\alpha/2)$. The following relations are easy consequences of the definitions:

$$T_\alpha T_\beta = T_{\alpha+\beta}, \quad S_\alpha S_\beta = T_{\alpha-\beta}.$$

The notation S_α for the reflections in $\mathbb{R}^2$ is of course just an alternative to the more general notation we already have. The relation between the two is given by the following correspondence:

$$R_a = S_\alpha \Leftrightarrow a = \pm(-\sin\alpha/2, \ \cos\alpha/2).$$

In any context we will freely use whichever notation we find most convenient.

Let us determine all possible finite reflection groups in $\mathcal{O}_2$. Choosing

$$a = (-\sin\alpha/2, \cos\alpha/2), \quad b = (-\sin\beta/2, \cos\beta/2)$$

we obtain

$$\begin{aligned} bR_a &= (-\sin\beta/2, \cos\beta/2)\begin{bmatrix} \cos\alpha & \sin\alpha \\ \sin\alpha & -\cos\alpha \end{bmatrix} \\ &= (\sin(\alpha-\beta/2), -\cos(\alpha-\beta/2)). \end{aligned} \tag{8.7.1}$$

Denote by Φ^* the set of angles corresponding to the vectors in a given root system Φ in Ω_2. From (8.7.1) we see that the two conditions for Φ being a root system transform into the following single condition for the angles:

$$\alpha, \beta \in \Phi^* \ \Rightarrow\ 2\alpha - \beta \in \Phi^*.$$

Recall that we require that Φ and thus Φ^* should be finite. Suppose Φ^* corresponds to a root system and choose positive constants $\alpha, \beta \in \Phi^*$ such that $\delta = \alpha - \beta > 0$ is minimal. Define

$$\theta_0 = \beta, \quad \theta_1 = \alpha = \delta + \beta,$$

and in general for $k \geq 1$:

$$\theta_{k+1} = 2\theta_k - \theta_{k-1}.$$

As Φ^* is finite we must have $\theta_q = \theta_0$ for some finite q and therefore Φ^* is given by

$$\Phi^* = \{k\delta + \beta : 0 \leq k < q\}.$$

This shows that any root system in Ω_2 is an evenly distributed set of points in Ω_2. It also follows that there is a reflection group corresponding to each such system. The group is of course the dihedral group $\mathcal{D}_q$ of order $2q$. The dihedral groups are the only reflection groups in $\mathcal{O}_2$.

Finding all reflection groups in $\mathcal{O}_n$ for general n is more complicated and requires some additional tools. It is clear that for a characterization of a reflection group $\mathcal{G} \in \mathcal{O}_n$ the set Φ is—in general—quite redundant. With the purpose of obtaining a more efficient description we introduce the following concept.

Definition 8.7.1 *Let $\Phi \in \Omega_n$ be a root system. We say that a subset $\Delta = \{\varphi_1, \varphi_2, \dots, \varphi_n\} \subseteq \Phi$ is* simple *if Δ is a base in $\mathbb{R}^n$ such that any $a \in \Phi$ has an expansion*

$$a = \sum_{i=1}^{n} a_i \varphi_i$$

with all non-zero coefficients a_i having the same sign (either all positive or all negative).

It turns out that if Φ is a root system for the group $\mathcal{G}$ and if Δ is a simple subsystem in Φ then the set $\Delta = \{\varphi_1, \varphi_2, \dots, \varphi_n\}$ determines the group $\mathcal{G}$. More precisely, $\mathcal{G}$ is *generated* by the reflections in Δ. We indicate this by writing $\mathcal{G} =< R_a : a \in \Delta >$ (Humphreys [133], Theorem 1.5, p. 11).

It is clear that a given set of reflections cannot generate a reflection with a root outside the linear span of the given roots. Thus, in a sense a simple set provides a most efficient description of a reflection group.

We say that two different simple systems Δ and Δ' are *geometrically equivalent* if one is obtained from the other by an orthogonal transformation. It turns out that any finite reflection group contains a simple subsystem within its set of roots and that all of the simple subsystems are geometrically equivalent. Moreover, all inner products between simple roots from a fixed simple system are non-positive. We summarize these facts in the following theorem.

Theorem 8.7.1 *Any root system Φ contains a simple subsystem Δ. Moreover, any pair of simple subsystems of Φ are geometrically equivalent and any simple subsystem $\Delta = \{\varphi_1, \varphi_2, \dots, \varphi_n\}$ is such that $\langle \varphi_i, \varphi_j \rangle \leq 0$ for $i \neq j$.*

Proof: See Humphreys [133], Chapter 1:3–1:4, pp. 8-10 or Grove–Benson [121], Chapter 4, pp. 34-52. □

It follows from the above that any finite reflection group is fully characterized by a set $\Delta = \{\varphi_1, \varphi_2, \ldots, \varphi_n\}$ of linearly independent unit norm vectors such that $\langle\varphi_i, \varphi_j\rangle \leq 0$ for $i \neq j$. However, it is *not* true that any such system defines a finite reflection group. We proceed to determine the additional conditions that must be fulfilled in order for a finite reflection group to be generated.

For any subset $\Delta = \{\varphi_1, \varphi_2, \ldots, \varphi_n\}$ of linearly independent vectors in Ω_n we define a *Gram matrix* G as follows:

$$G = [G_{i,j}], \quad G_{ij} = \langle\varphi_i, \varphi_j\rangle.$$

It is clear that any such Gram matrix is *positive definite*: for any nonzero $x = (x_1, x_2, \ldots, x_n) \in \mathbb{R}^n$ we have

$$x\,G\,x^{\perp} = \sum_{i,j} G_{ij}\,x_i x_j = \|\sum_{i=1}^{n} x_i\,\varphi_i\|^2 > 0.$$

By definition all diagonal elements in G equal 1. By Theorem 8.7.1, if Δ is a simple set in a root system Φ then all the off-diagonal elements are non-positive. In addition we now notice that if G is obtained from a *finite* reflection group then the elements G_{ij} of the matrix G must be of the specific form

$$G_{ij} = -\cos(\pi/p_{ij}); \quad i,j = 1,2,\ldots,n,$$

for some integers $p_{ij} \geq 2$. This follows from the fact that each pair (R_i, R_j) of reflections generates a dihedral subgroup $\mathcal{D}_q$, as we have noticed above. The integer $p_{ij} = q$ simply indicates the order of the subgroup $D_q =< R_i, R_j >$.

It turns out that the above necessary conditions are actually also sufficient. More precisely, to any positive definite matrix G with elements G_{ij} of the form $G_{ij} = -\cos(\pi/p_{ij})$ corresponds a finite reflection group. This remarkable fact was proved by Coxeter [72]. The proof consists of first determining all possible Gram matrices with the above properties and then explicitly demonstrating for each one of them the existence of a corresponding reflection group.

For carrying out this process and for the presentation of the result a final concept is very useful: the *Coxeter graph.* A Coxeter graph is a weighted graph $\Gamma = \{V, B, w\}$ with vertices $V = \{1, 2, \ldots, n\}$, edges $B = \{b_1, b_2, \ldots, b_t\}$ and an integer weight $w(b) \geq 3$ associated with each one of the edges $b \in B$. Let $\mathcal{G}$ be a finite reflection group with simple root system $\Delta = \{\varphi_1, \varphi_2, \ldots, \varphi_n\}$. We associate a Coxeter graph $\Gamma_\mathcal{G}$ to the group $\mathcal{G}$ by assigning a vertex i to each root φ_i and an edge b_k to any pair (i,j) for which $p_{ij} \geq 3$, where the quantities p_{ij} are defined by the equalities $\langle \varphi_i, \varphi_j \rangle = -\cos(\pi/p_{ij})$. The weights are defined so that $w(b_k) = p_{ij}$ whenever b_k is an edge connecting the vertices i and j. In this way a Coxeter graph $\Gamma_\mathcal{G}$ is associated with each finite reflection group $\mathcal{G}$. The fundamental question is: to which Coxeter graphs can we associate finite reflection groups?

We say that two *Coxeter graphs* are *equivalent* if one is obtained from the other by a permutation of the vertices and a corresponding relabeling of the edges and their weights. We say that two finite reflection groups are *geometrically equivalent* if one is obtained from the other by an orthogonal transformation of its roots. The following result reduces the study of finite reflection groups to the study of the corresponding Coxeter graphs.

Theorem 8.7.2 *Two finite reflection groups $\mathcal{G}_1$ and $\mathcal{G}_2$ are geometrically equivalent if and only if they have equivalent Coxeter graphs.*

Proof: See Humphreys [133], Chapter 2.1, pp. 29-30. □

Let $\Delta = \{\varphi_1, \varphi_2, \ldots, \varphi_n\}$ be a simple root system for the reflection group $\mathcal{G}$. Suppose there are two subsets $\Delta_1, \Delta_2 \subseteq \Delta$ such that the conditions $\varphi_i \in \Delta_1$ and $\varphi_j \in \Delta_2$ imply $\langle \varphi_i, \varphi_j \rangle = 0$. We say that the subsets Δ_1, Δ_2 are *orthogonal.* It is an immediate consequence of the definition that the subgraphs Γ_1 and Γ_2 generated by Δ_1 and Δ_2 respectively are not connected to each other. Suppose $\Delta = \Delta_1 \cup \Delta_2$. It is clear that any transformation $T \in \mathcal{G}$ can be represented in the form

$$T = \begin{bmatrix} T_1 & 0 \\ 0 & T_2 \end{bmatrix},$$

with T_1 operating on the subspace W_1 generated by Δ_1 and with T_2 operating on the subspace W_2 generated by Δ_2. It follows that

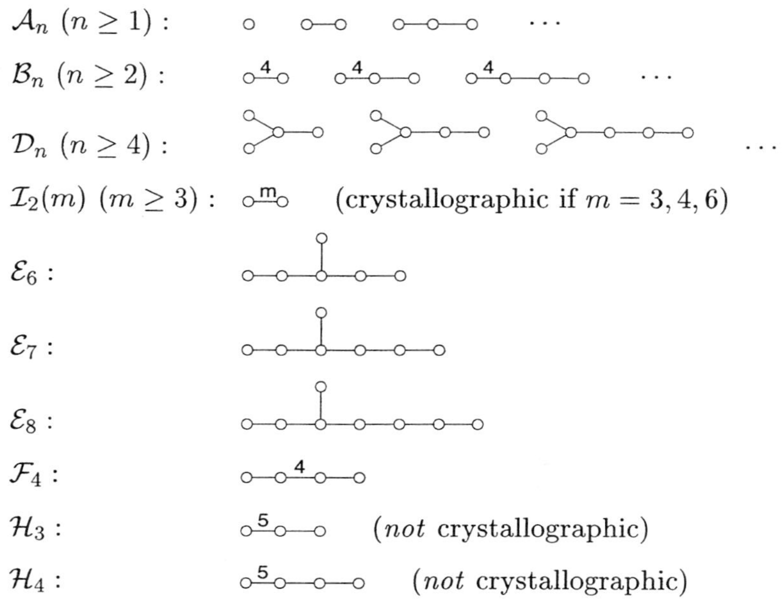

Figure 8.1: Coxeter graphs corresponding to all existing finite reflection groups. Labels indicate weights. Unlabeled edges are understood to have weight $w(b) = 3$.

the graph Γ corresponding to a finite reflection group $\mathcal{G}$ is *connected* if and only if $\mathcal{G}$ is *irreducible*. It also follows that there is always one connected subgraph in $\mathcal{G}$ for each irreducible component in $\mathcal{G}$. For the purpose of classifying the finite reflection groups it is therefore enough to determine the possible *irreducible* such groups. By the above observations this amounts to finding all possible connected Coxeter graphs corresponding to finite reflection groups. This approach is due to Coxeter [72].

Consider Figure 8.1. It shows four infinite classes of Coxeter graphs along with six additional individual graphs. We follow the notation used in Humphreys [133], Chapter 2.4, p. 32. It turns out that each one of these graphs corresponds to an irreducible finite reflection group, and that actually all such groups are generated by these graphs. The key tool for proving this result is the following lemma.

Lemma 8.7.1 *Let A be a symmetric matrix of order n, let A_1 be the submatrix obtained by deleting the first row and the first column in A and let A_2 be the submatrix obtained by deleting the first two rows and columns. Suppose in addition that the first row in A has the form $a = (1, \alpha, 0, \ldots, 0)$ (i.e. that only the first two components are non-zero). Then the following identity holds:*

$$\det A = \det A_1 - \alpha^2 \det A_2.$$

Proof: The proof follows by expanding $\det A$ along its first row. □

Theorem 8.7.3 (Coxeter, [72]) *A Coxeter graph Γ corresponds to an irreducible finite reflection group if and only if it is one of the graphs indicated in Figure 8.1.*

Proof: The proof is long and consists of considering separately a number of individual cases. For a complete proof see Humphrey [133], Chapter 2.7, pp. 36-38 and Chapters 5-6, Grove-Benson [121], Chapters 5.1–5.3, pp. 53-76, or see the original text by Coxeter [72]. Essentially the following steps are involved.

First it must be shown that each one of the graphs in Figure 8.1 has a positive definite Gram matrix. This is done by a repeated application of Lemma 8.7.1. By induction using this lemma one first shows that the graph $\mathcal{A}_n$ has determinant $\det \mathcal{A}_n = (n+1)2^{-n} > 0$ (for simplicity we use the same notation for the graph and for its Gram matrix). Then one observes that the k-th principal minor (using an appropriate basis) is given by $\det \mathcal{A}_k$, $k = 1, 2, \ldots, n$. From this follows that all $\mathcal{A}_n$ are positive definite. By minor modifications of this argument it can then be shown that also all the other graphs in Figure 8.1 have positive definite Gram matrices.

The next step is to show that there are no other Coxeter graphs with positive definite Gram matrices. This is done by a careful investigation of all possible cases, again employing Lemma 8.7.1 to show that any graph of a form different from those indicated in Figure 8.1 has a Gram matrix with zero determinant.

$\mathcal{G}$	$\mathcal{A}_n$ $(n \geq 1)$	$\mathcal{B}_n$ $(n \geq 2)$	$\mathcal{D}_n$ $(n \geq 4)$	$\mathcal{I}_2(m)$ $(m \geq 3)$
$\lvert\mathcal{G}\rvert$	$(n+1)!$	$2^n \cdot n!$	$2^{n-1} \cdot n!$	$2m$
$\lvert\Phi\rvert$	$n(n+1)$	$2 \cdot n^2$	$2n(n-1)$	$2m$

$\mathcal{G}$	$\mathcal{E}_6$	$\mathcal{E}_7$	$\mathcal{E}_8$	$\mathcal{F}_4$
$\lvert\mathcal{G}\rvert$	$2^7 \cdot 3^4 \cdot 5$	$2^{10} \cdot 3^4 \cdot 5 \cdot 7$	$2^{14} \cdot 3^5 \cdot 5^2 \cdot 7$	$2^7 \cdot 3^2$
$\lvert\Phi\rvert$	$72 = 2^3 \cdot 3^2$	$126 = 2 \cdot 3^3 \cdot 7$	$240 = 2^4 \cdot 3 \cdot 5$	$48 = 2^4 \cdot 3$

$\mathcal{G}$	$\mathcal{H}_3$	$\mathcal{H}_4$
$\lvert\mathcal{G}\rvert$	$2^3 \cdot 3 \cdot 5$	$2^6 \cdot 3^2 \cdot 5^2$
$\lvert\Phi\rvert$	$30 = 2 \cdot 3 \cdot 5$	$120 = 2^3 \cdot 3 \cdot 5$

Table 8.1: Sizes of all finite reflection groups $\mathcal{G}$ and their corresponding root systems Φ.

Finally it must be shown that to each one of the graphs in Figure 8.1 there actually corresponds a finite reflection group. This is done by giving in each case an explicit construction. □

In Table 8.1 we have listed the sizes of all the irreducible finite reflection groups along with the sizes of their corresponding root systems. Even with an explicit exhibition of the root system Φ the determination of the group orders is not always easy. We refer again to the specialized texts: Humphrey [133], Chapter 2.11, pp. 43–44 and Grove-Benson [121], Chapter 5.4, pp. 77–80. Notice that in general the order of the group is considerably larger than the size of the root system.

In Figure 8.2 we have indicated possible simple root systems for all the irreducible finite reflection groups. The roots are indicated as rows in various matrices. It is straightforward to check that the matrices in Figure 8.2 and the graphs in Figure 8.1 generate the same Gram matrices (up to equivalence).

We now turn to a discussion of the codes.

$$\frac{1}{\sqrt{2}}\begin{bmatrix} \sqrt{2} & 0 \\ -1 & 1 \end{bmatrix} \updownarrow \mathcal{A}_1 \quad \updownarrow \mathcal{B}_2 \qquad \frac{1}{\sqrt{2}}\begin{bmatrix} \sqrt{2} & 0 & 0 \\ -1 & 1 & 0 \\ 0 & -1 & 1 \end{bmatrix} \updownarrow \mathcal{A}_2 \quad \updownarrow \mathcal{B}_3$$

$$\frac{1}{\sqrt{2}}\begin{bmatrix} \sqrt{2} & 0 & 0 & 0 \\ -1 & 1 & 0 & 0 \\ 0 & -1 & 1 & 0 \\ 0 & 0 & -1 & 1 \end{bmatrix} \updownarrow \mathcal{A}_3 \quad \updownarrow \mathcal{B}_4 \qquad \mathcal{I}_2(m) \begin{bmatrix} 1 & 0 \\ -\cos\frac{\pi}{m} & \sin\frac{\pi}{m} \end{bmatrix} \quad m = 3, 4, \cdots$$

$$\frac{1}{\sqrt{2}}\begin{bmatrix} 1 & 1 & 0 & 0 \\ -1 & 1 & 0 & 0 \\ 0 & -1 & 1 & 0 \\ 0 & 0 & -1 & 1 \end{bmatrix} \updownarrow \mathcal{A}_3 \quad \updownarrow \mathcal{D}_4 \qquad \frac{1}{\sqrt{2}}\begin{bmatrix} 1 & 1 & 0 & 0 & 0 \\ -1 & 1 & 0 & 0 & 0 \\ 0 & -1 & 1 & 0 & 0 \\ 0 & 0 & -1 & 1 & 0 \\ 0 & 0 & 0 & -1 & 1 \end{bmatrix} \updownarrow \mathcal{A}_4 \quad \updownarrow \mathcal{D}_5$$

$$\frac{1}{\sqrt{2}}\begin{bmatrix} \frac{1}{2} & \frac{1}{2} & \frac{1}{2} & -\frac{1}{2} & -\frac{1}{2} & -\frac{1}{2} & -\frac{1}{2} & -\frac{1}{2} \\ 1 & -1 & 0 & 0 & 0 & 0 & 0 & 0 \\ 0 & 1 & -1 & 0 & 0 & 0 & 0 & 0 \\ 0 & 0 & 1 & -1 & 0 & 0 & 0 & 0 \\ 0 & 0 & 0 & 1 & -1 & 0 & 0 & 0 \\ 0 & 0 & 0 & 0 & 1 & -1 & 0 & 0 \\ 0 & 0 & 0 & 0 & 0 & 1 & -1 & 0 \\ 0 & 0 & 0 & 0 & 0 & 0 & 1 & -1 \end{bmatrix} \updownarrow \mathcal{E}_6 \quad \updownarrow \mathcal{E}_7 \quad \updownarrow \mathcal{E}_8$$

$$\frac{1}{2}\begin{bmatrix} -1 & -1 & -1 & -1 \\ 2 & 0 & 0 & 0 \\ -\sqrt{2} & \sqrt{2} & 0 & 0 \\ 0 & -\sqrt{2} & \sqrt{2} & 0 \end{bmatrix} \mathcal{F}_4 \qquad \frac{1}{2}\begin{bmatrix} 1 & \sigma & \tau & 0 \\ -1 & \sigma & -\tau & 0 \\ \sigma & \tau & 1 & 0 \\ \tau & 0 & -1 & -\sigma \end{bmatrix} \updownarrow \mathcal{H}_3 \quad \updownarrow \mathcal{H}_4$$

$$\sigma = 1/2 - \sqrt{5}/2 \qquad \tau = 1/2 + \sqrt{5}/2$$

Figure 8.2: Possible root systems for all irreducible finite reflection groups.

8.8 Codes from finite reflection groups

It is easy to see that a group code obtained from a reducible finite reflection group is the direct sum (as defined in Chapter 1.7) of group codes generated by its irreducible components. For this reason we will discuss only codes obtained from *irreducible* groups. As all such groups were obtained in the previous chapter all that remains is to determine the dominating codes for each one of them. This is the initial vector problem. It turns out that for the special case of finite reflection groups this problem can be fully solved.

Let $\mathcal{G}$ be a finite group in $\mathcal{O}_n$ (not necessarily a reflection group). By a *fundamental domain* we understand an *open* subset $D \subseteq \mathbb{R}^n$ with the following properties:

$$(i) \quad T_1, T_2 \in \mathcal{G},\ T_1 \neq T_2 \ \Rightarrow\ DT_1 \bigcap DT_2 = \emptyset,$$

$$(ii) \quad \bigcup_{T \in \mathcal{G}} \overline{D}T = \mathbb{R}^n$$

where $\overline{D}$ denotes the closure of D and where DT is the set in $\mathbb{R}^n$ into which D is transformed by $T \in \mathcal{G}$. In words: the sets DT, $T \in \mathcal{G}$, are disjoint and the union of their closures exhausts the space $\mathbb{R}^n$ in which $\mathcal{G}$ operates. Notice that condition (i) implies that all points in D belong to different orbits.

Let $\mathcal{G}$ be a finite reflection group in $\mathcal{O}_n$ and let Δ be a simple set of roots such that $\mathcal{G}$ is generated by the reflections corresponding to the roots in Δ. A fundamental domain D_Δ is given by the following definition (Humphreys [133], Lemma 1.12, p. 22):

$$D_\Delta \triangleq \{x \in \mathbb{R}^n : \langle a, x\rangle > 0, \quad a \in \Delta\}. \tag{8.8.1}$$

Notice that D_Δ is a convex cone ($x \in D_\Delta$ implies $\lambda x \in D_\Delta$ for any positive $\lambda \in \mathbb{R}$) generated by the hyperplanes $a^\perp$ corresponding to the roots a in Δ. All roots a in Δ are pointing *into* the domain D_Δ.

Now consider the code $\mathcal{X} = x_0\, \mathcal{G}$ generated by $\mathcal{G}$ and the initial vector x_0. As by definition all points in D_Δ correspond to different orbits it is clear that as long as we restrict the initial vector $x_0 \in \Omega_n$ to be inside some fundamental domain it is enough to let x_0 vary over the specific fundamental domain D_Δ. It is also clear that for each defining hyperplane $a^\perp$ there is a codeword $x_0 R_a$ and that the distance from x_0 to $x_0 R_a$ is exactly twice the distance from x_0 to the hyperplane $a^\perp$. It follows that a locally optimum position for x_0 in D_Δ—that is a position that maximizes the minimum distance in the spherical code $\mathcal{X} = x_0\mathcal{G}$—is determined by the condition that $x_0 \in \Omega_n$ should have the same distance to all the defining hyperplanes $a^\perp$, $a \in \Delta$. We can formulate this by the requirement $\langle a, x_0\rangle = \lambda$, $a \in \Delta$. It is easy to see that there will be exactly one codeword $x = x_0 T$ in each fundamental domain $D_\Delta T$ and that the fundamental domains of the group $\mathcal{G}$ coincide with the Voronoi regions of the code $\mathcal{X} = x_0\mathcal{G}$. One immediate consequence is that the size $M = |\mathcal{X}|$ of the code equals, in this case, the size $|\mathcal{G}|$ of the group: $M = |\mathcal{G}|$.

It remains to determine the optimal positions in those cases where the initial vector is located in one or several of the defining hyperplanes $a^\perp$, i.e. when the initial vector is located on the borders between the fundamental domains. We refer to Example 8.2.1 for a simple illustration of the ideas involved.

It is clear that if x_0 lays in all of the hyperplanes $a^\perp$ corresponding to the roots $a \in \Delta_P$ in $\mathcal{G}$ then x_0 is necessarily invariant with respect to any of the transformations T belonging to the subgroup $\mathcal{H} = < R_a : a \in \Delta_P >$ generated by these roots. In fact, it is also true that there are no other transformations in $\mathcal{G}$ under which x_0 is invariant (Humphreys [133], Theorem 1.12, p. 22). This means that $\mathcal{H}$ is the stabilizer for x_0 in $\mathcal{G}$ and that the size of the code is

$$M = \frac{|\mathcal{G}|}{|\mathcal{H}|}$$

in this case (cf. Section 8.2). It turns out that the locally optimum position for x_0 is determined by a simple modification of the previous condition. More precisely, let $\Delta_P \subseteq \Delta$ be a set of roots and consider the problem of finding an optimal location for $x_0 \in \Omega_n$ in the region $\cup_{a \in \Delta_P} a^\perp$. Denote by Δ_A the complement $\Delta \setminus \Delta_P$ of Δ_P in Δ. We will refer to Δ_A as the set of *active* roots while those in Δ_P will be referred to as *passive* roots. It turns out that the optimality criterion for $x_0 \in \Omega_n$ in the region $\cup_{a\in\Delta_P} a^\perp$ is given by

$$\langle a, x_0\rangle = \begin{cases} \lambda, & if \quad a \in \Delta_A, \\ 0, & if \quad a \in \Delta_P. \end{cases} \tag{8.8.2}$$

Notice that we have

$$D_\Delta T = D_{\Delta T} = \{x \in \mathbb{R}^n : \langle aT, x\rangle > 0, \quad a \in \Delta\}.$$

Denote by $S(x_0, \lambda)$ the open sphere of radius λ around x_0. The following result is crucial.

Theorem 8.8.1 (Lahtonen–Mittelholzer [171]) *Let $\mathcal{G}$ be a finite reflection group in $\mathcal{O}_n$, let Δ be a simple set of roots belonging to $\mathcal{G}$ and let D_Δ be the corresponding fundamental domain (as defined in (8.8.1)). Let Δ_A and Δ_P be complementary disjoint subsets in Δ and let x_0 satisfy the condition (8.8.2). Finally, let $\mathcal{H} =< R_a : a \in \Delta_P >$ be the stabilizer of x_0 in $\mathcal{G}$. Then the following inclusion holds:*

$$S(x_0, \lambda) \subseteq \bigcup_{T\in\mathcal{H}} \overline{D}_{\Delta T}$$

(the overbar indicates closure).

Proof: We must show that $\|x - x_0\| < \lambda$ implies that there is some $T \in \mathcal{H}$ such that $\langle aT, x\rangle \geq 0$ for all $a \in \Delta$. Consider first the case $a \in \Delta_P$ and define

$$\Delta_N = \{a \in \Delta_P : \langle a, x\rangle < 0\}; \quad T = \prod_{a\in\Delta_N} R_a.$$

It is clear that T belongs to the subgroup $\mathcal{H}$. By Humphrey [133] (Proposition 1.4, p. 10) follows that we have $\langle aT, x\rangle \geq 0, \ a \in \Delta_P$. Now let $a \in \Delta_A$. We have

$$\langle xT, a\rangle = \langle xT - x_0, a\rangle + \langle x_0, a\rangle \geq -\|x - x_0\|\|a\| + \langle x_0, a\rangle,$$

where we have used Schwarz' inequality and the fact that x_0 is invariant under $\mathcal{H}$. Now we have $\|x - x_0\| < \lambda$ by assumption and $\langle x_0, a\rangle = \lambda$ by the definition of x_0. The vector $a \in \Delta_A$ has unit norm. It follows that we have $\langle aT, x\rangle \geq 0$ for all $a \in \Delta$. □

It follows from the above discussion that the Voronoi region for x_0 is given by the interior of the union $\bigcup_{T\in\mathcal{H}} \overline{D}_{\Delta T}$. It remains to demonstrate that the condition (8.8.2) can always be satisfied and that the parameter λ is uniquely determined by this condition.

Let the simple set Δ be given by the vectors $\{\varphi_1, \varphi_2, \dots, \varphi_n\}$. Denote by N_A and N_P respectively the sets of indices defining the subsets Δ_A and Δ_P in Δ. For generality let us assume that the basis vectors $\varphi_i = \{\varphi_{i1}, \varphi_{i2}, \dots, \varphi_{iN}\}$ are described with respect to some arbitrary basis, possibly using an excessive number of coordinates. We clearly have $N \geq n$, where n denotes the dimension of the space spanned by $\{\varphi_1, \varphi_2, \dots, \varphi_n\}$.

We collect the vectors $\{\varphi_1, \varphi_2, \dots, \varphi_n\}$ in an $n \times N$ matrix R, having the vectors $\varphi_i = \{\varphi_{i1}, \varphi_{i2}, \dots, \varphi_{iN}\}$ as rows. The Gram matrix G is given by $G = RR^T$. By assumption G is a nonsingular $n \times n$ symmetric matrix.

Next let us define a vector $e_A = (e_1, e_2, \dots, e_n)$ by the following condition:

$$e_i = \begin{cases} 1, & if\ i \in N_A, \\ 0, & if\ i \in N_P, \end{cases}$$

$$i = 1, 2, \dots, n.$$

It follows that the optimality criterion (8.8.2) for x_0 can be written as

$$x_0 R^T = \lambda e_A.$$

It is required that the initial vector x_0 should be in the space spanned by $\{\varphi_1, \varphi_2, \dots, \varphi_n\}$. That means that x_0 should have the form $x_0 = yR$ for some vector $y = (y_1, y_2, \dots, y_n) \in \mathbb{R}^n$. Thus we have

$$x_0\, R^T = yRR^T = yG = \lambda e_A,$$

and therefore $y = \lambda e_A G^{-1}$. We have also the requirement that x_0 is a vector of unit Euclidean norm. That gives us

$$\begin{aligned} 1 = x_0 x_0^T &= yRR^T y^T \\ &= \lambda^2 e_A G^{-1} G G^{-1} e_A^T = \lambda^2 e_A G^{-1} e_A^T, \end{aligned}$$

where we used the fact that G^{-1} is also symmetric (recall that G is symmetric). As $\lambda = \langle a, x_0 \rangle$ denotes the Euclidean distance from x_0

to the hyperplane $a^\perp$ and as for each hyperplane $a^\perp$, where $a \in \Delta_A$, there is a reflection point $x_0 R_a$ on distance 2λ from x_0, it is clear that the squared minimum distance ρ satisfies $\rho \leq 4\lambda^2$. On the other hand, combining Theorem 8.8.1 above with Lemma 1.6.1 of Chapter 1.6 we see that we may conclude $\rho \geq 4\lambda^2$. It follows that we actually have $\rho = 4\lambda^2$. These observations imply that x_0 and λ are both uniquely determined by the condition (8.8.2) and that a solution always exists. Moreover, we have the following simple explicit formula for the squared minimum distance:

$$\rho = 4\lambda^2 = \frac{4}{e_A G^{-1} e_A^T} = \frac{4}{\sum_{i,j \in N_A} (G^{-1})_{i,j}}, \tag{8.8.3}$$

where $(G^{-1})_{i,j}$ denotes (i,j)-th element of the matrix G^{-1}. By this formula we can easily compute the minimum squared distance for any optimized code. The code and its parameters depend on two things: the group $\mathcal{G}$ and the set Δ_A of active roots. The Gram matrix G is immediately obtained from the Coxeter graph. All that we need to do is to invert this matrix and then sum over all the indices $i, j \in N_A$ corresponding to the active roots.

The size M is given by the formula $M = |\mathcal{G}|/|\mathcal{H}|$. The size of the group $\mathcal{G}$ can be obtained from Table 8.1. In order to find the size of the stabilizer $\mathcal{H} = < R_a : a \in \Delta_A >$ we need only observe that $\mathcal{H}$ is a direct product of irreducible Coxeter groups, say

$$\mathcal{H} = \mathcal{G}_1 \times \mathcal{G}_2 \times \ldots \times \mathcal{G}_k .$$

We therefore obtain

$$|\mathcal{H}| = |\mathcal{G}_1||\mathcal{G}_2| \ldots |\mathcal{G}_k|,$$

where each magnitude $|\mathcal{G}_i|$ can again be found from Table 8.1. We will illustrate the procedure by the aid of several examples, but first we summarize our results in the following theorem.

Theorem 8.8.2 (Lahtonen-Mittelholzer [171]) *Let $\mathcal{G}$ be an irreducible finite reflection group, let Δ be a set of simple roots and let*

G be the Gram matrix. For any subset $\Delta_A \subseteq \Delta$ *there is a group code* $\mathcal{X} = x_0\mathcal{G}$ *with an optimal initial vector*

$$x_0 \in \bigcap_{a \in \Delta_P} a^\perp, \quad \Delta_P = \Delta \setminus \Delta_A,$$

uniquely specified by the condition

$$\langle a, x_0 \rangle = \begin{cases} \lambda, & if \ \ i \in \Delta_A, \\ 0, & if \ \ i \in \Delta_P. \end{cases}$$

The stabilizer $\mathcal{H}$ *for* x_0 *in* $\mathcal{G}$ *is generated by the reflections* R_a *corresponding to the roots in* $\Delta_P : \mathcal{H} = < R_a : a \in \Delta_P >$. *The parameters of the code* $\mathcal{X} = x_0\mathcal{G}$ *are as follows:*

$$n = \dim(\Delta),$$

$$\rho = 4\lambda^2 = \frac{4}{e_A G^{-1} e_A^T} = \frac{4}{\sum\limits_{i,j \in N_A} (G^{-1})_{i,j}},$$

$$M = \frac{|\mathcal{G}|}{|\mathcal{H}|},$$

where e_A *is an indicator vector for the set* $\Delta_A \subseteq \Delta$ *and where* N_A *indicates the corresponding vertices in the Coxeter graph for* $\mathcal{G}$.

Remark 8.8.1:: Notice that in order to determine the parameters (n, ρ, M) of the optimal code

$$\mathcal{X} = x_0\mathcal{G} \ \text{ with } \ x_0 \in \bigcap_{a \in \Delta_P} a^\perp, \quad \Delta_P = \Delta \setminus \Delta_A,$$

it is *not* necessary to explicitly compute the optimal initial vector x_0.

8.9 Examples

We illustrate the above results with several examples. In Tables 8.2–8.4 we have indicated all the optimized codes obtained from the

Δ_A	Δ_P	$\mathcal{H}$	$\|\mathcal{H}\|$	M	ρ
$1,2,3$	$\emptyset$	1	1	120	0.0692
$2,3$	1	$\mathcal{A}_1$	2	60	0.1629
$1,3$	2	$\mathcal{A}_1$	2	60	0.2006
$1,2$	3	$\mathcal{A}_1$	2	60	0.1134
3	$1,2$	$\mathcal{I}_2(5)$	10	12	1.1056
2	$1,3$	$\mathcal{A}_1$	4	30	0.3820
1	$2,3$	$\mathcal{A}_2$	6	20	0.5093

Table 8.2: Optimized codes from $\mathcal{H}_3$ of order $|\mathcal{H}_3| = 2^3 \cdot 3 \cdot 5 = 120$. Notice that for the trivial group containing just the unity element we use "1" to indicate both the group and its order.

groups $\mathcal{H}_3$, $\mathcal{H}_4$, and $\mathcal{F}_4$. For convenience we have introduced an arbitrary enumeration of the vertices in the Coxeter graphs. The first two columns indicate the sets Δ_A and Δ_P, the third and forth columns indicate the stabilizer $\mathcal{H}$ and its order. Notice that any subgraph of a Coxeter graph is also a Coxeter graph. Therefore all the numbers needed for calculating the sizes $M = |\mathcal{G}|/|\mathcal{H}|$ of the various codes can be found in Table 8.1. The squared minimum distance ρ has in each case been calculated from formula (8.8.3).

We recognize several extremely good codes in these tables. In particular, in Table 8.2 we find the icosahedron (an optimal code) and in Table 8.3 we find the optimal code with parameters $(n, \rho, M) = (4, \sigma^2, 120)$, where $\sigma = (1 - \sqrt{5})/2$, corresponding to $\rho \approx 0.38196$. In Table 8.5 we have indicated four additional examples. Each one of them generates an optimal code. We have indicated the Coxeter graph for $\mathcal{E}_8$ along with an enumeration of the vertices. It is clear that $\mathcal{E}_7$ is a subgraph in $\mathcal{E}_8$. Choosing $\Delta_P = \mathcal{E}_7$ we obtain a code of size

$$M = \frac{|\mathcal{E}_8|}{|\mathcal{E}_7|} = 2^4 \cdot 3 \cdot 5 = 240.$$

If the root matrix is chosen as suggested in Figure 8.2 we obtain an optimal initial vector x_0 which, apart from a normalization, is given by $(1^7, -1)$. This is all indicated in the figure. Indicated is

Δ_A	Δ_P	$\mathcal{H}$	$\lvert\mathcal{H}\rvert$	M	ρ
$1,2,3,4$	$\emptyset$	1	1	14400	0.0061
$2,3,4$	1	$\mathcal{A}_1$	2	7200	0.0120
$1,3,4$	2	$\mathcal{A}_1$	2	7200	0.0146
$1,2,4$	3	$\mathcal{A}_1$	2	7200	0.0105
$1,2,3$	4	$\mathcal{A}_1$	2	7200	0.0079
$3,4$	$1,2$	$\mathcal{I}_2(5)$	10	1440	0.0464
$2,4$	$1,3$	$\mathcal{A}_1^2$	4	3600	0.0271
$2,3$	$1,4$	$\mathcal{A}_1^2$	4	3600	0.0174
$1,4$	$2,3$	$\mathcal{A}_2$	6	2400	0.0365
$1,3$	$2,4$	$\mathcal{A}_1^2$	4	3600	0.0220
$1,2$	$3,4$	$\mathcal{A}_1^2$	4	2400	0.0148
4	$1,2,3$	$\mathcal{H}_3$	120	120	0.3820
3	$1,2,4$	$\mathcal{I}_2(5) \times \mathcal{A}_1$	20	720	0.1056
2	$1,3,4$	$\mathcal{A}_1 \times \mathcal{A}_2$	12	1200	0.0486
1	$2,3,4,$	$\mathcal{A}_3$	24	600	0.0729

Table 8.3: Optimized codes from $\mathcal{H}_4$ of order $|\mathcal{H}_4| = 2^6 \cdot 3^2 \cdot 5^2 = 14400$.

Δ_A	Δ_P	$\mathcal{H}$	$\lvert\mathcal{H}\rvert$	M	ρ
$1,2,3,4$	$\emptyset$	1	1	1152	0.0374
$2,3,4$	1	$\mathcal{A}_1$	2	576	0.0541
$1,3,4$	2	$\mathcal{A}_1$	2	576	0.0817
$3,4$	12	$\mathcal{A}_2$	6	192	0.1429
$2,4$	13	$\mathcal{A}_1^2$	4	288	0.1464
$2,3$	$1,4$	$\mathcal{A}_1^2$	4	288	0.0858
$1,4$	$2,3$	$\mathcal{I}_2(4)$	8	144	0.2929
4	$1,2,3$	$\mathcal{B}_3$	48	24	1.0000
3	$1,2,4$	$\mathcal{A}_2 \times \mathcal{A}_1$	12	96	0.3333

Table 8.4: Optimized codes from $\mathcal{F}_4$ of order $|\mathcal{F}_4| = 2^7 \cdot 3^2 = 1152$.

	Δ_A	Δ_P	x_0	y	(n, ρ, M)
$\mathcal{E}_8$	8	$1-7$	$(1^7, -1)$	32465432	$(8, 1, 240)$
$\mathcal{E}_7$	7	$1-6$	$(1^6, -3, 3)$	3246543	$(7, 4/3, 56)$
$\mathcal{E}_6$	6	$1-5$	$(1^5, -5, 3^2)$	324654	$(6, 3/2, 27)$
$\mathcal{D}_5$	5	$1-4$	$(1^4, -7, 3^3)$	32465	$(5, 8/5, 16)$

Table 8.5: Optimal codes from finite reflection groups.

also a vector $y = (3, 2, 4, 6, 5, 4, 3, 2)$ (the simplified notation used in the figure should be obvious). This vector indicates how x_0 is obtained as a linear combination of the root vectors. Again apart from a normalization the optimal initial vector x_0 is given by $y\,R$, with R chosen as in Figure 8.2. The resulting code has parameters $(n, \rho, M) = (8, 1, 240)$.

Starting instead with $\mathcal{E}_7$ and choosing $\Delta_P = \mathcal{E}_6$ we obtain a code of size

$$M = \frac{|\mathcal{E}_7|}{|\mathcal{E}_6|} = 2^3 \cdot 7 = 56.$$

We obtain a valid root matrix by simply deleting the last row in the previous matrix (as indicated in Figure 8.2). The corresponding optimal initial vector x_0—and the vector y generating x_0 from the root matrix—is also indicated in the figure. Again the necessary normalization has been omitted.

The remaining codes in Table 8.5 are obtained in a similar way from successive subgraphs in $\mathcal{E}_8$. For each new code we simply omit the last vertex in the previous graph, generating the sequence $\mathcal{E}_8 - \mathcal{E}_7 - \mathcal{E}_6 - \mathcal{D}_5 - \mathcal{A}_4$. The last column of the table indicates the parameters of the resulting codes. We recall from Chapter 2 that all of these codes except the one obtained from $\mathcal{A}_4$ meet the Levenshtein bound and are therefore optimal. There are good reasons to believe that also the last code in this sequence is optimal, although so far no proof is available.

There are many further good codes obtained from $\mathcal{E}_8$ and its subgraphs. We will meet several of them again in the chapter on lattices.

The groups $\mathcal{A}_n$ and $\mathcal{B}_n$ generate the same codes as we encountered in Chapter 4 under the name of permutation codes: $\mathcal{A}_n$ corresponds to *Variant 1* while $\mathcal{B}_n$ corresponds to *Variant 2*. The groups $\mathcal{D}_n$ are similar to *Variant 2*, but with the additional constraint that the number of sign changes should always be even. Let us first consider the reflection groups $\mathcal{A}_n$.

The root system for the group $\mathcal{A}_n$ is most easily studied in a space with an excessive basis. Let V_n be the subspace in $\mathbb{R}^{n+1}$ defined as follows:

$$W_n = \{x \in \mathbb{R}^{n+1} : x_1 + x_2 + \cdots + x_{n+1} = 0\}.$$

Clearly the dimension of W_n is n. A root system for $\mathcal{A}_n$ is given by (cf Figure 8.2)

$$\varphi_i = \frac{1}{\sqrt{2}}\,(0^{i-1}, 1, -1, 0^{n-i}), \quad i = 1, 2, \ldots, n.$$

Denote by R_i the reflection corresponding to φ_i. It is easy to check that R_i sends $x = (x_1, x_2, \ldots, x_{n+1})$ into $y = (y_1, y_2, \ldots, y_{n+1})$, where for $j = 1, 2, \ldots, n$

$$y_j = \begin{cases} x_{i+1}, & if \quad j = i, \\ x_i, & if \quad j = i+1, \\ x_j, & otherwise. \end{cases}$$

In words this means that the reflections corresponding to the simple basis $\Delta = \{\varphi_1, \varphi_2, \ldots, \varphi_n\}$ perform transpositions of adjacent coordinates. It is well known that any permutation can be generated by successive such transpositions, and so it is clear that the reflections in the simple basis $\Delta = \{\varphi_1, \varphi_2, \ldots, \varphi_n\}$ generate the full symmetric group $\mathcal{S}_{n+1}$. In other words, the finite reflection group $\mathcal{G} = <R_i : i = 1, 2, \ldots, n>$ performs all possible permutations of coordinates on the initial vector x_0. This is precisely how permutation modulation *Variant 1* is defined (cf Chapter 4). If x_0 has the form

$$x_0 = (\mu_1, \mu_1, \ldots, \mu_1,\ \mu_2, \ldots, \mu_2,\ \mu_3, \ldots, \mu_K, \ldots, \mu_K)$$

with m_i successive components equal to μ_i, $i = 1, 2, \ldots, K$ it is clear that the set Δ_P of passive vertices is given by those roots φ_j

which perform a transposition within a set of constant coordinates. Equivalently the active set of indices is $N_A = \{m_1, m_1+m_2, \cdots, m_1+m_2+\ldots+m_{K-1}\}$. The stabilizer $\mathcal{H}$ is clearly a direct product of K irreducible groups of the form $\mathcal{A}_k$, where $k = m_i - 1,\ i = 1, 2, \ldots, K$. It follows that $\mathcal{H}$ has order $\prod_i m_i\,!$ and that the size of the resulting code is

$$M = \frac{(n+1)\,!}{\prod_{i=1}^{K} m_i\,!}\,,$$

in agreement with our derivation in Chapter 4. It is clear that an application of the formulas derived in the present chapter for reflection groups leads to exactly the same results as those we derived by entirely different means in Chapter 4. Notice that in our present approach the balance condition $x_1 + x_2 + \cdots + x_{n+1} = 0$ has been included from the outset.

From the root matrices indicated in Figure 8.2 it is clear that in $\mathcal{B}_n$ the reflection corresponding to the first root φ_1 performs a change of sign in the first coordinate while in $\mathcal{D}_n$ the corresponding reflection changes sign in the first two coordinates. As by the other reflections we can generate any permutation of coordinates it is clear that $\mathcal{B}_n$ is the same group as the one used in permutation modulation *Variant 2*, while $\mathcal{D}_n$ is subject to the constraint that sign changes must always occur in an even number of coordinates.

The only remaining reflection groups are $\mathcal{I}_2(m)$. They operate in $\mathbb{R}^2$ and generate the dihedral groups $\mathcal{D}_m$. The corresponding spherical codes are trivial.

8.10 Remarks on some specific codes

We conclude this long chapter with a few remarks regarding the group structure of some well known simple codes. We start with the optimal codes in Ω_3.

Recall that optimal codes in Ω_3 are known for $2 \le M \le 12$ and $M = 24$. We refer to Chapter 3 for a detailed description of all these codes. All finite groups in $\mathcal{O}_3$ are also known, see Grove-Benson [121], Chapter 2. The question is which of them generate optimal codes.

The cases $M = 2,\ 3$ are trivial: it is clear that the optimal codes are generated by the dihedral groups $\mathcal{I}_2(2)$ and $\mathcal{I}_2(3)$ respectively.

The cases $M = 4,\ 6,\ 12$ are well known. They correspond to the classical regular polyhedra: the tetrahedron, the octahedron and the icosahedron. It was observed already in antiquity that these are subject to remarkable symmetries. In fact, exactly these symmetries inspired and motivated many of the early investigations on groups. The groups $\mathcal{A}_3$, $\mathcal{B}_3$, $\mathcal{H}_3$ are actually the full groups of symmetries for these polyhedra. For the tetrahedron this is easy to see. Any group of transformations leaving the tetrahedron invariant must permute its vertices. The only orthogonal transformation leaving all four vertices unchanged is the identity. As there are only 4 vertices there are at most $4! = 24$ possibilities. We observed already in Chapter 4 that the tetrahedron is generated by $\mathcal{A}_3$ (although a different terminology was used in that chapter). As the order of $\mathcal{A}_3$ is precisely 24 it follows that $\mathcal{A}_3$ contains all the orthogonal transformations leaving the tetrahedron invariant.

Also for the octahedron the argument is fairly simple. There are 3 axes that can be permuted in $3! = 6$ different ways, and for each one of these permutations there is a choice of direction for each one of the axes, in total $2^3 = 8$ possible directions, and a total of $6 \cdot 8 = 48$ possible transformations. This is precisely the order of $\mathcal{B}_3$. It is easy to verify (and was actually implicitly demonstrated in Chapter 4) that the octaheder is generated by $\mathcal{B}_3$.

There are of course also other groups generating these polyhedra. For instance, the octahedron is also generated by $\mathcal{A}_3$. This is no wonder as $\mathcal{A}_3$ is a subgroup in $\mathcal{B}_3$: it is quite a normal thing that a group code can be generated also by a subgroup of its full group of symmetries.

In general it is a difficult problem to determine the full group of symmetries for a given spherical code, be it a group code or not. Even if a group leaving the code invariant has been found it might be difficult to prove that there is no larger group under which the code is also invariant. For a proof that $\mathcal{H}_3$ is actually the full group of symmetries for the icosahedron we refer to Grove-Benson [121], Section 2.5.

The optimal code of size $M = 8$ is also a group code, but in this case the generating group is not a reflection group. Recall that the optimal code is a square antiprism. It can be viewed as two squares placed one over the other and rotated 45 degrees with respect to each other. Consider the following matrices for $0 \leq i < 8$:

$$T_i = \begin{bmatrix} \cos(\pi i/4) & -\sin(\pi i/4) & 0 \\ \sin(\pi i/4) & \cos(\pi i/4) & 0 \\ 0 & 0 & (-1)^i \end{bmatrix}.$$

It is clear that they form a cyclic group of orthogonal matrices, and that they generate the squared antiprism by a proper choice of initial vector. This is *not* a reflection group: indeed, it does not contain any reflection at all! It is, of course, still conceivable that the squared antiprism is also generated by some reflection group containing this cyclic group as a subgroup. A complete evaluation of all cases (there are only 3 irreducible finite reflection groups in $\mathcal{O}_3$) reveals that this is not the case.

The optimal codes in cases $M = 7, 9, 10$ very obviously are *not* group codes. This follows by the fact that they are not regular.

The only remaining known optimal code in Ω_3 is the "snub cube", with $M = 24$. Recall that we described this code as a collection of twisted squares on each one of the six faces of a cube (Chapter 3). This fact suggest that the snub cube might have the same symmetries as the cube. This is almost true, but not quite.

The cube has the same group of symmetries as has the octahedron (this is a consequence of their being dual, see Coxeter [75], Chapter 3, p. 47 or Grove-Benson [121], Chapter 2), namely the reflection group $\mathcal{B}_3$ of order $2^3 \cdot 3! = 48$. This group contains a subgroup of order 24, consisting of all possible rotations around centers of opposite faces ($3 \cdot 3 = 9$, disregarding the identity), around the midpoints of opposite edges ($6 \cdot 1 = 6$, disregarding the identity), and around diagonally opposite corners ($4 \cdot 2 = 8$, disregarding the identity). This gives us a group of order $9 + 6 + 8 + 1 = 24$. It is easy to see that the snub cube is invariant under all transformations in this group, and that this group actually generates the code upon an appropriate choice of initial vector. It is also easy to see that the code is *not* invariant under the remaining transformations in $\mathcal{B}_3$. In particular, any reflection in a half plane parallel to any of the faces of the cube transforms the snub cube into its mirror image, which is obtained by twisting all the squares on the faces of the cube in opposite directions. Thus: the snub cube comes in two versions with opposite "parity", each of these two variants correspond to different orbits with respect to the rotation subgroup in $\mathcal{B}_3$. With respect to $\mathcal{B}_3$ these two orbits join to one single orbit of order 48.

A few words also about the *Simplex* codes and the *Biorthogonal* codes. We noticed in Chapter 4 (in our discussion on permutation codes) that the *Simplex* code is always generated as a permutation code *Variant 1*, i.e. as a reflection group generated by $\mathcal{A}_{n+1}$. By our informal description $(1^n, -n)^C$ follows immediately that it is also always generated by a *cyclic* group.

For the *Biorthogonal* code we have the representation $(0^{n-1}, \pm 1)^C$. It is easy to see that the required transformations can be obtained by a *cyclic* group if and only if n is odd. The code is a group code also for n even, but then the group is not cyclic.

We have seen that all known optimal codes in dimension $n \leq 8$ can be generated as group codes, often but not always with a reflection group as the generating group. In addition we have seen that all binary antipodal codes obtained from linear codes are also group codes. In these cases the groups are commutative, and therefore not reflection groups. We have also seen that polyphase codes generated from group codes over the ring $\mathbb{Z}_q^n$ are group codes in Ω_n.

Recently Sidelnikov [206], [207] described a new class of spherical codes generated directly as orbits from certain groups of orthogonal matrices. In particular for squared distance $\rho = 1$ this class contains codes with dimension n and size M according to the following table:

n	2	4	8	16	32	64
M	2	24	240	4320	146880	9694080

For dimensions $n = 4, 8, 16$ this coincides with the best known codes; the code in dimension $n = 8$ is even known to be unique [10] and optimal.

In all this means that many of the best and many of the practically most significant spherical codes are actually group codes. This fact is of great importance for analyzing the codes and in connection with decoding. We have also seen that some optimal codes are *not* group codes. However, even for such codes it is of great interest to determine the full group of symmetries. Although we have been able to say a few things about a number of special cases, the general problem of determining the full symmetry group for a given spherical code remains a challenge.

Chapter 9

Distance regular spherical codes

9.1 Introduction

Let $\mathcal{X}$ be a spherical code and let $\{\rho_0, \rho_1, \ldots, \rho_d\}$ be its set of squared distances. Assume the ordering $\{0 = \rho_0 < \rho_1 < \ldots < \rho_d\}$ and let $\{s_0 = 1 > s_1 > \ldots > s_d\}$ be the corresponding set of inner products. As usual we have $s_i = 1 - \rho_i/2$, $i = 0, 1, \ldots, d$. Define

$$A_i(x, y) = \begin{cases} 1, & if \quad \langle x, y \rangle = s_i, \\ 0, & if \quad \langle x, y \rangle \neq s_i. \end{cases}$$

We obtain a set of $d+1$ indicator matrices A_i, $i = 0, 1, \ldots, d$, each square of order M, where M is the size of the code. Notice that rows and columns are indexed by the codewords $x, y \in \mathcal{X}$. The matrices A_i satisfy the following conditions:

$$\begin{array}{ll} (i) & A_0 = I, \\ (ii) & A_i^T = A_i, \\ (iii) & \sum_{i=0}^{d} A_i = J, \end{array}$$

where I denotes the $M \times M$ identity matrix and J denotes the $M \times M$ matrix with ones in all positions. We notice that the (x, y)-th component $A_iA_j(x, y)$ of the matrix product A_iA_j denotes the number of codewords which are simultaneously on squared distance ρ_i from the codeword x and on squared distance ρ_j from the codeword y. In general, of course, this number depends on both x and y. Occasionally, however, it might be uniquely determined by the *distance* between x and y. Suppose this to be the case, and let it be p_{ij}^k when the squared distance is ρ_k. It follows that the product $A_i A_j$ can be expressed in the following way:

$$(iv) \quad A_i A_j = \textstyle\sum_{i=0}^{d} p_{ij}^k A_k \,.$$

In addition, as all the matrices A_i are symmetric, we also obtain the property

$$(v) \quad A_i A_j = A_j A_i.$$

A spherical code satisfying property (iv)—and then automatically also property (v)—is called a *distance regular* spherical code. The parameters p_{ij}^k are called *intersection numbers.* The matrices P_i defined by having the element p_{ij}^k in the j-th row of the k-th column are called *intersection matrices.*

Let V be an arbitrary set of size $v = |V|$. Any set of indicator matrices A_i defined over the product set $V \times V$ and satisfying the properties (i)–(v) is referred to as an *association scheme.* Each individual matrix A_i defines a subset of $V \times V$. The set of all these matrices defines a *partition* of $V \times V$ into $d+1$ subsets. The specific matrix A_0 indicates the set $\{(x,x) : x \in V\}$, which is often referred to as the *diagonal.* The non-diagonal subsets are usually referred to as *classes.* A scheme with $d+1$ indicator matrices is referred to as a scheme with d classes. Any distance regular spherical code defines an association scheme. As we shall see the opposite is also true.

Association schemes were introduced by *Bose–Shimamoto* in 1952, [41]. The coding theory aspects of the topic were extensively investigated by *Delsarte* in his thesis from 1973, [80]. He considered a slightly more general definition by replacing (ii) with the milder condition that for any matrix A_i in the scheme the transpose A_i^T is also an element in the same scheme, but not necessarily equal to A_i. With this more general definition the condition (v) is essential. In the more restricted original definition (v) is—as we just pointed out—a consequence of the other assumptions. For our purposes the simpler definition will be sufficient.

The original definition can alternatively be given in terms of *graphs.* The condition (ii) guarantees that each matrix A_i can be interpreted as the incidence matrix of a graph Γ_i over the given vertex set $V = \{1, 2, \ldots, v\}$. Recall that a *loop* is an edge connecting a given vertex with itself. The condition (i) implies that the special matrix A_0 corresponds to a graph Γ_0 having one loop for each vertex, and no other edges. Condition (iii) implies that there are no loops at all in any of the other graphs. Property (iii) also implies that the graphs Γ_i, $i = 1, 2, \ldots, d$ form a partition of the complete graph K_v. Conditions (i)—(iii) hold for any partition of K_v. The crucial property under which such a partition qualifies as an *association scheme* is property (iv). Delsarte's more general definition permits *directed* graphs.

Still another definition can be given in terms of *relations*. A relation R in a set V is simply a subset of $V \times V$. Usually one writes xRy rather than $(x, y) \in R$ and one says that x has relation R to y when the pair (x, y) belongs to R. A relation is *symmetric* if yRx holds whenever xRy holds. We see that the restricted definition induces a family $R = \{R_0, R_1, \ldots, R_d\}$ of symmetric relations while Delsarte's more general definition allows *non-symmetric* relations. In both cases A_0 corresponds to the identity relation R_0 defined by

$$xR_0y \quad \Leftrightarrow \quad x = y, \quad x, y \in V.$$

In order to make the concepts more concrete, let us right away consider some specific cases. We give three examples in dimension $n = 3$, see Figure 9.1.

Example 9.1.1 (Octahedron): The *octahedron* is distance regular with *two* classes. The intersection matrices are

$$P_0 = \begin{bmatrix} p_{00}^0 & p_{00}^1 & p_{00}^2 \\ p_{01}^0 & p_{01}^1 & p_{01}^2 \\ p_{02}^0 & p_{02}^1 & p_{02}^2 \end{bmatrix} = \begin{bmatrix} 1 & 0 & 0 \\ 0 & 1 & 0 \\ 0 & 0 & 1 \end{bmatrix},$$

$$P_1 = \begin{bmatrix} p_{10}^0 & p_{10}^1 & p_{10}^2 \\ p_{11}^0 & p_{11}^1 & p_{11}^2 \\ p_{12}^0 & p_{12}^1 & p_{12}^2 \end{bmatrix} = \begin{bmatrix} 0 & 1 & 0 \\ 4 & 2 & 4 \\ 0 & 1 & 0 \end{bmatrix},$$

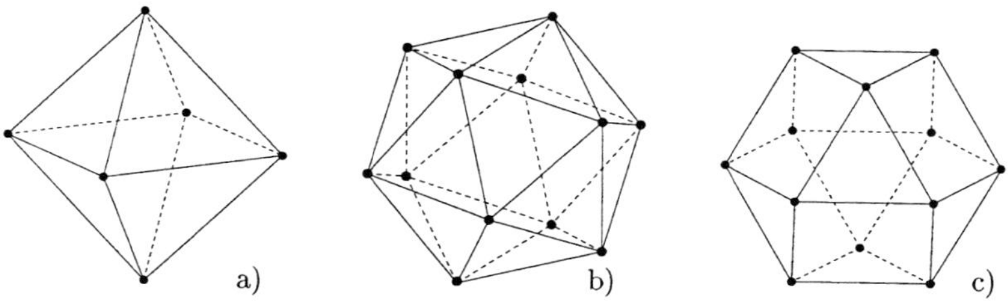

Figure 9.1: The octahedron (a), the icosahedron (b), and the cubeoctahedron (c).

$$P_2 = \begin{bmatrix} p_{20}^0 & p_{20}^1 & p_{20}^2 \\ p_{21}^0 & p_{21}^1 & p_{21}^2 \\ p_{22}^0 & p_{22}^1 & p_{22}^2 \end{bmatrix} = \begin{bmatrix} 0 & 0 & 1 \\ 0 & 1 & 0 \\ 1 & 0 & 0 \end{bmatrix}.$$

□

Example 9.1.2 (Icosahedron): The *icosahedron* is distance regular with three classes. The intersection matrices are

$$P_0 = \begin{bmatrix} p_{00}^0 & p_{00}^1 & p_{00}^2 & p_{00}^3 \\ p_{01}^0 & p_{01}^1 & p_{01}^2 & p_{01}^3 \\ p_{02}^0 & p_{02}^1 & p_{02}^2 & p_{02}^3 \\ p_{03}^0 & p_{03}^1 & p_{03}^2 & p_{03}^3 \end{bmatrix} = \begin{bmatrix} 1 & 0 & 0 & 0 \\ 0 & 1 & 0 & 0 \\ 0 & 0 & 1 & 0 \\ 0 & 0 & 0 & 1 \end{bmatrix},$$

$$P_1 = \begin{bmatrix} p_{10}^0 & p_{10}^1 & p_{10}^2 & p_{10}^3 \\ p_{11}^0 & p_{11}^1 & p_{11}^2 & p_{11}^3 \\ p_{12}^0 & p_{12}^1 & p_{12}^2 & p_{12}^3 \\ p_{13}^0 & p_{13}^1 & p_{13}^2 & p_{13}^3 \end{bmatrix} = \begin{bmatrix} 0 & 1 & 0 & 0 \\ 5 & 2 & 2 & 0 \\ 0 & 2 & 2 & 5 \\ 0 & 0 & 1 & 0 \end{bmatrix},$$

$$P_2 = \begin{bmatrix} p_{20}^0 & p_{20}^1 & p_{20}^2 & p_{20}^3 \\ p_{21}^0 & p_{21}^1 & p_{21}^2 & p_{21}^3 \\ p_{22}^0 & p_{22}^1 & p_{22}^2 & p_{22}^3 \\ p_{23}^0 & p_{23}^1 & p_{23}^2 & p_{23}^3 \end{bmatrix} = \begin{bmatrix} 0 & 0 & 1 & 0 \\ 0 & 2 & 2 & 5 \\ 5 & 2 & 2 & 0 \\ 0 & 1 & 0 & 0 \end{bmatrix},$$

$$P_3 = \begin{bmatrix} p_{30}^0 & p_{30}^1 & p_{30}^2 & p_{30}^3 \\ p_{31}^0 & p_{31}^1 & p_{31}^2 & p_{31}^3 \\ p_{32}^0 & p_{32}^1 & p_{32}^2 & p_{32}^3 \\ p_{33}^0 & p_{33}^1 & p_{33}^2 & p_{33}^3 \end{bmatrix} = \begin{bmatrix} 0 & 0 & 0 & 1 \\ 0 & 0 & 1 & 0 \\ 0 & 1 & 0 & 0 \\ 1 & 0 & 0 & 0 \end{bmatrix}.$$

□

Example 9.1.3 (Cubeoctahedron): The *cubeoctahedron* is distance regular with four classes. The intersection matrices are

$$P_0 = \begin{bmatrix} p_{00}^0 & p_{00}^1 & p_{00}^2 & p_{00}^3 & p_{00}^4 \\ p_{01}^0 & p_{01}^1 & p_{01}^2 & p_{01}^3 & p_{01}^4 \\ p_{02}^0 & p_{02}^1 & p_{02}^2 & p_{02}^3 & p_{02}^4 \\ p_{03}^0 & p_{03}^1 & p_{03}^2 & p_{03}^3 & p_{03}^4 \\ p_{04}^0 & p_{04}^1 & p_{04}^2 & p_{04}^3 & p_{04}^4 \end{bmatrix} = \begin{bmatrix} 1 & 0 & 0 & 0 & 0 \\ 0 & 1 & 0 & 0 & 0 \\ 0 & 0 & 1 & 0 & 0 \\ 0 & 0 & 0 & 1 & 0 \\ 0 & 0 & 0 & 0 & 1 \end{bmatrix},$$

$$P_1 = \begin{bmatrix} p_{10}^0 & p_{10}^1 & p_{10}^2 & p_{10}^3 & p_{10}^4 \\ p_{11}^0 & p_{11}^1 & p_{11}^2 & p_{11}^3 & p_{11}^4 \\ p_{12}^0 & p_{12}^1 & p_{12}^2 & p_{12}^3 & p_{12}^4 \\ p_{13}^0 & p_{13}^1 & p_{13}^2 & p_{13}^3 & p_{13}^4 \\ p_{14}^0 & p_{14}^1 & p_{14}^2 & p_{14}^3 & p_{14}^4 \end{bmatrix} = \begin{bmatrix} 0 & 1 & 0 & 0 & 0 \\ 4 & 1 & 2 & 1 & 0 \\ 0 & 1 & 0 & 1 & 0 \\ 0 & 1 & 2 & 1 & 4 \\ 0 & 0 & 0 & 1 & 0 \end{bmatrix},$$

$$P_2 = \begin{bmatrix} p_{20}^0 & p_{20}^1 & p_{20}^2 & p_{20}^3 & p_{20}^4 \\ p_{21}^0 & p_{21}^1 & p_{21}^2 & p_{21}^3 & p_{21}^4 \\ p_{22}^0 & p_{22}^1 & p_{22}^2 & p_{22}^3 & p_{22}^4 \\ p_{23}^0 & p_{23}^1 & p_{23}^2 & p_{23}^3 & p_{23}^4 \\ p_{24}^0 & p_{24}^1 & p_{24}^2 & p_{24}^3 & p_{24}^4 \end{bmatrix} = \begin{bmatrix} 0 & 0 & 1 & 0 & 0 \\ 0 & 1 & 0 & 1 & 0 \\ 2 & 0 & 0 & 0 & 2 \\ 0 & 1 & 0 & 1 & 0 \\ 0 & 0 & 1 & 0 & 0 \end{bmatrix},$$

$$P_3 = \begin{bmatrix} p_{30}^0 & p_{30}^1 & p_{30}^2 & p_{30}^3 & p_{30}^4 \\ p_{31}^0 & p_{31}^1 & p_{31}^2 & p_{31}^3 & p_{31}^4 \\ p_{32}^0 & p_{32}^1 & p_{32}^2 & p_{32}^3 & p_{32}^4 \\ p_{33}^0 & p_{33}^1 & p_{33}^2 & p_{33}^3 & p_{33}^4 \\ p_{34}^0 & p_{34}^1 & p_{34}^2 & p_{34}^3 & p_{34}^4 \end{bmatrix} = \begin{bmatrix} 0 & 0 & 0 & 1 & 0 \\ 0 & 1 & 2 & 1 & 4 \\ 0 & 1 & 0 & 1 & 0 \\ 4 & 1 & 2 & 1 & 0 \\ 0 & 1 & 0 & 0 & 0 \end{bmatrix},$$

$$P_4 = \begin{bmatrix} p_{40}^0 & p_{40}^1 & p_{40}^2 & p_{40}^3 & p_{40}^4 \\ p_{41}^0 & p_{41}^1 & p_{41}^2 & p_{41}^3 & p_{41}^4 \\ p_{42}^0 & p_{42}^1 & p_{42}^2 & p_{42}^3 & p_{42}^4 \\ p_{43}^0 & p_{43}^1 & p_{43}^2 & p_{43}^3 & p_{43}^4 \\ p_{44}^0 & p_{44}^1 & p_{44}^2 & p_{44}^3 & p_{44}^4 \end{bmatrix} = \begin{bmatrix} 0 & 0 & 0 & 0 & 1 \\ 0 & 0 & 0 & 1 & 0 \\ 0 & 0 & 1 & 0 & 0 \\ 0 & 1 & 0 & 0 & 0 \\ 1 & 0 & 0 & 0 & 0 \end{bmatrix}.$$

□

In all cases the intersection matrices can easily be derived by inspection.

Recall that a *graph* Γ is called *regular* if all vertices have the same *valency*. A *spherical code* with distance matrices $\Gamma_i, i = 0, 1, \ldots, d$ is said to be *regular* if all these graphs are regular in the sense of graphs. For such a code let k_i be the valency of the distance graph

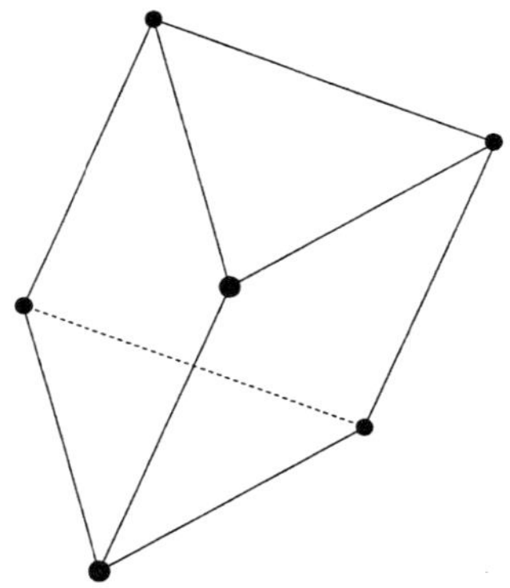

Figure 9.2: A triangular prism in $\mathbb{R}^3$.

$\Gamma_i, 1 \leq i \leq d$. Define separately $k_0 = 1$. The sequence $\{(k_i, \rho_i) : 0 \leq i \leq d\}$ is called the *distance distribution*. The numbers k_i indicate for any codeword x the number of codewords on squared distance ρ_i from x. We notice the identity $k_i = p_{ii}^0$, $i = 0, 1, \ldots, d$.

Each distance regular code is regular. The converse, however, is not true. A triangular prism such that each point has three close neighbors on squared distance 12/7 and two more distant neighbors on squared distance 24/7 offers a simple counterexample, see Figure 9.2.

A distance regular code does not necessarily have favorable properties in the sense of spherical codes. For instance, we noticed in Chapter 3 that the best possible configuration of 8 points on the unit sphere in three dimensions is not the cube but a *square antiprism.* The cube is distance regular, but the square antiprism is not. Both are regular.

The binary Hamming space H_n, which consists of all n-length binary sequences, forms an association scheme with n classes. The pair $\{x, y\}$ belongs to the i-th class if and only if $d_H(x, y) = i$. It follows that the spherical code obtained from H_n under the Y_2 construction is distance regular. The resulting code is referred to as a *hypercube.* All n-dimensional hypercubes induce association schemes.

The *Johnson space* $J(n, w)$, which consists of all binary sequences of length n and constant weight w, is also an association scheme. In this case the number of classes is $d = \min\{w, n - w\}$ and the pair $\{x, y\}$ belongs to the i-th class if and only if $d_H(x, y) = 2i$. Any balanced code generated by the Johnson space in accordance with the CW_2-construction described in Chapter 6 is distance regular.

Let v be an arbitrary positive integer and consider the set $\mathbb{Z}_v$ of integers modulo v. For $x \in \mathbb{Z}_v$ let $w_L(x)$ be the Lee weight of x. Define $d \triangleq \lfloor v/2 \rfloor$. For $i = 0, 1, \ldots, d$, let the matrices A_i be defined according to the following rule:

$$A_i(x, y) = \begin{cases} 1, & if \quad w_L(x - y) = i, \\ 0, & if \quad w_L(x - y) \neq i. \end{cases}$$

The matrices A_i, $i = 0, 1, \ldots, d$ form an association scheme. Schemes of this kind are generated by *polyphase codes.*

It turns out that several of the codes for which optimality has been established are distance regular, and so closely related to association schemes. We proceed with a brief discussion of the basic properties of association schemes.

9.2 Association schemes

Let $\mathcal{A} = \{A_0, A_1, \ldots, A_d\}$ be an association scheme with d classes. The matrices A_i are square of order v, with indices x, y from some given set V. A typical component in A_i is denoted $A_i(x, y)$. We assume a fixed but arbitrary ordering in the index set V. Consider the linear space $L(\mathcal{A})$ spanned by the matrices A_i. Property (iv) implies that this space is closed under matrix multiplication. That means that it is an *algebra*. Property (v) implies that the algebra is *commutative.* It is usually referred to as the *Bose–Mesner algebra*, [40]. The matrix A_0 is the *unit element* in this algebra. All matrices in the Bose–Mesner algebra are *symmetric.* It follows that the eigenvalues are *real* (Bellman [18], p. 35). Moreover, as the algebra is commutative all matrices in the algebra are jointly diagonalized by a single orthogonal matrix (Bellman [18], p. 56). Even more: all matrices in the algebra $L(\mathcal{A})$ have a simultaneous basis of eigenvectors (Cohn

[69], p. 203). A number of remarkable properties follow from this fact.

We start by observing that the intersection matrices P_i, $i = 0, 1, \ldots, d$ form an algebra isomorphic with $L(\mathcal{A})$. The key result is the following lemma.

Lemma 9.2.1 *Let $\mathcal{A}$ be an association scheme with intersection numbers p_{ij}^k, $i, j, k = 0, 1, \ldots, d$. Then the following identities hold:*

$$\sum_{l=0}^{d} p_{im}^{l} p_{jl}^{n} = \sum_{k=0}^{d} p_{ij}^{k} p_{km}^{n}. \tag{9.2.1}$$

Proof: Recall that $\{A_1, A_2, \ldots, A_d\}$ is a partitioning of the complete graph K_v, where v is the order of the matrices A_i. It follows that any branch in K_v can uniquely be assigned a number i, indicating to which subgraph Γ_i it belongs. Let x and y be two neighbors in Γ_n. The identity (9.2.1) follows by counting in two ways the number of ways by which x can be connected with y by a path in K_v of length 3 such the branches in the path have numbers j, i, m, in that order. □

Theorem 9.2.1 (Delsarte, [80]) *Let $\mathcal{A} = \{A_0, A_1, \ldots, A_d\}$ be an association scheme. The intersection matrices $\{P_0, P_1, \ldots, P_d\}$ corresponding to the scheme $\mathcal{A}$ generate a matrix algebra isomorphic with the Bose-Messner algebra $L(\mathcal{A})$.*

Proof: All we need to show is the identity

$$P_i P_j = \sum_{k=0}^{d} p_{ij}^{k} P_k. \tag{9.2.2}$$

This, however, is an immediate consequence of Lemma 9.2.1. The result follows by expressing both members of Equation (9.2.2) in terms of the intersection numbers p_{ij}^k. □

One consequence of Theorem 9.2.1 is that the eigenvalues of P_i are the same as those of A_i. This in turn implies that the Euclidean space $\mathbb{R}^v$ upon which the matrices A_i operate can be represented as a direct sum

$$\mathbb{R}^v = \bigoplus_{j=0}^{d} W_j$$

of subspaces W_j, where for any $A_i \in \mathcal{A}$ each vector u in W_j has the same eigenvalue $p_i(j)$. We refer to the subspaces W_j as the *eigenspaces* corresponding to the algebra $L(\mathcal{A})$.

Let E_i be the projection from $\mathbb{R}^v$ to W_i, $i = 0, 1, \ldots, d$. The matrices E_i play an important role in the theory of association schemes. They are usually referred to as the *principal idempotents.* We summarize some of the basic properties.

Theorem 9.2.2 *Let $\mathcal{A} = \{A_0, A_1, \ldots, A_d\}$ be an association scheme with d classes on a set V of cardinality v and let $W_0, W_1, \ldots, W_d$ be the corresponding eigenspaces. The projection matrices $E_i : \mathbb{R}^v \to W_i$ satisfy the following properties for any $i, j = 0, 1, 2, \ldots, d$:*

(i) $A_i E_j = p_i(j) E_j$,

(ii) $E_i E_j = \delta_{ij} E_i$,

(iii) *E_i is positive definite,*

(iv) $\sum_{i=0}^{d} E_i = I$,

(v) $f(A_i) = \sum_{j=0}^{d} f(p_i(j)) E_j$ *for any polynomial $f(t)$,*

(vi) $Tr(E_i) = dim(W_i)$.

Proof: Let u be any element in $\mathbb{R}^v$, represented as a column vector. The projection $u_j = E_j u$ of u on the subspace W_j has eigenvalue $p_i(j)$ with respect to the matrix A_i. Therefore we have

$$AE_j u = Au_j = p_i(j) u_j = p_i(j) E_j u.$$

As this holds for any $u \in \mathbb{R}^v$ property (i) follows.

Property (ii) reflects the fact that $\mathbb{R}^v$ is a direct sum of the orthogonal spaces E_i and property (iv) is an immediate consequence of the fact that the spaces W_i span $\mathbb{R}^v$. Property (iii) holds for any projection.

In order to establish (v) we notice that by multiplying equation (i) an appropriate number of times by A_i we obtain $A_i^k E_j = (p_i(j))^k E_j$ for any positive integer k. Using (iv) we obtain

$$A_i^k = \sum_{j=0}^{d} (p_i(j))^k E_j.$$

The identity (v) follows by linearity.

It remains to establish (vi). For that purpose we choose an orthonormal basis in $\mathbb{R}^v$ such that each subspace W_i is spanned by $m_i = dim(W_i)$ distinct base vectors. With respect to this basis each projection E_i has m_i diagonal elements equal to 1 while all other elements are equal to 0. The trace obviously equals m_i. Property (vi) follows as the trace is invariant under orthogonal transformations. □

It is easy to check that one of the principal idempotents is given by $(1/v)J$, where J as usual denotes the matrix with ones in all positions. It is an established convention to assign index 0 to this particular matrix: $E_0 = (1/v)J$.

Property (v) actually extends to any element in the algebra $L(\mathcal{A})$. Let $A = \sum_{i=0}^{d} \alpha_i A_i$. We get

$$AE_j = \sum_{i=0}^{d} \alpha_i A_i E_j = \sum_{i=0}^{d} \alpha_i p_i(j) E_j.$$

This means that the eigenvalue θ_j for the matrix A with respect to the eigenspace W_j is given by

$$\theta_j = \sum_{i=0}^{d} \alpha_i p_i(j).$$

Combining (v) with the identity $AE_j = \theta_j E_j$ we also obtain

$$A = \sum_{j=0}^{d} \theta_j E_j,$$

which shows that the matrices $E_0, E_1, \ldots, E_d$ span the linear space $L(\mathcal{A})$. Furthermore, by exactly the same argument as above we obtain

$$f(A) = \sum_{i=0}^{d} f(\theta_i) E_i, \tag{9.2.3}$$

where A is an arbitrary element in the algebra $L(\mathcal{A})$ and $f(t)$ is an arbitrary polynomial.

From properties (i) and (v) we have the expansion

$$A_i = \sum_{j=0}^{d} p_i(j) E_j,$$

by which the defining matrices A_i are expressed in terms of the principal idempotents E_j.

It turns out that there is also an inverse expansion by which the principal idempotents E_j are expressed in terms of the matrices A_i. Choose A in $L(\mathcal{A})$ with distinct eigenvalues $\theta_0, \theta_1, \ldots, \theta_d$, (for a proof that such an element A exists we refer to the specialized literature, [9], [40], [64], [80]). For each eigenvalue θ_i we define a polynomial $f_i(t)$ as follows:

$$f_i(t) = \prod_{j \neq i} (t - \theta_j).$$

Now we employ the expansion (9.2.3). We obtain

$$f_i(A) = \sum_{j=0}^{d} f_i(\theta_j) E_j = f_i(\theta_i) E_i, \tag{9.2.4}$$

where in the last step we have taken into account the fact that $f_i(\theta_j) = 0$ for $j \neq i$. Moreover, by construction we also have $f_i(\theta_i) \neq 0$.

Next we observe that $f_i(A)$ is an element in $L(\mathcal{A})$, and so has an expansion in the matrices A_i. Let it be

$$f_i(A) = \sum_{k=0}^{d} f_{ik} A_k. \tag{9.2.5}$$

Combining (9.2.4) and (9.2.5) we obtain

$$E_i = \frac{1}{f_i(\theta_i)} \sum_{k=0}^{d} f_{ik} A_k = \frac{1}{v} \sum_{k=0}^{d} q_i(k) A_k. \tag{9.2.6}$$

In the last step we have introduced the usual notation used for expanding the principal idempotents E_i in terms of the matrices A_k. The quantities $q_i(k)$ are usually called the *dual eigenvalues*, ([116], p. 226).

For our purposes the following little observation turns out to be of fundamental importance. Recall that all matrices in $L(\mathcal{A})$ are labeled by elements (x, y) in the set V, where x indicates row and y indicates column. Also recall that for each position (x, y) we have $A_i(x, y) = 1$ for one and only one index i, while $A_j(x, y) = 0$ for all indices $j \neq i$. In particular, $A_i(x, x) = 1$ for $i = 0$ while $A_i(x, x) = 0$ for all $i \neq 0$. Combining this little fact with (9.2.6) we obtain

$$E_i(x, x) = \frac{1}{v} q_i(0).$$

The point is that this holds for *any* $x \in V$, which means that the matrix E_i is *constant* along its main diagonal. As there are v diagonal elements we obtain $Tr(E_i) = v \cdot (1/v) \cdot q_i(0) = q_i(0)$. But we already know that the trace is also given m_i. It follows that we have $q_i(0) = m_i$. It also follows that the matrix

$$G_i \triangleq \frac{v}{m_i} E_i$$

is positive definite with all main diagonal elements $G_i(x, x)$ equal to 1. This means that each one of the matrices G_i defines a spherical code $\mathcal{X}_i$ having G_i as its Gram matrix. Alternatively it means that the code $\mathcal{X}_i$ has a codeword matrix X_i such that $X_i X_i^T = G_i$. One possible realization is

$$X_i = \sqrt{\frac{v}{m_i}} E_i\,.$$

Notice that it is the *rows* in X_i that constitute the words in the code $\mathcal{X}_i$, while the *columns* are elements in the eigenspace W_i. By Theorem 9.2.2, property (i) we have

$$A_i X_j = p_i(j) X_j, \quad i, j = 0, 1, \ldots d.$$

We remark that it may be that certain rows in X_i are repeated, corresponding to coinciding codewords. When this happens we can always create a code by choosing one representative from each equivalence class of coinciding rows in X_i.

We also remark that the representation above uses v columns for representing the codewords in $\mathcal{X}_i$, while $m_i = dim(\mathcal{X}_i)$ columns would actually be enough.

We illustrate with several examples.

Example 9.2.1 (Octahedron): Let $A = A_0, A_1, \ldots, A_d$ be the association scheme generated by the octahedron. We have $d = 2$, and the matrices can be taken to be

$$A_0 = \begin{bmatrix} 1 & 0 & 0 & 0 & 0 & 0 \\ 0 & 1 & 0 & 0 & 0 & 0 \\ 0 & 0 & 1 & 0 & 0 & 0 \\ 0 & 0 & 0 & 1 & 0 & 0 \\ 0 & 0 & 0 & 0 & 1 & 0 \\ 0 & 0 & 0 & 0 & 0 & 1 \end{bmatrix}, \qquad A_1 = \begin{bmatrix} 0 & 0 & 1 & 1 & 1 & 1 \\ 0 & 0 & 1 & 1 & 1 & 1 \\ 1 & 1 & 0 & 0 & 1 & 1 \\ 1 & 1 & 0 & 0 & 1 & 1 \\ 1 & 1 & 1 & 1 & 0 & 0 \\ 1 & 1 & 1 & 1 & 0 & 0 \end{bmatrix},$$

$$A_2 = \begin{bmatrix} 0 & 1 & 0 & 0 & 0 & 0 \\ 1 & 0 & 0 & 0 & 0 & 0 \\ 0 & 0 & 0 & 1 & 0 & 0 \\ 0 & 0 & 1 & 0 & 0 & 0 \\ 0 & 0 & 0 & 0 & 0 & 1 \\ 0 & 0 & 0 & 0 & 1 & 0 \end{bmatrix}.$$

The corresponding graphs are shown in Figure 9.3. We notice that the graph Γ_1 corresponding to A_1 is connected. It has an edge between any pair of vertices corresponding to codewords on minimum distance. The graph Γ_2, generated by codewords on squared distance

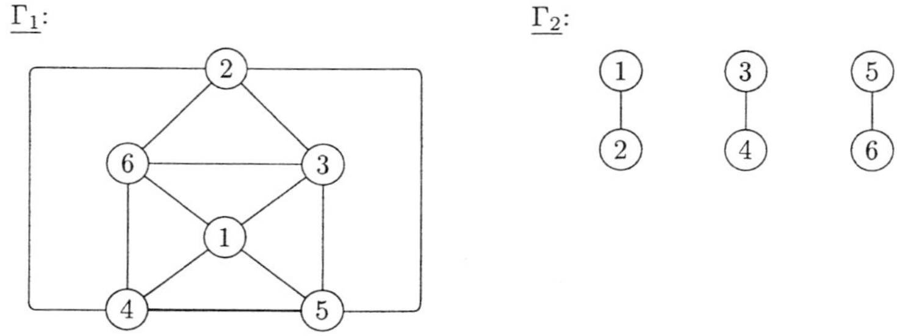

Figure 9.3: Distance graphs for the octahedron.

$\rho_2 = 4$, is disconnected. A possible representation of the spherical code has code matrix

$$X = \begin{bmatrix} 1 & 0 & 0 \\ -1 & 0 & 0 \\ 0 & 1 & 0 \\ 0 & -1 & 0 \\ 0 & 0 & 1 \\ 0 & 0 & -1 \end{bmatrix}.$$

It is easy to see that this matrix generates a common eigenspace to all the matrices A_i in the association scheme. The eigenvalues are $[\theta_0, \theta_1, \theta_2] = [1, 0, -1]$.

This, however, is not the only solution. There are two additional possibilities. One is given by $X = \mathbf{j}$, where $\mathbf{j}$ denotes a column vector with all ones. The eigenvalues are $[\theta_0, \theta_1, \theta_2] = [1, 4, 1]$. The other possibility is

$$X = \frac{1}{\sqrt{6}} \begin{bmatrix} 1 & 1 & -2 \\ 1 & 1 & -2 \\ 1 & -2 & 1 \\ 1 & -2 & 1 \\ -2 & 1 & 1 \\ -2 & 1 & 1 \end{bmatrix}, \qquad [\theta_0, \theta_1, \theta_2] = [1, -2, 1].$$

In the last two cases the solutions provide codes with *repeated codewords*. The code corresponding to $X = \mathbf{j}$ is trivial: it consists of a single point in dimension $n = 1$. In the other case we recognize a

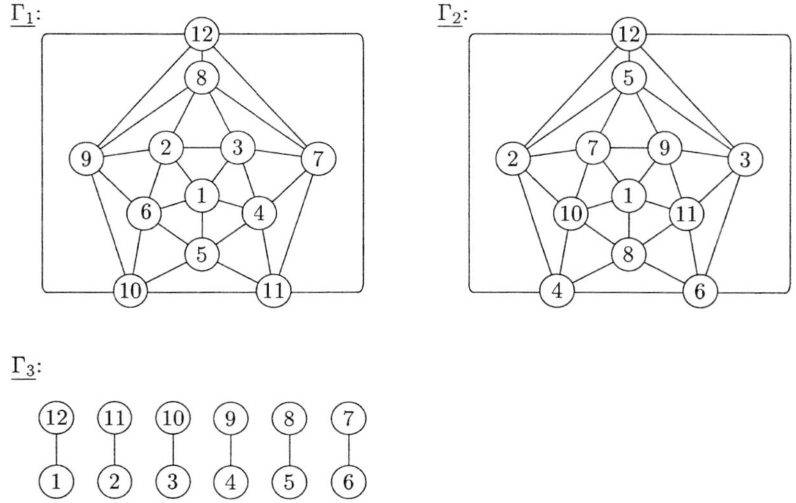

Figure 9.4: Distance graphs for the icosahedron.

repeated version of the simplex code in two dimensions, i.e. a triangle. In summary we have three spherical codes with the following parameters:

$$(n, \rho, M) = (3, 2, 6),\ \ (1, \infty, 1),\ \ (2, 3, 3).$$

We notice that the sum of the dimensions is $3+1+2=6=v$, which shows that the above solutions exhaust the possibilities. □

Example 9.2.2 (Icosahedron): The association scheme generated by the icosahedron has three classes and graphs as indicated in Figure 9.4. The graphs Γ_1 and Γ_2 are identical as graphs, although the enumeration of the vertices is different. The graph Γ_3 is disconnected. Altogether there are *four* spherical codes generated by this scheme: two versions of the icosahedron, the *Simplex* code in dimension five and the degenerated code with just one codeword. The parameters are

$$(n, \rho, M) = (3, 2-2/\sqrt{5}, 12),\ \ (1, \infty, 1),\ \ (5, 12/5, 6).$$

Again the sum of the dimensions is $3+3+1+5=12=v$. □

Example 9.2.3 (Cubeoctahedron): The graphs in the cubeoctahedral association scheme are shown in Figure 9.5. Again there

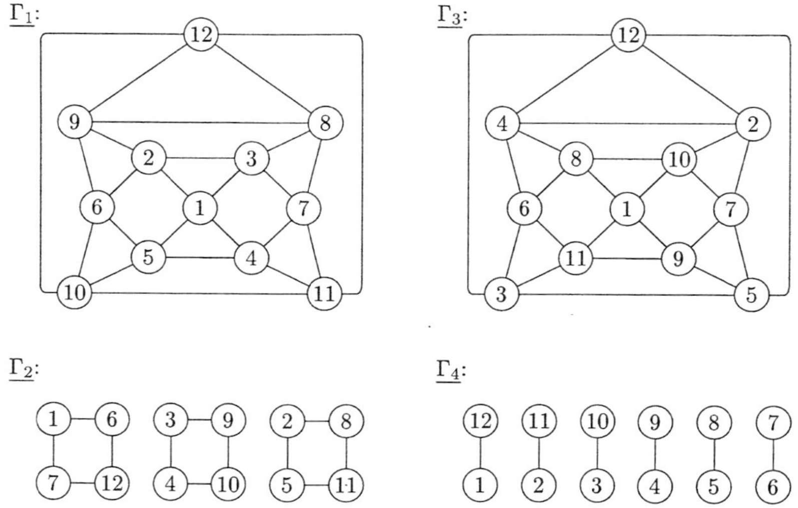

Figure 9.5: Distance graphs for the cubeoctahedron.

are two identical connected graphs. In addition there are two disconnected graphs. In this case we obtain two full size codes: the cubeoctahedron we started with and a code in dimension $n = 5$ with distance distribution

ρ_i	0	6/5	9/5	12/5	3
k_i	1	2	4	1	4

We can obtain this code by partitioning the cuboctahedron in three squares and add a suffix selected from the vertices of a triangle. By proper normalizations the above code is generated this way. The two remaining codes are the octahedron and the degenerated one-point code. The parameters of the four codes are

$$(n, \rho, M) = (3, 1, 12),\ (5, 6/5, 12),\ (1, \infty, 1),\ (3, 2, 6).$$

The sum of the dimensions is $3 + 5 + 1 + 3 = 12 = v$. □

The distances in the code $\mathcal{X}_i$ corresponding to the Gram matrix G_i are given by the dual eigenvalues $q_i(k)$. We have

$$G_i = \frac{1}{m_i} \sum_{k=0}^{d} q_i(k) A_k = \sum_{k=0}^{d} s_k A_k.$$

Therefore the inner products s_k corresponding to the code $\mathcal{X}_i$ generated by the eigenspace W_i are given by

$$s_k = \frac{1}{m_i} q_i(k), \quad k = 0, 1, \ldots, d.$$

The dimension is clearly $n = m_i$ and the size is $M = v$, provided there are no coinciding codewords.

Two codewords x and y have correlation s_i if and only if $A_i(x, y) = 1$. This means that the graph Γ_i connects the codewords on squared distance $\rho_i = 2 - 2s_i$. In each of the examples the graphs are enumerated so that the relations $\rho_1 < \rho_2 < \ldots < \rho_d$ hold for the spherical codes from which the schemes were *defined*. Notice, however, that for the other codes generated by the same scheme these relations do *not* necessarily hold. For instance, in the code with parameters $(n, \rho, M) = (5, 6/5, 12)$ generated by the cubeoctahedral scheme the smallest distance is the one corresponding to the graph Γ_3.

In all three examples we encountered codes with repeated codewords. In fact, $X = \mathbf{j}$ is generated by the projection $E_0 = (1/v)J$ and so occurs in *any* association scheme. We refer to this code as the trivial code. Our examples show that also non-trivial codes with repeated codewords are possible. Let us investigate a little closer the conditions under which this phenomenon occurs.

Let $\mathcal{A} = \{A_0, A_1, \ldots, A_d\}$ be an association scheme and let $\mathcal{X}$ be a corresponding code. Suppose that $\mathcal{X}$ has repeated codewords. It follows that we have $s_i = 1$ for some $i \neq 0$ and that neighboring codewords in the corresponding graph Γ_i coincide. If the graph Γ_i is connected this means that all codewords coincide. This leads to the degenerated one-point code. If Γ_i is disconnected the vertices in each one of its disjoint components are mapped on one and the same codeword. The components of a disconnected distance regular graph are necessarily isomorphic, and so the size of a component divides the size of the whole graph. It follows that the number of different codewords is always a factor of the size v of the vertex set V.

We have seen that for any association scheme $\mathcal{A} = \{A_0, A_1, \ldots, A_d\}$ there is one (possibly degenerated) spherical code for each one of the eigenspaces W_i, $i = 0, 1, \ldots, d$. A natural question to ask is whether there are any further spherical codes having the matrices in $\mathcal{A}$ as its distance graphs. More precisely: for a given association scheme $\mathcal{A} = \{A_0, A_1, \ldots, A_d\}$ can we characterize the set of *all* spherical codes having a Gram matrix G of the form

$$G = \sum_{i=0}^{d} s_i A_i,$$

i.e. having a Gram matrix G belonging to the space $L(\mathcal{A})$. The matrix G must be positive definite and have all its diagonal elements equal to 1. It follows that it must be a convex combination of the matrices G_i defined by the principal idempotents E_i, (recall that the matrices $E_i = (m_i/v)G_i$ form an orthogonal basis in $L(\mathcal{A})$). In other words: the Gram matrices corresponding to a given association scheme $\mathcal{A}$ are those located within a convex region in the space $L(\mathcal{A})$ having the matrices $G_i = (v/m_i)E_i$ as its extremal points.

For $i = 0, 1, \ldots d$ let $\mathcal{X}_i$ be a spherical code generated by the eigenspace W_i. Let X_i be a corresponding code matrix and let G_i be a corresponding Gram matrix. We assume that each matrix X_i is represented with the smallest possible number of columns. This means that each matrix X_i has v rows and $m_i = dim(\mathcal{X}_i)$ columns. We have

$$X_i X_i^T = G_i = (v/m_i)E_i.$$

Let $\alpha_0, \alpha_1, \ldots, \alpha_d$ be real numbers such that the squares sum to 1:

$$\alpha_0^2 + \alpha_1^2 + \cdots + \alpha_d^2 = 1.$$

Consider the following matrix:

$$X = [\alpha_0 X_0 | \alpha_1 X_1 | \ldots | \alpha_d X_d] .$$

We have

$$G = XX^T = \alpha_0^2 G_0 + \alpha_1^2 G_1 + \cdots + \alpha_d^2 G_d.$$

It is clear that the rows in X form a distance regular spherical code $\mathcal{X}$ corresponding to the given association scheme $\mathcal{A}$. Moreover *any* spherical code with distance graphs as defined by the scheme $\mathcal{A}$ must be equivalent to a spherical code having this form. In particular, choosing the coefficients α_i so as to *maximize* the squared minimum distance ρ in the resulting code $\mathcal{X}$ we obtain the direct sum $\mathcal{X} = \mathcal{X}_1 \oplus \mathcal{X}_2 \oplus \cdots \oplus \mathcal{X}_d$ as defined in Chapter 1.

The code matrices X_i are unique up to orthogonal transformations. Notice that the dimension of the code $\mathcal{X}$ is $\sum m_i$, where the summation extends over all indices i for which $\alpha_i \neq 0$. The best spherical codes are usually those corresponding to the *extremal* Gram matrices.

9.3 Metric schemes

Let Γ be a connected undirected graph with vertex set $V = \{1, 2, \ldots, v\}$. There is a natural function called *graph distance* defined on the product set $V \times V$. It assigns to any pair $\{x, y\} \in V \times V$ of vertices a number $d(x, y)$ equal to the number of edges in a shortest path connecting them. It is easy to show that this function satisfies all the usual metric axioms: it is *non-negative*, equals zero if and only if $x = y$, is *reflexive*, and satisfies the *triangle inequality*. The largest value d attained by this function in a given graph Γ is called the *diameter* of the graph.

For a given connected graph Γ with diameter d let us define indicator functions A_i by the rule

$$A_i(x, y) = \begin{cases} 1, & if \quad d(x, y) = i, \\ 0, & if \quad d(x, y) \neq i, \end{cases}$$

$$i = 0, 1, \ldots, d.$$

The definition parallels the one we employed for spherical codes. The difference is that now the metric used is *graph distance* rather than *Euclidean distance.*

We denote by Γ_i the graph formed by connecting by an edge all nodes $x, y \in V$ which are located on graph distance i. In this way we obtain a family of disjoint graphs Γ_i, one for each distance i, $i = 1.2. \ldots, d$. Together these graphs form a *partition* of the complete graph K_v. The matrix A_i is the incidence matrix of the graph Γ_i. In particular

$A = A_1$ is the incidence matrix of the graph $\Gamma = \Gamma_1$. Obviously all the matrices A_i are uniquely determined by A. Paralleling the corresponding definition for spherical codes we say that a connected graph Γ is *distance regular* if the matrices A_i define an association scheme.

Let Γ be distance regular and let p_{ij}^k denote the coefficient of A_k in the linear expansion of $A_i A_j$:

$$A_i A_j = \sum_{i=0}^{d} p_{ij}^k A_k.$$

It is not difficult to see that a distance regular graph generates an association scheme in the same way as does a distance regular spherical code. However, the present construction is more restricted: the association schemes for which there exists a generating distance regular graph constitute just a subclass of all association schemes. The schemes in this subclass are said to be *metric*.

Returning to our examples we see that the association schemes generated by the octahedron and the icosahedron are both metric. This can be seen by observing that in both cases there are obvious bijections between the graph distances and the Euclidean distances. The cubeoctahedron, on the other hand, is *not* metric. A simple inspection reveals that there are only *three* different non-zero *graph distances* but *four* different non-zero *Euclidean distances.*

The coefficients p_{ij}^k of a metric scheme exhibit all the usual properties enjoyed by the intersection numbers in general—and a few additional ones as well. In particular the following properties hold:

$$(i) \quad p_{ij}^k > 0, \quad \Rightarrow \quad k \le i + j, \ i \le k + j, \ j \le i + k;$$

$$(ii) \quad p_{i,1}^{i+1} > 0, \quad i = 0, 1, \ldots, d - 1.$$

The first property is an alternative statement of the triangle inequality for graph distance. The second property says that whenever two points x and y are on graph distance $i + 1$ there must be a neighbor z to y such that the graph distance between x and z is i. This is a special property of the graph distance, which follows directly from its definition. It can be shown (Cameron–van Lint, [61], p. 199) that whenever the intersection parameters of an association scheme satisfies these two properties, then the scheme is metric.

Let $\mathcal{A}$ be a metric scheme generated by the distance regular graph Γ. By (i) we have $p^i_{j,1} = 0$ unless $j = i-1, i, i+1$. It is customary to employ the following special notation:

$$a_i = p^i_{i,1}, \quad 0 \le i \le d,$$
$$b_i = p^i_{i+1,1}, \quad 0 \le i \le d-1,$$
$$c_i = p^i_{i-1,1}, \quad 1 \le i \le d.$$

Separately we define $c_0 = b_d = 0$.

It turns out that a metric scheme is completely specified by these parameters along with the single matrix $A = A_1$. More precisely, the following result is true.

Theorem 9.3.1 (Delsarte, [80]) *In any metric scheme there is a matrix A and a set of parameters $\{(a_i, b_i, c_i) : 0 \le i \le d\}$, such that the indicator matrices A_i can be expressed in the form $A_i = f_i(A)$, where $f_i(x)$ is a polynomial of degree i. The polynomials are obtained by the following recursion:*

$$f_0(x) = 1, \quad f_1(x) = x,$$
$$f_i(x) = (1/c_i) \cdot ((x - a_{i-1}) f_{i-1}(x) - b_{i-2} f_{i-2}(x)),$$
$$i = 2, 3, \ldots, d.$$

Proof: Let $\mathcal{A} = \{A_0, A_1, \ldots, A_d\}$ be a metric association scheme. Define $A = A_1$. We get $A_0 = f_0(A)$ and $A_1 = f_1(A)$. By the remarks above we have

$$A_{i-1}A_1 = \sum_{k=0}^{d} p^k_{i-1,1} A_k = b_{i-2}A_{i-2} + a_{i-1}A_{i-1} + c_i A_i,$$
$$i = 2, 3, \ldots, d.$$

Solving for A_i we find that the matrices A_i satisfy the same recursion as do the polynomials $f_i(x)$. It follows that we must have $A_i = f_i(A)$, $i = 0, 1, \ldots, d$. It is obvious from the definition that $c_i > 0$ for $i > 0$, so dividing by c_i is permissible. This completes the proof. □

We give three examples.

Example 9.3.1: Let $V = \mathbb{Z}_m^n$ be the set of n-length sequences of elements from the ring $\mathbb{Z}_m$ of integers modulo m. Define classes by the Hamming metric:

$$A_i(x,y) = \begin{cases} 1, & if \quad d_H(x,y) = i, \\ 0, & if \quad d_H(x,y) \neq i, \end{cases}$$

$$i = 0, 1, \ldots, n.$$

We obtain a metric scheme. The generating matrix A has diameter $d = n$ and valency $k = n(m-1)$. The intersection parameters are

$$a_i = i(m-2), \quad b_i = (n-i)(m-1), \quad c_i = i.$$

□

Example 9.3.2: Let $V = J(n, w)$ be the set of all binary sequences of Hamming weight w. This is the so called *Johnson space.* Notice that all Hamming distances in $J(n, w)$ are even. Define

$$A_i(x,y) = \begin{cases} 1, & if \quad d_H(x,y) = 2i, \\ 0, & if \quad d_H(x,y) \neq 2i, \end{cases}$$

$$i = 0, 1, \ldots, \min\{w, n-w\}.$$

Again we obtain a metric space. This time the generating matrix has diameter $d = \min\{w, n-w\}$ and valency $k = w(n-w)$. The intersection parameters are

$$a_i = i(n-2i), \quad b_i = (w-i)(n-w-i), \quad c_i = i^2.$$

□

Example 9.3.3: Let $V = E_2^n$ be the set of all binary n-length sequences with even weight. Let A_i be as in Example 9.3.2. We get a metric scheme such that the generating matrix has diameter $d = \lfloor n/2 \rfloor$ and valency $k = \binom{n}{2}$. The intersection parameters are

$$a_i = 2i(n-2i), \quad b_i = (1/2)\cdot(n-2i)(n-2i-1), \quad c_i = i(2i-1).$$

□

A distance regular graph is certainly regular. Let the valency of the generating graph Γ be k. The following identities are obvious:

$$a_i + b_i + c_i = k, \quad i = 0, 1, \ldots, d. \tag{9.3.1}$$

This means that the intersection parameters are not independent: given two of them the third one is always determined by the condition (9.3.1). It follows that we can characterize a metric scheme with the sequence $\{(b_i, c_i) : 0 \leq i \leq d\}$. This sequence is known as the *intersection array*. It turns out that most properties of the spherical codes generated by a metric association scheme can be obtained from the polynomials $f_i(x)$, and so from the intersection array alone.

As usual let k_i denote the valency of the graph Γ_i, $i = 1, 2, \ldots, d$, and define $k_0 = 1$. Let $\Gamma_i(x)$ denote the set of all vertices y on graph distance i from x. Counting in two ways the number of edges in Γ which connect $\Gamma_{i-1}(x)$ with $\Gamma_i(x)$ we obtain the relations

$$k_{i-1}b_{i-1} = k_i c_i, \quad i = 1, 2, \ldots, d.$$

From these identities the following explicit expression for the valencies k_i immediately follows:

$$k_i = \prod_{j=1}^{i} \frac{b_{j-1}}{c_j}, \quad i = 1, 2, \ldots, d. \tag{9.3.2}$$

Lemma 9.3.1 (Delsarte [80]) *Let $\{(b_i, c_i) : 0 \leq i \leq d\}$ be the intersection array to a metric association scheme. The following equalities are always satisfied:*

$$1 = c_1 \leq c_2 \leq \ldots \leq c_d; \quad k = b_0 \geq b_1 \geq \ldots \geq b_d = 0.$$

Proof: We prove only the first chain of inequalities. The second follows by an almost identical argument. The initial equality is trivial. Recall that the graph distance is the smallest number of edges connecting two vertices. It follows that for any i, $1 \le i \le d-1$ we can always find vertices x, u, y such that $d(x,u) = 1$, $d(u,y) = i$, $d(x,y) = i+1$. Consider $\Gamma_1(y) \cap \Gamma_{i-1}(u)$. This set is non-empty because Γ is connected. That means that we can choose a vertex v such that $d(u,v) = i-1$ and $d(v,y) = 1$. By the triangle inequality we have

$$d(x,y) - d(v,y) \le d(x,v) \le d(x,u) + d(u,v).$$

Evaluating the first and the last members in this chain we find the distance between the vertices x and v to be $d(x,v) = i$. This means that the set $\Gamma_1(y) \cap \Gamma_{i-1}(u)$ is a subset in $\Gamma_1(y) \cap \Gamma_i(x)$. Recalling the definitions of the intersection parameters we obtain

$$\begin{aligned} c_i = p^i_{i-1,1} &= |\Gamma_1(y) \cap \Gamma_{i-1}(u)| \le \\ &\le |\Gamma_1(y) \cap \Gamma_i(x)| = p^{i+1}_{i,1} = c_{i+1}, \\ i = 0, 1, \ldots, d-1. \end{aligned}$$

This completes the proof. □

With the aid of this lemma we can easily derive a very simple condition under which repeated codewords do *not* occur in a spherical code generated by a metric scheme.

We say that a distance regular graph Γ is *primitive* if all its constituting distance graphs $\Gamma_i, i = 1, 2, \ldots, d$ are connected. A distance regular graph which is not primitive is said to be *imprimitive.* By the remarks in the previous section a non-trivial spherical code generated from a primitive distance regular graph cannot have repeated codewords. The following result is true.

Theorem 9.3.2 *Let Γ be a distance regular graph with valency $k \ge 3$ and with diameter d. If Γ_2 and Γ_d are connected, then Γ is primitive.*

Proof: Let Γ be distance regular. By definition $\Gamma_1 = \Gamma$ is connected. Suppose that in addition also Γ_2 and Γ_d are connected. Under these conditions we will show that if Γ is imprimitive then necessarily the valency k cannot be larger than 2. If $d \leq 3$ all graphs Γ_i are connected already by assumption, so there is nothing to prove. So let $d \geq 4$ and suppose Γ is imprimitive. That means that there is at least one graph Γ_i, $3 \leq i \leq d-1$ which is not connected. Let Γ_i be such a graph. We will estimate $k = a_i + b_i + c_i$. We do that by estimating each one of the parameters a_i, b_i, c_i individually. The first observation is that Γ cannot contain an isosceles triangle with two sides of length i and one of length 1. If it did then for any pair of vertices x, y on distance 1 there would be a third vertex z on distance i from both. This is so because Γ is distance regular. But that would mean that any step of length 1 could be replaced by two steps of length i. This means that Γ_i would be connected because Γ_1 is connected. Thus, as Γ_i is disconnected it follows that we have $a_i = p^i_{i,1} = 0$.

The next two parameters require a slightly more complicated argument. We will apply Lemma 9.3.1.

As $3 \leq i \leq d-1$ both $b_{i-1} = p^{i-1}_{i,1}$ and $c_{i+1} = p^{i+1}_{i,1}$ are defined. We will show that if Γ_i is disconnected then both of these parameters must be strictly less than 2. For that purpose notice that if $b_{i-1} = p^{i-1}_{i,1} \geq 2$ then there exist distinct points x, y, u, v such that $d(x,u) = d(x,v) = i$; $d(u,y) = d(v,y) = 1$; $d(x,y) = i-1$. Similarly, if $c_{i+1} = p^{i+1}_{i,1} \geq 2$ there exist distinct points x, y, u, v such that $d(x,u) = d(x,v) = i$; $d(u,y) = d(v,y) = 1$; $d(x,y) = i+1$, (see Figure 9.6). In either case the triangle inequality implies $d(u,v) = 2$. It follows that $\{x,u,v\}$ is an isosceles triangle with two edges of length i and the third of length 1 or 2. By the same argument as before both of these possibilities are excluded since both $\Gamma = \Gamma_1$ and Γ_2 are connected. Thus both b_{i-1} and c_{i+1} must be strictly less than 2. By Lemma 9.3.1 the same conclusion holds also for b_i and c_i. As a consequence $k = a_i + b_i + c_i \leq 2$. □

Returning to our examples we notice that the graph generating the *octahedron* has diameter $d = 2$, with $\Gamma_2 = \Gamma_d$ disconnected. The graph generating the *icosahedron* has diameter $d = 3$, with $\Gamma_3 = \Gamma_d$ disconnected. The *cubeoctahedron* corresponds to a non-metric scheme and so is not generated by any distance regular graph.

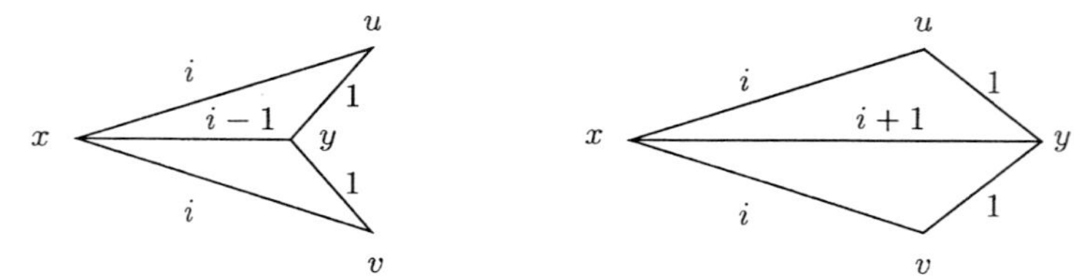

Figure 9.6: Illustration to the proof of Theorem 9.3.2

The restriction $k \geq 3$ is not severe. The only alternatives are $k = 1$ and $k = 2$. There is only one distance regular graph with valency $k = 1$, namely the trivial graph with just two connected vertices. A distance regular graph with valency $k = 2$ is a cycle.

Let X be a codeword matrix to a spherical code generated by a metric association scheme and let $\{\theta_0, \theta_1, \ldots, \theta_d\}$ be the corresponding set of eigenvalues, so that $A_i X = \theta_i X$, $i = 0, 1, \ldots, d$. The identities $A_i = f_i(A_1)$ imply $\theta_i = f_i(\theta_1)$. Denote $\alpha = \theta_1$ and notice that this is one of the eigenvalues corresponding to the incidence matrix $A = A_1$. Let the complete set of those eigenvalues be $\{\alpha_0, \alpha_1, \ldots, \alpha_d\}$, with $\alpha_0 = k$. Then we have $f_i(\alpha_j) = p_i(j)$, which means that the eigenvalue matrix $P = \{p_i(j)\}$ is also determined by the polynomials $f_i(x)$. Using the identity $\Sigma_{i=0}^d A_i = J$ we obtain

$$JE_j = \sum_{i=0}^{d} A_i E_j = \sum_{i=0}^{d} p_i(j) E_j = \sum_{i=0}^{d} f_i(\alpha_j) E_j.$$

Recall that the idempotents E_i are orthogonal and that $E_0 = (1/v)J$. We obtain

$$j = 0: \quad vE_0 = JE_0 = \sum_{i=0}^{d} f_i(k) E_0,$$

$$j \neq 0: \quad 0 = JE_j = \sum_{i=0}^{d} f_i(\alpha_j) E_j.$$

It follows that we have

$$\sum_{i=0}^{d} f_i(\alpha) = \begin{cases} v, & if \quad \alpha = k, \\ 0, & if \quad \alpha \neq k. \end{cases}$$

The eigenvalue $k = \alpha_0$ is given directly by the intersection parameters. The d remaining eigenvalues are solutions to the equation

$$\sum_{i=0}^{d} f_i(\alpha) = 0.$$

The left-hand side of this expression is a polynomial of degree d. It follows that this polynomial has d simple real roots. We collect our observations in the following theorem.

Theorem 9.3.3 (Delsarte [80]) *Let Γ be a distance regular graph with valency k and diameter d; let the generating polynomials be $f_i(x)$, $i = 0, 1, \ldots, d$, and let A be the incidence matrix of Γ. The following is true:*

(i) The eigenvalues α_i, $i = 1, 2, \ldots, d$, different from k, are obtained as roots of the polynomial $f_0(x) + f_1(x) + \cdots + f_d(x)$.

(ii) The components $p_i(j)$ of the eigenvalue matrix P are given by $p_i(j) = f_i(\alpha_j)$.

(iii) The multiplicity m_i of the i-th eigenvalue α_i is given by $m_i = q_i(0)$, $i = 0, 1, \ldots, d$, where $q_i(0)$ is the i-th component in the 0-th column of the matrix $Q = (1/v)P^{-1}$ of dual eigenvalues.

The above theorem contains all the information we need in order to determine the parameters of the different codes generated by a metric scheme $\mathcal{A}$. Except for the trivial code with matrix $X_0 = j$ corresponding to the eigenvalue $\alpha = k$ there are d additional codes. For each code $\mathcal{X}_i$ the number M_i of codewords is a factor of the order v of the graph Γ. If the graph is primitive and the valency k satisfies $k \geq 3$ the size equals v for all codes except $\mathcal{X}_0$. The dimension of the code $\mathcal{X}_j$ is given by the multiplicity m_j of the eigenvalue α_j. The inner products are $\{s_{ij} : 0 \leq i \leq d\}$, with $s_{ij} = f_i(\alpha_j)/k_i$, where k_i denotes the number of codewords having correlation s_i with a given codeword in $\mathcal{X}_j$. The numbers k_i—which are the same for all the codes—are directly obtained from the intersection parameters by the expression (9.3.2). We consider again our examples in dimension three.

Example 9.3.4 (Octahedron): The intersection parameters are:

i	a_i	b_i	c_i
0	0	4	0
1	2	1	1
2	0	0	4

We get

$$k_0 = 1, \quad k_1 = \frac{b_0}{c_1} k_0 = 4, \quad k_2 = \frac{b_1}{c_2} k_1 = 1.$$

The generating polynomials are

$$f_0(x) = 1, \qquad f_1(x) = x, \qquad f_2(x) = \frac{1}{4}x^2 - \frac{1}{2}x - 1.$$

Thus we have

$$\sum_{i=0}^{d} f_i(x) = \frac{1}{4}x(x+2),$$

with $d = 2$, which gives us $\{\alpha_0, \alpha_1, \alpha_2\} = \{4, 0, -2\}$. Evaluating $p_i(j) = f_i(\alpha_j)$ we obtain

$$P = \begin{bmatrix} 1 & 1 & 1 \\ 4 & 0 & -2 \\ 1 & -1 & 1 \end{bmatrix} \qquad Q = 6P^{-1} = \begin{bmatrix} 1 & 1 & 1 \\ 3 & 0 & -3 \\ 2 & -1 & 2 \end{bmatrix},$$

from which we conclude $(m_0, m_1, m_2) = (1, 3, 2)$. Finally, the inner products are given by $S = \{s_{ij}\}$, with $s_{ij} = p_i(j)/k_i$:

$$S = \begin{bmatrix} 1 & 1 & 1 \\ 1 & 0 & -\frac{1}{2} \\ 1 & -1 & 1 \end{bmatrix}.$$

□

Example 9.3.5 (Icosahedron): The intersection parameters are:

i	a_i	b_i	c_i
0	0	5	0
1	2	2	1
2	2	1	2
3	0	0	5

We get

$$k_0 = 1, \quad k_1 = \frac{b_0}{c_1} k_0 = 5, \quad k_2 = \frac{b_1}{c_2} k_1 = 5, \quad k_3 = \frac{b_2}{c_3} k_2 = 1.$$

The generating polynomials are

$$f_0(x) = 1, \quad f_1(x) = x, \quad f_2(x) = (1/2)x^2 - x - 5/2,$$
$$f_3(x) = (1/10)\,(x^3 - 4x^2 - 5x + 10).$$

We get

$$\begin{aligned}\textstyle\sum_{i=0}^{d} f_i(x) &= (1/10)\,(x^3 + x^2 - 5x - 5) \\ &= (1/10)\,(x+1)(x+\sqrt{5})(x-\sqrt{5}),\end{aligned}$$

which gives us $\{\alpha_0, \alpha_1, \ldots, \alpha_d\} = \{5, \sqrt{5}, -1, -\sqrt{5}\}$. Evaluating the quantity $p_i(j) = f_i(a_j)$ we obtain

$$P = \begin{bmatrix} 1 & 1 & 1 & 1 \\ 5 & \sqrt{5} & -1 & -\sqrt{5} \\ 5 & -\sqrt{5} & -1 & \sqrt{5} \\ 1 & -1 & 1 & -1 \end{bmatrix},$$

$$Q = 12P^{-1} = \begin{bmatrix} 1 & 1 & 1 & 1 \\ 3 & 3/\sqrt{5} & -3/\sqrt{5} & -3 \\ 5 & -1 & -1 & 5 \\ 3 & -3/\sqrt{5} & 3/\sqrt{5} & -3 \end{bmatrix},$$

from which we conclude $(m_0, m_1, m_2, m_3) = (1, 3, 5, 3)$. Finally, the inner products are all contained in the matrix

$$S = \begin{bmatrix} 1 & 1 & 1 & 1 \\ 1 & 1/\sqrt{5} & -1/5 & -1/\sqrt{5} \\ 1 & -1/\sqrt{5} & -1/5 & 1/\sqrt{5} \\ 1 & -1 & 1 & -1 \end{bmatrix}.$$

□

9.4 Strongly regular graphs

Let Γ be a regular graph of valency k over a vertex set of size v. Suppose that Γ is neither null nor complete. Let $\{x, y\}$ be a pair of distinct vertices and suppose the graph is such that the number of common neighbors to x and y depends only on whether x and y are neighbors or not. Let it be a in the former case and c in the latter case. A graph with this property is said to be *strongly regular*. It is said to have parameters (v, k, a, c). The following constraints are direct consequences of the definition:

$$1 \leq k \leq v - 2, \quad 0 \leq a \leq k - 1, \quad 0 \leq c \leq k. \tag{9.4.1}$$

If the graph is *connected* it is obviously distance regular with diameter $d = 2$. Conversely, any distance regular graph with diameter $d = 2$ is a connected strongly regular graph. A strongly regular graph is *disconnected* if and only if $c = 0$. In that case the graph is necessarily a disjoint union of identical complete graphs. For our purposes this case is of no interest. In the subsequent discussion we will always assume $c > 0$.

Strongly regular graphs were introduced by Bose, [39] (1963). An excellent overview of the general theory can be found in Cameron–van Lint, [61] (1991). Our presentation will focus on the connections with spherical codes. We give extensive references to the book by Cameron–van Lint.

Suppose Γ is distance regular with $d = 2$. The intersection parameters $\{(a_i, b_i, c_i)\} : 0 \leq i \leq 2\}$ are given by

$$\begin{array}{lll} a_0 = 0, & b_0 = k, & c_0 = 0, \\ a_1 = a, & b_1 = k - a - 1, & c_1 = 1, \\ a_2 = k - c, & b_2 = 0, & c_2 = c. \end{array}$$

We see that all intersection parameters are determined by $b_0 = k$, $a_1 = a$, and $c_2 = c$. In fact, also the size v is connected to these numbers. Let x be an arbitrary vertex in the graph Γ and let $\Gamma(x)$ denote the set of neighbors to x. By counting in two ways the number of edges $\{y, z\}$ with $y \in \Gamma(x)$, $z \notin \Gamma(x) \cup \{x\}$ we obtain the identity

$$k(k - a - 1) = (v - k - 1)c. \tag{9.4.2}$$

We get simple examples of strongly regular graphs by considering special cases of the examples from the previous section.

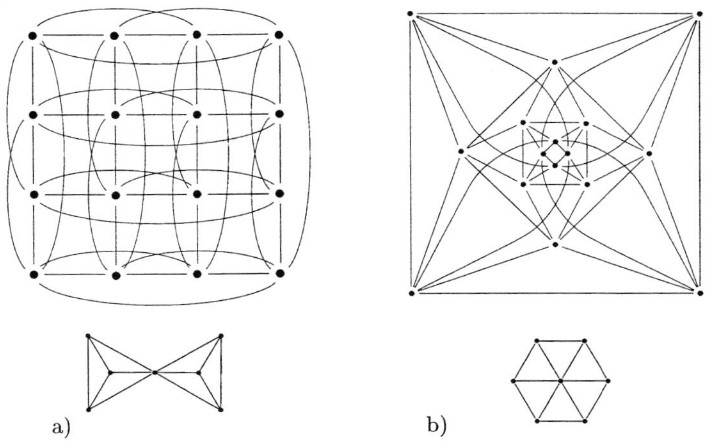

Figure 9.7: The square lattice $L_2(4)$ (a) and the Shrikhande graph (b) along with subgraphs indicating a set of neighbors.

Example 9.4.1: Put $n = 2$ in Example 9.3.1 from the previous section. We get a strongly regular graph with parameters $(v, k, a, c) = (m^2, 2m-2, m-2, 2)$, $m \geq 2$. Graphs of this kind are denoted $L_2(m)$ in Cameron–van Lint ([61], p. 34), and are referred to as *square lattice graphs.* □

Example 9.4.2: Consider the graph defined by the Johnson scheme $J(n, w)$ as defined in Example 9.3.2. Take $w = 2$. We get a strongly regular graph with parameters $(v, k, a, c) = [n(n-1)/2, 2(n-2), n-2, 4]$. Graphs of this kind are denoted $T(n)$ and are referred to as *triangular graphs* in Cameron–van Lint, [61], p. 33. □

Example 9.4.3: Consider the graph obtained from the set $V = E_2^n$ of all binary sequences of even weight, as considered in Example 9.3.3 of the previous section. Choose $n = 5$. We get a strongly regular graph with parameters $(v, k, a, c) = (16, 10, 6, 6)$. This graph is known as the *Clebsch graph*, (Cameron–van Lint, [61], p. 35, refer to the *complement* of this graph as the Clebsch graph). □

Strongly regular graphs with the same parameters (v, k, a, c) need not be isomorphic. In Figure 9.7 two graphs are displayed. One is the square lattice graph $L_2(4)$. The other is the so called *Shrikhande graph* (Cameron–van Lint, [61], p. 60). Both are strongly regular with parameters $(16, 6, 2, 2)$. They are not isomorphic. In $L_2(4)$ the set $\Gamma(x)$ of vertices adjacent to a given vertex x consists of two triangles, joined only in the vertex x; in the Shrikhande graph the set $\Gamma(x)$ is a cycle (see Figure 9.7). The Shrikhande graph can be obtained by connecting four square *antiprisms* in a ring, and so can be displayed as a plane graph on a torus.

The *complement* of a strongly regular graph is also strongly regular. The parameters $(\overline{v}, \overline{k}, \overline{a}, \overline{c}\})$ of the complement $\overline{\Gamma}$ are related to the parameters (v, k, a, c) of the graph Γ according to the following equations:

$$\overline{v} = v, \quad \overline{k} = v - 1 - k, \quad \overline{a} = v - 2k + c - 2, \quad \overline{c} = v - 2k + a.$$

As each parameter is nonnegative we obtain the following additional constraints on the parameters of a strongly regular graph:

$$v \geq 2k - c + 2, \quad v \geq 2k - a. \tag{9.4.3}$$

We notice that $\overline{k} = k$ if and only if $v = 2k + 1$ and that both of the relations $\overline{a} = a$ and $\overline{c} = c$ hold if and only if $c = a + 1$. A strongly regular graph with the property that all of these relations hold is called a *conference graph*. Inserting in (9.4.2) we find that the parameters of a conference graph have the form $(v, k, a, c) = (4a + 5, 2a + 2, a, a + 1)$. The following construction produces an infinite family of such graphs.

Example 9.4.4: Let q be a prime power such that $q \equiv 1 \bmod 4$ and let the vertex set V be identified with the finite field $\mathbb{F}_q$. We notice that $x - y$ is a square if and only if $y - x$ is a square. It follows that we can prescribe that x and y should be connected by an edge if and only if $x - y$ is a non-zero square. The resulting graph is a conference graph with parameters $(v, k, a, c) = (q, (q-1)/2, (q-5)/4, (q-1)/4)$. It is known as the *Paley* graph and is denoted $P(q)$ in Cameron–van Lint, [61], p. 35. □

We proceed to derive some properties of strongly regular graphs in general. By Theorem 9.3.1 we have

$$f_2(x) = \frac{1}{c}((x - a)x - k), \tag{9.4.4}$$

and so the eigenvalues different from k are the solutions to the equation

$$\sum_{i=0}^{d} f_i(x) = 1 + x + \frac{1}{c}((x-a)x - k) = 0.$$

We get

$$x^2 - (a-c)x - (k-c) = 0. \tag{9.4.5}$$

Let us define

$$\Delta = (a-c)^2 + 4(k-c).$$

The eigenvalues different from k are

$$\theta = (a - c + \sqrt{\Delta})/2, \quad \tau = (a - c - \sqrt{\Delta})/2.$$

As k is never less than c both roots are real. We notice the inequalities $\theta \geq 0$, $\tau < 0$.

The eigenvalue k has multiplicity 1. Let m_θ and m_τ be the multiplicities of θ and τ respectively. We get

$$v = m_\theta + m_\tau + 1.$$

Let A be the adjacency matrix of a strongly regular graph. Since $tr(A) = 0$ we have

$$0 = k + \theta m_\theta + \tau m_\tau.$$

Solving these two equations with respect to m_θ and m_τ and expressing the eigenvalues in terms of the parameters (v, k, a, c) we obtain

$$m_\theta,\ m_\tau = (1/2)\left[v - 1 \pm \frac{(v-1)(c-a) - 2k}{\sqrt{(c-a)^2 + 4(k-c)}}\right], \tag{9.4.6}$$

where the plus sign applies to m_θ and the minus sign applies to m_τ.

Apart from giving explicit expressions for the multiplicities these equations also provide us with a powerful test: a necessary condition for a strongly regular graph with parameters (v, k, a, c) to exist is that the right hand member of (9.4.6) evaluates to non-negative integers. This condition will be referred to as the *integrality condition.*

A strongly regular graph is *primitive* if both Γ and its complement $\overline{\Gamma}$ are connected. This definition is in accordance with the general definition of primitivity for distance regular graphs. The graph $\overline{\Gamma}$ is disconnected if and only if $\overline{c} = v - 2k + a$ equals zero. Inserting in (9.4.2) we find that this is the case if and only if c equals k. A graph satisfying this condition is necessarily a complete multipartite graph. For our purposes the most interesting cases are the so called *cocktail party graphs*, denoted $C(r)$, with parameters $(v, k, a, c) = (2r, 2r-2, 2r-4, 2r-2)$, $r \geq 2$, (Cameron–van Lint [61], p. 35). These graphs generate the *Biorthogonal codes*.

Let Γ be a primitive strongly regular graph. This means that it is also a primitive distance regular graph with diameter $d = 2$. It follows that it generates two nontrivial codes, both of size $M = v$. The dimension is $n = m_\theta$ or $n = m_\tau$ depending on the eigenvalue generating the code. Consider the code generated by θ. There are two different inner products in this code, given by $s_1 = \theta_1/k_1$ and $s_2 = \theta_2/k_2$. We have $\theta_1 = \theta$ and $\theta_2 = f_2(\theta)$, where $f_2(x)$ is given by equation (9.4.4). The *valencies* are $k_1 = (b_0/c_1) = k$ and $k_2 = (b_1/c_2) \cdot k_1 = v - k - 1$. Thus in a more explicit form the inner products can be expressed as follows:

$$s_1 = \frac{\theta}{k}, \quad s_2 = \frac{\theta^2 - a\theta - k}{c(v-k-1)} = -\frac{1+\theta}{v-1-k}.$$

The last expression for s_2 is obtained using (9.4.5). The inner products for the code generated by the negative eigenvalue τ are obtained by simply replacing θ with τ in these expressions. Let us define

$$\sigma_1 = -\frac{1+\tau}{v-1-k}, \quad \sigma_2 = \frac{\tau}{k}.$$

We have reversed the indices so as to maintain the relation $\sigma_1 > \sigma_2$. From the eigenvalue θ we obtain a spherical code with parameters $(n, \rho, M) = (m_\theta, 2-2s_1, v)$, while τ generates a code with parameters $(n, \rho, M) = (m_\tau, 2-2\sigma_1, v)$.

Let us consider the two codes generated from the complement $\overline{\Gamma}$. Simple calculations reveal the following identities:

$$\overline{\Delta} = \Delta, \quad \overline{\theta} = -1 - \tau, \quad \overline{\tau} = -1 - \theta, \quad m_{\overline{\theta}} = m_\tau, \quad m_{\overline{\tau}} = m_\theta,$$

where the meaning of the new symbols should be obvious. Further simple calculations reveal the following identities:

$$\overline{s}_1 = \sigma_1, \quad \overline{s}_2 = \sigma_2.$$

v	k	a	c	*Label*
5	2	0	1	$P(5)$
10	3	0	1	*Petersen*
9	4	1	2	$P(9)$, $L_2(3)$
16	5	0	2	$\overline{Clebsch}$
16	6	2	2	$L_2(4)$
15	6	1	3	$\overline{T_4(6)}$
13	6	2	3	$P(13)$
10	6	3	4	$T(5)$
50	7	0	1	*Hoffman–Singleton*
25	8	3	2	$L_2(5)$
17	8	3	4	$P(17)$
15	8	4	4	$T(6)$
28	9	0	4	$\nexists$
16	9	4	6	$\overline{L_2(4)}$
56	10	0	2	*Gewirtz*
36	10	4	2	$L_2(6)$
26	10	3	4	$S(13,3,2)$
21	10	5	4	$T(7)$
27	10	1	5	*Schläfli*
21	10	4	5	$\nexists$
21	10	3	6	$\overline{T(7)}$
16	10	6	6	*Clebsch*

Table 9.1: Parameters for tentative strongly regular graphs with parameters $1 \leq c < k \leq 10$.

In a similar way we also have

$$s_1 = \overline{\sigma}_1, \quad s_2 = \overline{\sigma}_2,$$

where again the meaning of the notation should be obvious. The interpretation is clear: the strongly regular graph Γ and its complement $\overline{\Gamma}$ generate the same pair of distance regular codes.

A quadruple (v, k, a, c) satisfying the conditions (9.4.1)–(9.4.3), and (9.4.6) is a *tentative* set of parameters for a strongly regular graph. In Table 9.1 we have listed all such tentative quadruples with $1 \leq c < k \leq 10$. In most of the cases a corresponding graph exists, and in most of the cases one or several of the constructions we have already considered generate such a graph. For two of the cases it can be shown that no graph exists. The non-existence of a graph with parameters $(28, 9, 0, 4)$ follows from the so called *absolute bound*, which will be established in Section 9.5. The non-existence of a graph with parameters $(21, 10, 4, 5)$ follows by a theorem of van Lint–Seidel (Cameron–van Lint, [61], p. 38). In four additional cases graphs do exist, but constructions different from those we have considered so far are needed. We proceed to derive these graphs—along with several additional graphs generated by the same constructions.

The Schläfli graph (27, 10, 1, 5)

Prescribing the parameters (k, a, c) to be of the form $(k, a, c) = (2u, 1, u)$ forces the order of the graph to be $v = 6u - 3$. The integrality conditions are met if and only if $u = 2, 3, 5, 11$. The case $u = 11$ is excluded because it would generate a spherical code violating the linear programming bound. The cases $u = 2$ and $u = 3$ correspond to graphs with parameters $(9, 4, 1, 2)$ and $(15, 6, 1, 3)$, which are both generated by the constructions we already know. This leaves us with the case $(v, k, a, c) = (27, 10, 1, 5)$. Such a graph exists and is known as the *Schläfli graph* (Cameron–van Lint, [61], p. 32). A construction is as follows.

Let $\mathbb{F}_2$ be the binary field and define $I \triangleq \{1, 2, 3, 4, 5\}$. Let $\mathcal{C}$ be the simple parity check code in $\mathbb{F}_2^5$:

$$\mathcal{C} = \{c = (c_1, c_2, \ldots, c_5) \in \mathbb{F}_2^5 : c_1 + c_2 + \cdots + c_5 = 0\}.$$

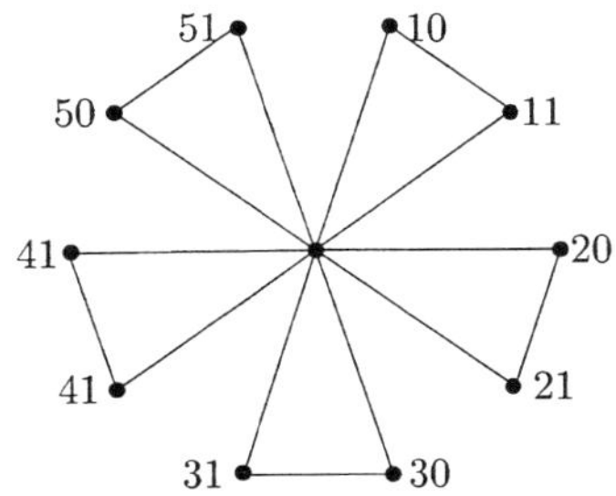

Figure 9.8: The point ∞ and the set $\Gamma(\infty) = I \times \mathbb{F}_2$, $I = \{1, 2, 3, 4, 5\}$ of neighbors in the Schläfi graph, with $(v, k, a, c) = (27, 10, 1, 5)$.

Notice that the distance distribution of $\mathcal{C}$ is

d_i	0	2	4
k_i	1	10	5

We construct a graph Γ by defining first a vertex set V and then giving rules for how the elements in V should be connected to form edges.

We let the vertex set V be the union of the following three sets: a singleton $\{\infty\}$ consisting of the special point ∞ alone, the Cartesian product $I \times \mathbb{F}_2$, and the code $\mathcal{C}$. The size is $v = |V| = 1+10+16 = 27$. We connect pairs of vertices as follows:

(i) The special point ∞ will be connected to all vertices $x \in I \times \mathbb{F}_2$, and to no other vertices.

(ii) The vertices $(i, x) \in I \times \mathbb{F}_2$ are pairwise connected so that $(i, 0)$ is connected to $(i, 1)$, $1 \leq i \leq 5$, (see Figure 9.8).

(iii) The vertex $c = (c_1, c_2, \ldots, c_5) \in \mathcal{C}$ is connected to each one of the vertices $(i, c_i) \in I \times \mathbb{F}_2$, $1 \leq i \leq 5$.

(iv) Each vertex $c \in \mathcal{C}$ is connected to all other vertices in $\mathcal{C}$ which are on Hamming distance 4 from c.

It is well known (and easy to show) that for any index i there are 8 codewords c with $c_i = 0$ and 8 codewords c with $c_i = 1$. It follows

in the three first cases. For $k = 2$ we have the pentagon with parameters $(5, 2, 0, 1)$. For $k = 3$ we have the so called *Petersen graph*, with parameters $(10, 3, 0, 1)$, see Figure 9.9. This graph is also obtained as the complement to the triangular graph $T(5)$. In the case $k = 7$ there is a graph known as the *Hoffman–Singleton graph*, with parameters $(50, 7, 0, 1)$. The existence of a strongly regular Moore graph with valency $k = 57$ is still an open question. We proceed with a construction of the Hoffman–Singleton graph.

Consider a set V of size 6, say $V = \{0, 1, 2, 3, 4, 5\}$. The complete graph K_6 over V contains $\binom{6}{2} = 15$ edges. Let the set of all these edges be denoted E. Notice that *any* pair of vertices defines an edge. A partition of the vertex set V into three disjoint edges will be called a *factor* (the proper graph-theoretic term is actually *1-factor*, but for our limited purposes the simplified term seems more appropriate). A partition of E into disjoint factors will be called a *factorization*. Let E' denote the set of all factors over V and let V' denote the set of all factorizations. It is not too difficult to see that there are altogether 15 factors and 6 factorizations. In other words: the number of *factors* equals the number of *edges* and the number of *factorizations* equals the number of *vertices*, so we have $|E'| = |E|$ and $|V'| = |V|$. This observation is essential in the construction. However, we need a few more observations.

In Figure 9.10 we have indicated all six different factorizations. Each factorization contains 5 factors. The factorizations have been named by the letters $a, b, \ldots, f$. Notice that each factor appears in precisely *two* factorizations. For instance, the factor 01 23 45 appears in the factorizations a and b and in no other factorization. Conversely, any pair of factorizations have exactly *one* factor in common. For instance, the factorizations d and f have the factor 05 14 23 in common, and no other common factor. This means that each factor (= element in E') uniquely determines a pair of factorizations (= pair of elements in V') and that each pair of factorizations (= pair of elements in V') uniquely determines a factor (= element in E'). It also means that if we choose to interpret V' as a vertex set we will have E' as the corresponding set of possible edges. Furthermore, if we continue to form the factors and factorizations over V' we find that each factor over V' corresponds to an edge over V and that each factorization

a		
$ab - 01$	23	45
$af - 02$	15	34
$ae - 03$	14	25
$ad - 04$	12	35
$ac - 05$	13	24

b		
$ab - 01$	23	45
$bc - 02$	14	35
$bd - 03$	15	24
$bf - 04$	13	25
$be - 05$	12	34

c		
$cd - 01$	25	34
$bc - 02$	14	35
$cf - 03$	12	45
$ce - 04$	15	23
$ac - 05$	13	24

d		
$cd - 01$	25	34
$de - 02$	13	45
$bd - 03$	15	24
$ad - 04$	12	35
$df - 05$	14	23

e		
$ef - 01$	24	35
$de - 02$	13	45
$ae - 03$	14	25
$ce - 04$	15	23
$be - 05$	12	34

f		
$ef - 01$	24	35
$af - 02$	15	34
$cf - 03$	12	45
$bf - 04$	13	25
$df - 05$	14	23

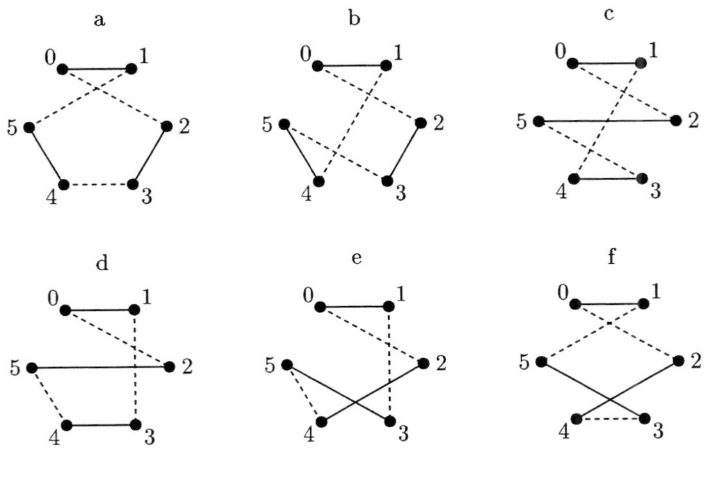

Figure 9.10: Factorizations of the hexagon.

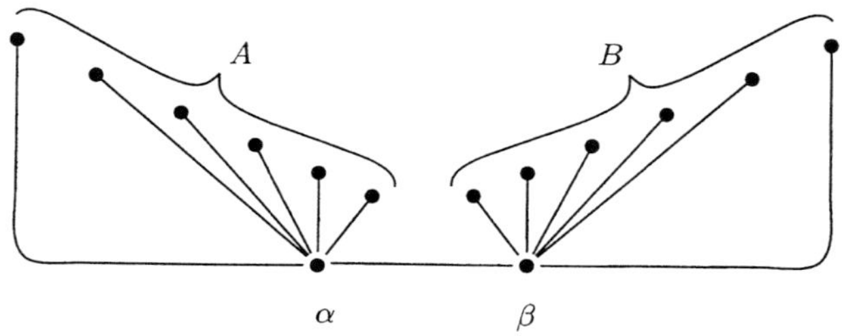

Figure 9.11: Two vertices, α and β, in the Hoffman–Singleton graph, along with their neighbors.

over V' corresponds to a vertex in V. This relation between the sets V and V' is referred to as a *duality* (Cameron–van Lint, [61], p. 83). It is the key element in the construction of the Hoffman–Singleton graph.

Our goal is to construct a strongly regular graph with parameters $(v, k, a, c) = (50, 7, 0, 1)$. Let α and β be two adjacent vertices in this graph and let $A = \Gamma(\alpha)\diagdown\{\beta\}$ and $B = \Gamma(\beta)\diagdown\{\alpha\}$ be the sets of additional neighbors, see Figure 9.11. The sets A and B are disjoint, each of size 6. There remains $50-2\cdot 7 = 36$ vertices. Each one of these must be connected to one vertex in A and one vertex in B. Identify the vertices in A with the set V and those in B with the set V'. Let the 36 remaining vertices be identified with $V \times V'$ and connect $(x, y) \in V \times V'$ with $x \in V$ and $y \in V'$. In this way each one of the elements in V and V' will be connected to precisely 7 other vertices. It remains to determine the connections within the set $V \times V'$. This is where the duality comes into play. Let (u, v) and (x, y) be two points in $V \times V'$. We connect them with an edge if and only if ux is an edge in the factor vy. The idea is illustrated by an example in Figure 9.12. In this way any vertex in $V \times V'$ is connected to precisely 5 other vertices in this set. In addition it is connected to one vertex in V and one vertex in V'. It follows that the construction gives a regular graph of valency $k = 7$ over a vertex set of size $v = 50$. By inspection we infer that any pair of adjacent vertices have no further common neighbor. It follows that the graph must be strongly regular, and so is a Moore graph with parameters (50, 7, 0, 1). This is the so called Hoffman–Singleton graph. A corresponding spherical code has parameters $(n, \rho, M) = (21, 40/21, 50)$. □

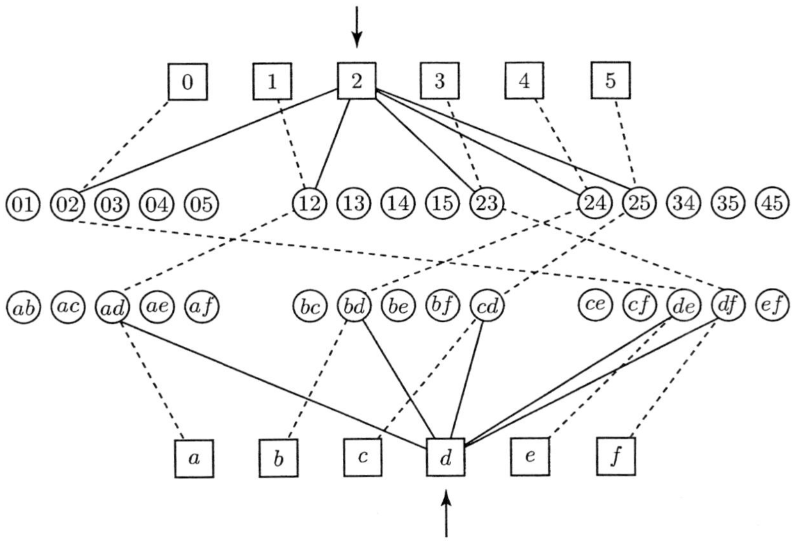

$$\Gamma(2d) = \{0e, 1a, 3f, 4b, 5c\}$$

Figure 9.12: Illustration to the construction of the Hoffman–Singleton graph.

Quasi-symmetric designs

The triangular graphs $T(n)$ can be generalized in two steps. Recall that those graphs can be viewed as the distance graphs of the Johnson schemes $J(n,2)$. The first generalization is obtained by replacing, if possible, the Johnson scheme with one of its possible subsets, namely a Steiner system $S(n,w,t)$ with $t=2$ and with $b=n(n-1)/(w(w-1))>n$. The condition $t=2$ implies that the corresponding constant weight code has at most two non-zero distances while the condition $b>n$ guarantees that the number of non-zero distances is at least two. A simple counting argument reveals that the distance graph is strongly regular with the following parameters (Cameron–van Lint [61], p. 69, Godsil [116], p. 185):

$$v=\frac{n(n-1)}{w(w-1)},\qquad k=\frac{w(n-w)}{w-1},$$

$$a=\frac{n-2w+1}{w-1}+(w-1)^2,\quad c=w^2.$$

The Steiner system $S(13,3,2)$ satisfies both of the conditions $t=2$ and $b>n$. The corresponding strongly regular graph has parameters $(v,k,a,c)=(26,15,8,9)$. The complement has parameters $(v,k,a,c)=(26,10,3,4)$ and is one of the graphs in Table 9.1 which is not generated by any of our previous constructions. The spherical codes generated by these two graphs have parameters $(n,\rho,M)=(12,26/15,26)$, $(13,8/5,26)$. Both are outperformed by other simple constructions.

From the point of view of spherical codes a further generalization suggested by Goethals–Seidel, [117], is more interesting. That construction is based on so called *quasi-symmetric designs*. A short digression is needed in order to establish the necessary definitions.

Recall that a t-design can be viewed as a constant weight code with the property that for any fixed set of t coordinates there is always a fixed number of codewords having ones in all those positions. If the code has length n and weight w and if the fixed number is λ_t, then we say that the design is a $t-(n,w,\lambda_t)$ design, also denoted $T(n,w,t,\lambda_t)$. In order not to infer with the notation already introduced for other purposes we employ a notation for t-designs that deviates slightly from the standard: the parameters n, w are usually denoted v, k.

A t-design is also an s-design for any s, $0 \leq s \leq t$. More precisely, if $\mathcal{C} \subseteq J(n,w)$ is a t-(n, w, λ_t) design then for any s, $0 \leq s \leq t$ it is also an s-(n, w, λ_s) design, where the parameters λ_t and λ_s are related according to the formula

$$\lambda_s \binom{w-s}{t-s} = \lambda_t \binom{n-s}{t-s},$$

see Cameron–van Lint [61], p. 3, MacWilliams–Sloane [167], p. 59. The following two special cases of this formula are particularly useful:

$$\lambda_0 = \lambda_t \frac{\binom{n}{t}}{\binom{w}{t}}, \qquad \lambda_{s+1} = \lambda_s \frac{w-s}{n-s}.$$

The parameters λ_0 and λ_1 are usually denoted b and r respectively.

A t-design with $t \geq 2$ satisfies Fisher's inequality: $b \geq n$. Designs fulfilling this bound with *equality* are called *symmetric designs.* Viewed as a code such a design is a binary constant weight code with only *one* non-zero distance. The distance graph has diameter $d = 1$ and is the complete graph. The code is a *Simplex.*

A design with *two* non-zero distances is called *quasi-symmetric*. Let $\mathcal{C}$ be the binary code generated by such a design and consider an arbitrary fixed codeword $c = (c_1, c_2, \ldots, c_n) \in \mathcal{C}$. The *support* of c is denoted *supp*(c). It indicates the set of those indices k for which $c_k = 1$. Let α and β be two disjoint subsets of *supp*(c). Denote by $N(\alpha, \beta)$ the set of codewords in $\mathcal{C}$ having ones in α and zeros in β. The inclusion-exclusion principle implies (Cameron–van Lint [61], pp. 21, 24-25) that the size of $N(\alpha, \beta)$ depends only on the sizes of α and β. It is customary to denote

$$\nu_{i,j} = |N(\alpha, \beta)|, \quad |\alpha| = j, \quad |\beta| = i - j, \quad 0 \leq j \leq i \leq w.$$

The following relations follow:

$$\nu_{i,i} = \begin{cases} \lambda_i, & 0 \leq i \leq t, \\ 1, & t+1 \leq i \leq w, \end{cases}$$

$$\nu_{i-1,j} = \nu_{i,j} + \nu_{i,j+1}.$$

With the aid of these formulas all the parameters $\nu_{i,j}$ can easily be computed from the parameters t-(n, w, λ) of the design $\mathcal{C}$. We notice that $\nu_{i,w}$ denotes the number of codewords which agree with c in the subset α of size $|\alpha| = i$ and disagree in the subset $\beta = supp(c) \setminus \alpha$. As the subset α can be selected in $\binom{w}{i}$ different ways we conclude that there are $\nu_{i,w}\binom{w}{i}$ codewords on distance $2w - 2i$ from the given codeword $c \in \mathcal{C}$. As this result holds for any codeword in $\mathcal{C}$ we conclude that $\mathcal{C}$ is *regular* and that the distance distribution is determined by the design parameters t-(n, w, λ_t).

For quasi-symmetric designs more can be said. The very definition suggests that such a design might generate a distance regular graph. Goethals–Seidel, [117] showed that so is actually the case. We refer to the proof given in Cameron–van Lint, [61], p. 70. The following explicit results are true.

Theorem 9.4.1 *Let $\mathcal{C} \subseteq J(n, w)$ be a quasi-symmetric design with parameters t-(n, w, λ_t) and such that any pair of distinct codewords have either x or y ones in common, where $x < y$. Then the distance graph of $\mathcal{C}$ is strongly regular with parameters (v, k, a, c) given by*

$$v = \lambda_t \frac{\binom{n}{t}}{\binom{w}{t}}, \qquad \lambda_1 = \frac{v\,w}{n}, \qquad \lambda_2 = \frac{v\,w\,(w-1)}{n\,(n-1)},$$

$$k = \frac{(\lambda_1 - 1)w - (v-1)x}{y - x}, \qquad \theta = \frac{\lambda_1 - \lambda_2 - w + x}{y - x},$$

$$\tau = -\frac{w - x}{y - x}, \qquad c = \theta\tau + k, \qquad a = \theta + \tau + c.$$

Remark 9.4.1: Recall that a strongly regular graph has precisely *three* different eigenvalues, the largest of which equals the valency k. The other two are given by the parameters θ and τ.

It is natural to display the parameters $\nu_{i,j}$ in a Pascal triangle, with $\nu_{0,0}$ at the top and with $\nu_{i,w}, 0 \leq i \leq w$ at the bottom. Such a triangle is referred to as an *intersection triangle*. In Figure 9.13 we

Intersection numbers	
759	0
506 · 253	1
330 · 176 · 77	2
210 · 120 · 56 · 21	3
130 · 80 · 40 · 16 · 5	4
78 · 52 · 28 · 12 · 4 · 1	5
46 · 32 · 20 · 8 · 4 · 0 · 1	6
30 · 16 · 16 · 4 · 4 · 0 · 0 · 1	7
30 · 0 · 16 · 0 · 4 · 0 · 0 · 0 · 1	8

Figure 9.13: Intersection triangle for the Steiner system S(24,8,5).

show such a triangle for the famous Steiner system $S(24,8,5)$, which can be found in the extended binary Golay code. We notice that the bottom row has only four non-zero entities. This implies that the derived designs $S(23,7,4)$, $T(22,7,3,4)$, $S(22,6,3)$, $T(21,7,2,12)$, and $T(21,6,2,4)$ all are quasi-symmetric. The corresponding strongly regular graphs have parameters (v,k,a,c) given by

v	k	a	c
253	140	87	65
176	105	68	54
77	60	47	45
120	77	52	44
56	45	36	36

All of these graphs generate quite good spherical codes.

The complement to the graph corresponding to the last entry in this table is a graph known as the *Gewirtz graph* (Cameron–van Lint [61], pp. 35, 72). It has parameters $(56,10,0,2)$. Having offered now a construction also for this graph we have actually offered constructions or statements on non-existence for all the graphs in Table 9.1. □

The Higman–Sims graph

The remarkable intersection triangle for the Steiner system $S(24,8,5)$ admits still further constructions. One of those leads to the so called *Higman–Sims graph* (Higman–Sims [127], 1968, Cameron–van Lint

[61], p. 107). This is a strongly regular graph with parameters $(v, k, a, c) = (100, 22, 0, 6)$ from which we obtain a spherical code with parameters $(n, \rho, M) = (22, 20/11, 100)$. A construction is as follows.

Let C be the constant weight code corresponding to the derived system $S(22, 6, 3)$. From the intersection triangle in Figure 9.13 we conclude that the design is quasi-symmetric. The code is regular with weight $w = 6$ and with distance distribution

d_i	0	8	12
k_i	1	60	16

The size is $v = 1 + 60 + 16 = 77$. We finally notice that for any index i, $1 \leq i \leq 22$ there are precisely $\lambda_1 = (vw)/n = (77 \cdot 6)/22 = 21$ codewords having a one in that position. With the aid of this code we construct a strongly regular graph with the parameters indicated above. As vertex set V we take the union of three different sets: a single point ∞, the index set $I = \{1, 2, \ldots, 22\}$ and the code C. The vertices $x \in V$ are connected as follows

(*i*) The vertex ∞ is connected to all vertices in the set I and to no other vertices.

(*ii*) In addition to the connection implied by (*i*) any vertex $\in I$ is connected to all vertices $y \in C$ such that $x \in supp(y)$.

(*iii*) In addition to the connections implied by (*ii*) any vertex $x \in C$ is connected also to those vertices $y \in C$ such that $supp(x) \cap supp(y) = \emptyset$.

The graph Γ generated by these rules is obviously regular with size $v = 1 + 22 + 77 = 100$ and with valency $k = 22$. It remains to show that it is strongly regular and to determine the parameters a and c.

The only non-trivial cases are those when both x and y belong to C. We notice that the subgraph of Γ generated by C is the complement to the third one of the graphs considered in the previous example. Thus, this subgraph is strongly regular with parameters $(77, 16, 0, 4)$. It follows that any neighbors in C have no further neighbors in C.

By construction they have no common neighbor in $I \bigcup \{\infty\}$. It follows that any neighbors in Γ have no common neighbor. Any non-neighbors in C have 4 common neighbors in C. As they overlap in precisely *two* positions it follows that in addition they also have two neighbors in $\{\infty\} \bigcup I$. Clearly ∞ is not a neighbor to any of them. It follows that any pair of non-neighbors in Γ has precisely 6 common neighbors in Γ. In conclusion it follows that the graph Γ is strongly regular with parameters $(v, k, a, c) = (100, 22, 0, 6)$. This is the Higman–Sims graph. □

Generalized quadrangles

An *incidence structure* is a triple (P, L, I), where P denotes a set of *points* and L denotes a set of *lines*, and where I is an *incidence relation* indicating which points belong to which lines. An incidence structure is called a *generalized quadrangle* if the following axioms hold:

- A1: Any line contains at least two points.
- A2: Any two distinct points are contained in at most one line.
- A3: For any $(p, l) \in P \times L$ such that $p \notin l$ there is a unique pair $(q, m) \in P \times L$ such that p belongs to m and q belongs to both l and m.
- A4: There is at least *one* pair of non-concurrent lines (i.e. lines with no common point).

To any generalized quadrangle $Q = (P, L, I)$ there is an associated *point graph* Γ, which is obtained by taking the points in the quadrangle as vertices in the graph and by letting two points in Q be connected by an edge in Γ whenever they are collinear.

Generalized quadrangles exist in abundance. Simple examples are provided by rectangular grids. One of those is shown in Figure 9.14.

In the present context we will restrict our attention to generalized quadrangles with the special property that all lines contain the same number of points and all points are incident with the same number of lines. It turns out that such quadrangles are closely related to strongly regular graphs. More precisely the following relation holds.

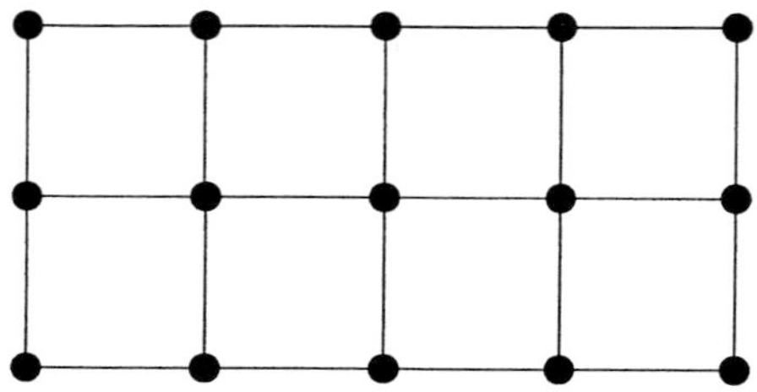

Figure 9.14: A rectangular grid with 15 points and 8 lines.

Theorem 9.4.2 (Bose [39]) *Suppose $Q = (P, L, I)$ is a generalized quadrangle such that each line contains $s+1$ points and each point is incident with $t+1$ lines. Then the associated point graph is strongly regular with parameters (v, k, a, c) given by*

$$v = (s+1)(st+1), \quad k = (t+1)s, \quad a = s-1, \quad c = t+1.$$

Proof: The number v of vertices in the graph Γ equals the number of points in the generalized quadrangle Q. Take an arbitrary line $l \in L$. There are $s+1$ points on l. Choose one of those, say $p \in l$. The point p is incident with t lines different from l, and each one of those contains s points different from p. By A3 all these lines have only the points on l in common and together they exhaust all the points in Q. These observations lead to the formula for v. To find the expression for k we simply observe that each point p is incident with $t+1$ lines, each containing s points except p. Recall that a denotes the number of common neighbors to two neighbors in the graph Γ. In the present case two points are neighbors if and only if they are collinear. If so there are by assumption precisely $s-1$ further points on that line, each being a neighbor to both of the two initial points. Finally, the formula for c follows directly from A3. □

Example 9.4.5: Consider a square grid, with $s+1$ points on each line. Any point is incident with precisely 2 lines. It follows that the point graph associated with the grid is a generalized quadrangle with $t = 1$. It is actually the square lattice graph $L_2(s+1)$ as defined in Example 9.4.1, with parameters (v, k, a, c) given by

$$v = (s+1)^2, \quad k = 2s, \quad a = s-1, \quad c = 2.$$

□

There are some famous examples usually referred to as the *classical* generalized quadrangles. The name is motivated by the fact that they occur naturally in connection with the so called *classical groups.* For a detailed account we refer to [226] or [180]. For our purposes the most interesting classical generalized quadrangles are those associated with the *orthogonal groups.* They can be defined as the sets of totally isotropic points and lines in the projective geometries $PG(3,q)$, $PG(4,q)$, or $PG(5,q)$ respectively, with respect to the following symmetric matrices, ([226], p. 290):

Fields of *odd* characteristic:

$$S_3 = \begin{bmatrix} 0 & I_2 \\ I_2 & 0 \end{bmatrix}, \quad S_4 = \begin{bmatrix} 0 & I_2 & \\ I_2 & 0 & \\ & & 1 \end{bmatrix},$$

$$S_5 = \begin{bmatrix} 0 & I_2 & & \\ I_2 & 0 & & \\ & & 1 & \\ & & & -z \end{bmatrix}.$$

Fields of characteristic 2:

$$S_3 = \begin{bmatrix} 0 & I_2 \\ & 0 \end{bmatrix}, \quad S_4 = \begin{bmatrix} 0 & I_2 & \\ & 0 & \\ & & 1 \end{bmatrix},$$

$$S_5 = \begin{bmatrix} 0 & I_2 & & \\ & 0 & & \\ & & \alpha & 1 \\ & & & \alpha \end{bmatrix}.$$

Here I_2 denotes the 2×2 unit matrix and z is a non-square element in $\mathbb{F}_q^*$ (q odd), while α is any element in $\mathbb{F}_q$ (q even) which is *not* of the form $x + x^2$ for any x in $\mathbb{F}_q$. The generalized quadrangles are usually denoted $Q(3,q)$, $Q(4,q)$, and $Q(5,q)$ respectively. The parameters are as follows:

	s	t
$Q(3,q)$	q	1
$Q(4,q)$	q	q
$Q(5,q)$	q	q^2

We recognize $Q(3,q)$ as a special case of Example 9.4.5.

Although we are not going to prove that the incidence structures as define above really do satisfy the axioms for a generalized quadrangle (the reader is referred to the specialized texts, for instance those mentioned above, [226] or [180]) we will at least briefly indicate how to determine the parameters, given the fact that the structures in question really are generalized quadrangles.

In order to determine s we first recall that points in $PG(n,q)$ are lines in $AG(n+1,q)$ on which the origin is excluded. More precisely, a point in $PG(n,q)$ can be viewed as a set $\{\lambda x : \lambda \in \mathbb{F}_q^*\}$, where x is a non-zero element in $AG(n+1,q)$. Similarly a *line* in $PG(n,q)$ corresponds to a *plane* in $AG(n+1,q)$, again excluding the origin. A simple combinatorial argument reveals that any line in $PG(n,q)$ is incident with $q+1$ points, and so $s=q$ is obvious.

Determining t requires a slightly more elaborate argument. We confine our discussion to $Q(5,q)$, which for our purposes is the most interesting case. The other two cases can be computed in exactly the same way. Although the result holds for *any* prime power q our argument will be restricted to the case when q is *not* a power of 2.

A totally isotropic *point* P in $PG(5,q)$ can be represented in $AG(6,q)$ by a vector $x=(x_1,x_2,\dots,x_6)\in\mathbb{F}_q^6$ such that $xS_5x^T=0$. Without loss of generality we choose $x=(1,0,0,0,0,0,0)$. The point P in $PG(5,q)$ is then given by $P=\{(\lambda,0,0,0,0,0)\in\mathbb{F}_q^6 : \lambda\neq 0\}$. A totally isotropic *line* in $PG(5,q)$ can be represented as a 2-dimensional subspace in $AG(6,q)$. We like to compute the number of lines in $PG(5,q)$ which are incident with P. In $AG(6,q)$ they correspond to the plane generated by the matrix

$$U=\begin{bmatrix} 1 & 0 & 0 & 0 & 0 & 0 \\ 0 & x_2 & x_3 & x_4 & x_5 & x_6 \end{bmatrix}.$$

Now in order for the line in $PG(5,q)$ corresponding to the matrix U to be totally isotropic it is required that the following identity is satisfied:

$$USU^T=0.$$

This puts a number of restrictions in the parameters. Performing the matrix multiplications we obtain

$$USU^T = \begin{bmatrix} 0 & x_3 \\ x_3 & \beta \end{bmatrix},$$

where $\beta = 2x_2x_4 + x_5^2 - zx_6^2$. It follows that we must have $x_3 = 0$. We also must have $\beta = 0$. Without loss of generality we may put $x_2 = 1/2$, (recall that q is a power of an *odd* prime, so 2 is invertible). Then we get $x_4 = zx_6^2 - x_5^2$. Clearly the parameters x_5 and x_6 can be chosen at will. There are q^2 possibilities, and so altogether q^2+1 lines through the point P. As the structure is a generalized quadrangle (this fact we have *not* shown) the same local structure is found around *all* totally isotropic points in $PG(5,q)$. It follows that we have $t = q^2$.

The point graph of $Q(5,q)$ is strongly regular with parameters

$$v = (q+1)(q^3+1), \quad k = q(q^2+1), \quad a = q-1, \quad c = q^2+1.$$

The negative eigenvalue τ gives rise to an infinite sequence of spherical codes which are optimal in the sense that they satisfy the Levenshtein bound with equality. The parameters are

$$n = q(q^2-q+1), \quad \rho = 2 - 2/q^2, \quad M = (q+1)(q^3+1).$$

The inner product distribution is

s_i	1	$1/q^2$	$-1/q$
k_i	1	q^4	q^3+q

Recall that q is a power of a prime. For the first few cases we obtain

q	n	ρ	M
2	6	3/2	27
3	21	16/9	112
4	52	15/8	325
5	105	48/25	756
7	301	96/49	2752
8	456	63/32	4617
9	657	160/81	7300
11	1221	240/121	15984
13	2041	336/169	30772
16	3856	255/128	69649

v	k	a	c	n	ρ	M	*Reference*
10	3	0	1	4	5/3	10	*Petersen*
16	5	0	2	5	8/5	16	*Clebsh*
27	10	1	5	6	3/2	27	*Schläfli*
56	10	0	2	20	28/15	56	*Gewirtz*
120	42	8	18	20	12/7	120	D^{pq}
50	7	0	1	21	40/21	50	*Hoffman–Singleton*
77	16	0	4	21	11/6	77	D_p
112	30	2	10	21	16/9	112	*1 subconst. McL*
162	56	10	24	21	12/7	162	*2 subconst. McL*
176	70	18	34	21	176/105	176	D^p
100	22	0	6	22	20/11	100	*Higman–Sims*
253	112	36	60	22	23/14	253	D
275	112	30	56	22	5/3	275	*McLaughlin*

Table 9.2: Parameters for selected spherical codes generated from strongly regular graphs. In all cases the codes are those obtained from the negative eigenvalue τ.

The first code we have already encountered several times: in Chapter 1, Example 1.2.5, it was defined as the vertices of the Hessian polytope and in the present chapter we constructed it from the Schläfli graph. In Appendix D we provide an independent construction also for the second code. Levenshtein [157] observed that *all* the codes obtained from the negative eigenvalues to the point graphs associated with $Q(5, q)$ are optimal. □

We conclude with a few general remarks on the interesting relation between spherical codes and strongly regular graphs.

Several of the best ones of all the known spherical codes—and in particular several of the codes known to be optimal—can be generated from strongly regular graphs. In Table 9.2 we have listed the parameters of some of best codes generated in this way. We have also indicated the parameters of corresponding graphs.

There is a *pairwise* correspondence between strongly regular graphs and spherical codes. As we have noted each strongly regular graph Γ generates *two* non-trivial spherical codes, one for each one of the two non-trivial eigenvalues θ and τ. Moreover, the *complement* $\overline{\Gamma}$ generates the same non-trivial spherical codes, with the roles of the eigenvalues interchanged. Similarly any distance regular spherical code with two non-zero distances generates *two* strongly regular graphs, one for each one of its two non-trivial distance graphs Γ_1 and Γ_2. It is easy to see that one is the complement of the other.

It turns out that the best codes are usually obtained from the negative eigenvalue τ of the graphs we have described in the text. In Table 9.2 only such codes are indicated.

9.5 The absolute bound

Beside the linear programming bound, which bounds the size of a spherical code in terms of its smallest non-zero distance, Delsarte–Goethals–Seidel, [81], [82] also derived a bound in terms of the number d of non-zero distances. This number is generally referred to as the *degree.* The bound is called the *absolute bound.*

Theorem 9.5.1 (Delsarte–Goethals–Seidel, [82]) *Let $\mathcal{X}$ be any spherical code with dimension n and degree d. Then the size M is bounded according to*

$$M \leq \binom{n+d-1}{d} + \binom{n+d-2}{d-1}.$$

Proof: Let the code be $\mathcal{X} = \{x_1, x_2, \ldots, x_M\}$ and let the set of non-zero inner products be $A = \{s_1, s_2, \ldots, s_d\}$. Define the sets Δ_i, $i = 2, 3, \ldots, M$, as follows:

$$\Delta_i = \{s \in A : \langle x_i, x_j \rangle = s;\ 1 \leq j < i\}.$$

Let $F_1(t) = 1$ and define

$$F_i(t) \triangleq \prod_{s \in \Delta_i} \frac{t-s}{1-s}$$

for $i = 2, 3, \ldots, M$, where t is a real variable. Finally define the multi-variable functions $f_i(z)$, where $z = (z_1, z_2, \ldots, z_n)$ is an n-dimensional real variable, by the following simple relations:

$$f_i(z) = F_i(\langle z, x_i \rangle), \; i = 1, 2, \ldots, M.$$

Clearly all functions f_i belong to $Pol(n, d)$. We have $f_i(x_j) = 0$ for $j < i$ and $f_i(x_i) = 1$, for $i = 1, 2, \ldots, M$, which implies that the functions f_i are linearly independent. Therefore we must have $M \leq dim\{Pol(n, d)\}$. The theorem follows now from the formula (2.2.2) in Chapter 2. □

Choosing $d = 1$ we get the bound $M \leq n + 1$. The *Simplex* codes satisfy this bound with equality.

For $d = 2$ the bound takes the form

$$M \leq n(n+3)/2; \quad n \geq 2.$$

For the first few values of n we have

n	2	3	4	5	6	7	8
$n(n+3)/2$	5	9	14	20	27	35	44

We notice that in addition to the somewhat trivial case $n = 2$ corresponding to the regular pentagon the bound is satisfied with equality also in the case $n = 6$: the spherical code generated by the *Schläfli* graph has parameters $(n, \rho, M) = (6, 3/2, 27)$ and only two non-zero squared distances: $\rho_1 = 3/2$ and $\rho_2 = 3$.

The *McLaughlin* graph is a strongly regular graph with parameters $(v, k, a, c) = (275, 162, 105, 81)$ (Cameron–van Lint [61], pp. 67, 112). The corresponding spherical code has parameters $(n, \rho, M) = (22, 5/3, 275)$ and is another example where the absolute bound is satisfied with equality (Delsarte et al [82], p 371).

We mentioned in Section 9.4 that a strongly regular graph with parameters $(v, k, a, c) = (28, 9, 0, 4)$ does not exist. We are now able to provide a simple argument why: if it did it would imply the existence of a spherical code with parameters $(M, n) = (28, 6)$, which would violate the absolute bound.

In a similar way Delsarte–Goethals–Seidel [82] employed the absolute bound to conclude the non-existence of a strongly regular graph with parameters $(276, 165, 108, 84)$.

For *antipodal* codes the following slightly stronger result holds.

Theorem 9.5.2 (Delsarte–Goethals–Seidel, [82]) *Let $\mathcal{X}$ be an* antipodal *spherical code with dimension n and degree d. Then the size M is bounded according to*

$$M \leq 2\binom{n+d-2}{d-1}.$$

Proof: An antipodal code $\mathcal{X}$ can be divided into two disjoint components $\mathcal{Y}$ and $-\mathcal{Y}$, so that $\mathcal{X} = \mathcal{Y} \bigcup -\mathcal{Y}$. Let $\mathcal{Y} = \{y_1, y_2, \ldots, y_N\}$, where $M = 2N$. Let A be the set of inner products in $\mathcal{X}$ and let A' be the set of inner products in $\mathcal{Y}$. Clearly A' is a subset in A. Moreover, we have $d' \triangleq |A'| < d \triangleq |A|$, because by definition -1 is an element in A but not in A'. We now define

$$F(t) \triangleq \prod_{s \in \Delta_i} \frac{t-s}{1-s}.$$

By letting $f_i(z) = F(\langle z, y_i \rangle)$ we get $f_i(y_j) = \delta_{ij}$. It follows that the functions $f_1, f_2, \ldots, f_N$ are linearly independent. They all have the same degree d'. The number N of these functions cannot exceed the dimension of the linear space $Pol(n, d')$. Consequently we have

$$M = 2N \leq 2\binom{n+d'-1}{d'} \leq 2\binom{n+d-2}{d-1}.$$

This completes the proof. □

For $d = 1$ this bound is met by the only antipodal code with degree $d = 1$, namely the trivial binary antipodal code. For $d = 2$ the bound reads $M \leq 2n$, $n \geq 2$, which is satisfied with equality by the *Biorthogonal* codes. For $d = 3$ we have $M \leq n(n+1)$, $n \geq 2$. The first few numerical values are

n	2	3	4	5	6	7	8
$n(n+1)$	6	12	20	30	42	56	72

The bound is met by the hexagon and the icosahedron. We recall that the unique spherical code with parameters $(n, \rho, M) = (7, 4/3, 56)$ has distance distribution

ρ_i	0	4/3	8/3	4
k_i	1	27	27	1

It follows that the bound is satisfied with equality by this code.

There is also a unique spherical code with parameters $(n, \rho, M) = (23, 8/5, 552)$, and with distance distribution given by

ρ_i	0	8/5	12/5	4
k_i	1	275	275	1

Also this code satisfies the bound with equality. Both codes are related to so called regular 2-graphs (Delsarte et al [82] p. 382).

There are three additional known cases where the bound is satisfied with equality. For those cases we have $(M, n, d) = (240, 8, 4)$, $(4\,600, 23, 4)$, $(196\,560, 24, 6)$. More will be said about these codes in the next section.

9.6 Spherical designs

Delsarte–Goethals–Seidel, [82] introduced the concept of *spherical design*. Like a code a spherical design is nothing else than a set of points on the unit sphere Ω_n. However, the characterization is different: while a code is characterized by its minimum distance a design is characterized by a parameter called *strength*. From another point of view this is, of course, just a new code parameter. The concept is illuminated by the following theorem, which actually embodies not less than *six* different but equivalent definitions.

Theorem 9.6.1 *Let $\mathcal{X} \subseteq \Omega_n$ be a spherical code of size $M = |\mathcal{X}|$, let $Q_i(t), i = 0, 1, \ldots$ denote the Gegenbauer polynomials, and let $\{v_{ij}, 1 \leq j \leq r_i, 0 \leq i \leq \tau\}$ be an orthonormal basis in the polynomial space Pol(n,τ). Let $f(x) = \Sigma_{i=0}^{\tau}\Sigma_{j=1}^{r_i} f_{ij}v_{ij}(x)$ be any multivariate polynomial in Pol(n,τ). The following six statements are equivalent:*

$$(i) \quad \sum_{x\in\mathcal{X}} f(xU) \quad \text{is independent of} \quad U \in \mathcal{O}_n,$$

$$(ii) \quad \int_{\Omega_n} f(x)d\mu(x) = \frac{1}{M}\sum_{x\in\mathcal{X}} f(x),$$

$$(iii) \quad \sum_{x\in\mathcal{X}} f(x) = Mf_{01},$$

$$(iv) \quad \sum_{x\in\mathcal{X}} v_{ij}(x) = 0, \quad j = 1, 2, \ldots, r_i, \; i = 1, 2 \ldots, \tau,$$

$$(v) \quad \sum_{x\in\mathcal{X}} Q_i(\langle x, y\rangle) = 0, \quad y \in \Omega_n, \; i = 1, 2 \ldots, \tau,$$

$$(vi) \quad \sum_{x,y\in\mathcal{X}} Q_i(\langle x, y\rangle) = 0, \; i = 1, 2 \ldots, \tau.$$

Proof: *(i)* ⇒ *(ii)*: Recall that μ denotes uniform Lebesgue measure on the sphere Ω_n. Suppose *(i)* holds. Let $F(x) = (1/M)\sum_{x\in\mathcal{X}} f(x)$. For any set $\{U_k : 1 \leq k \leq K\}$ of orthogonal transformations we have

$$F(x) = \frac{1}{K}\sum_{k=1}^{K} F(xU_k).$$

By proper choice of a large enough number of orthogonal transformations U_i we clearly can generate an arbitrarily close approximation to the integral in *(ii)*.

(ii) ⇒ *(iv)*: Recall that $Harm(n, 0)$ has dimension $r_0 = 1$ and that consequently we have $v_{01}(x) \equiv 1$. As a consequence we also have

$$(v_{ij}, 1) = \int_{\Omega_n} v_{ij}(x)d\mu(x) = \delta_{i,0}\delta_{j,1}.$$

Choosing $f(x) = v_{ij}(x)$ we see that *(ii)* implies *(iv)*.

(iv) $\Rightarrow$ *(iii)*: Expanding $f(x)$ in the basis $\{v_{ij}\}$ we see immediately that *(iv)* implies *(iii)*.

(iii) $\Rightarrow$ *(i)*: Consider the expansion of the functions $f(x)$ and $f(xU)$ in the basis $\{v_{ij}(x)\}$. Let the coefficients of the base function $v_{01}(x)$ be f_{01} and f'_{01} respectively. We clearly have $f_{01} = f'_{01}$. It follows that *(iii)* implies *(i)*.

(iv) $\Rightarrow$ *(v)* $\Rightarrow$ *(vi)* $\Rightarrow$ *(iv)*: All these implications are easy consequences of the definition of the Gegenbauer polynomials, see (2.2.11). This concludes the proof. □

Definition 9.6.1 *A spherical code satisfying the conditions of Theorem 9.6.1 will be called a τ-design. The* strength *is defined as the largest integer τ for which the conditions are satisfied.*

The following theorem offers an alternative characterization of designs of strength 1 or 2.

Theorem 9.6.2 *Let $\mathcal{X} \subseteq \Omega_n$ be a spherical code of size M with Gram matrix G. Let J be an $M \times M$ matrix with all entries equal to 1. The code $\mathcal{X}$ is a 1-design or more if and only if (i) holds. It is a 2-design or more if and only if both (i) and (ii) hold.*

$$\begin{array}{ll} (i) & JG = GJ = 0, \\ (ii) & G^2 = (M/n)\,G. \end{array}$$

Proof: Recall that the dimension of $Harm(n,1)$ is $r_1 = n$. Take $v_{1j}(x) = \sqrt{n}\,x_j$, $1 \le j \le n$, and notice that $\{v_{1i}(x)v_{1j}(x) : 1 \le i \le j \le n\}$ spans $Harm(n,2)$. The code $\mathcal{X}$ is at least a 1-design if and only if

$$\sum_{x\in\mathcal{X}} v_{1j}(x) = 0, \quad 1 \le j \le n.$$

With the given basis this means that we must have $\sum_{x\in\mathcal{X}} x_j = 0$ for all j, $1 \le j \le n$ or—what is the same—$JX = 0$. Recalling that the Gram matrix G is given by $G = XX^T$ the first statement follows easily.

The code $\mathcal{X}$ is at least a 2-design if and only if in addition to being a 1-design it also satisfies the condition

$$\sum_{x\in\mathcal{X}} v_{1i}(x)v_{1j}(x) = M\int_{\Omega_n} v_{1i}(x)v_{1j}(x)\,d\mu(x) = M\,\delta_{ij}.$$

With basis functions defined as above this condition alternatively can be written as

$$\sum_{x\in\mathcal{X}} x_i x_j = \frac{M}{n}\,\delta_{ij}.$$

Computing the (x,y)-component in the squared Gram matrix G^2 we obtain

$$\begin{aligned} G^2(x,y) &= \sum_{z\in\mathcal{X}} G(x,z)G(z,y) &= \sum_{z\in\mathcal{X}} \langle x,z\rangle\langle z,y\rangle \\ &= \sum_{z\in\mathcal{X}}\sum_{i=1}^{n} x_i z_i \sum_{j=1}^{n} y_j z_j &= \sum_{i,j=1}^{n} x_i y_j \sum_{z\in\mathcal{X}} z_i z_j. \end{aligned}$$

If $\mathcal{X}$ is a 2-*design* the last sum equals $(M/n)\,\delta_{ij}$ and we see that the relation $G^2 = (M/n)\,G$ prevails.

Conversely, let $\mathcal{X}$ be a 1-design or more with Gram matrix G satisfying (ii). Observing the identity $G(x,x)\equiv 1$ we obtain

$$G^2(x,x) = \sum_{y\in\mathcal{X}} \langle x,y\rangle^2 = M/n.$$

Summing also over x we get

$$\sum_{x,y\in\mathcal{X}} \langle x,y\rangle^2 = M^2/n\,.$$

The second Gegenbauer polynomial has the form

$$Q_2(t) = (nt^2-1)/(n-1).$$

By condition (vi) of Theorem 9.6.1 follows that $\mathcal{X}$ is indeed a 2-design (or possibly more). □

For *regular* codes a simple test is provided by the following theorem.

$$= \sum_{z\in\mathcal{X}} \frac{1}{r_i} \sum_{k=1}^{r_i} v_{ik}(x)v_{ik}(z) \frac{1}{r_j} \sum_{m=1}^{r_j} v_{jm}(z)v_{jm}(y) =$$

$$= \frac{1}{r_i r_j} \sum_{k=1}^{r_i} \sum_{m=1}^{r_j} v_{ik}(x)v_{jm}(y) \sum_{z\in\mathcal{X}} v_{ik}(z)v_{jm}(z).$$

If $\mathcal{X}$ is a τ-design and $0 \le i+j \le \tau$ then the last sum equals $M\delta_{ij}\delta_{km}$ and the whole expression collapses to

$$\frac{M}{r_i r_j}\delta_{ij} \sum_{k=1}^{r_i} v_{ik}(x)v_{ik}(y) = \frac{M}{r_i}\delta_{ij}Q_i(\langle x, y\rangle). \tag{9.6.4}$$

In summary we have arrived at the following set of equations:

$$\sum_{k,m=0}^{d} P_{km}(x,y)Q_i(s_k)Q_j(s_m) = \frac{M}{r_i}\delta_{ij}Q_i(\langle x, y\rangle). \tag{9.6.5}$$

We notice that $P_{k0}(x, y)$ and $P_{0k}(x, y)$ are equal to 1 if $\langle x, y\rangle = s_k$ and that they are zero otherwise. It follows that for any fixed pair $\{x, y\}$ there are at most d^2 unknown quantities $P_{k,m}(x, y)$. Suppose $\tau \ge 2d-2$ and let the indices i and j vary from 0 to $d-1$. The condition $0 \le i+j \le \tau$ is satisfied in all these cases, and so from (9.6.5) we obtain d^2 linear equation for the d^2 unknowns. The coefficient matrix is the tensor product $Q \otimes Q$ of the non-singular matrix Q with itself, and so is non-singular. It follows that the parameters $P_{k,m}(x, y)$ are uniquely determined, and so are independent of x and y. But this is exactly the condition required for the code $\mathcal{X}$ to be distance regular.□

On the other hand, there is a definite limit for how large the strength τ can be for a given degree d. The following theorem holds.

Theorem 9.6.5 (Delsarte et. al., [82]) *Let $\mathcal{X} \subseteq \Omega_n$ be a spherical code with degree d. The strength τ always satisfies the inequality $\tau \le 2d$; if $\mathcal{X}$ is antipodal the even stronger bound $\tau \le 2d-1$ holds.*

Proof: Let $f(x)$ be any polynomial with Gegenbauer expansion

$$f(t) = f_0 + f_1Q_1(t) + \cdots + f_kQ_k(t). \tag{9.6.6}$$

Then the following identity holds:

$$Mf(1) + \sum_{x \neq y} f(\langle x, y \rangle) = M^2 f_0 + \sum_{i=1}^{k} \frac{f_i}{r_i} \sum_{j=1}^{r_i} (\sum_{x \in \mathcal{X}} v_{ij}(x))^2. \quad (9.6.7)$$

The identity is obtained with the aid of the addition formula for Gegenbauer polynomials, (2.2.11) by summing $f(\langle x, y \rangle)$ over all x, y in $\mathcal{X}$. Let the inner products in $\mathcal{X}$ be $1 = s_0 > s_1 > s_2 > \ldots > s_d$ and choose

$$f(t) = \prod_{i=1}^{d} \frac{t - s_i}{1 - s_i} = f_0 + f_1 Q_1(t) + \cdots + f_d Q_d(t). \quad (9.6.8)$$

Let $g^{(j)}(t) = f(t)Q_j(t)$. We get $g_0^{(j)} = r_j f_j$, where $g_0^{(j)}$ denotes the zero coefficient in the Gegenbauer expansion of $g^{(j)}(t)$. Notice that $g^{(j)}(t) = f(t)Q_j(t)$ has degree $d + j$. Thus, if $\mathcal{X}$ is a τ-*design* we get from (9.6.8)

$$0 = M - M^2 g_0^{(j)} = M - M^2 r_j f_j, \quad 0 \le j \le \tau - d. \quad (9.6.9)$$

It follows that we have

$$f_j = \frac{1}{r_j M}, \quad 0 \le j \le \tau - d. \quad (9.6.10)$$

In particular we have $f_{\tau - d} \neq 0$, and so $\tau - d \le d$ or $\tau \le 2d$.

If $\mathcal{X}$ is antipodal the stronger inequality $\tau \le 2d - 1$ can be proved by a slight modification of the polynomial $f(t)$. This completes the proof. □

There are also bounds on the *size* that can be achieved by a spherical design of a given strength. Again the linear programming bound is useful, but this time it is used to formulate a *lower* bound.

Theorem 9.6.6 (Delsarte et. al., [82]) *Let $\mathcal{X}$ be a spherical code with inner products $\Sigma = \{s_1, s_2, \ldots, s_d\}$. Let the strength be τ and let $f(t)$ be a polynomial with Gegenbauer expansion $f(t) = f_0 + f_1 Q_1(t) + \cdots + f_k Q_k(t)$ satisfying the following properties:*

$$\begin{array}{ll} (i) & f(t) \ge 0 \quad \textit{for} \ -1 \le t \le 1, \\ (ii) & f_i \le 0, \quad i = \tau + 1, \tau + 2, \ldots, k. \end{array}$$

Then the size M of $\mathcal{X}$ satisfies the following inequality:

$$M \geq \frac{f(1)}{f_0}. \tag{9.6.11}$$

Equality holds if and only if $f(s) = 0$ for all $s \in \Sigma$.

Proof: We apply again the identity (9.6.7). By condition (i) the left hand side is lower bounded by $Mf(1)$. By condition (ii) the right hand side is upper bounded by $M^2 f_0$. The inequality follows. □

Theorem 9.6.7 (Delsarte et. al., [82]) *For $n \geq 3$ and for $\tau \geq 4$ the following bound holds for any τ-design:*

$$\tau = 2e: \quad M \geq \binom{n+e-1}{n-1} + \binom{n+e-2}{n-1}, \tag{9.6.12}$$

$$\tau = 2e+1: \quad M \geq 2\binom{n+e-1}{n-1}. \tag{9.6.13}$$

Proof: For $\tau = 2e$ choose

$$f(t) = \left(\sum_{i=0}^{e} r_i Q_i(t)\right)^2 = \sum_{i,j=1}^{e} r_i r_j Q_i(t) Q_j(t).$$

We get

$$f(1) = \left(\sum_{i=0}^{e} r_i\right)^2.$$

Recalling the identity $p_{ij}^0 = (1/r_i)\delta_{ij}$, (see (2.2.18)) we easily compute

$$f_0 = \sum_{i=0}^{e} r_i.$$

The first inequality now follows from the linear programming bound.

For $\tau = 2e+1$ choose

$$f(t) = (1+t)\left(\sum_{i=0}^{\lfloor e \rfloor} r_{e-2i} Q_{e-2i}(t)\right)^2.$$

By a similar evaluation and employing the linear programming bound the second inequality is established. □

The conditions for equality in these bounds is of great interest. Suppose $\mathcal{X}$ is a code with strength $\tau = 2e$ satisfying the bound (9.6.12) with equality. Let the inner products be $\Sigma = \{s_1, s_2, \cdots, s_d\}$. Theorem 9.6.6 implies that any valid test function $f(t)$ must satisfy $f(s) = 0$ for all $s \in \Sigma$. It follows that we must have $d \leq e$. On the other hand, by the absolute bound we have

$$M \leq \binom{n+d-1}{n-1} + \binom{n+d-2}{n-1}.$$

As the right hand side is an increasing function of d it follows that we must have $d \geq e$. The conclusion is that any code $\mathcal{X}$ of strength $\tau = 2e$ satisfying the linear programming bound for designs with equality must have degree $d = e$.

Similarly, if $\mathcal{X}$ is a code of strength $\tau = 2e+1$ satisfying the bound (9.6.13) with equality it follows from Theorem 9.6.6 that any corresponding test function $f(t)$ must satisfy $f(s) = 0$ for $s \in \Sigma \setminus \{-1\}$. It follows that we must have $d \leq e+1$. On the other hand, from the absolute bound for antipodal codes we have

$$M \leq 2\binom{n+d-2}{n-1}.$$

Again the right-hand member is an increasing function of d, and so if

$$M = 2\binom{n+e-1}{n-1}.$$

we must have $d-1 \geq e$. It follows that if $\mathcal{X}$ has strength $\tau = 2e+1$ it can satisfy the linear programming bound if and only if it is antipodal with degree d satisfying $d = e+1$.

Definition 9.6.2 *A design satisfying any of the two bounds in Theorem 9.6.7 with equality is said to be tight.*

n	M	d	τ
6	27	2	4
22	275	2	4
3	12	3	5
7	56	3	5
23	552	3	5
8	240	4	7
23	4600	4	7
24	196560	6	11

Table 9.3: All known tight designs of strengths $\tau \geq 4$ in dimensions $n \geq 3$.

Tight designs do exist. For $n = 2$ any regular $(\tau + 1)$-gon is a tight τ-design. In dimension $n \geq 2$ the *Simplex* codes are the only tight 2-designs and the *Biorhtogonal* codes are the only tight 3-designs. For $n \geq 3$ and $\tau \geq 4$ tight designs are more rare. Bannai and Damerell [9] have shown that in this range tight designs cannot exist except for $\tau = 4, 5, 7, 11$. They also showed that there is only *one* design in dimension $n = 11$. In dimensions $n = 4, 5, 7$ a complete classification is still lacking. Table 9.3 gives a list of all known tight designs of strengths $\tau \geq 4$ in dimensions $n \geq 3$. It is remarkable that all of these designs were mentioned already in the pioneering paper by Delsarte–Goethals–Seidel [82].

Non-tight designs exist in abundance. As we have already remarked, any spherical code is a design of some strength. Seymour and Zaslavsky [202] have shown that designs of any strength can be constructed in any dimension n, provided we allow the size M to be large enough. Sidelnikov, [208], [209], recently was able to construct infinite families of very good 7-designs and 11-designs.

9.7 Regular polytopes

The five regular polyhedra represent spherical codes of high regularity in three dimensions. We noticed already in Chapter 3 that three of them—the tetrahedron, the octahedron and the icosahedron—represent optimal spherical codes, while the remaining two—the cube and the dodecahedron—although equally regular are less impressive when interpreted as spherical codes.

Regular polyhedra and regular polygons are special cases of more general objects known as *regular polytopes.* The *Simplex code*, the *Biorthogonal code* and the vertices of the *hypercube* are examples of spherical codes that correspond to regular polytopes. As we know both the *Simplex* code and the *Biorthogonal* code are optimal spherical codes while the *hypercube* is *not*—except in the trivial case $n = 2$.

It turns out that in addition to the ones just mentioned there are three further regular polytopes. They all occur in dimension $n = 4$. One of them is the optimal configuration of 120 points which we encountered already in Chapter 1, Example 1.2.8. Another one is a regular polytope which is generated by those 24 points which provide the best known construction for contact number in four dimensions. The third regular polytope contains 600 vertices. Viewed as a spherical code this third polytope is *not* optimal or even best known. Nevertheless, as a geometrical object it is, of course, very interesting.

Although the number of regular polytopes is quite limited, and although we have already constructed most of them by other means, the extreme regularity and great beauty of these objects motivate a brief account of the general theory. The classical reference on this topic is the famous book by Coxeter [75].

A *polytope* is a finite convex region bounded by hyperplanes in Euclidean n-space. A more precise definition is as follows. Consider a family of linear inequalities:

$$b_{1k}x_1 + b_{2k}x_2 + \cdots + b_{nk}x_n \leq a_k, \quad k = 1, 2, \ldots, K.$$

The family is *consistent* if it has a non-empty set of solutions. It is *redundant* if one or more of the inequalities can be derived from the other ones. If the volume—in terms of the usual uniform Lebesgue measure—is *finite* the region corresponding to the set of solutions is called a *polytope.* For convenience we always assume that the inequalities are both consistent and non-redundant.

For a given polytope we obtain various subsets by requiring that some of the inequalities should be satisfied with *equality*. The collection of all non-empty ones of these subsets are referred to as the *elements* of the polytope. Clearly all the elements are also polytopes. This means that any polytope contains a number of smaller polytopes as its elements. This property is the fundamental geometric idea behind the concept of polytope. The elements defined by those points for which exactly *one* of the inequalities is satisfied with equality are called *cells*.

The *dimension* of a polytope is defined as the smallest dimension of any Euclidean space in which the polytope can be contained. An n-dimensional polytope has $(n-1)$-dimensional cells. The 2-dimensional elements are called *faces*, the 1-dimensional elements are called *edges* and the 0-dimensional elements—which of course consist of single points—are called *vertices*.

A *regular* polytope is defined recursively as follows. We first define a *regular polygon* as a polygon such that all edges have *equal length* and all angles between adjacent edges have *equal size*. Clearly the number of edges equals the number of vertices. The polygon is uniquely determined by this common number. If the number is p the polygon is called to as a *regular* p-gon and is referred to by the notation $\{p\}$. Regular p-gons exist for all $p \geq 3$.

A *regular polyhedron* is a polytope in three dimensions such that all its 2-dimensional elements—which of course are polygons—are regular and all its vertices have *regular equal surroundings*. The latter statement needs a clarification.

Consider a general polyhedron. For a given vertex let us join the midpoints of all edges connected to this vertex by new edges in such a way that each new edge is located in one of the faces adjacent to the given vertex. In this way a new polygon is generated for each vertex in the polyhedron. In general the vertices of these new polygons are not located in a plane, so in general the polygons will be so-called *skew* polygons. For the polyhedron to be *regular* we require not only that all these polygons are plane but also that they are *regular* and *equal*. This is what we mean by saying that the vertices in a regular polyhedron should have regular equal surroundings.

Alternatively, define the convex region spanned by the midpoints of all edges connected to given vertex as the *vertex figure*. A polyhedron is *regular* if all faces are equal regular polygons and all vertex figures are also regular polygons, all equal but not necessarily of the same kind as the faces.

For concreteness: the vertex figures of the tetrahedron, the cube, and the dodecahedron are equilateral triangles, while the vertex figures of the octahedron are squares and those of the icosahedron are regular pentagons. We notice that the vertex figures of a regular polyhedron belong to the same limited class of plane polygons as do the faces: equilateral triangles, squares and regular pentagons. We also notice that each regular polyhedron is uniquely determined by two numbers p and q indicating the types of the regular polygons of their faces and of their vertex figures respectively. The notation $\{p,q\}$ is employed to indicate a regular polyhedron having regular p-gons as faces and regular q-gons as vertex figures. As we know there are only five possible cases: $\{3,3\}$, $\{3,4\}$, $\{4,3\}$, $\{3,5\}$, $\{5,3\}$.

We also notice that the concept of vertex figure can be applied also to p-gons: in that case the vertex figure is simply the line segment connecting the midpoints of adjacent edges.

The general definition of a *regular polytope* is obtained by a generalization of the above ideas. A regular polytope in two dimensions is a regular p-gon; a regular polytope in three dimensions is a regular polyhedron. Consider dimension $n = 4$. In a regular polytope we require that the cells should be regular polyhedra. Again we define vertex figures as the convex regions spanned by the midpoints of all edges connected to a given vertex. For a polytope to be regular we require that also the vertex figures are regular polyhedra, all of the same kind.

As both the cells and the vertex figures in a regular 4-dimensional polytope are polyhedra it follows that for each one of these we have vertex figures and faces that are regular polygons. Moreover, the faces of those polyhedra that are vertex figures are vertex figures of those polyhedra that are cells in the polytope. It follows that if the cells are $\{p,q\}$'s then the vertex figures must be $\{q,r\}$'s for some positive integer r. As there are only five possible regular polyhedra this restriction limits considerably the possibilities for the regular 4-dimensional polytopes. Moreover, even if both a $\{p, q\}$ and a $\{q, r\}$ exist this fact alone does not imply that a corresponding regular polytope exists: the condition is just *necessary* and by no means *sufficient*. However, *if* a corresponding polytope exists it is uniquely determined by the numbers p, q, r and it will be referred to as a $\{p, q, r\}$.

In general the *vertex figure* to a vertex in an n-dimensional polytope is defined as the convex region spanned by the midpoints of all edges connecting a given vertex with all its neighbors. We say that an n-dimensional polytope is *regular* if all its cells are regular $(n-1)$-dimensional polytopes, all of the same kind, and all its vertex figures are also $(n-1)$-dimensional regular polytopes, all of the same kind. We emphasize that although it *is* required that all cells are polytopes of one and the same kind and all the vertex figures are also polytopes of one and the same kind it is *not* required that the cell polytopes should be of the same type as the vertex polytopes. We use the notation $\{p_1, p_2, \ldots, p_{n-1}\}$ to indicate a regular n-dimensional polytope such that the cells are $(n-1)$-dimensional polytopes of the type $\{p_1, p_2, \ldots, p_{n-2}\}$ and the vertex figures are $(n-1)$-dimensional polytopes of the type $\{p_2, p_3, \ldots, p_{n-1}\}$.

In order to illustrate the above concepts, and in order to see that the definitions are not void, let us consider the three infinite classes of regular polytopes we mentioned above, namely those corresponding to the *Simplex* codes, the *Biorthogonal* codes and the *hypercubes*.

Example 9.7.1: Consider the *Simplex* code $\mathcal{X} = S_n$. It consists of $M = n + 1$ equidistant codewords, x_1, x_2,...,x_M such that

$$\langle x_i, x_j \rangle = \delta_{ij}(1 + 1/n) - 1/n, \quad i, j = 1, 2, \ldots, M.$$

Consider the polytope defined by the inequalities

$$-1/n \leq \langle x, x_i \rangle \leq 1, \quad i = 1, 2, \ldots, M.$$

The extreme points are obviously the codewords of the code $\mathcal{X}$. The polytope is the set of all convex combinations of the extreme points.

Consider the cell defined by those points x for which we have $\langle x, x_i \rangle = -1/n$. It is clear that this cell is a convex region spanned by the points $\{x_j : j \neq i\}$. By a proper translation and scaling these points also form a *Simplex*. In the same way we find that the extreme points of the vertex figures can also be similarly transformed to simplex codes. We conclude that the polytope spanned by a simplex is a regular polytope of the type {3,3,...,3}. □

Example 9.7.2: For $x = (x_1, x_2, \ldots, x_n) \in \mathbb{R}^n$ consider the following set of inequalities

$$-1 \leq x_i \leq 1, \quad i = 1, 2, \ldots, n. \tag{9.7.1}$$

There are altogether $2n$ inequalities and it is clear that they are non-redundant. The set of solutions is referred to as an n-dimensional *hypercube*. The extreme points are those satisfying n of the equations with equality. Clearly, the n equations cannot be chosen arbitrarily. For each index $i = 1, 2, \ldots, n$ equality can hold in at most one of the above inequalities. There are 2^n different possible combinations. The extreme points are the words of the constant energy code $\mathcal{X} = (\pm 1^n)$. The hypercube is the set of all convex combinations of the extreme points.

The vertex figures are spanned by a set of n points, namely those n points which are located half-way towards each of the n neighbors to a given codeword in $\mathcal{X}$. All of these points are on equal distance from each other, i.e. each vertex figure is a regular *Simplex*.

The cells are those points which satisfy *one* of the defining inequalities with equality. Clearly each such cell is a hypercube of dimension $(n-1)$.

The above observations hold for any $n \geq 2$. For $n = 2$ the hypercube is a square, i.e. a regular polygon of type {4}. We conclude that in general the hypercubes are regular polytopes of the type {4,3,...,3}.□

Example 9.7.3: Let the vertices of the hypercube $\mathcal{X}$ be $x_1, x_2, \ldots, x_M$, with $M = 2^n$. Consider the 2^n inequalities

$$\langle x, x_i \rangle \leq 1, \quad i = 1, 2, \ldots, M.$$

Again it is easy to see that at most n relations can be satisfied with equality. There are precisely $2n$ extreme solutions. It is easy to see that these are given by the words of the *Biorthogonal* code $\mathcal{Y} = (\pm 1, 0^{n-1})^C$. Each cell is a regular *Simplex*. To see this, consider $x_1 = (1, 1, \ldots, 1)$. The cell defined by those points $y \in \mathcal{Y}$ satisfying $\langle y, x_1 \rangle = 1$ are orthogonal and all are located in the plane $\sum_{i=1}^n y_i = 1$. By proper translation they form a *Simplex* code. The vertex figures are easily seen to be hypercubes. We conclude that the convex regions spanned by the *Biorthogonal* codes are regular polytopes of the type {3,3,...,3,4}. A regular polytope of this kind is referred to as a *cross polytope*. □

We notice that the linear inequalities generated by the extreme points in a hypercube define a cross polytope and vice versa. This is a special case of a general phenomenon: if the coordinates of the extreme points to a regular polytope are taken as the coefficients in a set of linear inequalities the new polytope generated is also regular. Moreover, applying the same procedure using the extreme points of this new regular polytope the first one is obtained again. We express this fact by saying that the two polytopes are *reciprocal*. The regular simplex is reciprocal to itself. The cube and the octahedron are mutually reciprocal, and so are the icosahedron and the dodecahedron.

What further regular polytopes exist beyond those we have already encountered? We have already noticed that there are only five regular polyhedra, and that this fact limits the possibilities for regular polytopes in higher dimensions. Actually, there is a general condition formulated by Schläfli (see Coxeter, [75]) by which a complete classification is obtained.

Let $\{p_1, p_2, \ldots, p_{n-1}\}$ be a regular polytope centered at the point O_n. Consider one of its faces F, and let O_0 be one of the vertices contained in F. Let O_1 and O_1' be two points located at the midpoints of the two edges in F connected to O_0, see Figure 9.15. By definition O_1 and O_1' are both vertices in the vertex figure V defined by O_0 and by definition the vertex figure is also a regular polytope. The center is a point O_{n-1} which can be defined as the common orthogonal projection of all the vertices in V on the line connecting O_0 and O_n. Let r_1 denote the Euclidean distance between the points O_0 and O_n,

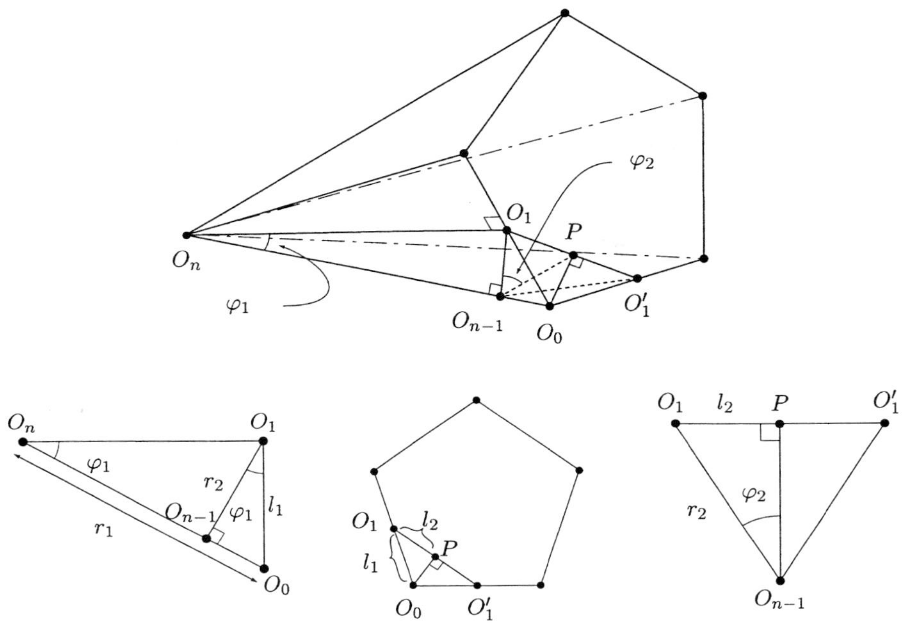

Figure 9.15: One face of a regular polytope $\{p_1, p_2, \ldots, p_{n-1}\}$. The figure illustrates the case $n = 3$ with $p_1 = 5$.

let r_2 denote the distance between O_1 and O_{n-1}, and let l_1 denote the distance between O_0 and O_1. Consider the triangle $O_0O_1O_n$. Let φ_1 be the angle at O_n. By elementary trigonometry we have:

$$l_1 = r_1 \sin\varphi_1, \quad r_2 = l_1 \cos\varphi_1. \tag{9.7.2}$$

The vertex figure at O_0 is a $\{p_2, p_3, \ldots, p_{n-1}\}$ with center O_{n-1}. Let P be the midpoint of the edge O_1O_1'. The triangle O_1PO_{n-1} is a counterpart to $O_0O_1O_n$. The angle φ_2 at O_{n-1} is a counterpart to φ_1. Denoting by l_2 the length of the half-edge O_1P we obtain the following counterpart to the first one of the equations in (9.7.2):

$$l_2 = r_2 \sin\varphi_2.$$

We now notice that O_0, O_1, and P all occur in the face F, which is a $\{p_1\}$. Again by an elementary trigonometric argument we have

$$\cos(\pi/p_1) = \frac{l_2}{l_1} = \frac{r_2 \sin\varphi_2}{r_2/\cos\varphi_1} = \sin\varphi_2 \cos\varphi_1.$$

This expression in turn can easily be transformed to the following equation:

$$\sin^2 \varphi_1 = 1 - \frac{\cos^2(\pi/p_1)}{\sin^2 \varphi_2} \,. \tag{9.7.3}$$

Now if a $\{p_1, p_2, \ldots, p_{n-1}\}$ exists then obviously the angle φ_1 must satisfy $0 < \varphi_1 < \pi/2$. This implies that the expression (9.7.3) is strictly positive. Moreover, the same argument can be applied successively for all the regular polytopes $\{p_i, p_{i+1}, \ldots, p_{n-1}\}$, $i = 1, 2, \ldots, n-1$ which can be generated from $\{p_1, p_2, \ldots, p_{n-1}\}$. Let φ_i, $i = 1, 2, \ldots, n-1$ be the successive center angles in a sequence of regular polytopes generated in this way. It is clear that all these angles must satisfy $0 < \varphi_i < \pi/2$, $i = 1, 2, \ldots, n-1$. Notice that we have $\varphi_{n-1} = \pi/p_{n-1}$. We obtain the following sequence of *necessary* conditions:

$$\sin^2 \varphi_i = 1 - \frac{\cos^2(\pi/p_i)}{\sin^2 \varphi_{i+1}} > 0, \quad i = 1, 2, \ldots, n-2. \tag{9.7.4}$$

It turns out that for $n \geq 3$ these conditions are also *sufficient*. In order to establish this fact let us investigate the first few cases in detail.

<u>$n = 3$:</u>

For $n = 3$ we have $\{p_1, p_2, \ldots, p_{n-1}\} = \{p, q\}$. The general conditions (9.7.4) imply the following inequality:

$$\sin^2 \varphi_1 = 1 - \frac{\cos^2(\pi/p)}{\sin^2 \varphi_2} > 0 \,.$$

Taking into account the equality $\varphi_2 = \pi/q$ we obtain the following condition:

$$\sin^2(\pi/p) + \sin^2(\pi/q) > 1.$$

This condition is equivalent with the more familiar inequality $1/p + 1/q > 1/2$. The only possibilities are $\{p, q\} = \{3, 3\}$, $\{3, 4\}$, $\{4, 3\}$, $\{3, 5\}$, $\{5, 3\}$. It is well known that all the corresponding polyhedra exist.

$n = 4$:

In dimension $n = 4$ we have $\{p_1, p_2, \ldots, p_{n-1}\} = \{p, q, r\}$. The conditions (9.7.4) give us

$$\sin^2 \varphi_1 = 1 - \frac{\cos^2(\pi/p)}{\sin^2 \varphi_2} > 0,$$

$$\sin^2 \varphi_2 = 1 - \frac{\cos^2(\pi/q)}{\sin^2 \varphi_3} > 0.$$

Inserting $\varphi_3 = \pi/r$ the following inequality can be derived:

$$\sin(\pi/p)\sin(\pi/r) > \cos(\pi/q). \tag{9.7.5}$$

For a $\{p, q, r\}$ to exist the parameters must be such that both a $\{p, q\}$ and a $\{q, r\}$ exist. Combining that restriction with those given by (9.7.5) leaves us with the following remaining possibilities

$$\{p, q, r\} = \{3, 3, 3\}, \{3, 3, 4\}, \{4, 3, 3\}, \{3, 4, 3\}, \{3, 3, 5\}, \{5, 3, 3\}.$$

It turns out that there is a regular polytope corresponding to each of these cases. The first three we have already encountered as the regular simplex, the cross polytope and the hypercube, respectively. We will see shortly that $\{3, 4, 3\}$ is a polytope corresponding to the 24 points which provide a best known construction for contact number in dimension $n = 4$, while $\{3, 3, 5\}$ is a polytope generating the optimal code of size $M = 120$. The final polytope $\{5, 3, 3\}$ is the reciprocal of $\{3, 3, 5\}$. The existence of this polytope follows from the existence of $\{3, 3, 5\}$. The spherical code corresponding to this last regular polytope has size $M = 600$. The minimum distance is not impressive.

$n = 5$:

For the case $n = 5$ we employ the notation $\{p_1, p_2, \ldots, p_{n-1}\} = \{p, q, r, s\}$. From the conditions (9.7.4) we derive the following inequality:

$$\frac{\cos^2(\pi/q)}{\sin^2(\pi/p)} + \frac{\cos^2(\pi/r)}{\sin^2(\pi/s)} < 1.$$

Observing also the condition that the cells and the vertex figures must belong to the class of admissible four-dimensional regular polyhedra we find that the only possibilities are $\{3, 3, 3, 3\}$, $\{3, 3, 3, 4\}$, and $\{4,3,3,3\}$. In all cases corresponding polytopes do exist. Indeed, we recognize again the regular *Simplex*, the *cross polytope* and the *hypercube*. In dimension $n = 5$ there are no other regular polytopes.

We already know that these three polytopes—the regular *Simplex*, the *cross polytope* and the *hypercube*—exist in any dimension. As they are the only regular polytopes in dimension $n = 5$ it follows that they are the only polytopes also in any higher dimension. It follows that we have arrived at a complete classification of the regular polytopes.

In dimension $n = 2$ any regular polygon is a trivial special case of a regular polytope.

In dimension $n \geq 3$ we have three infinite classes of regular polytopes:

- the regular *Simplex*: $\{3^{n-1}\}$,
- the regular *cross polytope*: $\{3^{n-2}, 4\}$,
- the *hypercube*: $\{4, 3^{n-2}\}$.

In addition there are five isolated regular polytopes, namely

- $n = 3$: $\{3,5\}$, $\{5,3\}$,
- $n = 4$: $\{3,4,3\}$, $\{3,3,5\}$, $\{5,3,3\}$.

The regular polytopes $\{3,5\}$, $\{5,3\}$ in dimension $n = 3$ are the icosahedron and the dodecahedron respectively. These are familiar objects which we have already encountered several times. The fact that they are regular polytopes is well known and easy to establish. We have already mentioned that the vertices of $\{3,4,3\}$ constitute the well known best construction for kissing number in dimension $n = 4$ and that the vertices of $\{3,3,5\}$ is the optimal code of size $M = 120$. Both these codes were described already in Chapter 1 (Examples 1.2.6 and 1.2.8 respectively).

The only object that is new to us is the regular polytope $\{5,3,3\}$. The existence of this polytope follows—as we have already pointed out—directly from the existence of the polytope $\{3,3,5\}$. Of course the vertices constitute a spherical code, but as we have also mentioned the parameters are not impressing. We will not discuss this code any further. So what remains for us to establish is the fact that the polytopes generated by the two other codes are actually regular.

Consider first the code of size $M = 24$. In Chapter 1 we gave two constructions of this code:

$$\mathcal{X} = (1/\sqrt{2})(\pm 1, \pm 1, 0, 0)^S,$$

$$\mathcal{Y} = (\pm 1, 0, 0, 0)^C \cup (1/2)(\pm 1, \pm 1, \pm 1, \pm 1).$$

It is to easy convince oneself that $\mathcal{X}$ and $\mathcal{Y}$ are equivalent spherical codes, both with distance distribution

ρ_i	0	1	2	3	4
k_i	1	8	6	8	1

We need to establish that the cells are octahedra and that the vertex figures are cubes. We notice that the codewords in $\mathcal{Y}$ define the hyperplanes containing the cells in $\mathcal{X}$. Consider the hyperplane defined by $y = (1, 0, \ldots, 0) \in \mathcal{Y}$. The corresponding cell in $\mathcal{X}$ is spanned by the vectors $(1, (\pm 1, 0, 0)^C)$, where the first component is fixed and the three last components span an octahedron. A similar octahedral cell in $\mathcal{X}$ is defined by each of the codewords $y \in \mathcal{Y}$.

For the rest of the argument we prefer to use the coordinate representation given by $\mathcal{Y}$. The vertex figure corresponding to the codeword $y = (1, 0, \ldots, 0) \in \mathcal{Y}$ is clearly given by vectors of the form $(1/2)(1, \pm 1, \pm 1, \pm 1)$, having the first component fixed to 1 and having arbitrary signs in the three remaining positions. Clearly those vectors span a cube, and clearly the same is true for the vertex figures of all other codewords.

By definition these two facts regarding cells and vertex figures together imply that the code defined either by $\mathcal{X}$ or by $\mathcal{Y}$ span a regular polytope of type $\{3, 4, 3\}$.

Next consider the optimal code of size $M = 120$. In Chapter 1 it was defined as the union of the following three codes:

$$\mathcal{X} = (\pm 1, 0, 0, 0)^C, \quad \mathcal{Y} = (1/2)(\pm 1, \pm 1, \pm 1, \pm 1),$$
$$\mathcal{Z} = (1/2)(0, \pm 1, \pm \sigma, \pm \tau)^A,$$

where σ and τ are the two roots to the equation $x^2 = x+1$, and where the exponents indicate cyclic permutation and even permutations, respectively. We take $\sigma = 1/2 - \sqrt{5}/2 \approx -0.6180$ and $\tau = 1/2 + \sqrt{5}/2 \approx 1.6180$. In this case we need to establish that each cell is a simplex and that each vertex figure is an icosahedron. Again we can show—with some effort—that the code is *regular* in the sense of spherical codes, i.e. that all codewords have the same inner product distribution. We find

s_i	1	$\tau/2$	$1/2$	$-\sigma/2$	0	$\sigma/2$	$-1/2$	$-\tau/2$	-1
k_i	1	12	20	12	30	12	20	12	1

The vertex figure corresponding to $x = (0, 0, 0, 1)$ is spanned by the codewords having the form $(1/2)((0, \pm 1, \pm \sigma)^A, \tau)$. We recognize the icosahedron.

We notice that one of the hyperplanes containing a cell is defined by the vector $a = (1, 1, 0, 0)$. The corresponding cell is spanned by the following six vectors:

$$(1, 0, 0, 0), \ (0, 1, 0, 0), \ (1/2)(1, 1, \pm 1, \pm 1).$$

Subtracting the mean $(1/2)(1, 1, 0, 0)$ and then normalizing by multiplying with $\sqrt{2}$ we obtain the six vectors

$$(1/\sqrt{2})(1, -1, 0, 0), \ (1/\sqrt{2})(-1, 1, 0, 0), \ (1/\sqrt{2})(0, 0, \pm 1, \pm 1),$$

which we easily identify as a fancy representation of the biorthogonal code. In conclusion we have established that the vertices of the optimal code with $M = 120$ span a regular polytope of the form $\{3, 3, 5\}$.

It is interesting to note that although the five regular *polyhedra* were known already in antiquity the remaining regular *polytopes* in higher dimensions were not discovered until in the nineteenth century. The major work on this matters was performed by the Swiss mathematician Ludwig Schläfli [191]. An extensive account of his work and much more is provided by Coxeter [75].

Chapter 10

Lattices

10.1 Introduction

A lattice is a regular, unbounded discrete pointset in Euclidean n-space; a sphere packing is a discrete pointset without any specific constraints on regularity. A lattice is the counterpart to a group code in Ω_n, while a sphere packing is the counterpart in $\mathbb{R}^n$ to the general concept of spherical code. The relations between lattices and sphere packings in $\mathbb{R}^n$ and codes in Ω_n extend far beyond the formal similarity: many spherical codes of importance are actually best described as shells or unions of shells in various lattices or sphere packings.

The present chapter is devoted to the theory of lattices and sphere packings. Our treatment is short and focused on those aspects that are of direct importance for spherical codes. For a full account of the general theory we refer to the specialized texts, in particular to the extensive texts by Conway and Sloane [71] and by Zong [242].

10.2 Lattices

A *lattice* is a discrete pointset $\mathbb{L}$ in a Euclidean space $\mathbb{R}^N$ with the property that if one point is located at the origin then the set of vectors describing the points in $\mathbb{L}$ is closed under vector addition. "Discrete" means that the points in $\mathbb{L}$ can be surrounded by disjoint open spheres; "closed" means that the points of $\mathbb{L}$ form an additive group under vector addition. From the latter property follows that each point has the same distance distribution to all other points and that all the surrounding spheres can be given the same radius. The maximal such radius is a parameter of major interest. It is called the *packing radius*.

Let $\{v_1, v_2, \ldots, v_n\}$ be a set of linearly independent vectors in $\mathbb{R}^N$, $n \leq N$. The set $\mathbb{L}$ of all linear combinations $x = u_1 v_1 + u_2 v_2 + \cdots + u_n v_n$ with integer coefficients u_i is obviously a lattice. Conversely, any lattice can be represented in this form. A proof of this fact can be found in Cassels [65], Section II-4, p. 78.

The vectors v_i are usually collected in a matrix V called the *generator matrix* and defined by the property that the vectors v_i form the rows in V. By collecting the coefficients u_i in a vector $u = (u_1, u_2, \ldots, u_n)$ we may write the generating equation in the form $x = u\,V$. Using the notation $\mathbb{Z}$ for the set of integers we obtain

$$\mathbb{L} = \mathbb{Z}^n V \triangleq \{zV : z \in \mathbb{Z}^n\}.$$

We see that the problem of defining a lattice in $\mathbb{R}^N$ reduces to that of defining a matrix V of rank n.

In principle there is no loss of generality in always assuming $N = n$. However, in the same way as it is sometimes convenient to use an excessive number of coordinates for describing spherical codes it turns out to be sometimes convenient to define a lattice using a generator matrix V with a number N of columns exceeding the number n of rows. Throughout the present text it will always be assumed that the rows are linearly independent, so the rank of the generating matrix is always equal to the number of rows. This number n will be called the *dimension* of the lattice.

The set $\mathbb{Z}^n$ is of course in itself a simple example of a lattice. The formula $\mathbb{L} = \mathbb{Z}^n V$ indicates that any lattice $\mathbb{L}$ in $\mathbb{R}^n$ can be seen as a linear transformation of the simple lattice $\mathbb{Z}^n$.

Recall that any real number r has a unique decomposition in an integer and a fractional part, $r = u + \vartheta$, where $u \in \mathbb{Z}$ and $\vartheta \in [0,1)$. For $r \in \mathbb{R}^n$ the same kind of decomposition applies, with $u \in \mathbb{Z}^n$ and $\vartheta \in [0,1)^n$. For any integer n the set $[0,1)^n$ is an example of a *fundamental region* for the lattice $\mathbb{Z}^n$.

More generally, let $\mathbb{L}$ be an arbitrary lattice in $\mathbb{R}^n$. Any set D such that the sets $D + x : x \in \mathbb{L}$ form a *partitioning* of $\mathbb{R}^n$ is called a *fundamental region* for $\mathbb{L}$. More explicitly the requirements are

$$(i) \quad \forall\, x, y \in \mathbb{L},\ x \neq y : \ (D + x) \bigcap (D + y) = \emptyset,$$

$$(ii) \quad \bigcup_{x \in \mathbb{L}} (D + x) = \mathbb{R}^n .$$

Clearly, any element $z \in \mathbb{R}^n$ has a unique decomposition $z = x + \vartheta$, with $x \in \mathbb{L}$ and $\vartheta \in D$.

To a lattice $\mathbb{L}$ of the form $\mathbb{L} = \mathbb{Z}^n V$ we normally associate the fundamental region

$$D = [0,1)^n\, V \triangleq \{\theta_1 v_1 + \theta_2 v_2 + \cdots, \theta_n v_n : 0 \leq \theta_i < 1\}.$$

Assume for the moment (and without any loss of generality) that the matrix V is square. As the rows are linearly independent it is clear that we have $\det V \neq 0$. By the standard formula for change of variables in integration we obtain

$$|D| = |\int_D dy| = |\det\, V| \int_{[0,1)^n} dx = |\det\, V|.$$

It is clear that any other fundamental region has the same volume. Thus, $|D| = |\det\, V|$ is an important invariant of the lattice.

If V is not square the above formulas must be modified. A useful quantity is the *Gram matrix*, which for any V—square or not—is defined as $G = VV^T$. The Gram matrix is always square and the volume of the fundamental region can always be obtained by the formula

$$|D| = (\det\, G)^{1/2}. \tag{10.2.1}$$

Another parameter of great importance is the *minimum distance* d between points in the lattice:

$$d \triangleq \min\{\|x - y\| : \ x, y \in \mathbb{L};\ x \neq y\}.$$

Recall Lemma 1.6.1. By this lemma we can surround the lattice points x by disjoint spheres $S_n(x, r)$ with radius r as long as $r < d/2$. We say that the lattice $\mathbb{L}$ defines a *packing* of spheres with radius $r = d/2$. Unlike the fundamental regions the spheres do not fill the space (except in the trivial case $n = 1$). A basic problem is to find lattices $\mathbb{L}$ such that the space between the spheres constitutes as small a fraction as possible of the total space.

Recall that the volume of a sphere with radius r is $V_n(r) = k_n\, r^n/n$, where k_n/n is the volume of the unit sphere (see Appendix B). As a measure of the packing density of a lattice $\mathbb{L}$ with minimum distance $d = 2r$ we use the quantity

$$\Delta \triangleq \frac{V_n(r)}{|D|},$$

defined as the quotient between the volume of the largest disjoint spheres and the fundamental region. Clearly both of these quantities are proportional to r^n, so the quantity Δ, called the *density* of the lattice, is independent of r. We like to have lattices with the density as *large* as possible.

It turns out often to be more convenient to use a slightly different parameter called the *center density*. This alternative parameter is denoted δ and is defined as

$$\delta \triangleq \frac{r^n}{|D|}. \tag{10.2.2}$$

A lattice is a *geometrical* object and as such it should be independent of the coordinate system used for the description. More precisely: any parameter of interest should be invariant under distance preserving transformations. We say that two lattices $\mathbb{L}$ and $\tilde{\mathbb{L}}$ that are related by a distance preserving transformation are *congruent*. If also a change of scale is involved we say that the lattices are *equivalent*.

Recall that the distance preserving transformations in $\mathbb{R}^n$ are of two kinds: translations and orthogonal transformations. A *translation* $x \mapsto \tilde{x} = x + a$ transforms the lattice $\mathbb{L}$ into a *congruent* lattice $\tilde{\mathbb{L}} = \mathbb{L} + a$. A transformation of the form $x \mapsto \tilde{x} = cx + a$, where c is a non-zero real number, transforms the lattice $\mathbb{L}$ into an *equivalent* lattice $\tilde{\mathbb{L}} = c\mathbb{L} + a$. The description $\mathbb{L} = \mathbb{Z}^n V$ presupposes that the coordinate system is selected in such a way that the origin is a point in the lattice.

An *orthogonal* transformation T transforms any vector $x \in \mathbb{R}^n$ into $\tilde{x} = xT \in \mathbb{R}^n$ and the lattice $\mathbb{L} = \mathbb{Z}^n V$ into $\tilde{\mathbb{L}} = \mathbb{Z}^n \tilde{V}$, with $\tilde{V} = VT$. Thus, for any orthogonal matrix T of order n the matrices V and $\tilde{V} = VT$ define *congruent* lattices; the weaker concept of *equivalence* requires only the relation $\tilde{V} = cVT;\ c \neq 0$.

The description $\mathbb{L} = \mathbb{Z}^n V$ involves not only a coordinate system for the rows of the matrix V but also an enumeration of the points in the generating lattice $\mathbb{Z}^n$. We notice that we have $\mathbb{Z}^n = \mathbb{Z}^n U$ for any non-singular square matrix U with the property that both U and U^{-1} have only integer entries. (It is easy to see that the inverse to a non-singular integer matrix U is integer if and only if $\det U = \pm 1$). The generating matrices V and $\tilde{V} = UV$ therefore produce congruent lattices if U is of this form.

In summary we have:

- Two matrices V and $\tilde{V}$, each having linearly independent rows, generate *congruent* lattices if they are related as $\tilde{V} = UVT$, where U is a non-singular integer matrix with $\det U = \pm 1$, and where T is an orthogonal matrix.
- Two such matrices generate *equivalent* lattices if they are related as $\tilde{V} = cUVT$, where c is any non-zero real-valued constant.

Suppose $\tilde{V} = UVT$ and let v and $\tilde{v} = vU^{-1}$ belong to $\mathbb{Z}^n$. We have

$$|\det \tilde{V}| = |\det(UVT)| = |\det U| \cdot |\det V| \cdot |\det T| = |\det V|$$

and

$$\|\tilde{v}\tilde{V}\| = \|vU^{-1}UVT\| = \|v\,V\|.$$

It follows that we also have

$$|\tilde{D}| = |\det \tilde{V}| = |\det V| = |D|$$

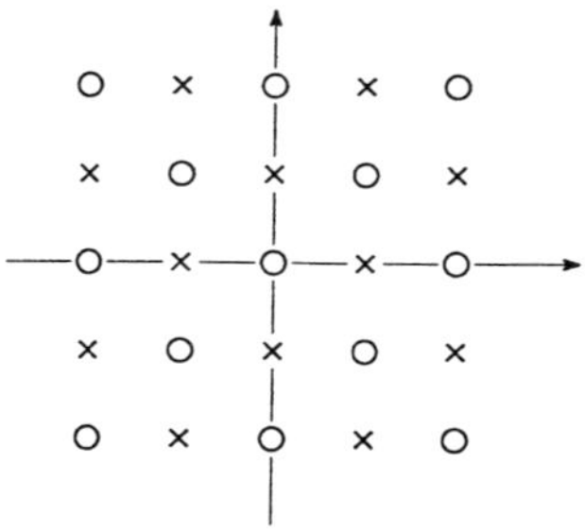

Figure 10.1: Illustrating Example 10.2.1. The rings indicate $\mathbb{L}_0 = \{(x, y) \in \mathbb{Z}^2 : x + y \equiv 0 \bmod 2\}$ while the crosses indicate $\mathbb{L}_1 = \{(x, y) \in \mathbb{Z}^2 : x + y \equiv 1 \bmod 2\}$.

and

$$\begin{aligned}\tilde{d} &= \min\left\{\|\tilde{v}\tilde{V}\| : \tilde{v} \in \mathbb{Z}^n \setminus \{0\}\right\} \\ &= \min\{\|vV\| : v \in \mathbb{Z}^n \setminus \{0\}\} = d.\end{aligned}$$

This means that the *volume* of the fundamental region and the *minimum distance* are invariant under *congruence*. As $|D|$ is proportional to d^n it follows that both the density and the central density are invariant under equivalence.

Before proceeding we illustrate the concepts introduced so far by some simple examples.

Example 10.2.1: Let $\mathbb{L} = \mathbb{Z}^2$ be the lattice of all integer points in the Euclidean plane and define

$$\mathbb{L}_0 = \{(x, y) \in \mathbb{Z}^2 : x + y \equiv 0 \bmod 2\},$$
$$\mathbb{L}_1 = \{(x, y) \in \mathbb{Z}^2 : x + y \equiv 1 \bmod 2\}.$$

In Figure 10.1we have used rings indicate the points in $\mathbb{L}_0$ and crosses to indicate those in $\mathbb{L}_1$. The lattice $\mathbb{L}$ is of course the union of $\mathbb{L}_0$ and $\mathbb{L}_1$. It is easy to see that all three point-sets constitute lattices. Choosing

$$V = I, \quad V_0 = \begin{bmatrix} 1 & -1 \\ 1 & 1 \end{bmatrix},$$

we can represent them as follows:

$$\mathbb{L} = \mathbb{Z}V, \quad \mathbb{L}_0 = \mathbb{Z}V_0, \quad \mathbb{L}_1 = \mathbb{Z}V_0 + (0, 1).$$

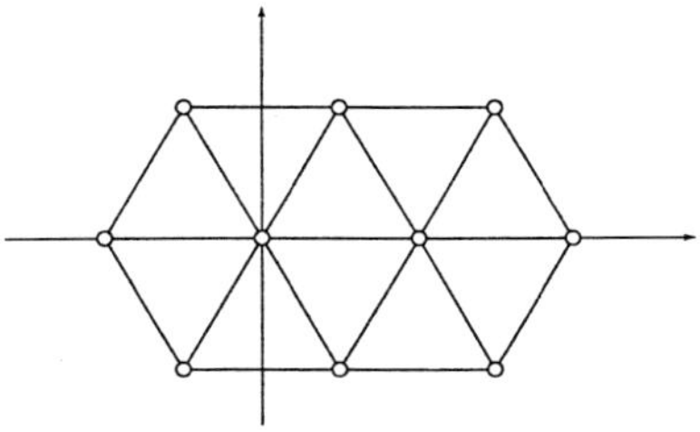

Figure 10.2: The hexagonal lattice.

We have $|D| = |\det V| = 1$ and $|D_0| = |D_1| = |\det V_0| = 2$. A simple consideration also reveals $d = 1$, $d_0 = d_1 = \sqrt{2}$. Thus we have $\delta = \delta_0 = \delta_1 = 1/4$. We have $V_0 = c\,U\,V\,T$ with

$$c = \sqrt{2}, \quad U = I, \quad T = \begin{bmatrix} \cos(\pi/4) & -\sin(\pi/4) \\ \cos(\pi/4) & \sin(\pi/4) \end{bmatrix}.$$

All three lattices are *equivalent*; the lattices $\mathbb{L}_0$ and $\mathbb{L}_1$ are even *congruent*. □

Example 10.2.2: The hexagonal lattice (see Figure 10.2) is generated by the matrix

$$V = \begin{bmatrix} 1 & 0 & 0 \\ -1/2 & \sqrt{3}/2 & 0 \end{bmatrix}.$$

In this description the Gram matrix is

$$G = VV^T = \begin{bmatrix} 1 & -1/2 \\ -1/2 & 1 \end{bmatrix},$$

and the volume of the fundamental domain is $|D| = \sqrt{\det G} = \sqrt{3}/2$. The minimum distance is $d = 1$. The center density is $\delta = 1/(2\sqrt{3})$. An alternative description is given by

$$\tilde{V} = \begin{bmatrix} 1 & -1 & 0 \\ 0 & 1 & -1 \end{bmatrix},$$

with the Gram matrix

$$\tilde{G} = \tilde{V}\tilde{V}^T = \begin{bmatrix} 2 & -1 \\ -1 & 2 \end{bmatrix}.$$

In this case the volume of the fundamental domain is $|\tilde{D}| = \sqrt{\det G} = \sqrt{3}$. The minimum distance is $\tilde{d} = \sqrt{2}$. The equivalence between the two descriptions is given by $\tilde{V} = c\,U\,V\,T$, with

$$c = \sqrt{2}, \quad U = I, \quad T = \frac{1}{\sqrt{6}} \begin{bmatrix} \sqrt{3} & -\sqrt{3} & 0 \\ 1 & 1 & -2 \\ \sqrt{2} & \sqrt{2} & \sqrt{2} \end{bmatrix}.$$

□

10.3 The root lattices

A lattice $\mathbb{L}$ can be regarded as an orbit of an additive group in $\mathbb{R}^n$. A coordinate description is simplest if the lattice is located so that the origin of the coordinate system is a point in the lattice. Let us suppose this to be the case. Denote by $Symm(\mathbb{L})$ the symmetry group of $\mathbb{L}$. This is the group of all distance preserving transformation under which $\mathbb{L}$ is invariant. The translations in $Symm(\mathbb{L})$ form a subgroup which is isomorphic with $\mathbb{L}$ itself; as a group code $\mathbb{L}$ is trivially generated by its translation subgroup.

What further elements could there be in the group $Symm(\mathbb{L})$? It is clear that there cannot be any further translations, so all other transformations must be orthogonal. In particular, let us investigate the conditions under which a *rotation* might belong to $Symm(\mathbb{L})$. Let the lattice be $\mathbb{L} = \mathbb{Z}^n V$ and let us suppose that V is square. Let $x = uV$ be an element in $\mathbb{L}$ and suppose T is a rotation belonging to $Symm(\mathbb{L})$. Then $y = xT$ belongs to $\mathbb{L}$, so we have $y = vV$ for some $v \in \mathbb{Z}^n$. It follows that we also have $uVT = vV$. By assumption the generator matrix is square and so invertible, and so we may define $U = VTV^{-1}$. This means that the vector v generating y is related to the vector u generating x according to $v = uU$. As this is true for any vector u, and as both u and v have only integer components, it follows that the matrix U also has only integer components. Taking the trace we obtain

$$Tr(U) = Tr(V^{-1}U\,V) = Tr(T).$$

As U is an integer matrix it follows that $Tr(T) = Tr(U)$ must be an integer. Recall that the trace is independent of the basis. With respect to some appropriately chosen basis the matrix T has the form

$$T = \begin{bmatrix} A & 0 \\ 0 & I \end{bmatrix},$$

where I is an $(n-2) \times (n-2)$ identity matrix, where A has the form

$$A = \begin{bmatrix} \cos\theta & -\sin\theta \\ \sin\theta & \cos\theta \end{bmatrix},$$

and where θ is the angle of rotation. It follows that we have $Tr(T) = 2\cos\theta + n - 2$. This quantity is an integer if and only if the angle θ has the form $\theta = 2\pi/p$, where $p = 1, 2, 3, 4, 6$. It follows that these are the only possible rotation angles under which a lattice $\mathbb{L}$ can possibly be invariant.

The restrictions regarding possible rotations in $Symm(\mathbb{L})$ also imply restrictions on the possible reflections. Let R_i and R_j be two reflections in $Symm(\mathbb{L})$. The product R_iR_j is a rotation with angle $2\pi/p_{ij}$, where p_{ij} must be one of the above five numbers.

Recall (from Chapter 8) that any reflection group can be decomposed into irreducible components, and that the irreducible reflection groups are precisely those indicated in Figure 8.1. The restrictions regarding possible angles for the rotations in $Symm(\mathbb{L})$ imply that the parameters p_{ij} occurring in the Gram matrix must be restricted as above. This condition is usually referred to as the *crystallographic condition.* An irreducible reflection group for which the Gram matrix satisfies the crystallographic condition is said to be a *crystallographic reflection group.*

The argument above implies that the crystallographic condition is a necessary condition for a reflection group to be a subgroup in $Symm(\mathbb{L})$. In Figure 8.1 we have explicitly indicated those irreducible reflection groups which satisfy this condition. As seen it is satisfied by all except $\mathcal{H}_3$, $\mathcal{H}_4$, and $\mathcal{I}_2(p)$ for $p \neq 1, 2, 3, 4, 6$.

It turns out that for each one of the crystallographic reflection groups there is actually a lattice $\mathbb{L}$ having this group as a subgroup in its symmetry group. In fact, it turns out that the corresponding lattices are unique up to equivalence and that they have generator matrices which are given by just minor modifications of the root matrices. The lattices generated in this way are called *root lattices.* They are usually denoted by the same letters as the corresponding reflection groups. For our purposes the most important ones of the root lattices are A_n, D_n, and E_n. In dimension $n \leq 8$ these lattices have the largest possible densities. They also generate—as we shall see shortly—a number of extremely good spherical codes. For $n > 8$ other lattices have larger densities and produce better spherical codes.

Because of the above remark we limit our discussion of root lattices to the cases A_n, D_n, and E_n. In all of these cases the generator matrix for the lattice can be taken directly as the root matrix of the corresponding reflection group, without any modification at all. We will follow the established convention of using the same letters for both the lattices and the reflection groups.

Theorem 10.3.1 *Each one of the lattices A_n, D_n, E_n is invariant under its corresponding reflection group.*

Proof: Let $v_1, v_2, \ldots, v_n$ be the set of vectors defining both the roots of the reflection group and the generator matrix of the corresponding lattice. This means that any reflection R_i in the reflection group has the form

$$R_i : x \mapsto x - 2\, v_i \langle v_i, x \rangle$$

and that each lattice point has the form $x = k_1 v_1 + k_2 v_2 + \cdots + k_n v_n$, $k_i \in \mathbb{Z}$. For each one of the cases under consideration we have $\langle v_i, v_j \rangle \in \{0, -1/2, -1\}$. It follows that any lattice point x is transformed to another lattice point under any one of the possible reflections. As the reflection group is generated by its reflections this means that the lattice $\mathbb{L}$ is invariant under the reflection group. □

Remark 10.3.1: The same proof applies with just minor modifications also to the remaining root lattices. The new thing that has to be taken into account is that in some cases we now also have $\langle v_i, v_j \rangle = -1/\sqrt{2}$. In order to obtain the proper lattices the generator matrices must be defined by roots of *different* lengths. For

n	1	2	3	4	5	6	7	8
$\mathbb{L}$	A_1	A_2	D_3	D_4	D_5	E_6	E_7	E_8
$1/\delta$	2	$2\sqrt{3}$	$4\sqrt{2}$	8	$8\sqrt{2}$	$8\sqrt{3}$	16	16

Table 10.1: The list of all known densest possible lattices for which the optimality has been proven.

instance, the reflection group B_2 was defined in Figure 8.2 by the roots $v_1 = (1,0)$, $v_2 = (1/\sqrt{2})(-1,1)$. A proper generator matrix for the corresponding lattice is

$$V = \begin{bmatrix} 2 & 0 \\ -1 & 1 \end{bmatrix}.$$

If all the rows in the generator matrix V of a root lattice $\mathbb{L}$ have unit length—as is the case for A_n, D_n, E_n—it is easy to see that the minimum distance is $d = 1$. It is also easy to see that the Gram matrix of the lattice equals—under these assumptions—the Gram matrix for the corresponding reflection group. Using Lemma 8.7.1 the following results are easily obtained, where by $\det(\mathbb{L})$ we mean the determinant of the corresponding Gram matrix:

$$\begin{aligned} \det A_n &= (n+1)2^{-n}; \quad n = 1,2,\ldots, \\ \det D_n &= 4 \cdot 2^{-n}; \qquad\quad n = 3,4,\ldots, \\ \det E_n &= (9-n)2^{-n}; \quad n = 6,7,8. \end{aligned}$$

From (10.2.1) and (10.2.2) we obtain

$$1/\delta = 2^n \sqrt{\det \mathbb{L}},$$

from which Table 10.1 can be computed.

It has been shown that these lattices are the densest possible. The subject has a long history. The case $n = 3$ (the first non-trivial case) was established by Gauss in 1831 [113]; the cases $n = 4$, 5 were established by Korkine-Zolotareff in 1872, [144] and 1877, [145] respectively. The cases $n = 6,7,8$ were established by Blichfeldt in 1935 [34], although the constructions were known much earlier. The lattice E_8 was actually constructed by Gosset [119] as early as 1900, see Coxeter [75], Section 11.8. It is usually referred to as the Gosset lattice.

For the lattices A_n and D_n we have the following general results:

$$A_n : 1/\delta = \sqrt{n+1} \cdot 2^{n/2}, \quad n = 1, 2, \dots ,$$
$$D_n : 1/\delta = 2 \cdot 2^{n/2}, \quad n = 3, 4, \dots .$$

However, for large n there are lattices much denser than these ones. For $n = 3$ the two formulas produce the same value. This is no coincidence: in this case the lattices A_n and D_n are actually equivalent.

10.4 Sphere packings and packing bounds

It should be emphasized that although the above root lattices have been proven to be the densest possible *lattice packings* there still remains the possibility that there might be even denser *non-lattice packings*. No such packing is known, but neither is there a proof that none exists, except for the trivial cases $n = 1, 2$. According to a famous conjecture, usually attributed to Kepler (Bezdek [21]), the face-centered cubic lattice A_3 is optimal also within the wider class of general sphere packings. A proof of this conjecture was recently claimed (Hsiang [132]), but doubts have been expressed as to the completeness of the proof. However, in this and all remaining cases upper bounds are known, although none of them is strong enough to establish the optimality of the above constructions in a wider sense, i.e. with the lattice condition relaxed.

We will devote the present paragraph to a short discussion of the known results on bounds for packings.

Any discrete pointset $\mathbb{P}$ in $\mathbb{R}^n$ defines a sphere packing. The parameters of interest are the same as for lattices, although the definitions must be modified. The minimum distance d is defined as an infimum over all distances within the packing:

$$d \triangleq \inf\{\|x - y\| : x, y \in \mathbb{P}, \ x \neq y\}.$$

Notice that this definition involves a limiting process. Also the density must be defined by some limiting process. The simplest way to proceed is the following. First, without loss of generality, we normalize the minimum distance to $d = 2$, corresponding to a packing of unit spheres. Then we denote by $N(r)$ the number of points from $\mathbb{P}$ that are inside the sphere $S_n(0, r)$. The volume of the sphere is $V_n(r) = k_n\, r^n$. Finally we define the *center density* as

$$\delta = \delta_n(\mathbb{P}) \triangleq \limsup_{r \to \infty} N(r)/V_n(r). \tag{10.4.1}$$

The *density* $\Delta = \Delta_n(\mathbb{P})$ is defined as $\Delta_n(\mathbb{P}) = \delta_n(\mathbb{P})k_n/n$. It should be clear that in case $\mathbb{P}$ is a lattice our new definitions coincide with the old ones.

Let us remark that the choice of *spheres* in the definitions above is somewhat arbitrary. There are many other bounded regions that could have been used. However, by a famous theorem by Hlawka, [128], (see also Zong, [242], p. 4), any convex body leads to an equivalent definition.

All of the non-lattice packings we shall consider are so called *periodic packings*. The definition is as follows.

Let $\mathbb{L}$ be a given lattice in $\mathbb{R}^n$ and let Q be an arbitrary finite subset of $\mathbb{R}^n$. The set

$$\mathbb{P} \triangleq \{Q + x : x \in \mathbb{L}\}$$

is a periodic packing associated with the lattice $\mathbb{L}$ and the set Q. Any fundamental region D associated with $\mathbb{L}$ will also be referred to as a valid fundamental region for the periodic packing $\mathbb{P}$. In this case there are $q \triangleq |Q|$ points associated with each fundamental region. The minimum distance has to be determined by some special argument.

It is clear that the concept of lattice is a special case of the concept of sphere packing. Therefore any upper bound on the density of sphere packings is automatically also an upper bound for the density of lattices. Conversely, any lower bound for the density of lattice packings is also a lower bound for the density of sphere packings. We will give a short account of the known results for both kinds of bounds.

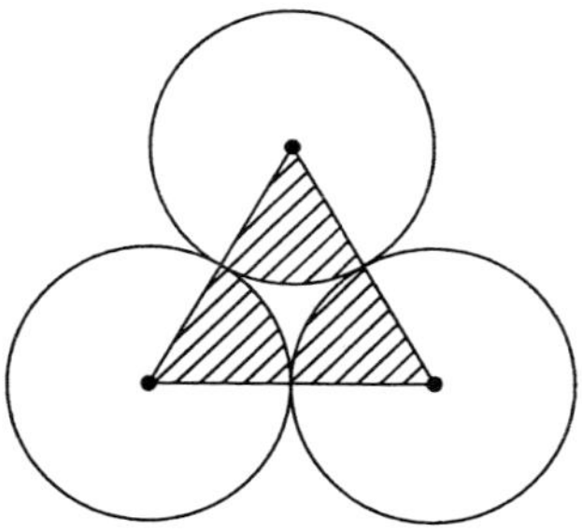

Figure 10.3: Graphical illustration to the derivation of Rogers' bound. The region B_n is the cross-hatched part of the triangle.

There are two major upper bounds, closely related to the two major general upper bounds for spherical codes. The first bound was derived by Rogers [186] and is usually referred to as Rogers' bound. It is the sphere packing counterpart to the Coxeter–Böröczky extension of the Fejes Tóth bound. The second bound belongs to Kabatianskii–Levenshtein [140]. It is directly related to the linear programming bound for spherical codes.

We consider first Rogers' bound. It is conceptually very simple, but unfortunately the numerical evaluation is quite complicated.

Let $\mathcal{S}_n$ be the *Simplex* code in Ω_n. As we have seen several times this code has parameters $(n, \rho, M) = (n, 2 + 2/n, n + 1)$. It has also the property that the distance is the same between any pair of codewords. Consider a sphere packing $\mathbb{P}$ in $\mathbb{R}^n$. A key step in the proof of Rogers' bound is to establish the intuitively natural fact that any $n+1$ neighboring points in $\mathbb{P}$ cannot be closer to each other than are the $n+1$ points in the *Simplex* code $\mathcal{S}_n$. Denote by C_n the convex region spanned by $\mathcal{S}_n$ and let $|C_n|$ be its volume. Let us form a sphere with radius $r = (1/2)\sqrt{2 + 2/n}$ around each of the codewords in $\mathcal{S}_n$. Let B_n be the intersection of all these spheres with the convex set C_n and denote by $|B_n|$ the volume of this set. Rogers [186] showed that the density $\Delta_n = k_n \delta_n / n$ for any sphere packing in $\mathbb{R}^n$ cannot exceed the quotient of these two volumes:

$$\Delta_n \leq \frac{|B_n|}{|C_n|}.$$

The idea is illustrated for the case $n = 2$ in Figure 10.3.

Unfortunately it is usually quite difficult to really compute the volumes $|B_n|$ and $|C_n|$. Let the functions $F_n(\alpha)$, $n = 1, 2, \ldots$, be defined as in Theorem 3.6.1 of Section 3.6. Define the functions $f_n(x)$ by the relation

$$f_n\Big(\frac{1}{\cos 2\alpha}\Big) \triangleq F_n(\alpha).$$

Rogers derived the following form of the bound

$$\delta_n \leq \delta_R(n) = \frac{(n+1)^{1/2}(n!)^2}{2^{3n/2}} f_n(n).$$

For the first two cases we obtain $\delta_1 = 1/2$, $\delta_2 = 1/2\sqrt{3}$, proving that A_1 and A_2 are densest possible. For $n = 3$ there is no known packing meeting Rogers' bound.

Next we consider the Kabatianskii–Levenshtein bound. It is obtained by a combination of two inequalities. In the first step the density Δ_n of any sphere packing in $\mathbb{R}^n$ is upper bounded by the maximal size $A(n+1, \rho)$ of an appropriately selected spherical code in $\mathbb{R}^{n+1}$. In the second step the quantity $A(n+1, \rho)$ is upper bounded by the linear programming bound. The idea in the first step is the same as the one used in the Bassalygo–Elias bound for binary codes (Bassalygo [16], see also MacWilliams–Sloane [166], p. 558). For packings in Euclidean space this idea was first employed by Yaglom [229] already in 1958.

Theorem 10.4.1 (Yaglom) *Let Δ_n be the density of a sphere packing $\mathbb{P}$ in $\mathbb{R}^n$ and let $A(n+1, \rho)$ denote the largest possible size of a spherical code in Ω_{n+1} with squared minimum distance ρ. Then for any $r > 0$ the following bound holds:*

$$\Delta_n \leq \frac{1}{r^n} A(n+1, 4/r^2).$$

Proof: As in the general definition (10.4.1) we normalize the minimum distance to $d = 2$ and denote by $N(r)$ the number of points from $\mathbb{P}$ that are inside the sphere $S_n(a, r)$ in $\mathbb{R}^n$. Consider now a sphere

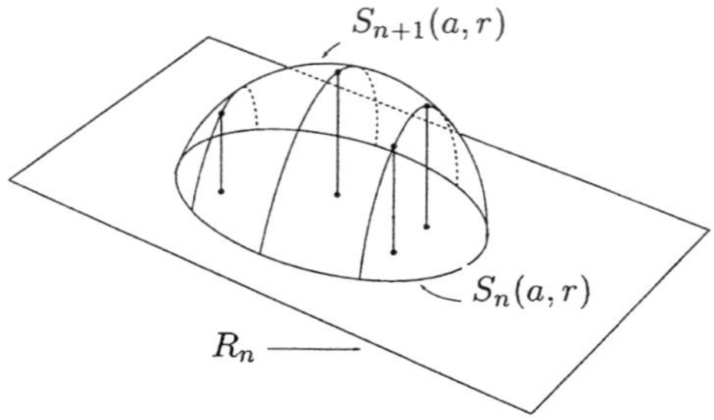

Figure 10.4: Graphical illustration to the Yaglom construction.

$S_{n+1}(a,r)$ in $\mathbb{R}^{n+1}$ such that $S_n(a,r)$ is a projection of $S_{n+1}(a,r)$ on an n-dimensional hyperplane in $\mathbb{R}^n$. By a proper selection of the center a we have $N(r) \geq \Delta_n r^n$. By projecting each point from the packing $\mathbb{P}$ located *inside* the sphere $S_n(a,r)$ to the *surface* of $S_{n+1}(a,r)$ we obtain a constant energy code in $\mathbb{R}^{n+1}$ with minimum distance $d_E \geq d = 2$, (Figure 10.4 illustrates the idea). By definition we have

$$\Delta_n \, r^n \leq N(r) \leq A(n+1, 4/r^2).$$

□

The bound is implicit inasmuch as it is expressed in terms of the unknown function $A(n+1, 4/r^2)$. In order to get an explicit bound we need to insert some further bound on $A(n+1, 4/r^2)$. The natural bound to insert is the linear programming bound. This leads to the Kabatianskii–Levenshtein bound. However, any other bound will also do. In order to obtain the best possible result out of Theorem 10.4.1 we should choose the best bound available and then minimize with respect to the parameter $r > 0$. Notice that we have expressed the bound in terms of the density $\Delta_n = k_n \, \delta_n / n$.

According to Conway–Sloane ([71], p. 20) Rogers' bound is the best of the two for $n \leq 42$, while for $n > 42$ the Kabatianskii–Levenshtein bound is better. For the special case $n = 3$ a bound by Lindsey [161] is slightly better than both.

The classical *lower* bound is a bound by Minkowski [169] according to which there exists for any n a lattice packing with density Δ_n

satisfying

$$\Delta_n \geq \frac{\zeta(n)}{2^{n-1}},$$

where $\zeta(n)$ is the Riemann zeta-function:

$$\zeta(n) = \sum_{k=1}^{\infty} k^{-n}.$$

The bound is actually a special case of a more general result usually referred to as the Minkowski–Hlawka theorem. Minkowski [169], 1905, proved the result for packing of spheres and conjectured that the same should hold for a much wider class of convex bodies. Hlawka [129] gave the first proof of the general form in 1943. For small n the bound is quite weak in as much as there are explicit constructions giving much higher densities. Asymptotically as n tends to infinity it is, however, the best result known. Define

$$\varepsilon \triangleq \limsup_{n\to\infty} \frac{1}{n} \log \Delta_n.$$

From the Minkowski–Hlawka theorem follows that we have $\varepsilon \geq -1$, and that the lower bound can be achieved by a lattice packing. From the Kabatianskii–Levenshtein bound it can be shown [140] that we must have $\varepsilon \leq -0.5990$ for any packing, lattice or not.

10.5 Sphere packings and codes

There are several constructions by which sphere packings can be constructed from known lattices and codes. If the codes are linear the resulting sphere packings are usually lattices. We describe here one such construction and one of its modifications. Both were devised by Leech [152]. Together they generate all the best known packings in dimensions $n \leq 24$ (see Conway–Sloane [71], Chapter 1).

Our formulation differs slightly from the one originally given by Leech, and is actually slightly more general. The basic idea, however, is exactly the same.

Let $\mathbb{L}$ be a given lattice in $\mathbb{R}^m$ and let $Q = \{a_1, a_2, \ldots, a_q\}$ be a finite pointset in the fundamental region of $\mathbb{L}$. Let $\mathbb{F}_q$ be the finite field of size q and let $f : \mathbb{F}_q \to Q$ be a mapping associating a point $a = f(\alpha) \in Q$ with each field element $\alpha \in \mathbb{F}_q$. Let $\mathcal{C}$ be a code in $\mathbb{F}_q^N$, possibly but not necessarily a linear code. Finally, denote by $f_N : \mathbb{F}_q^N \to Q^N$ the coordinate-wise extension of the mapping f. By this mapping any codeword $c = (c_1, c_2, \ldots, c_N)$ in $\mathcal{C}$ is mapped on a vector $f_N(c) = (f(c_1), f(c_2), \ldots, f(c_N))$ in $Q^N \subseteq \mathbb{R}^n$, where $n = Nm$. Denote by $f_N(\mathcal{C})$ the pointset in $Q^N \subseteq \mathbb{R}^n$ corresponding to the code $\mathcal{C} \subseteq \mathbb{F}_q^N$.

Definition 10.5.1 *For any lattice $\mathbb{L} \subseteq \mathbb{R}^m$ and any code $\mathcal{C} \subseteq \mathbb{F}_q^N$ we define*

$$\mathbb{P}(\mathbb{L}, \mathcal{C}) \triangleq \mathbb{L}^N + f_N(\mathcal{C}) = \{x = z + f_N(c) \in \mathbb{R}^n : z \in \mathbb{L}^N,\ c \in \mathcal{C}\}.$$

Let $D_{\mathbb{L}}$ be a fundamental region for the lattice $\mathbb{L}$. The Cartesian product $\mathbb{L}^N$ is a lattice with fundamental region $D_{\mathbb{L}}^N$. The packing $\mathbb{P}(\mathbb{L}, \mathcal{C})$ puts $M = |\mathcal{C}|$ points in each one of the fundamental domains of $\mathbb{L}^N$. We define

$$\begin{aligned}
d_{\mathbb{L}} &= \min\{\|x - y\| : x, y \in \mathbb{L},\ x \neq y\},\\
d_Q &= \min\{\|z + x - y\| : z \in \mathbb{L},\ x, y \in Q,\ x \neq y\},\\
d_{\mathcal{C}} &= \min\{d_H(u, v) : u, v \in \mathcal{C},\ u \neq v\}.
\end{aligned}$$

The minimum distance $d_{\mathbb{P}}$ in the packing $\mathbb{P}(\mathbb{L}, \mathcal{C})$ is

$$d_{\mathbb{P}}^2 = \min\{d_{\mathcal{C}} d_Q^2,\ d_{\mathbb{L}}^2\} = d_{\mathbb{L}}^2\, d^2;$$

where in the last step we have introduced the quantity

$$d \triangleq \min\{\frac{\sqrt{d_{\mathcal{C}}}\, d_Q}{d_{\mathbb{L}}}, 1\}.$$

Theorem 10.5.1 *The center density $\delta_{\mathbb{P}}$ of the packing $\mathbb{P}(\mathbb{L}, \mathcal{C})$ is given by*

$$\delta_{\mathbb{P}} = \delta_{\mathbb{L}}^N\, M\, d^n.$$

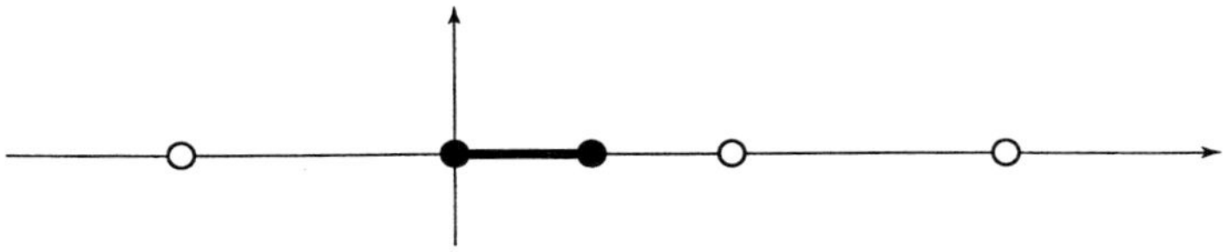

Figure 10.5: The lattice $\mathbb{L} = \mathbb{Z}$ and the set $Q = \{0, 1/2\}$.

Proof: Recall (see equation (10.2.2)) that the center density of the lattice $\mathbb{L}$ is

$$\delta_{\mathbb{L}} = \frac{(d_{\mathbb{L}}/2)^m}{|D_{\mathbb{L}}|} .$$

The center density of the N-fold Cartesian product $\mathbb{L}^N$ is $\delta_{\mathbb{L}}^N$. We have noticed that the packing $\mathbb{P}(\mathbb{L}, \mathcal{C})$ puts M points in each fundamental region $D_{\mathbb{L}}^N$ of $\mathbb{L}^N$. We have also noticed that the minimum distance in $\mathbb{P}(\mathbb{L}, \mathcal{C})$ is $d_{\mathbb{P}}^2 = d_{\mathbb{L}}^2\, d^2$. The theorem follows by these observations. □

The packing $\mathbb{P}(\mathbb{L}, \mathcal{C})$ gives alternative constructions of all the root lattices considered above. We illustrate this fact by considering several examples.

Example 10.5.1: Let the lattice $\mathbb{L}$ be given by $\mathbb{L} = \mathbb{Z}$ and take $Q = \{0, 1/2\}$ (see Figure 10.5). We obtain $d_{\mathbb{L}} = 1$, $d_Q = 1/2$, which gives us $d = \min\{(1/2)\sqrt{d_{\mathcal{C}}}, 1\}$. Taking $\mathcal{C}$ as a simple binary parity check code we obtain $d = 1/\sqrt{2}$, which by Theorem 10.5.1 gives us

$$\delta_{\mathbb{P}} = \left(\frac{1}{2}\right)^n \cdot 2^{n-1} \cdot \left(\frac{1}{\sqrt{2}}\right)^n = 2^{-1-n/2}.$$

This is the same density as we have already derived for the lattices D_n. In order to see that the new construction actually produces the same lattice we first notice that with $\mathbb{L}$ and $\mathcal{C}$ as above we have

$$\mathbb{P}(\mathbb{L}, \mathcal{C}) = \frac{1}{2}\{x \in \mathbb{Z}^n : x_1 + x_2 + \cdots + x_n \equiv 0 \bmod 2\}.$$

Next we recall that a basis for the lattice D_n is given by

$$V = \begin{bmatrix} 1 & 1 & 0 & \dots & 0 & 0 \\ -1 & 1 & 0 & \dots & 0 & 0 \\ 0 & -1 & 1 & \dots & 0 & 0 \\ \cdot & \cdot & \cdot & \cdots & \cdot & \cdot \\ 0 & 0 & 0 & \dots & -1 & 1 \end{bmatrix}.$$

With $u = (u_1, u_2, \ldots, u_n)$ and $x = u\,V$ we obtain

$$\begin{aligned}
x_1 &= u_1 - u_2, \\
x_2 &= u_1 + u_2 - u_3, \\
x_3 &= u_3 - u_4, \\
&\ldots \\
x_{n-1} &= u_{n-1} - u_n, \\
x_n &= u_n.
\end{aligned}$$

That gives us

$$\begin{aligned}
2u_1 &= x_1 + x_2 + \cdots + x_n, \\
2u_2 &= -x_1 + x_2 + \cdots + x_n, \\
u_3 &= x_3 + x_4 + \cdots + x_n, \\
&\ldots \\
u_{n-1} &= x_{n-1} + x_n, \\
u_n &= x_n.
\end{aligned}$$

It is easy to check that these equations have an integer solution if and only if the condition $x_1 + x_2 + \cdots + x_n \equiv 0 \bmod 2$ is satisfied. This means that the root lattice generated by the matrix V and the lattice $\mathbb{P}(\mathbb{L}, \mathcal{C})$ are equivalent. □

Example 10.5.2: With $\mathbb{L}$ and Q as in Example 10.5.1 and with a binary code having parameters $[n, k, d]_q = [8, 4, 4]_2$ (extended Hamming code) we obtain $d = 1$ and the following center density:

$$\delta_{\mathbb{P}} = \left(\frac{1}{2}\right)^8 \cdot 2^4 \cdot 1 = \frac{1}{16}.$$

This is an alternative definition of the E_8-lattice. □

Example 10.5.3: With $\mathbb{L}$ and Q again the same as in Example 10.5.2 and with a binary code having parameters $[n, k, d]_q = [7, 3, 4]_2$ (binary *Simplex code*) we obtain

$$\delta_{\mathbb{P}} = \left(\frac{1}{2}\right)^7 \cdot 2^3 \cdot 1 = \frac{1}{16}.$$

This is the same density as for the lattice E_7 and indeed it can be shown that the present construction is just another definition. □

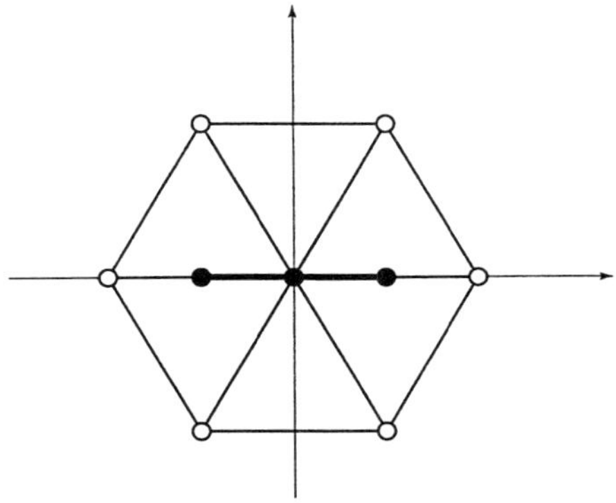

Figure 10.6: The hexagonal lattice $\mathbb{L}_{hex}$ along with the set $Q = \{(0,0),(\pm 1/\sqrt{3},0)\}$.

Example 10.5.4: Let $\mathbb{L}$ be the hexagonal lattice defined by the generator matrix

$$V = \begin{bmatrix} 1 & 0 \\ -1/2 & \sqrt{3}/2 \end{bmatrix}$$

and let the set Q be $Q = \{(0,0),(\pm 1/\sqrt{3},0)\}$, see Figure 10.6. We obtain $d_{\mathbb{L}} = 1, d_Q = 1/\sqrt{3}$ and thus $d = \min\{\sqrt{d_{\mathcal{C}}/3}, 1\}$. Choosing $\mathcal{C}$ as a linear ternary repetition code with parameters $[n,k,d]_q = [3,1,3]_3$ we obtain

$$\delta_{\mathbb{P}} = \left(\frac{1}{2\sqrt{3}}\right)^3 \cdot 3 \cdot 1 = \frac{1}{8\sqrt{3}}.$$

Again equality actually prevails. We recognize the lattice E_6. □

Example 10.5.5: With $\mathbb{L}$ and Q as in Example 10.5.4 but choosing the code $\mathcal{C}$ with parameters $[n,k,d]_q = [4,2,3]_3$ (a ternary Hamming code) we obtain

$$\delta_{\mathbb{P}} = \left(\frac{1}{2\sqrt{3}}\right)^4 \cdot 3^2 \cdot 1 = \frac{1}{16}.$$

This is still another construction of the E_8-lattice. □

Example 10.5.6: Let $\mathbb{L}$ still be the hexagonal lattice, but let the set Q be

$$Q = \{(0,0),(1/2,0),(-1/4,\pm\sqrt{3}/4)\},$$

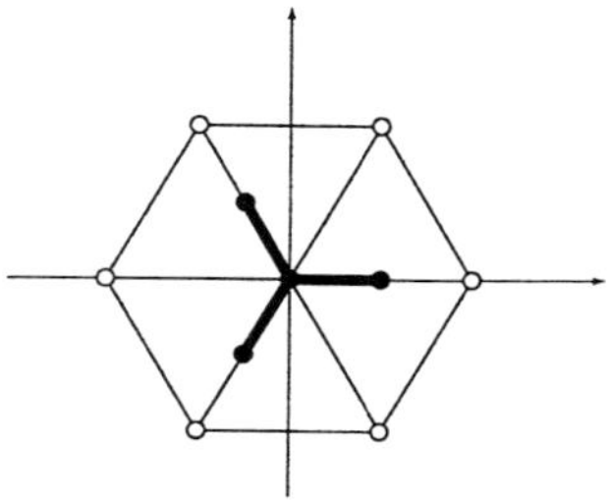

Figure 10.7: The hexagonal lattice $\mathbb{L}_{hec}$ along with the set $Q = \{(0,0),(1/2,0),(-1/4,\pm\sqrt{3}/4)\}$.

see Figure 10.7. Choose $\mathcal{C}$ as a quaternary code with parameters $[n,k,d]_q = [6,3,4]_4$, (a quaternary MDS code due to Bush [52], sometimes referred to as the "hexacode", [71], p. 82, pp. 300-303). We obtain

$$\delta_{\mathbb{P}} = \left(\frac{1}{2\sqrt{3}}\right)^6 \cdot 4^3 \cdot 1 = \frac{1}{27}.$$

The construction produces a lattice known as the Coxeter–Todd lattice, [76]. □

10.6 Lattices and codes

As a special case of the construction above it might occur that $\mathbb{L}+Q = \{x+a : x \in \mathbb{L}, a \in Q\}$ is in itself a lattice. As a matter of fact, this happens to be the case in all of the examples considered above. In all such cases the construction can be formulated in a slightly different way. The alternative form is less general, but it is better suited for exhibiting the algebraic properties.

Let $\mathbb{L}$ be a lattice. As we know from the definition if we choose a coordinate system such that one lattice point is located at the origin the lattice $\mathbb{L}$ forms an infinite abelian group with respect to any such coordinate system. Let $\mathcal{G}$ be a finite group and suppose there is a map f from $\mathbb{L}$ onto $\mathcal{G}$ which is an *homomorphism*, so that $\mathcal{G}$ is an *homomorphic image* of $\mathbb{L}$. Let the order of $\mathcal{G}$ be $|\mathcal{G}| = q$.

We denote by f_N the mapping $\mathbb{L}^N \to \mathcal{G}^N$ obtained by applying the homomorphism f individually to each component x_i in a vector $x = (x_1, x_2, \ldots, x_N) \in \mathbb{L}^N$. Clearly f_N is also an homomorphism:

$$\forall x,\, y \in \mathbb{L}^N :\; f_N(x+y) = f_N(x) + f_N(y).$$

Let $\mathcal{C}$ be a subset in $\mathcal{G}^N$. Typically $\mathcal{G}$ is the additive group in a finite field $\mathbb{F}_q$ and $\mathcal{C}$ is a linear code in $\mathbb{F}_q^n$, but for the general definition it is only the group structure in $\mathcal{G}$ that is important, and $\mathcal{C}$ can in principle be any subset in $\mathcal{G}^N$. Given the lattice $\mathbb{L}$, the finite homomorphic image $\mathcal{G}$, and the subset $\mathcal{C} \subseteq \mathcal{G}^N$ we define

$$\mathbb{P}_*(\mathbb{L}, \mathcal{C}) \triangleq f_N^{-1}(\mathcal{C}) = \{x \in \mathbb{L}^N : f_N(x) \in \mathcal{C}\}.$$

By construction it is obvious that the packing $\mathbb{P}_*(\mathbb{L}, \mathcal{C})$ is a lattice whenever $\mathcal{C}$ is a subgroup in $\mathcal{G}^N$. In particular this is the case if $\mathcal{C}$ is a linear code over a finite field $\mathbb{F}_q$ and $\mathcal{G}$ is the additive group of this field. Let us see how this new construction is related to the previous one.

For any $\alpha \in \mathcal{G}$ we define

$$\mathbb{L}_\alpha \triangleq f^{-1}(\alpha) = \{a \in \mathbb{L} : f(a) = \alpha\}.$$

Denote by 0 the neutral element in $\mathcal{G}$. It is clear that $\mathbb{L}_0$ is a subgroup in $\mathbb{L}$ and that $\mathbb{L}_\alpha$—for each $\alpha \neq 0$—is a coset in the quotient group $\mathbb{L}/\mathbb{L}_0$. It is also clear that $\mathbb{L}/\mathbb{L}_0$ is isomorphic with the group $\mathcal{G}$. All of the sets $\mathbb{L}_\alpha$ are equivalent lattices. Let $Q = \{a_1, a_2, \ldots, a_q\}$ be a set of coset representatives. The packing $\mathbb{P}_*(\mathbb{L}, \mathcal{C})$ is nothing else than the old construction applied to $\mathbb{L}_0$ and Q:

$$\mathbb{P}_*(\mathbb{L}, \mathcal{C}) = \mathbb{P}(\mathbb{L}_0, \mathcal{C}).$$

For reproducing the examples considered in Section 10.5 three different groups are needed, corresponding to the additive groups in the finite fields $\mathbb{F}_2$, $\mathbb{F}_3$ and $\mathbb{F}_4$ respectively. Let us denote by $(\mathbb{F}_q, +)$ the additive group in $\mathbb{F}_q$.

In Examples 10.5.1–10.5.3 we used binary codes. For an alternative definition using the construction $\mathbb{P}_*(\mathbb{L}, \mathcal{C})$ we employ the obvious homomorphism $f : \mathbb{Z} \to (F_2, +)$ given by $a \mapsto a \bmod 2$.

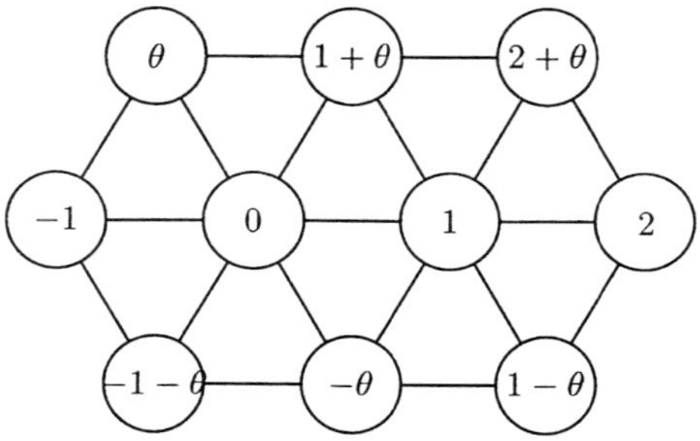

Figure 10.8: The hexagonal lattice with Eisenstein coordinates.

In the remaining examples two different constructions using the hexagonal lattice A_2 along with the fields $\mathbb{F}_3$ and $\mathbb{F}_4$ are employed. The homomorphisms needed for a reformulation in terms of the construction $\mathbb{P}_*(\mathbb{L}, \mathcal{C})$ are best described using the *Eisenstein integers.*

Let θ be a root to the equation $1 + \theta + \theta^2 = 0$ over $\mathbb{C}$, say

$$\theta = -1/2 + i\sqrt{3}/2,$$

where $i^2 = -1$. The Eisenstein integers are defined as

$$\mathbb{E} = \{u + \theta v : (u, v) \in \mathbb{Z}^2\}.$$

It is clear that the Eisenstein integers offer a complex representation of the hexagonal lattice A_2. Alternatively we can simply regard the pair (u, v) as a vector enumerating the lattice points in A_2 under the mapping $(x, y) = (u, v)V$, with V as in Example 10.5.4:

$$V = \begin{bmatrix} 1 & 0 \\ -1/2 & \sqrt{3}/2 \end{bmatrix}.$$

In Figure 10.8 a small part of A_2 is displayed in the notation of Eisenstein integers.

There are two simple homomorphisms mapping $\mathbb{E} = A_2$ onto $(\mathbb{F}_3, +)$ and $(\mathbb{F}_4, +)$ respectively. A mapping onto $(\mathbb{F}_3, +)$ is given by

$$f : \mathbb{E} \to (\mathbb{F}_3, +); \quad u + \theta v \mapsto (u + v) \bmod 3.$$

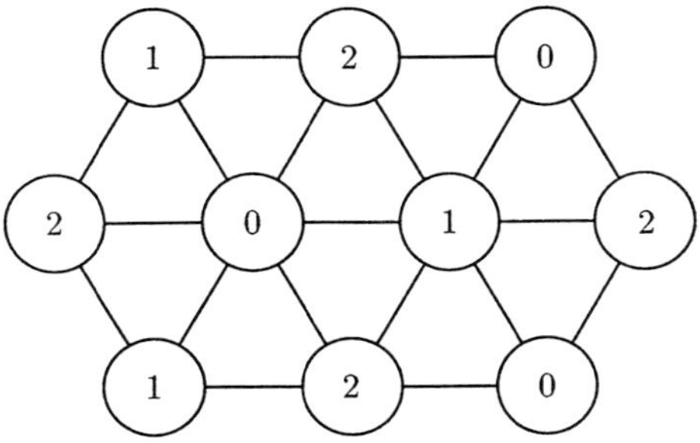

Figure 10.9: Illustrating the homomorphism $f : u + \theta \cdot v \mapsto (u + v) \bmod 3$.

This map is illustrated in Figure 10.9. Notice that all three corners in any triangle are mapped on different elements on $\mathbb{F}_3$. Also notice that the three cosets corresponding to the three different elements in $\mathbb{F}_3$ define congruent lattices. In this case all three sublattices are also equivalent with the original lattice $\mathbb{E}$. The minimum distance in any sublattice is a factor $\sqrt{3}$ larger than in the original lattice. Consider a triangle defined by three neighboring zeros. There is one point from the original lattice in the middle of each such triangle.

Let ω be a primitive third root of unity over the binary field, i.e. a root to the irreducible polynomial $1 + x + x^2$ over $\mathbb{F}_2$. Any element in $\mathbb{F}_4$ can be represented in the form $\alpha + \beta\,\omega$, $\alpha, \beta \in \mathbb{F}_2$. Identify $\mathbb{F}_2$ with the set $\mathbb{Z}_2$ of integers modulo 2 and denote by $(a)_2$ the residue class of integers equivalent to the integer a under reduction modulo 2: $(a)_2 \equiv a \bmod 2$. An homomorphism $f : \mathbb{E} \to (\mathbb{F}_4, +)$ can be defined as

$$f : \mathbb{E} \to (\mathbb{F}_4, +); \quad u + \theta \cdot v \mapsto (u)_2 + \omega \cdot (v)_2.$$

This mapping is illustrated in Figure 10.10. Again the various cosets define congruent lattices which are all equivalent with the original lattice. The minimum distance in any of the sublattices is a factor $2 = \sqrt{4}$ larger than in the original lattice. Consider a triangle of three neighboring zeros. There is one point from the original lattice on each edge of any such triangle.

We finally notice that the dimension m of the hexagonal lattice is

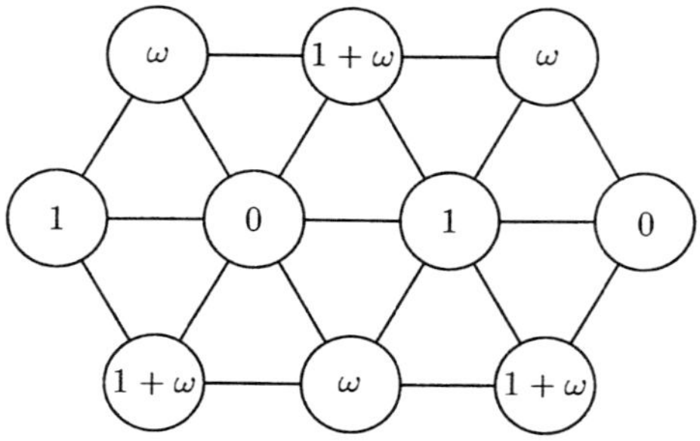

Figure 10.10: Illustrating the homomorphism $f : u+\theta\cdot v \mapsto (u)_2+\omega\cdot(v)_2$.

$m = 2$ and that in both of the above cases we have

$$\left(\frac{d_{\mathbb{L}_0}}{d_{\mathbb{L}}}\right)^m = q.$$

10.7 Expurgated constructions

Let $\mathbb{L} \subset \mathbb{R}^n$ be a lattice with minimum distance d and fundamental region D. Recall that the center density is defined by the formula

$$\delta = \frac{(d/2)^n}{|D|}\ .$$

By deleting certain points in the lattice $\mathbb{L}$ we might affect both the minimum distance d and the fundamental region D. Under certain circumstances this might lead to a new lattice or packing with larger density. This is the basic idea in the following constructions.

Let $\mathcal{C}$ be a binary code with parameters $(n, d_{\mathcal{C}}, M)$. A simple construction (called construction A [153], see also [71], p. 137) is the following:

$$\mathbb{P}_A(\mathcal{C}) \triangleq \{x \in \mathbb{Z}^n : x \bmod 2 \in \mathcal{C}\}.$$

The minimum distance of this packing is $d = \min\{2, \sqrt{d_{\mathcal{C}}}\}$. For $d_{\mathcal{C}} = 4$ we have $d = 2$ and the central density is

$$\delta_A = M2^{-n}.$$

If $\mathcal{C}$ is a linear code then the packing $\mathbb{P}_A(\mathcal{C})$ is a lattice. The construction $\mathbb{P}_A(\mathcal{C})$ is equivalent with construction $\mathbb{P}(\mathbb{L}, \mathcal{C})$ in the case $\mathbb{L} = 2\mathbb{Z}$.

A second construction (called construction B [153], see also [71], p. 141) is the following. Let $\mathcal{C} \subseteq \mathbb{F}_2^n$ be a binary code with the property that all codewords have *even weight.* Define

$$\mathbb{P}_B(\mathcal{C}) \triangleq \{x \in \mathbb{Z}^n : (i)\ x \bmod 2 \in \mathcal{C};\ (ii)\ x_1+x_2+\cdots+x_n \equiv 0 \bmod 4\}.$$

It turns out that equal performance can be obtained from a slight generalization, which we call B^*.

Definition 10.7.1 *For an arbitrary binary code $\mathcal{C}$ (no restrictions on weights) the packing $\mathbb{P}_{B^*}(\mathcal{C})$ is defined according to the formula*

$$\mathbb{P}_{B^*}(\mathcal{C}) \triangleq \{x \in \mathbb{Z}^n : (i)\ x \bmod 2 \in \mathcal{C},\ (ii) \sum_{i=1}^{n} x_i - (x_i)_2 \equiv 0 \bmod 4\},$$

where $(x_i)_2$ denotes the modulo 2 reduction of x_i, interpreted as an integer.

We notice that if $x \bmod 2$ belongs to $\mathcal{C}$, where $\mathcal{C}$ is a binary code having even weight codewords only, then we have

$$\sum_{i=1}^{n} (x_i)_2 \equiv 0 \bmod 4,$$

which immediately shows that the B^*-construction is a true generalization of the B-construction.

The density of the B^*-construction is given in the following theorem.

Theorem 10.7.1 *Let $\mathcal{C}$ be a binary code and let its parameters be (n, d_c, M). The packing $\mathbb{P}_{B^*}(\mathcal{C})$ has center density δ given by*

$$\delta = M d^n 2^{-(2n+1)}.$$

where $d^2 = \min\{d_{\mathcal{C}}, 8\}$.

Remark 10.7.1: We notice that for the special case $d_{\mathcal{C}} = 8$ we have $d^2 = 8$, which gives us

$$\delta = M \cdot 2^{-(n/2+1)}.$$

Proof: The components x_i in any vector $x = (x_1, x_2, \ldots, x_n) \in \mathbb{Z}^n$ can be represented as

$$x_i = (x_i)_2 + 2\,k_i,$$

where $(x_i)_2$ is the modulo 2 reduction of x_i and where k_i is an integer. Let us form the vector $k = (k_1, k_2, \ldots, k_n) \in \mathbb{Z}^n$. We conclude that we have

$$\sum_{i=1}^{n} x_i - (x_i)_2 = 2 \sum_{i=1}^{n} k_i\,.$$

In order to satisfy property (*ii*) the sum of the components of the vector k must be even, which means that k is an element in the lattice D_n.

Consider two distinct points x and x' from $\mathbb{P}_{B^*}(\mathcal{C})$. Let us represent them as $x = c + 2k$ and $x' = c' + 2k'$ respectively, where c and c' are elements in the binary code $\mathcal{C}$ and where k and k' are elements in the lattice D_n. If $c = c'$ then k and k' are distinct elements in D_n. It follows that we have $\|x - x'\|^2 \geq 4 \cdot d(D_n) = 8$. We notice that the lower bound is attained. Assume $c \neq c'$. If $k = k'$ we immediately conclude $\|x - x'\|^2 \geq d_{\mathcal{C}}$. Again the bound is attained. Next we notice that for any index i such that $k_i \neq k_i'$ we have $|x_i - x_i'| = |c_i - c_i' + 2(k_i - k_i')| \geq 1$. Therefore the distance cannot be smaller if $k \neq k'$. Thus the minimum distance d in the packing $\mathbb{P}_{B^*}(\mathcal{C})$ is given by

$$d^2 = \min\{d_{\mathcal{C}}, 8\}.$$

The fundamental region in $2\,D_n$ has volume 2^{n+1}. In the packing $\mathbb{P}_{B^*}(\mathcal{C})$ each fundamental region from $2\,D_n$ contains $M = |\mathcal{C}|$ points. It follows that the center density is

$$\delta = M \cdot \frac{\left(\frac{1}{2}d\right)^n}{2^{n+1}} = M d^n 2^{-(2n+1)}.$$

□

We offer three examples.

Example 10.7.1: Taking $\mathcal{C}$ as the binary repetition code of length $n = 8$ we obtain $M = 2$ and $d_\mathcal{C} = 8$. The packing $\mathbb{P}_{B^*}(\mathcal{C})$ is a lattice with density $\delta = 2 \cdot 2^{-4-1} = 1/16$. This is still another definition of the Gosset lattice E_8, [119]. □

Example 10.7.2: Taking $\mathcal{C}$ as the Reed-Muller code $R(r, m) = R(1, 4)$ we obtain $M = 16$, $d_\mathcal{C} = 8$. Again $\mathbb{P}_{B^*}(\mathcal{C})$ is a lattice. The density is $\delta = 2^{5-8-1} = 1/16$. This is the celebrated Barnes–Wall lattice, [15]. □

Example 10.7.3: Taking $\mathcal{C}$ as the extended binary Golay code $\mathcal{G}_{24}$ with parameters $[n, k, d] = [24, 12, 8]$ we obtain $M = 4096$, $d_\mathcal{C} = 8$. The construction $\mathbb{P}_{B^*}(\mathcal{G}_{24})$ produces a lattice with density $\delta = 2^{12-12-1} = 1/2$. This is *half* of the so called *Leech lattice*, [152], which we discuss in more detail in the next section. □

10.8 The Leech lattice

The last example in the previous section is of particular interest. It employs the extended binary Golay code and was first constructed by Leech in a paper published 1964, [151]. Soon after [152] Leech discovered that two translated copies of this lattice could be fitted together without decreasing the minimum distance, thus creating a lattice in 24 dimensions with density $\delta = 1$. This lattice is today referred to as the Leech lattice. It possesses a large number of remarkable properties (see [71]) and although being unique it has a number of different constructions. The simplest is probably the following.

Denote by $\mathbb{L}_e$ the lattice defined in Example 10.7.3 above:

$$\mathbb{L}_e = \mathbb{P}_0(\mathcal{G}_{24}),$$

where $\mathcal{G}_{24}$ denotes the extended binary Golay code with parameters $[n, k, d]_q = [24, 12, 8]_2$. Consider next the translate

$$\mathbb{L}_o = \mathbb{L}_e + ((1/2)^{23}, -3/2).$$

We notice that any pair of lattice points $x \in \mathbb{L}_e$ and $y \in \mathbb{L}_o$ differ by squared Euclidean distance at least

$$23 \cdot (1/2)^2 + (3/2)^2 = 8.$$

This follows because each codeword in the Golay code has even weight, so there must always be at least one coordinate where x and y differ by $3/2$. But 8 is the minimum squared distance also in each one of $\mathbb{L}_e$ and $\mathbb{L}_o$. It follows that the union

$$\mathbb{L} = \mathbb{L}_e \cup \mathbb{L}_o$$

has the same minimum distance as each one of the components $\mathbb{L}_e$ and $\mathbb{L}_o$ individually, and hence twice the density, i.e. $\delta = 1$. A special argument is of course required to show that $\mathbb{L}$ is actually a lattice, but this fact is of less importance for our needs.

It turns out that a number of extremely good spherical codes can be obtained from various shells in the Leech lattice—as also from several of the other lattices we have defined. We will discuss these matters in some detail in the next few sections. For the moment we just notice that the set of points in the first shell in the Leech lattice—the points which are closest to the origin—form an optimal spherical code. The Golay code has 759 codewords of minimum weight 8 (the so called *octads*, forming the famous Steiner system $S(24, 8, 5)$ found by Witt in 1938 [227]). For each one of these there are $2^7 = 128$ possible combinations of signs (observing the constraint $\sum x_i \equiv 0 \bmod 4$). All of these are on squared Euclidean distance 8 from the origin, i.e. smallest possible. In addition there are $\binom{24}{2} \cdot 2^2 = 1104$ points in $\mathbb{L}_e$ of the form $(0^{22}, (\pm 2)^2)$. All of these are also on squared Euclidean distance 8 from the origin. Finally, from $\mathbb{L}_o$ we have $24 \cdot 2^{12} = 98304$ points of the form $((\pm 1/2)^{23}, \pm 3/2)$ of minimum squared norm, corresponding to the 2^{12} codewords in the Golay code and all the 24 possible locations of the $\pm 3/2$-coordinate. Altogether this gives us

$$759 \cdot 128 + 1104 + 98304 = 196560$$

points, all on the sphere of squared radius 8 and all with the squared distance at least 8 from one to another (see Chapter 5, Example 5.9.4, or Ericson-Zinoviev [100]). By normalization this gives us a spherical code found by Leech [152] with parameters

$$(n, \rho, M) = (24, 1, 196560).$$

Inserting the polynomial

$$f(t) = (t+1)(t+\frac{1}{2})^2(t+\frac{1}{4})^2\, t^2(t-\frac{1}{4})^2(t-\frac{1}{2})^2$$

in the Levenshtein bound it can be shown (see Levenshtein [156], [157] or Odlyzko-Sloane [176], reprinted in Conway-Sloane [71], pp. 337-339) that this number coincides with the upper bound. It follows that the code is best possible. It has been shown ([10]) that this code is *unique*, up to equivalence.

10.9 Theta functions

Let $\mathbb{P}$ be a given sphere packing in $\mathbb{R}^n$ and denote by $N_{\mathbb{P}}$ the set of all possible norms $u = ||x||^2$; $x \in \mathbb{P}$. For any $u \in N_{\mathbb{P}}$ denote by $\mathcal{X}_u$ the set of elements $x \in \mathbb{P}$ with norm $||x||^2 = u$. Let us use A_u as a notation for the size of $\mathcal{X}_u$: $A_u = |\mathcal{X}_u|$. It is clear that X_u is a constant energy code in $\mathbb{R}^n$ and that the corresponding normalized spherical code has parameters (n, ρ, M), where $\rho \geq d^2/u$, $M = A_u$. Thus any sphere packing in $\mathbb{R}^n$ defines a sequence of spherical codes, one for each element in the set $N_{\mathbb{P}}$. It turns out that some of the densest lattices also produce in this way a number of very good spherical codes.

A useful tool for the investigation of the spherical codes associated with sphere packings is the *theta function*. Given a sphere packing $\mathbb{P} \in \mathbb{R}^n$ there is a corresponding theta function $\theta_{\mathbb{P}}(w)$ defined as follows:

$$\theta_{\mathbb{P}}(w) \triangleq \sum_{u \in N_{\mathbb{P}}} A_u \omega^u = \sum_{x \in \mathbb{P}} \omega^{||x||^2}.$$

As seen the theta function is nothing else than a generating function for the sequence $\{A_u : u \in N_{\mathbb{P}}\}$, a book-keeping device by which we represent the numbers A_u. The nature of the variable ω is nonessential for our purposes.

We shall derive the theta functions for several of the lattices and sphere packings which generate spherical codes with interesting properties. For that purpose a few fundamental facts are useful.

Let $\mathbb{P}$ be a sphere packing in $\mathbb{R}^n$. We denote by $m\mathbb{P}$ the packing obtained by multiplying each element x in $\mathbb{P}$ by the constant m:

$$m\mathbb{P} = \{mx \in \mathbb{R}^n : x \in \mathbb{P}\}.$$

The following relation is obvious:

$$\theta_{m\mathbb{P}}(\omega) = \theta_{\mathbb{P}}(\omega^{m^2}). \tag{10.9.1}$$

If $\mathbb{P}_1$ and $\mathbb{P}_2$ are two sphere packings in $\mathbb{R}^n$ such that for any $x \in \mathbb{P}_1$ and any $y \in \mathbb{P}_2$ we have $\langle x, y \rangle = 0$, then we say that $\mathbb{P}_1$ and $\mathbb{P}_2$ are orthogonal. We denote this fact by writing $\mathbb{P}_1 \perp \mathbb{P}_2$. By $\mathbb{P}_1 + \mathbb{P}_2$ we mean all elements $x + y$, with any $x \in \mathbb{P}_1$ and any $y \in \mathbb{P}_2$. The following result is easy to establish:

$$\theta_{\mathbb{P}_1+\mathbb{P}_2}(\omega) = \theta_{\mathbb{P}_1}(\omega)\theta_{\mathbb{P}_2}(\omega), \quad \mathbb{P}_1 \perp \mathbb{P}_2. \tag{10.9.2}$$

Consider $\mathbb{P} = \mathbb{Z} + a$, where $a \in [0, 1)$. For $a = 0$ we have $N_{\mathbb{P}} = \{0, 1, 4, 9, \ldots\}$ (the set of squares of integers), and the function A_u is given by

$$A_u = \begin{cases} 1, & if \quad u = 0, \\ 2, & if \quad u = 1, 4, 9, 16, \ldots \end{cases}.$$

Thus we have

$$\theta_{\mathbb{Z}}(\omega) = 1 + 2\omega + 2\omega^4 + 2\omega^9 + 2\omega^{16} + \cdots. \tag{10.9.3}$$

For $a = 1/2$ we have $\mathbb{P} = \mathbb{Z} + 1/2$ (the same lattice, but described in a coordinate system with the origin halfway between two lattice points). We have $N_{\mathbb{P}} = \{1/4, 9/4, 25/4, \ldots\}$. In this case the function A_u is identically equal to 2 and we have

$$\theta_{\mathbb{Z}+1/2}(\omega) = 2\omega^{1/4} + 2\omega^{9/4} + 2\omega^{25/4} + \cdots. \tag{10.9.4}$$

In general we have

$$\theta_{\mathbb{Z}+a}(\omega) = \omega^{a^2} + \omega^{(a-1)^2} + \omega^{(a+1)^2} + \omega^{(a-2)^2} + \cdots. \tag{10.9.5}$$

Except for the two cases $a = 0, 1/2$ all exponents are different.

The formulas (10.9.1)–(10.9.5) produce the theta functions for all possible lattices in the one dimensional Euclidean space $\mathbb{R}$. For instance, by (10.9.1) and (10.9.4) we obtain

$$\theta_{2\mathbb{Z}+1}(\omega) = 2\omega + 2\omega^9 + 2\omega^{25} + \cdots .$$

The formula (10.9.2) is a key for deriving theta functions in higher dimensions. For the square lattice (10.9.2) and (10.9.3) immediately give us

$$\theta_{\mathbb{Z}^2}(\omega) = (\theta_{\mathbb{Z}}(\omega))^2 = 1 + 4\omega + 4\omega^2 + 4\omega^4 + \cdots .$$

The *hexagonal lattice* is slightly more complicated. Let the generator matrix be

$$V = \begin{bmatrix} 1 & 0 \\ -1/2 & \sqrt{3}/2 \end{bmatrix}$$

and let the lattice be $\mathbb{L}_{hex} = \{uV : u \in \mathbb{Z}^2\}$. We obtain

$$\|x\|^2 = uVV^T u^T = u_1^2 - u_1u_2 + u_2^2 .$$

The presence of the middle term prevents us from immediately applying the formula (10.9.2). The following trick resolves the difficulties. Let T be the matrix

$$T = \begin{bmatrix} 1 & 0 \\ 1 & 2 \end{bmatrix}.$$

and define

$$\mathbb{P}_0 = \{vT : v \in \mathbb{Z}^2\},$$

$$\mathbb{P}_1 = \{vT + (0,1) : v \in \mathbb{Z}^2\}.$$

Notice that we have $\mathbb{Z}^2 = \mathbb{P}_0 \cup \mathbb{P}_1$. This relation implies

$$\mathbb{L}_{hex} = \mathbb{P}_0 V \cup \mathbb{P}_1 V.$$

This also means that in terms of the variables v_1 and v_2 of the vector $v = (v_1, v_2)$ we can express the lattice as a union of two direct sums. We have

$$u = vT : \quad \|x\|^2 = v_1^2 + 3v_2^2,$$

$$u = vT + (0,1) : \quad \|x\|^2 = (v_1 - 1/2)^2 + 3(v_2 + 1/2)^2.$$

From these observations we deduce

$$\begin{aligned}\theta_{hex}(\omega) &= \textstyle\sum_{x\in\mathbb{L}_{hex}} \omega^{\|x\|^2} \\ &= \textstyle\sum_{v\in\mathbb{Z}^2}\left(\omega^{v_1^2+3v_2^2} + \omega^{(v_1-1/2)^2+3(v_2+1/2)^2}\right) \\ &= \theta_{\mathbb{Z}}(\omega)\theta_{\mathbb{Z}}(\omega^3) + \theta_{\mathbb{Z}-1/2}(\omega)\theta_{\mathbb{Z}+1/2}(\omega^3).\end{aligned}$$

Inserting (10.9.3) and (10.9.4) and observing that $\mathbb{Z}-1/2$ is the same as $\mathbb{Z}+1/2$ we obtain

$$\begin{aligned}\theta_{hex}(\omega) &= (1+2\omega+2\omega^4+\cdots)(1+2\omega^3+2\omega^{12}+\cdots)+ \\ &+(2\omega^{1/4}+2\omega^{9/4}+\cdots)(2\omega^{3/4}+2\omega^{27/4}+\cdots) = \\ &= 1+6\omega+6\omega^3+6\omega^4+12\omega^7+\cdots\ .\end{aligned}$$

For future purposes we also need the theta series for those representations having the origin at the middle of an edge or at the middle of a face (see Figures 10.6 and 10.7). Exactly the same ideas may be employed. Let $\mathbb{P}_a$ be the packing obtained by translating the lattice $\mathbb{L}_{hex}$ by the vector $a : \mathbb{P}_a = \mathbb{L}_{hex}+a$. Let $y = x+a$ where $a = (a_1, a_2)$. We obtain

$$\begin{aligned}u = vT: &\quad \|y\|^2 = (v_1+a_1)^2 + 3\left(v_2+a_2/\sqrt{3}\right)^2, \\ u = vT+(1,0): &\quad \|y\|^2 = [v_1+(1/2)(2a_1-1)]^2+ \\ &\quad + 3\left[v_2+(1/2)\left(1+2a_2/\sqrt{3}\right)\right]^2.\end{aligned}$$

Notice that for the special case $a = 0$ we have the same expressions as before. For the two other cases of main interest we deduce

$a = (1/2, 0)$, (*middle of an edge*):

$$\begin{aligned}\theta_{edge}(\omega) &= \textstyle\sum_{v\in\mathbb{Z}^2}[\omega^{(v_1+1/2)^2+3v_2^2} + \omega^{v_1^2+3(v_2+1/2)^2}] \\ &= \theta_{\mathbb{Z}+1/2}(\omega)\theta_{\mathbb{Z}}(\omega^3) + \theta_{\mathbb{Z}}(\omega)\theta_{\mathbb{Z}+1/2}(\omega^3) \\ &= 2\omega^{1/4}+2\omega^{3/4}+4\omega^{7/4}+4\omega^{9/4}+\cdots\ .\end{aligned}$$

$a = (1/2, 1/2\sqrt{3})$, *(middle of a face)*:

$$\theta_{face}(\omega) = \sum_{v \in \mathbb{Z}^2} (\omega^{(v_1+1/2)^2+3(v_2+1/6)^2} + \omega^{v_1^2+3(v_2+2/3)^2})$$

$$= \theta_{\mathbb{Z}+1/2}(\omega)\theta_{\mathbb{Z}+1/6}(\omega^3) + \theta_{\mathbb{Z}}(\omega)\theta_{\mathbb{Z}+2/3}(\omega^3)$$

$$= 3\omega^{1/3} + 3\omega^{4/3} + 6\omega^{7/3} + 6\omega^{10/3} + \cdots .$$

10.10 Spherical codes from lattices

We are now finally ready for the main topic of this chapter—the main reason for our considering sphere packings at all—namely the spherical codes generated from various lattices and packings. By a *shell* in a packing or lattice we mean all the points on a given fixed distance from the origin. It is clear that any shell defines a spherical code. The main tool for analyzing such spherical codes is the theta functions. The basic result is a simple relation between theta functions and weight distribution functions for codes over finite alphabets.

Consider the constructions $\mathbb{P}(\mathbb{L}, \mathcal{C})$ and $\mathbb{P}_*(\mathbb{L}, \mathcal{C})$. They both define lattices from codes. For simplicity we assume that the code $\mathcal{C}$ is defined over some finite field $\mathbb{F}_q$. For each element $a \in \mathbb{F}_q$ and for any code vector $c \in \mathcal{C}$ let $w_a(c)$ denote the number of occurrences of the letter a in the vector c. Denote by $w(c)$ the collection of all these functions: $w(c) = \{w_a(c) : a \in \mathbb{F}_q\}$. We call $w(c)$ the composition of c. Let $\{z_a\}$ be a set of variables, one for each element a in the alphabet $\mathbb{F}_q$. For any set $\{u_a\}$ of integers u_a define

$$z^u \triangleq \prod_{a \in \mathbb{F}_q} z_a^{u_a}.$$

As a special case, if $w(c)$ is the composition of the vector $c \in \mathcal{C}$ we have

$$z^{w(c)} \triangleq \prod_{a \in \mathbb{F}_q} z_a^{w_a(c)}.$$

In this way we have, for each codeword $c \in \mathcal{C}$, a function $z^{w(c)}$ describing the composition of c. The complete weight enumeration function for the code $\mathcal{C}$ is simply the sum of all these functions:

$$W(z) \triangleq \sum_{c \in \mathcal{C}} z^{w(c)}.$$

It is clear that for binary codes the complete weight distribution is just the Hamming weight distribution.

In all cases we shall consider the size q of the alphabet $\mathbb{F}_q$ will be small. In those cases it is more convenient to avoid the indices and instead use different letters for the variables: $x, y, z, \ldots$. For instance, the binary extended Hamming code with parameters $[n,k,d]_q = [8,4,4]_2$ has weight distribution

$$W(x,y) = x^8 + 14x^4y^4 + y^8,$$

indicating that there is one codeword with weight 0 (indicated by x^8), 14 codewords of weight 4, and one codeword of weight 8. As another example, the ternary Hamming code with parameters $[n,k,d]_q = [4,2,3]_3$ has complete weight distribution

$$W(x,y,z) = x(x^3 + y^3 + z^3) + 3xyz(y+z),$$

indicating that there is one codeword consisting of 4 zeros, there are 2 codewords with 1 zero and three other letters, all of the same kind, and finally 6 codewords with all 3 letters represented, with the letter 0 occurring only once.

Now consider the construction $\mathbb{P}(\mathbb{L}, \mathcal{C}) = \mathbb{L}^N + f_N(\mathcal{C})$, where $\mathbb{L}$ is a lattice in $\mathbb{R}^m$ and where f_N denotes a coordinate-wise mapping from $\mathbb{F}_q^N$ into $\mathbb{R}^n$, $n = Nm$. Let the complete weight distribution of $\mathcal{C} \in \mathbb{F}_q^N$ be $W(z)$ and for each $a \in \mathbb{F}_q^N$ let $\gamma_a(\omega)$ be the theta function of the lattice $\mathbb{L} + f(a)$.

Theorem 10.10.1 *Let the sphere packing $\mathbb{P}(\mathbb{L}, \mathcal{C})$ be generated by the code $\mathcal{C} \in \mathbb{F}_q^N$ and let the function f_N be as described above; let $\gamma_a(\omega)$ be the theta function for the lattice $\mathbb{L} + f(a)$ and denote*

$$\gamma(\omega)^{w(c)} = \prod_{a \in \mathbb{F}_q} \gamma_a(\omega)^{w_a(c)}.$$

The theta function for the packing $\mathbb{P}(\mathbb{L}, C)$ *is given by*

$$\theta(\omega) = W(\gamma(\omega)),$$

where $W(z)$ *is the complete weight distribution for the code* C*.*

Proof: By definition we have

$$\theta(\omega) = \sum_{c \in C} \sum_{x \in \mathbb{L}^N} \omega^{||x + f_N(c)||^2}.$$

Expanding the inner sum we obtain

$$\sum_{x \in \mathbb{L}^N} \omega^{||x+f_N(c)||^2} = \sum_{x \in \mathbb{L}^N} \prod_{i=1}^{N} \omega^{||x_i + f(c_i)||^2}$$

$$= \prod_{i=1}^{N} \sum_{x_i \in \mathbb{L}} \omega^{||x_i + f(c_i)||^2} = \prod_{i=1}^{N} \gamma_{f(c_i)}(\omega)$$

$$= \prod_{a \in \mathbb{F}_q} \gamma_a(\omega)^{w_a(c)} = \gamma(\omega)^{w(c)}.$$

Summing over $c \in C$ we obtain the result. □

We illustrate the theorem by applying it to the root lattices $D_3 - D_5$ and $E_6 - E_8$.

Example 10.10.1: Recall that the lattices D_n are generated by simple parity check binary codes:

$$D_n = \{x \in \mathbb{Z}^n : x \bmod 2 \in C\},$$

where

$$C = \{c \in \mathbb{F}_2^n : c_1 + c_2 + \cdots + c_n = 0\}.$$

The weight distribution of the simple parity check code, i.e. the binary code $[n, n-1, 2]_2$, is

$$W(x, y) = x^n + \sum_{i=1}^{\lfloor n/2 \rfloor} \binom{n}{2i} x^{n-2i} y^{2i}.$$

With

$$\gamma_0(\omega) = \theta_{2\mathbb{Z}}(\omega) = 1 + 2\omega^4 + 2\omega^{16} + 2\omega^{36} + \cdots ,$$

$$\gamma_1(\omega) = \theta_{2\mathbb{Z}+1}(\omega) = 2\omega + 2\omega^9 + 2\omega^{25} + 2\omega^{49} + \cdots ,$$

u	$A_u(n)$	$A_u(3)$	$A_u(4)$	$A_u(5)$
0	1	1	1	1
2	$4\binom{n}{2}$	12	24	40
4	$2n+16\binom{n}{4}$	6	24	90
6	$A_6(n)$	24	96	240
8	$A_8(n)$	12	24	200

Table 10.2: The cardinalities of the first shells in the lattices D_n. The coefficients $A_6(n)$ and $A_8(n)$ are given in the expression for $\theta(\omega)$ above.

we obtain

$$\theta(\omega) = W(\gamma_0(\omega), \gamma_1(\omega)) = (1 + 2\omega^4 + 2\omega^{16} + \cdots)^n +$$

$$+\binom{n}{2}(1 + 2\omega^4 + 2\omega^{16} + \cdots)^{n-2}(2\omega + 2\omega^9 + \cdots)^2 + \cdots$$

$$= 1 + 4\binom{n}{2}\omega^2 + [2n + 16\binom{n}{4}]\,\omega^4 +$$

$$+[8(n-2)\binom{n}{2} + 64\binom{n}{6}]\omega^6 + 4[\binom{n}{2} +$$

$$+8(n-4)\binom{n}{4} + 64\binom{n}{8}]\,\omega^8 + \cdots .$$

The values of the coefficients $A_u(n)$ for the cases $n = 3, 4, 5$ are given in Table 10.2. □

Example 10.10.2: From Example 10.5.2 we recall that the lattice E_8 is obtained by the same construction using the code $\mathcal{C} : [8, 4, 4]_2$ (the extended binary Hamming code of length 8). The weight distribution is

$$W(x, y) = x^8 + 14x^4y^4 + y^8.$$

We obtain

$$\theta(\omega) = W(\gamma_0(\omega)\gamma_1(\omega)) = (1 + 2\omega^4 + 2\omega^{16} + \cdots)^8 +$$

$$+14(1 + 2\omega^4 + 2\omega^{16} + \cdots)^4 (2\omega + 2\omega^9 + 2\omega^{25} + \cdots)^4 +$$

$$+(2\omega + 2\omega^9 + 2\omega^{25} + \cdots)^8 =$$

$$= 1 + 240\omega^4 + 2160\omega^8 + 6720\omega^{12} + 17520\omega^{16} + \cdots .$$

□

Example 10.10.3: The lattice E_7 has a definition using the same construction with the Hadamard code $\mathcal{C} : [7, 3, 4]_2$ (see Example 10.5.3). The weight distribution is

$$W(x, y) = x^7 + 7x^3y^4.$$

We obtain

$$\theta(\omega) = W(\gamma_0(\omega), \gamma_1(\omega)) = (1 + 2\omega^4 + 2\omega^{16} + \cdots)^7 +$$

$$+ 7\,(1 + 2\omega^4 + 2\omega^{16} + \cdots)^3 (2\omega + 2\omega^9 + \cdots)^4 =$$

$$= 1 + 126\omega^4 + 756\omega^8 + 2072\omega^{12} + 4158\omega^{16} + \cdots .$$

□

We notice that in all of these examples the minimum distance in the lattice is $d = 2$. The exponents indicate the squared radius E and the coefficients indicate the size M of the spherical code. It follows that the successive shells define spherical codes with squared normalized distance ρ such that

$$\rho = \frac{d^2}{E} \geq 1,\ 1/2,\ 1/3,\ 1/4,\ \ldots .$$

Example 10.10.4: In Example 10.5.4 we defined the lattice E_6 from the hexagonal lattice $\mathbb{L}_{hex}$ with minimum distance $d = 1$ and the point-set Q consisting of the origin and two deep holes. As code we used the ternary repetition code $\mathcal{C}$ with parameters $[3, 1, 3]_3$ over the field $\mathbb{F}_3$, with 0 mapped on the origin and the elements 1 and 2 mapped on the two other points. The corresponding theta functions are

$$\gamma_0(\omega) = \theta_{hex}(\omega) = 1 + 6\omega + 6\omega^3 + 6\omega^4 + 12\omega^7 + \cdots ,$$

$$\gamma_1(\omega) = \gamma_2(\omega) = \theta_{face}(\omega) = 3\omega^{1/3} + 3\omega^{4/3} + 6\omega^{7/3} + \cdots .$$

The ternary repetition code has complete weight distribution

$$W(x, y, z) = x^3 + y^3 + z^3.$$

We obtain the following theta function for the lattice E_6:

$$\theta(\omega) = (1 + 6\omega + 6\omega^3 + \cdots)^3 + 2(3\omega^{1/3} + 3\omega^{4/3} + 6\omega^{7/3} + \cdots)^3 =$$

$$= 1 + 72\omega + 270\omega^2 + 720\omega^3 + 936\omega^4 + 2160\omega^5 + 2214\omega^6 + \cdots .$$

The spherical codes have normalized distances ρ satisfying

$$\rho \geq 1, 1/2, 1/3, \ldots .$$

□

In Table 10.3 we have summarized the parameters for the spherical codes generated by the first few shells in the densest lattices in dimensions $1 \leq n \leq 8$. We have indicated by ρ_0 the lower bound to the normalized squared minimum distances ρ as given by $\rho_0 = d^2/E$, where d is the minimum distance in the lattice and where E is the squared radius of the shell in question. Often this bound is tight, but for instance in the hexagonal lattice it gives a very crude bound for most of the shells.

The case $\rho = 1$ corresponds to the contact number problem. We see that of the five known cases for which the optimal contact number is known four occur in the above list: $\tau_1 = 2$, $\tau_2 = 6$, $\tau_3 = 12$, $\tau_8 = 240$. We have also seen that the fifth known optimal contact number $\tau_{24} = 196560$ appears in the Leech lattice. Thus all of them are found in lattices.

Also many of the other spherical codes generated by the shells in various lattices are quite good. Indeed, quite often they represent best known spherical codes.

ρ_0	Z	A_2	A_3	D_4	D_5	E_6	E_7	E_8
1	2	6	12	24	40	72	126	240
1/2			6	24	90	270	756	2160
1/3		6	24	96	240	720	2072	6720
1/4		6	12	24	200	936	4158	17520
1/5			24	144	560	2160	7560	30240
1/6			8	96	400	2214	11592	60480
1/7		12	48	192	800	3600	16704	82560
1/8			6	24	730	4590	24948	140400
1/9		6	36	312	1240	6552	31878	181680
1/10			24	144	712	5184	39816	272160

Table 10.3: Spherical codes generated from the first few shells in the densest lattices in dimensions $1 \leq n \leq 8$.

As we mentioned in Section 10.4 all the lattices occurring in Table 10.3 are densest possible. Of course this fact in itself does not imply that they also must generate best possible spherical codes. There are many cases in higher dimensions where a less dense lattice generates better spherical codes than a denser one.

Example 10.10.5: In Example 10.5.5 we noticed that the E_8-lattice can also be generated by the hexagonal lattice and the same set Q as in Example 10.10.4 above. The generating code is the ternary Hamming code with parameters $[4, 2, 3]_3$ and with complete weight distribution

$$W(x, y, z) = x(x^3 + y^3 + z^3) + 3xyz(y + z).$$

By this construction we obtain the following alternative form of the theta function for the lattice E_8:

$$\theta(\omega) = \theta_{hex}(\omega)[\theta_{hex}^3(\omega) + 2\theta_{face}^3(\omega)] + 6\theta_{hex}(\omega)\theta_{face}^3(\omega) =$$

$$= 1 + 240\omega + 2160\omega^2 + 6720\omega^3 + 17520\omega^4 + \cdots .$$

The difference in the exponents as compared with Example 10.10.2 occurs because in the present case we realize the lattice in a form with a minimum distance $d = 1$, while the realization in Example 10.10.2 has $d = 2$. □

Example 10.10.6: In Example 10.5.6 we generated the *Coxeter–Todd* lattice using the hexagonal lattice and a set Q consisting of the origin and three points located at the edges of some fundamental triangles. The code employed is the so called *Hexacode* with parameters $[n, k, d]_q = [6, 3, 4]_4$ over the quaternary field $\mathbb{F}_4$. The zero element is mapped on the origin while the non-zero elements are mapped on the points in Q located on the triangle edges. The Hamming weight distribution for the *Hexacode* is

$$W(x, y) = x^6 + 45x^2y^4 + 18y^6,$$

with the variable x corresponding to zero and the variable y corresponding to any non-zero element in $\mathbb{F}_4$. We obtain

$$\gamma_0(\omega) = \theta_{hex}(\omega) = 1 + 6\omega + 6\omega^3 + \cdots ,$$

$$\gamma_1(\omega) = \gamma_a(\omega) = \theta_{edge}(\omega) = 2\omega^{1/4} + 2\omega^{3/4} + 4\omega^{7/4} + \cdots ,$$

which combined with the weight enumerator gives

$$\theta(\omega) = 1 + 756\omega + 4032\omega^{3/2} + 20412\omega^2 + 60480\omega^{5/2} + 139860\omega^3 + \cdots .$$

□

10.11 Theta functions for expurgated constructions

The expurgated constructions A, B, B^* lead to theta functions slightly different from those we have seen so far. Recall the definitions:

$$\mathbb{P}_A(\mathcal{C}) \triangleq \{x \in \mathbb{Z}^n : (i)\ x \bmod 2 \in \mathcal{C}\},$$

$$\mathbb{P}_B(\mathcal{C}) \triangleq \{x \in \mathbb{Z}^n : (i)\ x \bmod 2 \in \mathcal{C},\quad (ii)\ x_1 + \cdots + x_n \equiv 0 \bmod 4\},$$

$$\mathbb{P}_{B^*}(\mathcal{C}) \triangleq \{x \in \mathbb{Z}^n : (i)\ x \bmod 2 \in \mathcal{C},\quad (ii)\ \textstyle\sum_{i=1}^n x_i - (x_i)_2 \equiv 0 \bmod 4)\}.$$

Let C be a binary code and let the lattice $\mathbb{L}$ be the set of even numbers: $\mathbb{L} = 2\mathbb{Z}$. Under these conditions the constructions $\mathbb{P}_A(C)$ and $\mathbb{P}(\mathbb{L}, C)$ produce the same packings. Therefore, as a special case of Theorem 10.10.1 we have the following result.

Theorem 10.11.1 *Let C be a binary code with weight distribution $W(x, y)$. Define*

$$\gamma_0(\omega) \triangleq \theta_{2\mathbb{Z}}(\omega), \quad \gamma_1(\omega) \triangleq \theta_{2\mathbb{Z}+1}(\omega). \tag{10.11.1}$$

The theta function for the packing $\mathbb{P}_A(C)$ is given by

$$\theta(\omega) = W(\gamma_0(\omega), \gamma_1(\omega)). \tag{10.11.2}$$

The constructions B and B^* are slightly more complicated. As B is a special case of B^* it is enough to consider the latter construction. Recall that a point x is in $\mathbb{P}_{B^*}(C)$ if and only if it has the form $x = c+2k$, with $c = (c_1, c_2, \ldots, c_n) \in C$ and $k = (k_1, k_2, \ldots, k_n) \in D_n$. By definition the theta function is

$$\theta(\omega) = \sum_{c\in C} \sum_{k\in D_n} \omega^{||c+2k||^2} =$$

$$= \sum_{c\in C} \sum_{k\in \mathbb{Z}^n} \left(\frac{1}{2}\,\omega^{||c+2k||^2} + \frac{1}{2}\,(-1)^{\sum_i k_i}\,\omega^{||c+2k||^2}\right).$$

In the last parenthesis the two terms either cancel or coincide depending on whether k is in D_n or not. Define

$$\gamma_0(\omega) = \theta_{2\mathbb{Z}}(\omega); \quad \gamma_1(\omega) = \theta_{2\mathbb{Z}+1}(\omega).$$

For the first term we obtain

$$\sum_{k\in\mathbb{Z}^n} \omega^{||c+2k||^2} = \prod_{i=1}^{n} \sum_{k_i\in\mathbb{Z}} \omega^{(c_i+2k_i)^2}$$

$$= \gamma_0(\omega)^{n-w_H(c)}\,\gamma_1(\omega)^{w_H(c)}.$$

For the second term we obtain

$$\sum_{k\in\mathbb{Z}^n} (-1)^{\sum_i k_i}\,\omega^{||c+2k||^2} = \prod_{i=1}^{n} \sum_{k_i\in\mathbb{Z}} (-1)^{k_i} \omega^{(c_i+2k_i)^2}.$$

Define

$$\gamma_*(\omega) \triangleq \sum_{k\in\mathbb{Z}} (-\omega)^{k^2} = 1 - 2\omega + 2\omega^4 - 2\omega^9 + \cdots.$$

There are two possibilities:

(*i*) $c_i = 0$:

$$\sum_{k_i\in\mathbb{Z}} (-1)^{k_i}\omega^{(c_i+2k_i)^2} = \sum_{k_i\in\mathbb{Z}} (-1)^{k_i^2}\omega^{4k_i^2} = \sum_{k_i\in\mathbb{Z}} (-\omega^4)^{k_i^2} = \gamma_*(\omega^4).$$

(*ii*) $c_i = 1$:

$$\sum_{k_i\in\mathbb{Z}} (-1)^{k_i}\omega^{(c_i+2k_i)^2} = \sum_{k_i\in\mathbb{Z}} (-1)^{k_i}\omega^{(1+2k_i)^2} =$$

$$= \sum_{i=0}^{\infty} (-1)^i\omega^{(1+2i)^2} + \sum_{j=0}^{\infty} (-1)^{-j-1}\omega^{(-1-2j)^2} = 0.$$

In summary we have

$$\sum_{k\in\mathbb{Z}^n} (-1)^{\sum_i k_i}\omega^{\|c+2k\|^2} = \begin{cases} \gamma_*(\omega^4)^n, & c = (c_1, c_2, \ldots, c_n) = 0, \\ 0, & c = (c_1, c_2, \ldots, c_n) \neq 0. \end{cases}$$

Combining the above observations we arrive at the following result.

Theorem 10.11.2 *Let $\mathcal{C}$ be a binary code with weight distribution $W(x, y)$. Define*

$$\gamma_0(\omega) \triangleq \theta_{2\mathbb{Z}}(\omega), \quad \gamma_1(\omega) \triangleq \theta_{2\mathbb{Z}+1}(\omega), \quad \gamma_*(\omega) \triangleq \sum_{k\in\mathbb{Z}} (-\omega)^{k^2}.$$

The theta function for the packing $\mathbb{P}_{B^*}(\mathcal{C})$ is given by

$$\theta(\omega) = \frac{1}{2} W(\gamma_0(\omega), \gamma_1(\omega)) + \frac{1}{2}\gamma_*(\omega^4)^n. \tag{10.11.3}$$

Example 10.11.1: In Example 10.7.1 we constructed the E_8-lattice by the construction $\mathbb{P}_B(\mathcal{C})$, taking $\mathcal{C}$ as the trivial repetition code $\mathcal{C} : [8,1,8]_2$. The weight distribution of $\mathcal{C}$ is $W(x,y) = x^8 + y^8$. We obtain the following alternative form of the theta function for the E_8-lattice:

$$\theta(\omega) = \frac{1}{2}\,\theta_{2\mathbb{Z}}(\omega)^8 + \frac{1}{2}\,\theta_{2\mathbb{Z}+1}(\omega)^8 + \frac{1}{2}\,\gamma_*(\omega^4)^8.$$

□

Example 10.11.2: The *Barnes–Wall* lattice was constructed in Example 10.7.2 as $\mathbb{P}_B(\mathcal{C})$ with $\mathcal{C}$ chosen as a first order Reed–Muller code with parameters $[16,5,8]_2$. The weight enumeration function is

$$W(x,y) = x^{16} + 30x^8y^8 + y^{16}.$$

We obtain

$$\theta(\omega) = \frac{1}{2}\,(1 + 2\omega^4 + \cdots)^{16} + 15(1 + 2\omega^4 + \cdots)^8(2\omega + 2\omega^9 + \cdots)^8$$

$$+\frac{1}{2}\,(2\omega + 2\omega^9 + \cdots)^{16} + \frac{1}{2}\,(1 - 2\omega^4 + \cdots)^{16}$$

$$= 1 + 4320\,\omega^8 + 61440\,\omega^{12} + 522720\,\omega^{16} + 2211840\,\omega^{20} + \cdots\,.$$

□

Example 10.11.3: In Example 10.7.3 we constructed a half of the *Leech lattice* using the extended Golay code with parameters $[24,12,8]_2$. It has weight enumeration function

$$W(x,y) = x^{24} + 759\,x^{16}y^8 + 2576\,x^{12}y^{12} + 759\,x^8y^{16} + y^{24}.$$

By Theorem 10.11.2 we obtain after some calculations

$$\theta(\omega) = 1 + 98256\,\omega^8 + 8384512\,\omega^{12} + 198289488\,\omega^{16} + \cdots\,.$$

□

10.12 Unions of shells

Any shell in any packing defines a spherical code. As we have seen many good codes are generated in this way. In particular several of the best known constructions for *contact number* occur as *inner shells* in well known lattices.

In general, however, we usually get better codes by taking *unions* of several successive shells. The key result for the analysis is the following theorem.

Theorem 10.12.1 *Let $\mathbb{P}$ be a packing with minimum distance d. Let $\mathcal{X}$ be the spherical code obtained by projecting onto the unit sphere all points x in $\mathbb{P}$ such that $E_1 \leq \|x\|^2 \leq E_2$. The squared minimum distance ρ of the code $\mathcal{X}$ satisfies the following bound:*

$$\rho \geq \min\{\frac{d^2}{E_2}, \frac{d^2 - (\sqrt{E_2} - \sqrt{E_1})^2}{\sqrt{E_1 E_2}}\}. \tag{10.12.1}$$

Proof: Let $x'$and y' be two distinct points in $\mathbb{P}$ such that $\|x'\|^2 = a$ and $\|y'\|^2 = b$, where $E_1 \leq a, b \leq E_2$. The projections x and y on the unit sphere are given by

$$x = x'/\sqrt{a}, \quad y = y'/\sqrt{b}.$$

By assumption we have

$$d^2 \leq \|x' - y'\|^2 = a + b - 2\langle x', y'\rangle.$$

From this inequality we derive the following bound for the inner product between x and y:

$$\langle x, y\rangle = \frac{\langle x', y'\rangle}{\sqrt{ab}} \leq \frac{a + b - d^2}{2\sqrt{ab}}.$$

This inequality in turn immediately gives us the following estimate for the squared minimum distance of the code $\mathcal{X}$:

$$\rho \geq \min\{\frac{d^2 - (\sqrt{b} - \sqrt{a})^2}{\sqrt{ab}}\},$$

where the minimum is taken over all possible squared norms a, b such that $\{E_1 \leq a, b \leq E_2\}$. A straightforward analysis of this expression reveals that the minimum is given by the expression as stated in the theorem. □

For *lattices* the bound (10.12.1) is *tight*, provided only that both of the parameters E_1 and E_2 correspond to the squared radius of an existing shell. A simple calculation reveals that the minimum distance is determined by the outer shell provided the following inequality holds:

$$E_1 \geq \frac{(E_2 - d^2)^2}{E_2}.$$

This means that if to a code generated by a shell with squared radius E_2 we add all shells from the same lattice having squared radius a such that

$$\frac{(E_2 - d^2)^2}{E_2} \leq a < E_2\,, \tag{10.12.2}$$

then the minimum distance will *not* decrease. For outer shells we can often find shells satisfying this condition.

In Tables 10.4-10.8 we have indicated a few of the spherical codes that can be generated as unions of shells from some well known lattices. Except for the inner shells all these codes are generated by two or more shells.

n	ρ_{exact}	$\rho_{numer.}$	M	E_1	E_2
4	$2-7/\sqrt{15}$	0.19261	264	3	5
4	$1/5$	0.20000	168	4	5
4	$2-\sqrt{3}$	0.26795	144	1	3
4	$1/3$	0.33333	120	2	3
4	$1/2$	0.50000	48	1	2
4	1	1.00000	24	1	1
5	$2-3\sqrt{3}/(2\sqrt{2})$	0.16288	1160	4	6
5	$2-7/\sqrt{15}$	0.19261	1000	3	5
5	$1/5$	0.20000	760	4	5
5	$2-5/(2\sqrt{2})$	0.23223	530	2	4
5	$1/4$	0.25000	440	3	4
5	$2-\sqrt{3}$	0.26795	370	1	3
5	$1/3$	0.33333	330	2	3
5	$1/2$	0.50000	130	1	2
5	1	1.00000	40	1	1

Table 10.4: Unions of shells in the lattices D_4 and D_5.

n	ρ_{exact}	$\rho_{numer.}$	M	E_1	E_2
6	$2-3\sqrt{3}/(2\sqrt{2})$	0.16288	5310	4	6
6	$1/6$	0.16667	4374	5	6
6	$2-7/\sqrt{15}$	0.19261	3816	3	5
6	$1/5$	0.20000	3096	4	5
6	$2-5/(2\sqrt{2})$	0.23223	1926	2	4
6	$1/4$	0.25000	1656	3	4
6	$2-\sqrt{3}$	0.26795	1062	1	3
6	$1/3$	0.33333	990	2	3
6	$1/2$	0.50000	342	1	2
6	1	1.00000	72	1	1
7	$2-3\sqrt{3}/(2\sqrt{2})$	0.16288	23310	4	6
7	$1/6$	0.16667	19152	5	6
7	$2-7/\sqrt{15}$	0.19261	13790	3	5
7	$1/5$	0.20000	11718	4	5
7	$2-5/(2\sqrt{2})$	0.23223	6986	2	4
7	$1/4$	0.25000	6230	3	4
7	$2-\sqrt{3}$	0.26795	2954	1	3
7	$1/3$	0.33333	2828	2	3
7	$1/2$	0.50000	882	1	2
7	1	1.00000	126	1	1

Table 10.5: Unions of shells in the lattices E_6 and E_7.

n	ρ_{exact}	$\rho_{numer.}$	M	E_1	E_2
8	$2-3\sqrt{3}/(2\sqrt{2})$	0.16288	108240	4	6
8	$1/6$	0.16667	90720	5	6
8	$2-7/\sqrt{15}$	0.19261	54480	3	5
8	$1/5$	0.20000	47760	4	5
8	$2-5/(2\sqrt{2})$	0.23223	26400	2	4
8	$1/4$	0.25000	24240	3	4
8	$1/3$	0.33333	8880	2	3
8	$1/2$	0.50000	2400	1	2
8	1	1.00000	240	1	1
12	$1/5$	0.20000	3999492	14	20
12	$2-4/\sqrt{5}$	0.21115	2199456	10	18
12	$2/9$	0.22222	2138976	12	18
12	$2-5/(2\sqrt{2})$	0.23223	1199772	8	16
12	$1/4$	0.25000	1179360	10	16
12	$2-8/\sqrt{21}$	0.25426	551378	6	14
12	$1/3$	0.33333	224784	6	12
12	$2/5$	0.40000	85680	4	10
12	$1/2$	0.50000	25200	4	8
12	$2/3$	0.66667	4788	4	6
12	1	1.00000	756	4	4

Table 10.6: Unions of shells in the lattice E_8 and the Coxeter–Todd lattice K_{12}.

n	ρ_{exact}	$\rho_{numer.}$	M	E_1	E_2
16	$2-20/(3\sqrt{13})$	0.15100	3924500160	18	26
16	$2/13$	0.15384	3789332160	20	26
16	$2-3\sqrt{3}/(2\sqrt{2})$	0.16288	2219970720	16	24
16	$1/6$	0.16667	2152816320	18	24
16	$2-16/\sqrt{77}$	0.17663	1095551520	14	22
16	$2/11$	0.18181	1072327200	16	22
16	$2-7/\sqrt{15}$	0.19261	554316960	12	20
16	$1/5$	0.20000	545356320	14	20
16	$2-4/\sqrt{5}$	0.21115	236719200	10	18
16	$2/9$	0.22222	234507360	12	18
16	$2-5/(2\sqrt{2})$	0.23223	102073920	8	16
16	$1/4$	0.25000	101551200	10	16
16	$2-8/\sqrt{21}$	0.25426	34980960	6	14
16	$2/7$	0.28571	34919520	8	14
16	$1/3$	0.33333	11756640	6	12
16	$2/5$	0.40000	2800320	4	10
16	$1/2$	0.50000	588480	4	8
16	$2/3$	0.66667	65760	4	6
16	1	1.00000	4320	4	4

Table 10.7: Unions of shells in the Barnes–Wall lattice Λ_{16}.

n	ρ_{exact}	$\rho_{numer.}$	M	E_1	E_2
24	$2-20/(3\sqrt{13})$	0.15100	279932381033760	18	26
24	$2/13$	0.15384	276956829545760	20	26
24	$2-3\sqrt{3}/(2\sqrt{2})$	0.16288	110816165481840	16	24
24	$1/6$	0.16667	110001285707040	18	24
24	$2-16/\sqrt{77}$	0.17663	40517418417840	14	22
24	$2/11$	0.18181	40329928482480	16	22
24	$22-7/\sqrt{15}$	0.19261	13498890153840	12	20
24	$1/5$	0.20000	13464472497840	14	20
24	$2-4\sqrt{5}$	0.21115	4016968235280	10	18
24	$2/9$	0.22222	4012338854160	12	18
24	$2-5/(2\sqrt{2})$	0.23223	1041814781280	8	16
24	$1/4$	0.25000	1041416747280	10	16
24	$2-8/\sqrt{21}$	0.25426	226951779600	6	14
24	$2/7$	0.28571	226935006480	8	14
24	$1/3$	0.33333	39461844240	6	12
24	$2/5$	0.40000	5044384800	4	10
24	$1/2$	0.50000	415003680	4	8
24	$2/3$	0.66667	16969680	4	6
24	1	1.00000	196560	4	4

Table 10.8: Unions of shells in the Leech lattice Λ_{24}.

Chapter 11

Decoding

11.1 Introduction

In applications decoding is crucial. Often this is the most difficult part in the implementation of a code. Depending on the channel the decoding problem takes many different forms. A complete discussion of all the various cases is far beyond the scope of this book. We will confine ourselves to the special case when the received signal belongs to the same Euclidean space as does the code. The problem will be formulated entirely in terms of Euclidean metric. This is a robust approach which can be justified in many cases. It is particularly adequate if the main disturbance is additive white Gaussian noise.

We will confine our discussion to spherical codes based on one or several codes from Hamming spaces. We will assume that decoding algorithms with respect to the Hamming metric are already available for those codes. The problem of interest to us is the adaptation needed for the Euclidean space.

Decoding in the Euclidean space is often referred to as "soft decoding". The case of binary antipodal encoding (construction Y_2) is of particular interest. It is by far the most thoroughly investigated case and in a sense it is the simplest one. Our discussion will focus on that case, with just minor indications regarding extensions to more general cases.

In order to employ the digital algorithm to the analog problem we must convert the given received sequence $y = (y_1, y_2, \ldots, y_n) \in \mathbb{R}^n$ into a sequence of binary symbols and erasures. We assume that this conversion is performed in the simplest possible way using a component-wise mapping. Define

$$\Psi_\theta(y_i) \triangleq \begin{cases} 0, & if \quad y_i > 1 - \sqrt{\theta}, \\ \mathcal{E}, & if \quad |y_i| \leq 1 - \sqrt{\theta}, \\ 1, & if \quad y_i < -1 + \sqrt{\theta}, \end{cases} \tag{11.3.1}$$

where θ is a real-valued parameter in the interval $0 \leq \theta \leq 1$. The function $\Psi_\theta(y_i)$ plays a fundamental role in the so called GMD algorithm, (GMD = Generalized Minimum Distance decoding, to be described in Section 4 of the present chapter). It will be referred to as the *GMD quantizer.* We assume that $y = (y_1, y_2, \ldots, y_n) \in \mathbb{R}^n$ is converted into a sequence of zeroes, ones and erasures using this mapping successively on all components y_i, $i = 1, 2, \ldots, n$. We also define

$$t_\theta(a, y) \triangleq |\{i : \psi_\theta(y_i) = a_i \oplus 1\}|$$

$$r_\theta(y) \triangleq |\{i : \psi_\theta(y_i) = \mathcal{E}\}|,$$

where $\oplus$ denotes addition modulo 2. This means that $t_\theta(a, y)$ denotes the number of errors that are made in the component-wise decisions if $x \in \mathcal{X}$ corresponding to $a \in \mathcal{A}$ is transmitted and $y \in \mathbb{R}^n$ is received, while $r_\theta(y)$ denotes the number of erasures. The codeword $a \in \mathcal{A}$ can be restored provided the following inequality is satisfied:

$$2t_\theta(a, y) + r_\theta(y) < d. \tag{11.3.2}$$

It is clear that the set $B_a \subseteq \mathbb{R}^n$ of vectors $y = (y_1, y_2, \ldots, y_n) \in \mathbb{R}^n$ that will be decoded into the codeword $a \in \mathcal{A}$ depends on the parameter θ. We proceed to investigate this dependence in some detail. We begin with the following simple observation.

Lemma 11.3.1 *Let $\mathcal{A}$ be a binary code with minimum distance d and let the spherical code $\mathcal{X} \subseteq \Omega_n$ be obtained from $\mathcal{A}$ by the Y_2-construction. Then, for all $y \in \mathbb{R}^n$ and for all θ, $0 \leq \theta \leq 1$, there is at most one (and possibly none) codeword $a \in \mathcal{A}$ for which the inequality $2t_\theta(a, y) + r_\theta(y) < d$ is fulfilled.*

Proof: Suppose (11.3.2) holds for $a \in \mathcal{A}$ and let a' be an element in $\mathcal{A}$ different from a. By assumption a and a' differ in at least d positions. Therefore for any $y = (y_1, y_2, \ldots, y_n) \in \mathbb{R}^n$ we have $t_\theta(a, y) + t_\theta(a', y) \geq d$, and consequently we also have

$$2\, t_\theta(a', y) + r_\theta(y) \geq 2d - 2\, t_\theta(a, y) + r_\theta(y) > d + 2\, r_\theta(y) \geq d.$$

□

An immediate consequence is that the following function is well defined:

$$\Phi_\theta(y) \triangleq \begin{cases} a \in \mathcal{A}, & if \;\; 2t_\theta(a, y) + r_\theta(y) < d, \\ \mathcal{E}, & otherwise, \end{cases}$$

$$0 \leq \theta \leq 1.$$

The function $\Phi_\theta(y)$ defines a decoder $\mathbb{R}^n \rightarrow \mathcal{A} \cup \{\mathcal{E}\}$. Figure 11.1 illustrates this function in the simple case $\mathcal{A} = \{00, 11\}$, which implies $\mathcal{X} = \{(1, 1), (-1, -1)\}$. In this case $d = 2$, so (11.3.2) is satisfied if and only if $(t, r) = (0, 0)$ or $(0, 1)$. In the figure we have indicated the decision regions for three different values of the parameter θ. Indicated in the figure are also the largest spheres that are contained in the decoding regions. As seen the size of the spheres is constrained by two different conditions depending on the size of θ. The diagram in the middle indicates the case when those two constraints coincide. It is easy to show that this is the case where the spheres have largest possible radius.

In general, the smallest size $\|w\|^2$ of a vector $w = (w_1, w_2, \ldots, w_n)$ such that $y = x + w$ cannot be correctly decoded is determined by the pair (t, r) minimizing $\|w\|^2 = t(2 - \sqrt{\theta})^2 + r\theta$ under the condition $2t + r \geq d$. A simple calculation reveals that this *minimal* quantity with respect to the pair (t, r) is *maximized* with respect to the parameter θ by

$$\sqrt{\theta} = 2(\sqrt{2} - 1).$$

The maximum is given by

$$\|w\|^2/d = 4(3 - 2\sqrt{2}) \approx 0.6863.$$

These equations determine the best possible decoder in the sense that the smallest noise vector that might cause an error is largest possible.

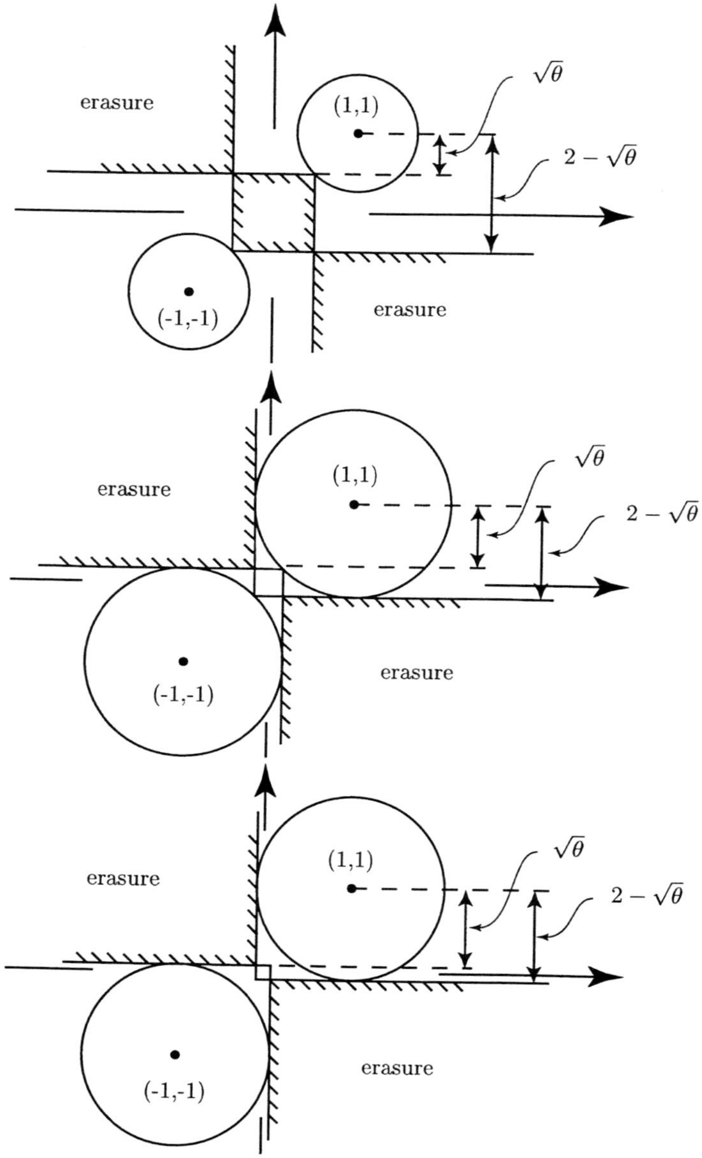

Figure 11.1: Decoding regions for the function $\Phi_\theta(y)$ and the code $\mathcal{X} = \{(1,1),(-1,-1)\}$ for three different thresholds θ, $0 \leq \theta \leq 1$. The circles indicate the largest possible spheres—centered at the codewords—that are contained within the decoding regions.

It follows that a *single* decoder of this form cannot decode correctly for *all* errors w of size $\|w\|^2 < (d_E/2)^2 = d$. However, for the simple example illustrated in Figure 11.1 it is easy to see that although none of the decoders $\Phi_\theta(y)$ *alone* is able to decode all errors w of size $\|w\|^2 < (d_E/2)^2 = d$, *each* of those error vectors is decoded by *some* decoder $\Phi_\theta(y)$, $0 \le \theta \le 1$. This observation suggests the idea of employing simultaneously *several different* decoders. We proceed to develop this idea.

11.4 Generalized minimum distance decoding

The idea of using error-erasure decoding with several different erasure thresholds as indicated in the previous section was first developed by Forney [107] under the name of *Generalized Minimum Distance decoding* (*GMD*). It appears in the literature in many different forms and under many different names. Our discussion is based on [93] and is closely related to the results developed by Blokh-Zyablov [35], Zinoviev–Zyablov [240], Dumer–Zinoviev–Zyablov [88], Zinoviev [238] and Zinoviev–Litzyn–Portnoi [239]. A few additional variants will also be briefly discussed.

The key to the analysis is the selection of an efficient method for estimating the size of the vectors w that generate errors or erasures. For that purpose we define the following function.

$$\Gamma_\theta(w_i) \triangleq \begin{cases} 0, & if \quad |w_i| < \sqrt{\theta}, \\ 1, & if \quad \sqrt{\theta} \le |w_i| \le 2 - \sqrt{\theta}, \\ 2, & if \quad 2 - \sqrt{\theta} < |w_i|. \end{cases} \tag{11.4.1}$$

This function has two useful properties which we formulate in two lemmas.

Lemma 11.4.1 *Let $x \in \mathcal{X} = Y_2(\mathcal{A})$ correspond to $a \in \mathcal{A}$. For all $w \in \mathbb{R}^n$ and for all $\theta \in [0, 1]$ the following inequality holds:*

$$\sum_{i=1}^{n} \Gamma_\theta(w_i) \ge 2t_\theta(a, x + w) + r_\theta(x + w).$$

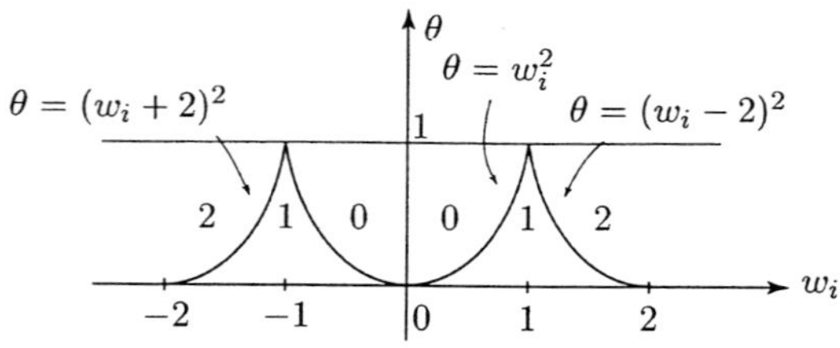

Figure 11.2: The function $\Gamma_\theta(w_i)$, $0 \le \theta \le 1$, $w_i \in \mathbb{R}$. The figure displays the domain of the function along with an indication of the values attained at the various parts of the domain.

Proof: With Ψ_θ defined as in (11.3.1) we have the following obvious implications:

$$\Psi_\theta(x_i + w_i) = a_i \oplus 1 \;\Rightarrow\; |w_i| > 2 - \sqrt{\theta},$$
$$\Psi_\theta(x_i + w_i) = \mathcal{E} \;\Rightarrow\; |w_i| \ge \sqrt{\theta}.$$

A comparison of the definitions (11.3.1) and (11.4.1) immediately reveals the result. □

Lemma 11.4.2 *For any $w_i \in \mathbb{R}$ the following inequality holds:*

$$\int_0^1 \Gamma_\theta(w_i)d\theta \le w_i^2.$$

Proof: The result follows easily by evaluating the integral separately in the following three cases, namely $|w_i| < 1$, $1 \le |w_i| \le 2$, and $|w_i| > 2$, (see Figure 11.2). □

These two lemmas lead to a third one, which is actually the key result.

Lemma 11.4.3 *Let $x \in \mathcal{X} = Y_2(\mathcal{A})$ correspond to $a \in \mathcal{A}$. If the inequality $2t_\theta(a, x+w) + r_\theta(x+w) \ge d$ holds for all θ, $0 \le \theta \le 1$, then the noise vector $w = (w_1, w_2, \ldots, w_n) \in \mathbb{R}^n$ satisfies the inequality $\|w\|^2 \ge d = (d_E/2)^2$.*

Proof: The result is obtained through a series of inequalities:

$$\|w\|^2 =_{(i)} \sum_{i=1}^{n} w_i^2 \geq_{(ii)} \sum_{i=1}^{n} \int_0^1 \Gamma_\theta(w_i) d\theta$$

$$=_{(iii)} \int_0^1 \sum_{i=1}^{n} \Gamma_\theta(w_i) d\theta \geq_{(iv)} \int_0^1 (2t_\theta(a, x+w) + r_\theta(x+w)) d\theta$$

$$\geq_{(v)} \min_{0 \leq \theta \leq 1} \{2t_\theta(a, x+w) + r_\theta(x+w)\} \geq_{(vi)} d.$$

The successive steps are motivated as follows:

(i) : by definition,
(ii) : by Lemma 11.4.2,
(iii) : by linearity of integration,
(iv) : by Lemma 11.4.1,
(v) : by standard estimation of an integral,
(vi) : by assumption.

□

We notice that Lemma 11.4.3 is precisely the property we promised: if the distortion w satisfies $\|w\|^2 < (d_E/2)^2 = d$ then *at least one* of the decoders $\Phi_\theta(y)$ will perform a correct decoding.

A property that is *not* illustrated in the simple example shown in Figure 11.2 is that even if *one* decoder $\Phi_\theta(y)$ performs a correct decoding there might be several other decoders (corresponding to other values of the parameter θ) that produce *false* decisions. For a simple illustration, let us modify the code in our example to $\mathcal{X} = \{x_0, x_1\}$, where $x_0 = (1,1,1)$, $x_1 = (-1,-1,-1)$, with $(d_E/2)^2 = 3$. Let the transmitted codeword be $x = x_1$ and let the distortion be $w = (\alpha, \alpha, 2\alpha)$, where $2/3 < \alpha < 1/\sqrt{2}$. For $\theta = 4/9$ we get $\sqrt{\theta} < \alpha < 2 - \sqrt{\theta} < 2\alpha$. Thus the two first symbols are erased and the last one is erroneously detected as $+1$, see Figure 11.3. Clearly the decoder Φ_θ, $\theta = 4/9$, will fail. Still we have $\|w\|^2 = 6\alpha^2 < 3 = (d_E/2)^2$.

In order to find a decoder that produces a correct decoding there is no need to consider *all* values of θ, $0 \leq \theta \leq 1$; as a matter of fact, it is enough to consider only d different values of θ. For $y =$

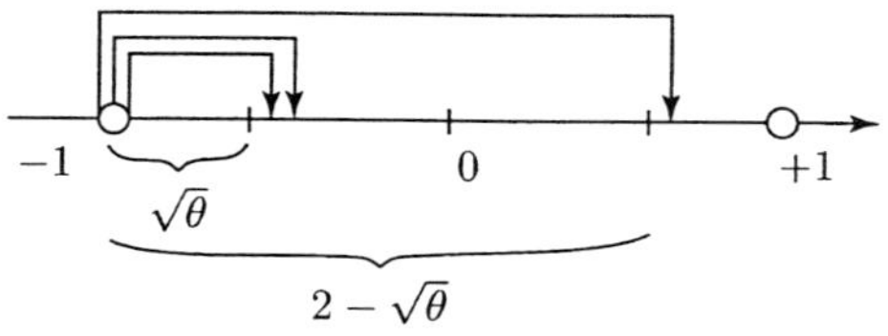

Figure 11.3: Illustrating the fact that a small noise vector might cause an erroneous decision in an error-erasure decoder.

$(y_1, y_2, \ldots, y_n) \in \mathbb{R}^n$, let $(i_1, i_2, \ldots, i_n)$ be a permutation of the indices $(1, 2, \ldots, n)$ such that $0 \leq |y_{i_1}| \leq |y_{i_2}| \leq \ldots \leq |y_{i_n}|$. Define

$$\theta_0 \triangleq 1, \quad \theta_d \triangleq 0,$$

$$\theta_j \triangleq (1 - |y_{i_j}|)^2, \quad j = 1, 2, \ldots, d-1.$$

We get

$$r_\theta(y) = constant \geq j; \quad \theta_j \geq \theta > \theta_{j+1}, \quad j = 0, 1, \ldots, d-1.$$

Thus we have the following lemma.

Lemma 11.4.4 *Whenever the inequality*

$$\|w\|^2 < (d_E/2)^2 = d$$

is satisfied we also have the inequality:

$$\min_{0 \leq j \leq d-1} [2t_{\theta_j}(a, x+w) + r_{\theta_j}(x+w)] \leq d,$$

where $x \in \mathcal{X}$ corresponds to $a \in \mathcal{A}$.

The significant thing here is, of course, that in order to perform the minimization over the domain $0 \leq \theta \leq 1$ we need in fact to consider only the d cases θ_j, $j = 0, 1, \ldots, d-1$. This observation leads to the following decoding algorithm, referred to as the *GMD* algorithm.

GMD algorithm:

1. Sort the components in the observed vector $y = (y_1, y_2, \ldots, y_n) \in \mathbb{R}^n$ such that $0 \leq |y_{i_1}| \leq |y_{i_2}| \leq \ldots \leq |y_{i_n}|$.
2. Perform d error-erasure decoding attempts by erasing successively the $d-1$ least reliable components in $y = (y_1, y_2, \ldots, y_n)$. Let the resulting tentative decisions be $\hat{x}_0, \hat{x}_1, \ldots, \hat{x}_{d-1}$.
3. Find the codeword $x \in \mathcal{X} = Y_2(\mathcal{A})$ closest to the observed vector y by directly computing all distances $||y - \hat{x}_j||^2$ and select the smallest.

Theorem 11.4.1 *The GMD algorithm maps $y \in \mathbb{R}^n$ into $x \in \mathcal{X}$ whenever the condition $||y - x||^2 < (d_E/2)^2$ is fulfilled.*

Proof: The theorem is a direct consequence of the preceding lemmas. □

A few comments are in order.

Remark 11.4.1: The complexity is algebraic: the number of distances $||y - \hat{x}_j||$ that need to be computed is d rather than $M = |\mathcal{X}|$ and the number of comparisons needed for the sorting is at most d^2.

Remark 11.4.2: The last step (3) is needed because—as we have seen by example—some of the tentative decisions might be erroneous. Other decisions might result in erasures. All we can guarantee is that *at least one* tentative decision is right.

Remark 11.4.3: It is worthwhile observing that although the algorithm is a so called *soft decision algorithm*—i.e. an algorithm that deals with the *analog* values of the observed vector y—most of the processing is actually performed on a *digital* basis using the assumed error-erasure decoding algorithm for the Hamming metric.

11.5 The Chase decoder

A simple modification of the above ideas is employed in a decoder suggested by Chase [67] and usually referred to as the *Chase decoder.* This decoder is simpler inasmuch as no initial erasures are produced, i.e. the received sequence $y = (y_1, y_2, \ldots, y_n) \in \mathbb{R}^n$ is converted into a set of binary sequences only. This means that the subsequent digital decoding is performed using a pure binary algorithm. We assume that such an algorithm is available for the code $\mathcal{A}$, and we assume that t binary errors are corrected by this algorithm whenever $t < d/2$. As we shall see the performance of the resulting decoder for $\mathcal{X} = Y_2(\mathcal{A})$ is at least as good as for the GMD algorithm.

Actually, Chase in his original paper discussed *three* different versions of his basic idea. However, only *one* of these cases leads to a decoder with *algebraic* complexity. We discuss only the algebraic variant.

For the analysis of the Chase decoder we need a new quantizer $\tilde{\Psi}_\theta(y_i)$, slightly different from the GMD quantizer $\Psi_\theta(y_i)$. We will refer to the new quantizer $\tilde{\Psi}_\theta(y_i)$ as the *Chase quantizer.* We start by defining $\tilde{\Psi}_1(y_i)$ according to

$$\tilde{\Psi}_1(y_i) \triangleq \begin{cases} 0, & if \quad y_i \geq 0, \\ 1, & if \quad y_i < 0. \end{cases}$$

For $0 \leq \theta < 1$ we define

$$\tilde{\Psi}_\theta(y_i) \triangleq \begin{cases} \tilde{\Psi}_1(y_i), & if \quad |y_i| > 1 - \sqrt{\theta}, \\ \tilde{\Psi}_1(y_i) \oplus 1, & if \quad |y_i| \leq 1 - \sqrt{\theta}. \end{cases}$$

In words this means that $\tilde{\Psi}_\theta(y_i)$ reverses the decisions made by $\tilde{\Psi}_1(y_i)$ whenever $|y_i| \leq 1 - \sqrt{\theta}$, i.e. it reverses the most uncertain ones of the preliminary component-wise decisions.

We also define

$$\tilde{t}_\theta(a, y) \triangleq |\{i : \tilde{\Psi}_\theta(y_i) \neq a_i\}|.$$

This function denotes the number of errors performed by the Chase quantizer $\tilde{\Psi}_\theta(y_i)$ when $a \in \mathcal{A}$ is transmitted and $y \in \mathbb{R}^n$ is received.

We notice that the *GMD* quantizer $\Psi_\theta(y_i)$ and the *Chase* quantizer $\tilde{\Psi}_\theta(y_i)$ produce the same output (0 or 1) whenever $|y_i| > 1 - \sqrt{\theta}$. For $|y_i| \leq 1 - \sqrt{\theta}$ the *GMD* quantizer produces an erasure while the *Chase* quantizer produces either 0 or 1. By definition we have

$$\tilde{\Psi}_1(y_i) \oplus \tilde{\Psi}_\theta(y_i) = \begin{cases} 1, & if \quad |y_i| \leq 1 - \sqrt{\theta}, \\ 0, & if \quad |y_i| > 1 - \sqrt{\theta}. \end{cases}$$

Suppose $a \in \mathcal{A}$ is the correct word and that $y \in \mathbb{R}^n$ is the received vector. Consider a fixed θ. We first notice that any error made in the *GMD* quantizer Ψ_θ will also be an error in the *Chase* quantizer $\tilde{\Psi}_1$. Denote by $\alpha_\theta(a, y)$ the number of *additional* errors made in $\tilde{\Psi}_1$. By definition the following relation holds:

$$\tilde{t}_1(a, y) = t_\theta(a, y) + \alpha_\theta(a, y).$$

Now consider the Chase quantizer $\tilde{\Psi}_\theta$. Still all the errors made in the GMD quantizer Ψ_θ also occur as errors in the Chase quantizer $\tilde{\Psi}_\theta$. Denote by $\beta_\theta(a, y)$ the number of *erasures* in Ψ_θ that occurs as *errors* in $\tilde{\Psi}_\theta$. Then we have

$$\tilde{t}_\theta(a, y) = t_\theta(a, y) + \beta_\theta(a, y)$$

Any erasure in $\tilde{\Psi}_\theta$ will be an error in either $\tilde{\Psi}_1$ or in Ψ_θ, but never in both. Therefore we have

$$\alpha_\theta(a, y) + \beta_\theta(a, y) = r_\theta(y).$$

The following lemma is the key to the analysis (for simplicity of notation the arguments a, y are suppressed).

Lemma 11.5.1 *Whenever the inequality* $2t_\theta + r_\theta < d$ *holds then also the following inequality holds:*

$$\min\{\tilde{t}_1, \tilde{t}_\theta\} \leq \left\lfloor \frac{d-1}{2} \right\rfloor.$$

Proof: From the three preceding equations we immediately have

$$\tilde{t}_1 + \tilde{t}_\theta = 2t_\theta + r_\theta < d.$$

The lemma follows directly from this. □

Now let θ_j, $j = 0, 1, \ldots, d$ be defined as in the previous section. Notice that this means that we are employing only thresholds θ_j at which two or more of the components y_i in the received vector have the same magnitude. The following result is immediate.

Corollary 11.5.1 *Whenever the inequality*

$$\|w\|^2 < (d_E/2)^2 = d$$

is satisfied the following inequality also necessarily holds:

$$\min_{0 \le j \le d-1} \{\tilde{t}_{\theta_j}\} \le \left\lfloor \frac{d-1}{2} \right\rfloor .$$

Proof: The result follows by combining Lemma 11.5.1 with Lemma 11.4.4. □

Denote by $z^{(i)}$, $0 \le i \le d-1$, the binary sequence generated by component-wise application of the function $\tilde{\Psi}_{\theta_i}(y_i)$ on the n successive components y_i of the received vector y:

$$z^{(i)} \triangleq (\tilde{\Psi}_{\theta_i}(y_1), \tilde{\Psi}_{\theta_i}(y_2), \ldots, \tilde{\Psi}_{\theta_i}(y_n)), \quad 0 \le i \le d-1.$$

Notice that this means that we successively reverse the binary decisions corresponding to the $d-1$ least reliable components in the received vector y. The corollary says that whenever the size $\|w\|$ of the noise vector w is less than half the minimum distance (non-squared Euclidean distance) of the code $\mathcal{X} = Y_2(\mathcal{A})$, then at least one of the d binary sequences $z^{(i)}$ is within Hamming distance $\lfloor (d-1)/2 \rfloor$ of the codeword $a \in \mathcal{A}$ which generated the transmitted codeword $x \in \mathcal{X} = Y_2(\mathcal{A})$. This in turn means that at least one of the binary sequences $z^{(i)}$ is decoded into the correct codeword $a \in \mathcal{A}$. As with the previous decoder we have to test each one of the d tentative codewords individually against the received sequence $y = (y_1, y_2, \ldots, y_n) \in \mathbb{R}^n$.

The *Chase* decoder is based upon exactly the above considerations. It first generates the binary sequences $z^{(i)}$, $0 \le i \le d-1$. Then it performs a digital decoding upon each one of them. Finally it generates the codewords in $\mathcal{X}$ corresponding to each one of the tentative decisions obtained in the decoding of the sequences $z^{(i)}$. The final decision is in favor of that tentative codeword which is closest to the received vector $y \in \mathbb{R}^n$. It should be clear that these steps are achieved by the following algorithm.

The Chase decoding algorithm:

1. Sort the components in the observed vector $y = (y_1, y_2, \ldots, y_n) \in \mathbb{R}^n$ such that $0 \leq |y_{i_1}| \leq |y_{i_2}| \leq \cdots \leq |y_{i_n}|$.

2. Form the binary sequence $z^{(0)} = (z_1^{(0)}, z_2^{(0)}, \ldots, z_n^{(0)}) \in \mathbb{F}_2^n$ using the simple rule $y_i > 0 \Rightarrow z_i^{(0)} = 0$, $y_i < 0 \Rightarrow z_i^{(0)} = 1$, $i = 1, 2, \ldots, n$. (If $y_i = 0$ either decision will do.)

3. Form the additional binary sequences $z^{(i)} = (z_1^{(i)}, z_2^{(i)}, \ldots, z_n^{(i)}) \in \mathbb{F}_2^n$, $i = 1, 2, \ldots, d-1$, by successively reverting those symbols in $z^{(0)} = (z_1^{(0)}, z_2^{(0)}, \ldots, z_n^{(0)}) \in \mathbb{F}_2^n$ corresponding to the smallest magnitudes in the received vector $y = (y_1, y_2, \ldots, y_n) \in \mathbb{R}^n$, according to the sorting performed in step 1.

4. Apply the digital decoding algorithm upon each one of the binary sequences $z^{(i)} = (z_1^{(i)}, z_2^{(i)}, \ldots, z_n^{(i)}) \in \mathbb{F}_2^n$, $i = 0, 1, \ldots, d-1$. Let the resulting tentative decisions be denoted $\hat{a}^{(i)}$, $i = 0, 1, \ldots, d-1$.

5. Form all the corresponding codewords $\hat{x}^{(i)}$ and compute the distances $\|y - \hat{x}^{(i)}\|$, $i = 0, 1, \ldots, d-1$.

6. Choose as the final decision the codeword $\hat{x}^{(i)}$ minimizing the distance $\|y - \hat{x}^{(i)}\|$.

From the discussion above follows that the Chase decoder performs a correct decision whenever the *GMD* decoder does so, and in particular whenever the noise vector w satisfies $\|w\| < d_E/2$. So, like the *GMD* decoder the *Chase* decoder performs so called *bounded minimum distance decoding*. Although in principle the *Chase* decoder is

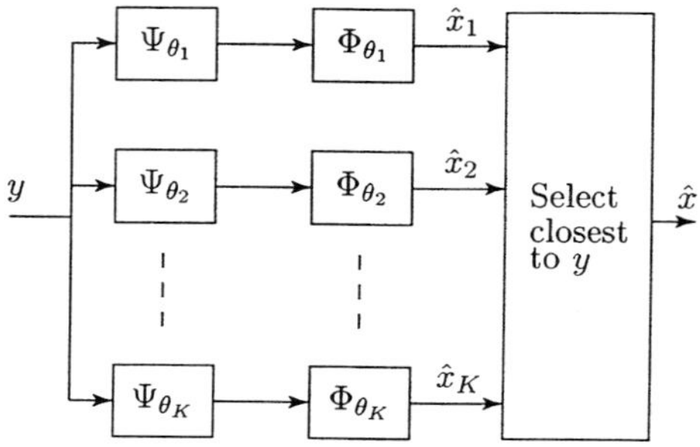

Figure 11.4: Parallel decoding.

slightly better than the *GMD* decoder, in practice the performance is essentially the same. With respect to implementation the *Chase* decoder has the advantage that the digital algorithm involved is a pure error correction algorithm, while the *GMD* decoder relies upon a digital algorithm handling both errors and erasures.

11.6 Parallel decoding

Another modification that naturally suggests itself is to use a fixed set of parallel branches corresponding to some appropriately selected set of different values of the threshold θ. This idea was first investigated by Zyablov [243] for decoding of concatenated codes according to Hamming distance, then by Dumer–Zinoviev–Zyablov [88] for *GMD* decoding of generalized concatenated codes and later by Hassan–Stark [124] for decoding of concatenated codes according to Euclidean distance. The problem of interest is to find, for any given number K of parallel branches, the best possible thresholds. We proceed with a discussion of that problem.

Consider a decoder with K parallel branches, where each branch i contains a *GMD* quantizer Ψ_{θ_i} followed by a digital error-erasure decoder $\Phi_{\theta_i}(y)$, $i = 1, 2, \ldots, K$, see Figure 11.4. We recall the definitions from Section 3:

$$\Psi_\theta(y_i) \triangleq \begin{cases} 0, & if \quad y_i > 1-\sqrt{\theta}, \\ \mathcal{E}, & if \quad |y_i| \le 1-\sqrt{\theta}, \\ 1, & if \quad y_i < -1+\sqrt{\theta}, \end{cases}$$

$$\Phi_\theta(y) \triangleq \begin{cases} a \in \mathcal{A}, & if \quad 2t_\theta(a,y) + r_\theta(y) < d, \\ \mathcal{E}, & otherwise, \end{cases}$$

$$0 \le \theta \le 1.$$

We define $z_i = \sqrt{\theta_i}$ and we assume $z_1 > z_2 > \cdots > z_K$. We still assume that the code $\mathcal{X} \subseteq \Omega_n$ is generated from a binary code $\mathcal{A} \subseteq \mathbb{F}_2^n$ using the construction Y_2 and the alphabet $\mathcal{L} = \{\pm 1\}$. Without loss of generality we also assume that $\mathcal{X}$ contains the codeword $x_0 = (-1, -1, \ldots, -1)$ and that this particular codeword has been transmitted. The decision levels of the various quantizers Ψ_{θ_i} are indicated in Figure 11.4. We need a notation for the number of components y_i from the observed vector $y = (y_1, y_2, \ldots, y_n) \in \mathbb{R}^n$ that belong to the various intervals between the various quantization levels. We define

$$v_1 = |\{j : -1 + z_1 \le y_j \le 1 - z_1\}|,$$

$$v_i = |\{j : -1 + z_i \le y_j < -1 + z_{i-1}\}|, \quad i = 2, 3, \ldots, K.$$

$$u_i = |\{j : 1 - z_i < y_j \le 1 - z_{i+1}\}|, \quad i = 1, 2, \ldots, K-1,$$

$$u_K = |\{j : 1 - z_K < y_j\}|.$$

We want first to determine a smallest possible vector $w = (w_1, w_2, \ldots, w_n)$ that causes an erroneous decision in each one of the K branches, and then to find thresholds $z_1, z_2, \ldots, z_K$ such that this smallest noise-vector is as large as possible. With $y_j = -1 + w_j$ we alternatively have

$$u_i = |\{j : 2 - z_i < w_j \le 2 - z_{i+1}\}|, \quad i = 1, 2, \ldots, K-1,$$

$$u_K = |\{j : 2 - z_K < w_j\}|,$$

$$v_1 = |\{j : z_1 \le w_j \le 2 - z_1\}|,$$

$$v_i = |\{j : z_i \le w_j < z_{i-1}\}|, \quad i = 2, 3, \ldots, K.$$

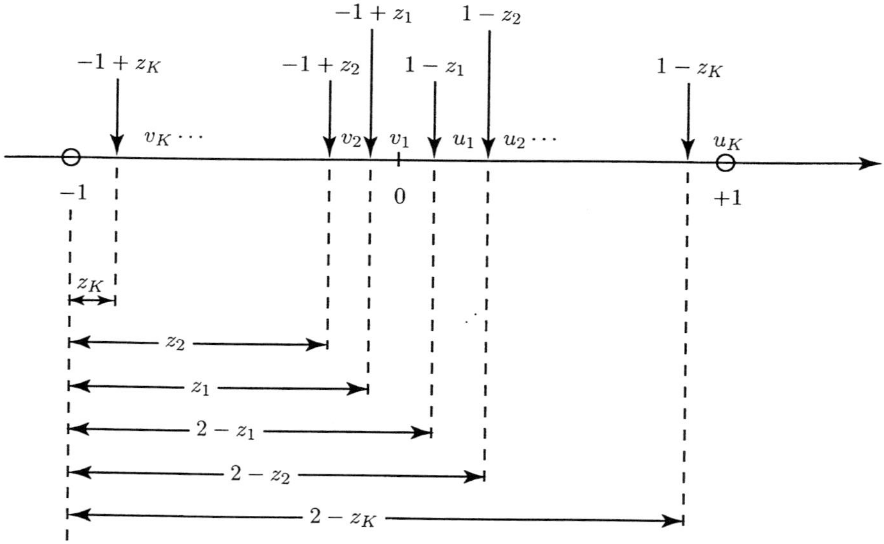

Figure 11.5: Thresholds $z_1, z_2, \ldots, z_K$ in a K-branch parallel decoder.

Denote by t_i and r_i the *total* numbers of errors and erasures made in Ψ_{θ_i}. The following equalities are evident:

$$\begin{array}{ll} t_1 = u_1 + u_2 + \cdots + u_K, & r_1 = v_1, \\ t_2 = u_2 + u_3 + \cdots + u_K, & r_2 = v_1 + v_2 + u_1, \\ t_3 = u_3 + u_4 + \cdots + u_K, & r_3 = v_1 + v_2 + v_3 + u_1 + u_2, \\ \cdots & \cdots \\ t_K = u_K, & r_K = v_1 + \cdots + v_K + u_1 + \cdots + u_{K-1}. \end{array}$$

It is clear that for an error-generating noise-vector w of smallest possible size each one of the relations $2t_i + r_i \geq d$, $i = 1, 2, \ldots, K$, must be satisfied with equality. For $i = 2, 3, \ldots, K$ we obtain

$$v_i = r_i - r_{i-1} - u_{i-1} = d - 2t_i - d + 2t_{i-1} - u_{i-1}$$

$$= 2(t_{i-1} - t_i) - u_{i-1} = u_{i-1}, \tag{11.6.1}$$

while for $i = 1$ we get

$$v_1 = r_1 = d - 2t_1 = d - 2(u_1 + u_2 + \cdots + u_K). \tag{11.6.2}$$

For a given set $\{t_1, t_2, \ldots, t_K; r_1, r_2, \ldots, r_K\}$ the infimum of all possible sizes $\|w\|^2$ of vectors w that produce this set of errors and erasures is given by

$$\|w\|^2 = u_1(2-z_1)^2 + u_2(2-z_2)^2 + \cdots + u_K(2-z_K)^2$$
$$+v_1 z_1^2 + v_2 z_2^2 + \cdots + v_K z_K^2.$$

Inserting (11.6.1) and (11.6.2) we get after some simple calculations

$$\begin{aligned}\|w\|^2 = dz_1^2 &+ u_1\left((2-z_1)^2 - 2z_1^2 + z_2^2\right) + \\ &+ u_2\left((2-z_2)^2 - 2z_1^2 + z_3^2\right) + \\ &\ldots \\ &+ u_{K-1}\left((2-z_{K-1})^2 - 2z_1^2 + z_K^2\right) + \\ &+ u_K\left((2-z_K)^2 - 2z_1^2\right). \end{aligned} \tag{11.6.3}$$

This equation gives the infimum of $\|w\|^2$ in terms of the non-negative integers $u_1, u_2, \ldots, u_K$, which are subject to the constraint $u_1 + u_2 + \cdots u_K \leq n$ and in terms of the real-valued variables $z_1, z_2, \ldots, z_K$, which are subject to the constraints $0 < z_K < z_{K-1} < \ldots < z_1 < 1$. For any allowable set $\{z_1, z_2, \ldots, z_K\}$ of thresholds the performance is characterized by the minimum of (11.6.3) with respect to the integers $\{u_1, u_2, \ldots, u_K\}$. We would like to select the thresholds $\{z_1, z_2, \ldots, z_K\}$ so that this minimum is as large as possible. We proceed to show that it is possible to choose them so that all the coefficients of the variables u_i become zero. This makes the expression (11.6.2) identically equal to dz_1, independently of the variables $\{u_1, u_2, \ldots, u_K\}$.

Setting all coefficients in (11.6.1) equal to zero gives us the following set of equations:

$$z_i^2 = 2z_1^2 - (2-z_{i-1})^2, \quad i = 2, 3, \ldots, K, \tag{11.6.4}$$

$$z_K = 2 - \sqrt{2}z_1. \tag{11.6.5}$$

For $K = 1$ only (11.6.5) applies. This is the case we encountered in Section 11.3. The solution is

$$z_1 = \sqrt{\theta} = 2(\sqrt{2}-1) \approx 0.8284.$$

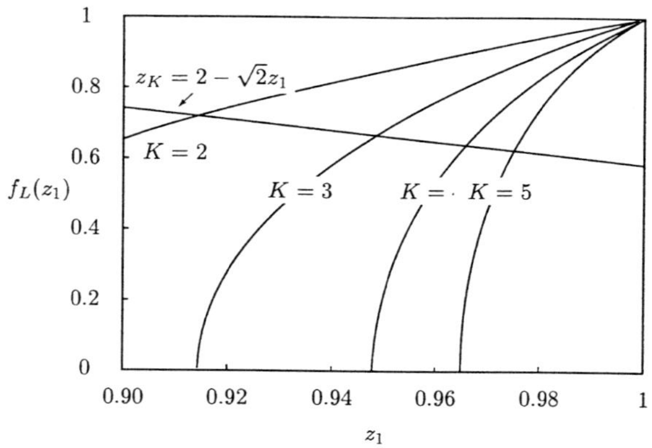

Figure 11.6: The function $z_K = f_K(z_1)$ and the line $z_K = 2 - \sqrt{2}$, $K = 2, 3, 4, 5$.

K	z_1	z_2	z_3	z_4	z_5	z_6
1	0.8284					
2	0.9152	0.7058				
3	0.9483	0.8320	0.6590			
4	0.9649	0.8891	0.7925	0.6355		
5	0.9745	0.9207	0.8571	0.7701	0.6219	
6	0.9806	0.9403	0.8947	0.8376	0.7563	0.6137

Table 11.1: Optimal thresholds for parallel decoding.

In general, for $K \geqq 2$, the solution can be computed as follows. First notice that (11.6.3) defines z_K as a function of z_1, say $z_K = f_K(z_1)$. In Figure 11.6 we have plotted this function for the first few values of K. The value of z_1 corresponding to the solution of our problem is obtained by the intersection between the function $f_K(z_1)$ and the line defined by (11.6.5). This fact is also illustrated in Figure 11.6. Once z_1 has been determined all the other parameters z_i are easily computed from equation (11.6.4). The first few cases are given in Table 11.1. Recall that the minimum Euclidean distance is $d_E^2 = 4d$ and that the GMD decoder decodes correctly whenever $\|w\|^2 < (d_E/2)^2 = d$. For parallel decoding correct decisions can be guaranteed only if $\|w\|^2 < dz_1$. Thus the parameter z_1 can be interpreted as a measure of efficiency for the parallel decoder. From the table we see that the performance is quite close to that of GMD decoding, even for rather small values of K.

11.7 Decoding Y_3

The ideas for decoding Y_2 can be extended to all the Y_K- and all the CW_K-constructions. We consider briefly how this can be done for codes from the Y_3-construction.

Recall that each codeword x in $Y_3(\mathcal{C}, \mathcal{A})$ depends on two binary codewords a and c, where a is selected from an unconstrained code $\mathcal{A} \subseteq \mathbb{F}_2^w$ and c is selected from a constant weight code $\mathcal{C} \subseteq \mathbb{F}_2^{n,w}$. The mapping is defined by

$$x_i = \begin{cases} 0, & if \quad c_i = 0, \\ 1, & if \quad c_i = 1, \ a_{\varphi(c,i)} = 1, \\ -1, & if \quad c_i = 1, \ a_{\varphi(c,i)} = 0, \end{cases}$$

where the function $\varphi(c,i)$ is defined as $\varphi(c,i) \triangleq \sum_{j=1}^{i} c_j$. We have chosen to represent the code in the alphabet $\mathcal{L} = \{0, \pm 1\}$, which gives us the energy $E = w$ and the following bound on the minimum distance:

$$d_E^2 \geq \min\{d_{\mathcal{C}}, 4d_{\mathcal{A}}\}.$$

It was shown first by Dumer–Zinoviev–Zyablov [88] and later by Zinoviev–Litsyn–Portnoi [239] (for codes in Euclidean space) that

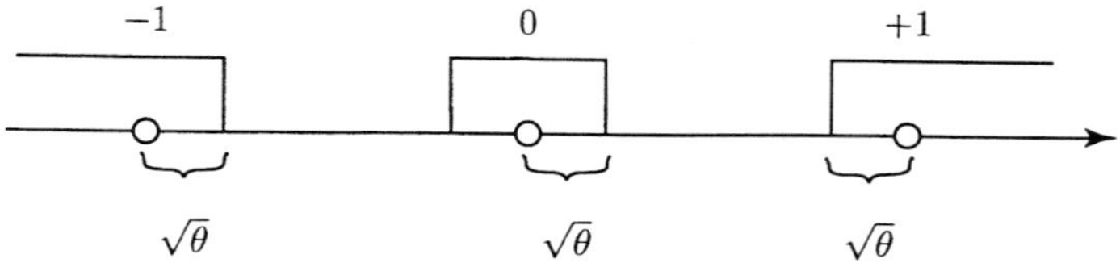

Figure 11.7: Decoding regions used for decoding $c \in \mathcal{C}$ in the spherical code $\mathcal{X} = Y_3(\mathcal{C}, \mathcal{A})$.

for concatenated codes in the generalized sense, involving more than one inner code, GMD decoding can always be achieved by *successive decoding*. This means that the various outer codes are successively decoded according to their position in the code hierarchy, starting at the root of the tree and successively proceeding further out in the branches. In each new step it is assumed that all the previous decisions were correct. For the construction Y_3 successive decoding means that we first try to find the codeword $c \in \mathcal{C}$ and then—provided that a decision was possible—try to decode $a \in \mathcal{A}$. As we shall see correct decoding of both a and c can be guaranteed provided the error w satisfies

$$\|w\|^2 < (d_E^*/2)^2,$$

where $d_E^*/2$ denotes the estimate of the minimum distance of $Y_3(\mathcal{C}, \mathcal{A})$ obtained by the product bound:

$$(d_E^*)^2 \triangleq \min\{d_{\mathcal{C}}, 4d_{\mathcal{A}}\}.$$

We assume that digital decoding algorithms are available for both $\mathcal{C}$ and $\mathcal{A}$ so that each of them can be decoded up to half its minimum Hamming distance.

The first step is to decode $c \in \mathcal{C}$. For that purpose we need as before a family of functions $\Psi_\theta^{\mathcal{C}}$ for converting the components y_i of the received vector $y = (y_1, y_2, \ldots, y_n) \in \mathbb{R}^n$ into a sequence of zeros, ones and erasures. For $0 \leq \theta \leq 1/4$ we define

$$\Psi_\theta^{(\mathcal{C})}(y_i) \triangleq \begin{cases} 0, & if \quad |y_i| < \sqrt{\theta}, \\ \mathcal{E}, & if \quad \sqrt{\theta} \leq |y_i| \leq 1 - \sqrt{\theta}, \\ 1, & if \quad 1 - \sqrt{\theta} < |y_i|. \end{cases}$$

The arguments that follow are almost exactly the same as those employed in the discussion of the decoding of Y_2. However, there are a few subtle differences of importance. The first one of them is that the range of the parameter θ is different: the present range is $0 \leq \theta \leq 1/4$, while in our discussion of Y_2 we had the range $0 \leq \theta \leq 1$. We refer to Figure 11.7, where the various decoding regions defined by $\Psi_\theta^{(C)}$ are indicated. The reason for the new domain for θ is that the minimum distance in the present alphabet $\mathcal{L} = \{0, \pm 1\}$ is smaller than the minimum distance in the alphabet $\{\pm 1\}$ employed for Y_2. The choice of alphabet is, as we have seen, to a certain degree arbitrary. We could, of course, have chosen the alphabet for Y_3 so as to achieve the same minimum distance as in the previous case. That would have simplified somewhat the present argument, but it would have led to a disadvantage in the forthcoming discussion of decoding of $a \in \mathcal{A}$.

We also define

$$\begin{aligned} &t_\theta^{(C)}(c, y) \triangleq |\{i : \Psi_\theta^{(C)}(y_i) = c_i \oplus 1\}|, \\ &r_\theta^{(C)}(y) \triangleq |\{i : \Psi_\theta^{(C)}(y_i) = \mathcal{E}\}, \\ &0 \leq \theta \leq 1/4. \end{aligned}$$

The following result is proved in exactly the same way as was the corresponding result for Y_2:

Lemma 11.7.1 *For any $y = (y_1, y_2, \ldots, y_n) \in \mathbb{R}^n$ and for any $\theta \in [0, 1/4]$ there is at most one (and possibly none) $c \in C$ such that*

$$2t_\theta^{(C)}(c, y) + r_\theta^{(C)}(y) < d_C.$$

As before this lemma is needed in order to guarantee that the following family of decoders is well defined:

$$\Phi_\theta^{(C)}(y) \triangleq \begin{cases} c \in C, & if \quad 2t_\theta^{(C)}(c, y) + r_\theta^{(C)}(y) < d_C, \\ \mathcal{E}, & otherwise, \end{cases}$$

where $0 \leq \theta \leq 1/4$.

The next step is to demonstrate that at least one of these decoders will produce a correct decision provided $\|w\|^2 < d_{\mathcal{C}}/4$. For that purpose we need again a function which conveniently relates the size of the noise-vector w and the numbers of errors and erasures it causes. We define

$$\Gamma_\theta^{(\mathcal{C})}(w_i) \triangleq \begin{cases} 0, & if \quad |w_i| < \sqrt{\theta}, \\ 1, & if \quad \sqrt{\theta} \leq |w_i| \leq 1 - \sqrt{\theta}, \\ 2, & if \quad 1 - \sqrt{\theta} < |w_i|. \end{cases}$$

This is almost exactly the same as our previous definition (11.4.1), but not quite: the previous bound $2 - \sqrt{\theta}$ has been replaced by the new bound $1 - \sqrt{\theta}$. The reason is the new domain employed for θ. The following three lemmas are again proved in exactly the same way as were the corresponding earlier versions.

Lemma 11.7.2 *Let $x \in Y_3(\mathcal{C}, \mathcal{A})$ be generated by $(c, a) \in \mathcal{C} \times \mathcal{A}$. Then for all $w \in \mathbb{R}^n$ and for all $\theta \in [0, 1/4]$ the following inequality applies:*

$$\sum_{i=1}^{n} \Gamma_\theta^{(\mathcal{C})}(w_i) \geq 2t_\theta^{(\mathcal{C})}(c, x + w) + r_\theta^{(\mathcal{C})}(x + w).$$

Lemma 11.7.3 *For all $w = (w_1, w_2, \ldots, w_n) \in \mathbb{R}^n$ the following inequalities hold:*

$$\int_0^{1/4} \Gamma_\theta^{(\mathcal{C})}(w_i) d\theta \leq w_i^2, \quad i = 1, 2, \ldots, n.$$

Lemma 11.7.4 *Let $x \in Y_3(\mathcal{C}, \mathcal{A})$ be generated by $(c, a) \in \mathcal{C} \times \mathcal{A}$. If the inequality $2t_\theta^{(\mathcal{C})}(c, x + w) + r_\theta^{(\mathcal{C})}(x + w) \geq d_{\mathcal{C}}$ holds for all θ, $0 \leq \theta \leq 1/4$, then the noise-vector $w = (w_1, w_2, \ldots, w_n) \in \mathbb{R}^n$ satisfies the inequality $\|w\|^2 \geq d_{\mathcal{C}}/4$.*

The only essential difference in comparison to the previous lemmas is the appearance of the factor 1/4 at various places. The reason is always the same: the minimum distance in the alphabet $\mathcal{L} = \{0, \pm 1\}$ is 1; the square of half the minimum distance is 1/4.

Lemma 11.7.4 implies that whenever the noise w satisfies the inequality $\|w\|^2 < d_{\mathcal{C}}/4$ at least one of the decoders $\Phi_\theta^{(\mathcal{C})}$ will make a correct decision. Again the correct decision can be found by successively erasing the $d_{\mathcal{C}} - 1$ least reliable components in the received vector $y = (y_1, y_2, \ldots, y_n) \in \mathbb{R}^n$. Let us now see how $a \in \mathcal{A}$ can be decoded.

For the decoding of $a \in \mathcal{A}$ we will assume that $c \in \mathcal{C}$ has already been decoded—and that it was decoded correctly. If the decoding attempt for $c \in \mathcal{C}$ resulted in an erasure—i.e. if no decision was possible—we abstain from any further efforts regarding the decoding of $a \in \mathcal{A}$. If the decoding of $c \in \mathcal{C}$ was erroneous—this, of course, the decoder is necessarily unaware of—nevertheless we proceed according to the procedure to be described, but in those cases the decoding of $a \in \mathcal{A}$ will usually also be erroneous.

Assume that $c \in \mathcal{C}$ is available and denote by $supp(c)$ the set of indices i for which c_i is non-zero. Clearly $a \in \mathcal{A}$ depends only on those components y_i in $y = (y_1, y_2, \ldots, y_n) \in \mathbb{R}^n$ for which $i \in supp(c)$. Therefore, for $y \in \mathbb{R}^n$ such that the inequality

$$\min_{0 \le \theta \le 1/4} \{2t_\theta^{(\mathcal{C})}(c, y) + r_\theta^{(\mathcal{C})}(y)\} < d_{\mathcal{C}}$$

is satisfied for some $c \in \mathcal{C}$ and for $i \in supp(c)$ we define

$$\Psi_\theta^{(\mathcal{A},c)}(y_i) \triangleq \begin{cases} 0, & if \quad |y_i| < \sqrt{\theta}, \\ \mathcal{E}, & if \quad \sqrt{\theta} \le |y_i| \le 2 - \sqrt{\theta}, \\ 1, & if \quad 2 - \sqrt{\theta} < |y_i|, \end{cases}$$

$$0 \le \theta \le 1.$$

By a slight modification of Lemma 11.3.1 we have that the inequality

$$\min_{0 \le \theta \le 1/4} \{2t_\theta^{(\mathcal{C})}(c, y) + r_\theta^{(\mathcal{C})}(y)\} < d_{\mathcal{C}}$$

is satisfied for at most *one* codeword $c \in \mathcal{C}$. As a consequence we have that for any $y = (y_1, y_2, \ldots, y_n) \in \mathbb{R}^n$ at most *one* of the functions $\Psi_\theta^{(\mathcal{A},c)}$ is defined, and—of course—for many vectors y *none* of them is. Notice also that the present definition coincides almost exactly with the one we used in our discussion of decoding of Y_2. The reason is, of course, that in the decoding of $a \in \mathcal{A}$ the effective alphabet—i.e.

the subset $\mathcal{L} = \{0, \pm 1\}$ distinguished by the components in $a \in \mathcal{A}$—is exactly the same as the alphabet we used in our representation of Y_2, namely $\{\pm 1\}$. We also define

$$t_\theta^{(\mathcal{A},c)}(a,y) \triangleq |\{i \in supp(c) : \Psi_\theta^{(\mathcal{A},c)}(y_i) = a_{\varphi(c,i)} \oplus 1\},$$
$$r_\theta^{(\mathcal{A},c)}(y) \triangleq |\{i \in supp(c) : \Psi_\theta^{(\mathcal{A},c)}(y_i) = \mathcal{E}\}.$$

The new versions of the lemmas we need read as follows.

Lemma 11.7.5 *For any $y \in \mathbb{R}^n$ such that*

$$\min_{0 \le \theta_c \le 1/4} \{2t_{\theta_c}^{(\mathcal{C})}(c,y) + r_{\theta_c}^{(\mathcal{C})}(y)\} < d_{\mathcal{C}}$$

for some $c \in \mathcal{C}$ there is, for any θ_a, $0 \le \theta_a \le 1$, at most one—and possibly none—$a \in \mathcal{A}$ such that

$$2t_{\theta_a}^{(\mathcal{A},c)}(a,y) + r_{\theta_a}^{(\mathcal{A},c)}(y) < d_{\mathcal{A}}.$$

By this lemma the following definition is valid for any $y \in \mathbb{R}^n$ such that

$$\min_{0 \le \theta_c \le 1/4} \{2t_{\theta_c}^{(\mathcal{C})}(c,y) + r_{\theta_c}^{(\mathcal{C})}(y)\} < d_{\mathcal{C}}$$

for some $c \in \mathcal{C}$:

$$\Phi_{\theta_a}^{(\mathcal{A},c)}(y) \triangleq \begin{cases} a \in \mathcal{A}, & if \quad 2t_{\theta_a}^{(\mathcal{A},c)}(a,y) + r_{\theta_a}^{(\mathcal{A},c)}(y) < d_{\mathcal{A}}, \\ \mathcal{E}, & otherwise, \end{cases}$$

where $0 \le \theta_a \le 1$.

Next we define

$$\Gamma_{\theta_a}^{(\mathcal{A})}(w_i) \triangleq \begin{cases} 0, & if \quad |w_i| < \sqrt{\theta_a}, \\ 1, & if \quad \sqrt{\theta_a} \le |w_i| \le 2 - \sqrt{\theta_a}, \\ 2, & if \quad 2 - \sqrt{\theta_a} < |w_i|, \end{cases}$$

$0 \le \theta_a \le 1$.

Notice that this definition coincides exactly with the definition (11.4.1) used in our discussion of decoding for Y_2. The corresponding lemmas are as follows.

Lemma 11.7.6 *For any $x \in Y_3(\mathcal{C}, \mathcal{A})$ corresponding to a pair $(c, a) \in \mathcal{C} \times \mathcal{A}$, for any $w \in \mathbb{R}^n$, and for any $\theta_a \in [0, 1]$:*

$$\sum_{i=1}^{n} \Gamma_{\theta_a}^{(\mathcal{A})}(w_i) \geq 2t_{\theta_a}^{(\mathcal{A},c)}(a, x + w) + r_{\theta_a}^{(\mathcal{A},c)}(x + w).$$

Combining now these new lemmas with Lemma 11.7.4 we obtain the following result:

Lemma 11.7.7 *Let $x \in Y_3(\mathcal{C}, \mathcal{A})$ be generated by $(c, a) \in \mathcal{C} \times \mathcal{A}$ and suppose*

$$2t_{\theta_c}^{(\mathcal{C})}(c, x + w) + r_{\theta_c}^{(\mathcal{C})}(x + w) < d_{\mathcal{C}}$$

holds for some $c \in \mathcal{C}$ and some θ_c, $0 \leq \theta_c \leq 1/4$. Also suppose that the inequality

$$2t_{\theta_a}^{(\mathcal{A},c)}(a, x + w) + r_{\theta_a}^{(\mathcal{A},c)}(x + w) \geq d_{\mathcal{A}}$$

holds for all θ_a, $0 \leq \theta_a \leq 1$. Under these circumstances the noise-vector $w = (w_1, w_2, \ldots, w_n) \in \mathbb{R}^n$ satisfies the inequality

$$\|w\|^2 \geq d_{\mathcal{A}}.$$

Combining the above lemmas with the results derived in Section 11.3 we are led to the following decoding algorithm for Y_3.

Decoding $Y_3(\mathcal{C}, \mathcal{A})$:

1. First decode $c \in \mathcal{C}$ by error-erasure decoding, where the $d_{\mathcal{C}} - 1$ least reliable components in the received vector $y \in \mathbb{R}^n$ are successively erased. For each component y_i reliability is determined by its distance to the closest point in the alphabet $\mathcal{L} = \{0, \pm 1\}$. The selection between different tentative decisions for $c \in \mathcal{C}$ should be based on the criterion function $\min\{\|y - x\|^2\}$, where for given c and y the minimization is over all $x \in Y_3(\mathcal{C}, \mathcal{A})$ corresponding to different possible choices of $a \in \mathcal{A}$. Clearly $c \in \mathcal{C}$ should be chosen so as to minimize $\min\{\|y - x\|^2\}$.

2. If the decoding of $c \in \mathcal{C}$ fails, i.e. if it leads to an erasure, stop the decoding efforts. If the decoding succeeds—i.e. if a vector $c \in \mathcal{C}$ satisfying $2t_{\theta_c}^{(\mathcal{C})}(c, x + w) + r_{\theta_c}^{(\mathcal{C})}(x + w) < d_\mathcal{C}$ is found—then perform a new error-erasure decoding for $a \in \mathcal{A}$ by successively erasing the $d_\mathcal{A} - 1$ least reliable components in $supp(c)$. Again the selection between different tentative decisions should be based on the criterion function $\|y - x\|^2$, but this time x is always computed using the $c \in \mathcal{C}$ we have already decided upon.

We have deliberately left out a number of logical steps, trying to emphasize the principal ideas rather than the mathematical details. It appears that clarity in these matters is more related to finding appropriate notation and distinguishing between various cases rather than in mastering the mathematics needed for the analysis. The details left out should be easy to fill in.

We summarize our observations by noticing the above algorithm decodes correctly whenever the inequality $\|w\|^2 < \min\{d_\mathcal{C}/4, d_\mathcal{A}\}$ is satisfied. This is precisely the product bound for the minimum distance of $Y_3(\mathcal{C}, \mathcal{A})$. Thus by applying the *GMD* algorithm successively first for $c \in \mathcal{C}$ and then for $a \in \mathcal{A}$ we are guaranteed to obtain correct decoding up to half the estimated minimum distance.

Appendix A

Algebraic codes and designs

For the convenience of the reader we give a brief description of some of the most important ones of the codes and designs used in the constructions of spherical codes.

1. Binary unrestricted codes

Tables A1-A2 give the parameters (n, d, M) for the best binary codes known in the range $5 \leq n \leq 24$, $d = 4, 6, 8, 10, 12, 14$. The parameters indicate *length*, *size* and *Hamming distance*. The list is based upon Brouwer et al [51], see also Conway–Sloane, [71], p. 248. We will indicate briefly the various constructions leading to codes with these parameters. It is well known that a binary code with parameters (n, d, M), d odd, exists if and only if a code with parameters $(n + 1, d + 1, M)$ exists. Therefore we consider only even distances. The case $d = 2$ is trivial.

A binary code which is described as a linear subspace in the space $\mathbb{F}_2^n$ is called a *linear* code. The size M of such a code is always a power of 2, $M = 2^k$ for some k, $0 \leq k \leq n$. The exponent k is called the *dimension* of the code. For linear codes we use the notation $[n, k, d]$ to indicate a code of length n, minimum distance d and size $M = 2^k$.

1.1 Hadamard and conference matrix codes:

For small lengths all the best codes known are generated from so called *Hadamard* matrices. A square matrix H is said to be Hadamard if all entries are $+1$ or -1 and if the rows are *orthogonal*, i.e. if the matrix satisfies the condition $HH^T = nI$, where T denotes transponate, n is the order of the matrix and I denotes the $n \times n$ identity matrix. It is easy to show that with the exception of the two cases $n = 1$ and $n = 2$ an Hadamard matrix can exist only if n is a multiple of four. For the cases $n = 1$ and $n = 2$ we have

$$H_1 = [1], \quad H_2 = \begin{bmatrix} 1 & 1 \\ 1 & -1 \end{bmatrix}.$$

Both of these matrices are Hadamard. For larger values of n, n a multiple of four, there are many different constructions producing Hadamard matrices. It is a famous conjecture that a Hadamard matrix exists whenever n is a multiple of four, but this still has not been proved. For our purposes two different constructions will be enough. Both give infinite sequences of Hadamard matrices. In particular they produce Hadamard matrices for all orders $n = 1, 2, 4, 8, 12, \ldots, 32$.

The Sylvester construction: Let H_1 and H_2 be as above and for any order n of the form $n = 2^m$, $m = 2, 3, 4, \ldots$, define H_n recursively according to the formula

$$H_{2n} = \begin{bmatrix} H_n & H_n \\ H_n & -H_n \end{bmatrix}.$$

It is easy to see that H_1, H_2, H_4, H_8, $H_{16}, \ldots$ is an infinite sequence of Hadamard matrices.

The Paley construction: Let $q = p^m$ be an odd prime power and let $\mathbb{F}_q = \{\alpha_1, \alpha_2, \ldots, \alpha_q\}$ be a finite field. Denote by $\mathcal{Q}$ the set of squares (quadratic residues) and denote by $\mathcal{N}$ the set of non-squares (non-residues; the element 0 is classified as neither a square nor a non-square). The Legendre symbol $\chi(\alpha_i)$ is defined as follows:

$$\chi(\alpha_i) = \begin{cases} 0, & if \quad \alpha_i = 0, \\ 1, & if \quad \alpha_i \in \mathcal{Q}, \\ -1, & if \quad \alpha_i \in \mathcal{N}. \end{cases}$$

The matrix $Q = [Q_{ij}]$, with $Q_{ij} \triangleq \chi(\alpha_i - \alpha_j)$, is called the *Jakobsthal* matrix. We define

$$H \triangleq \begin{bmatrix} 1 & \mathbf{j} \\ \mathbf{j}^T & Q - I \end{bmatrix}.$$

where $\mathbf{j}$ denotes a row-vector of q ones and where I is an $q \times q$ identity matrix. It is well known (MacWilliams-Sloane [167], p. 47) that this construction produces a Hadamard matrix whenever $q + 1$ is a multiple of four, i.e. when $q = p^m \equiv 3 \bmod 4$. The first few cases are $n = 4, 8, 12, 20, 24, 28, 32$.

Several different classes of binary codes can be obtained from the Hadamard matrices. They are all known as *Hadamard codes*. Let d be even and suppose there is a Hadamard matrix H of order $n = 2d$. Let each row in H generate two codewords in $\mathbb{F}_2^n$ by replacing the elements $+1$ and -1 by either of the two possible rules:

$$\begin{aligned} +1 &\mapsto 0, \qquad & +1 &\mapsto 1, \\ -1 &\mapsto 1, \qquad & -1 &\mapsto 0. \end{aligned}$$

The resulting binary code will be denoted $\mathcal{C}_d$. It has parameters $(n, d, M) = (2d, d, 4d)$. It is also known as the *Biorthogonal* code. In this code half of the codewords begin with 0 and one fourth of the codewords begin with 00. By taking only the subsets of codewords in $\mathcal{C}_d$ specified by these conditions and deleting the initial one or two zeros we obtain two further codes $\mathcal{A}_d$ and $\mathcal{B}_d$ with parameters $(n, d, M) = (2d - 2, d, d)$, $(2d - 1, d, 2d)$ respectively.

It turns out that all codes in the range $n \leq 2d$ of Tables A1-A2 can be generated form these three classes of Hadamard codes. The key idea is a simple construction by which a new code $\mathcal{C}$ with parameters (n, d, M) is generated from two codes $\mathcal{C}_1$ and $\mathcal{C}_2$ with parameters (n_1, d_1, M_1) and (n_2, d_2, M_2) respectively. Let $M = \min\{M_1, M_2\}$. The codewords in $\mathcal{C}$ are obtained by forming M binary sequences of length $n = n_1 + n_2$ with a prefix from $\mathcal{C}_1$ and a suffix from $\mathcal{C}_2$ in such a way that a given prefix is always followed by the same suffix and vice versa. It is clear that the resulting code $\mathcal{C}$ has parameters $(n, d, M) = (n_1 + n_2, d \geq d_1 + d_2, \min\{M_1, M_2\})$. We use the notation $\mathcal{C} = \mathcal{C}_1 + \mathcal{C}_2$ in order to indicate that $\mathcal{C}$ is generated from $\mathcal{C}_1$ and $\mathcal{C}_2$ by this construction. A sum of m copies of the same code $\mathcal{C}$ is denoted $m\,\mathcal{C}$.

In Tables A1-A.2 we have indicated how the codes in the range $n \leq 2d$ can be obtained in this way from the Hadamard codes. It turns out that all these codes are optimal inasmuch as they meet the *Plotkin bound*. This bound says that the maximal number $A(n,d)$ of codewords in a code of length n and with even distance d satisfies the following inequality:

$$A(n,d) \leq \begin{cases} 2\left\lfloor \dfrac{d}{2d-n} \right\rfloor, & if \quad n < 2d, \\ \\ 4d, \quad if \quad n = 2d. \end{cases}$$

Actually, by Levenshtein's theorem (MacWilliams–Sloane, [167], p. 49) codes meeting the Plotkin bound can be obtained for all $n \leq 2d$ by the above construction, provided only that the conjecture on existence of Hadamard matrices is correct.

A conference matrix of order r is an $r \times r$ matrix C with 0 in the diagonal positions and with $+1$ or -1 in all other positions. In addition the matrix C must satisfy the following condition:

$$CC^T = (r-1)I.$$

Without loss of generality we may assume it to have the form

$$C = \begin{bmatrix} 0 & \mathbf{j} \\ \mathbf{j}^T & S. \end{bmatrix}.$$

In terms of the matrix S the defining condition takes the following form:

$$SS^T = (r-1)I - J, \quad SJ = 0,$$

where I is the identity matrix and J is a square matrix with $+1$ in all positions. We use "0" to denote an all-zero matrix.

Suppose C and S are symmetric. Then $r \equiv 2 \bmod 4$ (MacWilliams–Sloane, [167], p. 55) or—what is the same—$d \triangleq (1/2)(r-2)$ is even. Moreover, the matrices $(1/2)(S+I+J)$ and $(1/2)(-S+I+J)$ have all entries equal to 0 or 1. Taking all $2 \times (r-1)$ rows in these two matrices and in addition the all zero vector and the all one vector we obtain a code $\mathcal{CM}_d$ with parameters $(n,d,M) = (r-1,(r-2)/2,2r)$.

Alternatively we may write $(n, d, M) = (2d+1, d, 4d+4)$, with d even.

Codes generated from conference matrices in this way are referred to as *conference matrix codes.* When they exist they offer codes one unit longer than the longest Hadamard code of the same distance, with four additional codewords.

For $q \equiv 1 \bmod 4$, q a power of a prime, a possible choice for the matrix S is $S = Q$. This construction is due to Paley.

The first few conference matrix codes have parameters $(n, d, M) = (5, 2, 12)$, $(9, 4, 20)$, $(13, 6, 28)$, $(17, 8, 36)$. From Tables A1-A2 we see that two of these are best codes known.

1.2 Hamming codes:

Let $\mathbb{F}_2$ be the binary field, let $n = 2^m - 1$, $m = 2, 3, 4, \ldots$, and let $H = [h_1, h_2, \ldots, h_n]$ be an $m \times n$ matrix with columns h_i consisting of all possible different non-zero column vectors of length m. It is well known—and easy to show—that

$$\mathcal{H}_m \triangleq \{x \in \mathbb{F}_2^n : Hx^T = 0\}$$

is a *linear* code with parameters $[n, k, d] = [2^m - 1, 2^m - m - 1, 3]$. It is known as the *Hamming code.* Adding an extra parity check symbol we get the *extended Hamming code* $\mathcal{H}_m^*$. It has the following parameters:

$$\mathcal{H}_m^* : \ [n, k, d] = [2^m, 2^m - m - 1, 4], \quad m = 2, 3, \ldots .$$

For $m = 2, 3, 4$ we get $[n, k, d] = [4, 1, 4]$, $[8, 4, 4]$, $[16, 11, 4]$, respectively. The first two codes are identical with the Hadamard codes $2\mathcal{A}_2$ and $\mathcal{C}_4$ respectively.

All Hamming codes are best possible. The shorter versions, having minimum distance $d = 3$, satisfy the sphere packing bound with equality and are examples of so called *perfect codes.*

Any linear code has half of its codewords beginning with 0 and half of them beginning with 1. Taking any of these sets of codewords and deleting the initial symbol we get a new code one unit shorter and with half the size. The code obtained from the codewords beginning with 0 in the original code is again linear. In this way we get from the even distance Hamming codes the following complete set of codes with distance $d = 4$:

$$[n, k, d] = [n, n - m - 1, 4], \quad 2^{m-1} + 1 \leq n \leq 2^m, \quad m = 2, 3, 4, \dots .$$

These are the best *linear* codes with distance 4. However, for $2^{m-1} + 1 \leq n \leq 2^{m-1} + 2^{m-2}$ we can always find better *nonlinear* codes. In Table A1.1 we see that the Hamming parameters occur for $5 \leq n \leq 8$ and $13 \leq n \leq 16$, while for all other lengths we have larger sizes, reached by various non-linear codes. We now discuss how some of these can be obtained.

1.3 Some nonlinear codes with distance $d = 4$:

Consider again the Hadamard code $\mathcal{C}_6$ with parameters $(n, d, M) = (12, 6, 24)$. It has a subcode with parameters $(n, d, M) = (12, 6, 12)$ obtained by selecting only those codewords having an initial 0. The modulo 2 sum of any pair of different codewords in this subcode is a vector of weight 6. Taking all these pairs and all complements to such pairs we get a code S_{12} with parameters $(n, d, M) = (12, 4, 132)$, and with the property that all codewords have weight 6. This is already better than the parameters $(12, 4, 128)$ of the shortened Hamming code of the same length. However, further improvements are possible. It is easy to see that there are codes with parameters $(n, d, M) = (12, 4, 6)$ and with all codewords having weight 2 or 10 respectively. Both these codes have distance 4 to S_{12}, while they have distance 8 to each other. Thus, taking the union of all three codes we get a nonlinear code $\mathcal{C}_{12}^*$ with parameters $(n, d, M) = (12, 4, 144)$.

By appropriately shortening the code $\mathcal{C}_{12}^*$ we get codes $\mathcal{C}_{11}^*$, $\mathcal{C}_{10}^*$, $\mathcal{C}_9^*$ with parameters $(n, d, M) = (11, 4, 72)$, $(10, 4, 38)$, $(9, 4, 20)$. From Table A1.1 we see that two of these actually are best possible. The last code has the same parameters as one of the conference matrix codes we discussed above.

The code $\mathcal{C}_{12}^*$ is uniquely determined by the above construction, but for the remaining codes there are a few slightly different possibilities.

Possible weight distributions are as follows:

$$\mathcal{C}_{12}^*: \ 2^6 6^{132} 10^6, \qquad \mathcal{C}_{11}^*: \ 1^1 5^{66} 9^5,$$
$$\mathcal{C}_{10}^*: \ 1^1 5^{36} 9^1, \qquad \mathcal{C}_{9}^*: \ 1^1 5^{18} 9^1.$$

For length $n = 10$ an even better code is available. Best, [20] found a code with parameters $(n, d, M) = (10, 4, 40)$. It can be obtained as the union of the cyclic shifts of the vectors 01 00 00 00 11, 00 11 11 11 01, 11 00 10 11 00, 00 01 01 01 11. Although nonlinear it has the remarkable property that all codewords have the same distance distribution $0^1 4^{22} 6^{12} 8^5$. It has been shown that no binary code with parameters $(n, d) = (10, 4)$ can have more than 40 codewords, so this code is really the "Best" code.

It has recently been established, [177], that also the codes with parameters given by $(n, d, M) = (11, 4, 72)$ and $(n, d, M) = (12, 4, 144)$ are optimal.

1.4 The Nordstrom–Robinson code:

The *Nordstrom–Robinson* code $\mathcal{NR}$, [175], (the same code was independently found also by Semakov–Zinoviev, [198]), has parameters $(n, d, M) = (16, 6, 256)$. Although the size is a power of 2 this is *not* a linear code. It has recently been observed, however, that the code has a simple description as a *group code* over the alphabet $\mathbb{Z}_4$. This fact explains the remarkable property that all codewords have the same distance distribution, $0^1 6^{112} 8^{30} 10^{112} 16^1$, (first pointed out in [198], see also [200]). Shortening the code we get a sequence of nonlinear codes with parameters $(n, d, M) = (15, 6, 128)$, $(14, 6, 64)$, $(13, 6, 32)$. All of these codes—as well as the Nordstrom–Robinson code itself—are optimal, see [138]. The last code in this series outperforms the conference matrix code with parameters $(13, 6, 28)$.

The Hamming code $\mathcal{H}_4^*$ with parameters $(16, 4, 2048)$ has the code $\mathcal{C}_8$ with parameters $(16, 8, 32)$ as a linear subcode (both of these can be described as Reed–Muller codes). There are $2^{11-5} = 64$ cosets of $\mathcal{C}_8$ in H_4^*. The Nordstrom–Robinson code can be formed as the union of $\mathcal{C}_8$ and seven of these cosets, i.e. as the union of eight different versions of $\mathcal{C}_8$, [198]. Furthermore, it is easy to see that we can form

8 disjoint such unions of cosets in $\mathcal{H}_4^*$, so the code $\mathcal{H}_4^*$ is a union of 8 disjoint versions of the Nordstrom–Robinson code:

$$\mathcal{H}_4^* = \bigcup_{i=1}^{8} \{\mathcal{NR} + a_i\}$$

for some appropriately selected set $\{a_1, a_2, \dots, a_8\} \subseteq \mathcal{H}_4^*$, [198]. If we add a tail from the simple parity check code with parameters $(n, d, M) = (4, 2, 8)$ to each codeword in $\mathcal{H}_4^*$ so that all codewords c in the same subcode $\mathcal{NR} + a_i$ get the same tail we increase the length by 4 and the minimum distance by 2, thus generating a code $\mathcal{H}\backslash\mathcal{NR}$ with parameters $(n, d, M) = (20, 6, 2048)$, [232]. This code is also best known. By successively shortening this code we generate a sequence of codes with parameters $(n, d, M) = (19, 6, 1024)$, $(18, 6, 512)$, $(17, 6, 256)$, all of them best known, [232].

The Nordstrom–Robinson code is a first member of two infinite sequences of codes: the *Preparata* codes, $\mathcal{P}(m)$, with parameters

$$\mathcal{P}(m): \; (n, d, M) = (2^m, 6, 2^{(2^m - 2m)}),$$

and the *Kerdock* codes, $\mathcal{K}(m)$, with parameters

$$\mathcal{K}(m): \; (n, d, M) = (2^m, 2^{m-1} - 2^{(m-2)/2}, 2^{2m}).$$

In both cases we have $m = 4, 6, 8, \dots$. All of these codes (or closely related codes with the same parameters) have recently been found to be *group codes* with very simple descriptions over the ring $\mathbb{Z}_4$, [123]. This fact explains the remarkable observation—first made in [200]—that the Preparata and the Kerdock codes are *formally dual* to each other. The Nordstrom–Robinson code is *formally self-dual.*

1.5 The Golay code:

The binary *Golay* code $\mathcal{G}_{24}$ is a linear code with parameters $[n, k, d] = [24, 12, 8]$. There are many different constructions. One of the simplest is the *Turyn* construction, [224]. Let $\mathcal{H}_3^*$ be the extended Hamming code of length 8 and denote by $\bar{\mathcal{H}}_3^*$ the code obtained by reversing all codewords in $\mathcal{H}_3^*$. The Golay code can be defined as follows (see for instance MacWilliams–Sloane, [167], p. 588, or van Lint, [162], p. 43)

$$\mathcal{G}_{24} = \{(a+x, b+x, a+b+x) \in \mathbb{F}_2^{24} : a, b \in \mathcal{H}_3^*,\ x \in \bar{\mathcal{H}}_3^*\}.$$

The odd distance version of the Golay code has parameters $[n,k,d] = [23,12,7]$. This code along with the Hamming codes are the only nontrivial perfect binary codes (the trivial ones are the whole space, a single point, and the binary repetition code of odd length).

The Nordstrom–Robinson code can be obtained from the Golay code by deleting eight coordinates from an appropriately selected subset of the codewords in $\mathcal{G}_{24}$, [198].

The Golay code and the shortened linear codes obtained from it constitute a sequence of very good codes. The seven first members in this class are best codes known. The parameters are $[n,k,d] = [24-i, 12-i, 8]$, $i = 0, 1, \ldots, 6$. The four first codes have been proved to be best possible. For these codes the weight distributions are as follows:

n	k_0	k_8	k_{12}	k_{16}	k_{24}
24	1	759	2576	759	1
23	1	506	1288	253	
22	1	330	616	77	
21	1	210	280	21	

1.6 The Reed–Muller codes

The Reed–Muller codes constitute a large class of codes, among which several very good codes can be found. In addition the properties are comparatively well understood for most of the codes in this class. Several definitions are possible. A simple and compact one is the following.

Let $\mathbb{F}_2^m = \{a_1, a_2, \ldots, a_n\}$, $n = 2^m$ denote the space of all binary m-tuples, with the elements 0 and 1 interpreted as elements of the binary field $\mathbb{F}_2$, and denote by $P(r,m)$ the set of all polynomials of degree at most r in m variables over $\mathbb{F}_2$. The Reed–Muller code $\mathcal{R}(r,m)$ is obtained as the evaluation space of $P(r,m)$ over $\mathbb{F}_2^m$:

$$\mathcal{R}(r,m) = \{(f(a_1), f(a_2), \ldots, f(a_n)) \in \mathbb{F}_2^n : f \in P(r,m)\},$$

$$0 \leq r \leq m.$$

The code $\mathcal{R}(r,m)$ is a linear code with parameters

$$[n,k,d] = [2^m, k(r,m), 2^{m-r}],$$

where

$$k(r,m) = \sum_{i=0}^{r} \binom{m}{i}$$

is the dimension of the code.

The Reed–Muller codes have a number of interesting properties. In particular they are nested: $\mathcal{R}(r,m)$ is a subcode in $\mathcal{R}(r+1,m)$. Moreover, $\mathcal{R}(m-r-1,m)$ is the dual of $\mathcal{R}(r,m)$, $0 \leq r \leq m-1$, (MacWilliams–Sloane, [167], p. 375).

An alternative, iterative definition is as follows:

$$\mathcal{R}(r+1,m+1) = \{(u,u+v) : u \in \mathcal{R}(r+1,m), v \in \mathcal{R}(r,m)\}.$$

Some of the codes we have already discussed can also be seen as Reed-Muller codes: $\mathcal{C}_d = \mathcal{R}(1,m)$ for $d = 2^{m-1}$ (provided $\mathcal{C}_d$ is linear), $\mathcal{H}_m^* = \mathcal{R}(m-2,m)$, $m = 2,3,\ldots$.

1.7 Other binary codes:

The reader might have observed that we have not mentioned one of the most famous classes of codes: the *BCH* codes. They are not forgotten, but we feel that they are best defined as subfield subcodes of Reed-Solomon codes. Therefore we will discuss them in connection with non-binary codes. Moreover, the best *BCH* codes occur at large code lengths. For a discussion of Tables A1-A.2 we can actually do entirely without them.

2. Binary constant weight codes

A code with the property that all codewords have the same weight is called a *constant weight* code. Many constructions of spherical codes rely on such codes. Brouwer et al, [51], p. 1337, contains an extensive list of best known binary constant weight codes. Our list over spherical codes relies heavily on this list. We describe a few of the most important constant weight codes.

The set $J(n,w) \subseteq \mathbb{F}_2^n$ of all binary vectors of length n and weight w is called the *Johnson space.* Any subset $\mathcal{C}$ in $J(n,w)$ is an (n,w)-constant weight code. As with unrestricted codes we are interested in the size $T = |\mathcal{C}|$ and the minimum distance $d = \min\{d_H(x,y) : x \neq y;\ x,y \in \mathcal{C}\}$. We characterize the code $\mathcal{C}$ with the quadruple (n,w,d,T).

The *correlation* $\lambda(x,y)$ between two vectors $x,y \in J(n,w)$ is the number of positions i such that $x_i = y_i = 1$. We have $d_H(x,y) = 2w - 2\lambda(x,y)$, so the distance is always an *even* number.

We notice the obvious fact that to any code with parameters (n,w,d,T) there is another code with parameters $(n,n-w,d,T)$ obtained by reversing zeros and ones in the first code.

Denote by $T(n,w,d)$ the maximal possible size of a constant weight code with parameters (n,w,d). There are two important simple bounds on this function, known as Johnson's first and second bounds

Johnson 1:

$$T(n,w,d) \leq \left\lfloor \frac{nd}{nd - 2w(n-w)} \right\rfloor, \quad n < \frac{2w^2}{2w-d}.$$

Johnson 2:

$$T(n,w,d) \leq \frac{\binom{n}{w-d/2+1}}{\binom{w}{w-d/2+1}}.$$

(Both bounds can actually be tightened slightly, see MacWilliams–Sloane, [167], pp. 526-527).

The first bound is the counterpart in $J(n,w)$ to the Plotkin bound in $\mathbb{F}_2^n$. We consider two special cases. For $d = 2w$ the bound takes the form $T \leq \lfloor n/w \rfloor$. In this case the codewords are orthogonal and it is easy to see that we can always find codes satisfying the bound with equality. Thus we have

$$T(n, w, 2w) = \lfloor n/w \rfloor.$$

By deleting the all-zero codewords from the Hadamard codes $\mathcal{A}_d$ and $\mathcal{B}_d$ we get constant weight codes with parameters $(n, w, d, T) = (2d-2, d, d, d-1)$, $(2d-1, d, d, 2d-1)$. By reversing zeros and ones we get codes with parameters $(n, w, d, T) = (2d-2, d-2, d, d-1)$, $(2d-1, d-1, d, 2d-1)$. By a slight modification of the code A_d we get a constant weight code with parameters $(n, w, d, T) = (2d-2, d-1, d, d)$. All of these satisfy the first Johnson bound with equality. By deleting the all-zero codeword and the all-one codeword from the Hadamard code $\mathcal{C}_d$ we get a constant weight code with parameters $(n, w, d, T) = (2d, d, d, 4d-2)$. Codes with these parameters exist by the Sylvester construction whenever d is a power of 2 and by the Paley construction whenever $2d-1$ is a prime power.

In a similar way we obtain constant weight codes from conference matrix codes by deleting the all-zero and the all-one codewords. The parameters are $(n, w, d, T) = (2d+1, d+1, d, 4d+2)$. By reversing zeros and ones we alternatively get codes with parameters $(n, w, d, T) = (2d+1, d, d, 4d+2)$. These codes exist whenever $2d+1$ is a prime power.

Codes satisfying the second bound with equality are of particular interest to us. They correspond to so called *Steiner systems.* A definition is as follows.

Let $X = \{x_1, x_2, \ldots, x_v\}$ be an arbitrary set with v elements and let $S = \{B_1, B_2, \ldots, B_b\}$ be a family of b subsets, each containing exactly k elements from X. The family S is a Steiner system $S(v, k, t)$ if any subset $A \subseteq X$ of size $|A| = t \geq 2$ is contained in one and only one of the sets B_i. Defining $c(B_i) = (c_1(B_i), c_2(B_i), \ldots, c_v(B_i)) \in \mathbb{F}_2^v$ by the rule

$$c_j(B_i) = \begin{cases} 1, & if \quad x_j \in B_i, \\ 0, & if \quad x_j \notin B_i, \end{cases}$$

we see that $\mathcal{C} = \{c(B_1), c(B_2), \ldots, c(B_b)\}$ is a constant weight code with parameters $(n, w, d, T) = (v, k, 2(k-t+1), b)$.

It can be shown that whenever a Steiner system $S(v, k, t)$ exists the parameters (v, k, t, b) are related as follows:

$$b\binom{k}{t} = \binom{v}{t}.$$

Inserting the parameters for the induced constant weight code we see that it satisfies Johnson's second bound with equality. Conversely it is easy to see that any constant weight binary code satisfying Johnson's second bound with equality induces a Steiner system, [198], [199].

It is clear that for given values of (v, k, t) a Steiner system $S(v, k, t)$ can exist only if the number b obtained from the equation above is an integer. However, although necessary this condition is certainly not sufficient. For instance, the quadruple $(v, k, t, b) = (15, 5, 4, 273)$ satisfies the equation, but no Steiner system $S(15, 5, 4)$ exists, [70].

It is easy to see that whenever a system $S(v, k, t)$, $t \geq 3$ exists, so does a system $S(v - 1, k - 1, t - 1)$. It is clear that for non-trivial systems we always have $2 \leq t < k < v$.

Table A3 contains a complete list of all known Steiner systems of size $v \leq 24$, along with an indication of all undecided cases in this range. We see that there are nine different families, six of which are closed in the sense that the chain of systems cannot be further extended. For three of the families the question of possible extensions is undecided. For instance, the system $S(24, 6, 5)$ exists but $S(37, 19, 18)$ does not. Between these extremes there are 12 quadruples of parameters satisfying the equation for which it is not known whether a Steiner system exists or not. The same is true for the extreme pairs $\{(20, 4, 3),\ (33, 17, 16)\}$ and $\{(16, 4, 3),\ (25, 13, 12)\}$.

The known Steiner systems with $v \leq 24$ give rise to constant weight codes with parameters as shown in Table A4.

There are certain relations to the unrestricted codes we discussed above. Most remarkable is the large number of best constant weight codes that can be found in the Golay code and the first few shortened versions of it. The codewords of weights $w = 8, 12, 16$ in the codes of lengths $n = 21, 22, 23, 24$ constitute constant weight codes of largest sizes known. Reversing zeros and ones in the sets of codewords of weight $w = 16$ we get all the constant weight codes corresponding to the Steiner family $S(21+r, 5+r, 2+r)$, $0 \leq r \leq 3$.

The Steiner systems $S(8,4,3)$ and $S(16,4,3)$ generate constant weight codes that can be found as the sets of codewords of weight 4 in the Hamming codes $\mathcal{H}_3^*$ and $\mathcal{H}_4^*$ respectively. The Steiner system $S(12,6,5)$ is induced by the constant weight code S_{12} used in the construction of the code $\mathcal{C}_{12}^*$. In the same way the Steiner system $S(11,5,4)$ is induced by the codeword of weight 5 in the shortened code $\mathcal{C}_{11}^*$. Both occur also in the ternary Golay codes as supports of codewords.

One further constant weight code deserves to be mentioned. It has parameters $(n, w, d, T) = (16, 6, 6, 112)$ and is largest possible, given length, weight and minimum distance, ([51]). It is formed by all codewords of weight 6 in the Nordstrom–Robinson code (represented in a way so as to contain the all-zero codeword), and was first described by Semakov and Zinoviev, [198]. These authors also showed the remarkable fact that this code can be partitioned into seven so called *orthogonal 2-designs*. The designs constitute constant weight codes with parameters $(n, w, d, T) = (16, 6, 8, 16)$. They satisfy the first Johnson bound with equality, and so are also optimal. We use this partitioning in several constructions for spherical codes.

3. Non-binary codes

Although in general non-binary codes are much less understood than binary codes there are several known classes of good non-binary codes. We mention a few.

3.1 Non-binary Hamming codes:

The Hamming codes can be constructed over any field. For the finite field $\mathbb{F}_q$, (with q a power of a prime), let H be a matrix over $\mathbb{F}_q$ containing all possible non-zero column vectors of length m, with the restriction that the first non-zero entry in each column must be 1. There are $n = (q^m - 1)/(q-1)$ such m-vectors, and the code

$$\mathcal{H}_{q,m} = \{c \in \mathbb{F}_q^n : Hc^T = 0\}$$

is called a *q-ary Hamming code.* It is a linear perfect code with parameters $[n,k,d]_q = [(q^m-1)/(q-1), n-m, 3]_q$.

Of particular interest for the construction of spherical codes is the ternary code of length $n=4$. The parameters are $[n,k,d]_q = [4,2,3]_3$. The non-zero codewords are

$$\begin{matrix} 0111 & 1012 & 1201 & 1120 \\ 0222 & 2021 & 2102 & 2210 \end{matrix}$$

The complete weight enumerator is

$$W(x,y,z) = x(x^3+y^3+z^3) + 3xyz(y+z),$$

while the Hamming weight enumerator is

$$W(x,y) = x^4 + 8xy^3.$$

This code is sometimes referred to as the *Tetracode.*

3.2 The ternary Golay codes:

There is also a pair of ternary *Golay* codes: a perfect two-error-correcting code with parameters $[n,k,d]_q = [11,6,5]_3$ and an extended version with parameters $[n,k,d]_q = [12,6,6]_3$. Both are linear. The extended code is self-dual. The complete weight enumerator is

$$W(x,y,z) = x^{12} + y^{12} + z^{12} + 22(x^6y^6 + x^6z^6 + y^6z^6) +$$
$$+ 220x^3y^3z^3(x^3+y^3+z^3).$$

The Steiner systems $S(11,5,4)$ and $S(12,6,5)$ occur as the supports of the codewords of weight 5 and 6 respectively in the Golay codes $[11,6,5]_3$ and $[12,6,6]_3$.

3.3 MDS codes and the Singleton bound:

Let C be a code with parameters $(n,d,M)_q$. The *Singleton bound* says that the following inequality must be satisfied:

$$M \le q^{n-d+1}.$$

Codes satisfying this bound with equality are said to be *maximum distance separable*, (MDS). In all known cases where an MDS code exists there is always a *Reed-Solomon* code (or a slightly modified such code) with the same parameters. The Reed-Solomon codes (RS codes) can be defined as follows.

Let $\{a_1, a_2, \ldots, a_{q-1}\}$ be the non-zero elements of the finite field $\mathbb{F}_q$. For any r, q, $1 \le r \le q-1$ we define the following *truncated Vandermonde matrix*:

$$V(r,q) = \begin{bmatrix} 1 & 1 & \ldots & 1 \\ a_1 & a_2 & \ldots & a_{q-1} \\ \vdots & \vdots & \ldots & \vdots \\ a_1^{r-1} & a_2^{r-1} & \ldots & a_{q-1}^{r-1} \end{bmatrix}.$$

Using this matrix as the parity check matrix we obtain a linear code with parameters $[n,k,d]_q = [q-1, q-r-1, r+1]_q$. Choosing $a_i = \alpha^{i-1}$, $1 \le i \le q-1$, where α is a primitive element in $\mathbb{F}_q$, we get a linear *cyclic* code. This is one of the standard definitions of the *Reed-Solomon* code. The dimension is $k = q-r-1$, because the rows in $V(r,q)$ are linearly independent. The minimum distance is *at least* $r+1$ because any set of r columns are linearly independent; it is *at most* $r+1$ because of the Singleton bound.

Let $u = [1, 0, \ldots, 0]^T$ and $v = [0, 0, \ldots, 0, 1]^T$ be column vectors, both of length r. The parity check matrix $H = [V(r,q)|u|v]$ defines a linear MDS code. By this construction we get linear MDS codes with parameters $[n,k,d]_q = [q+1, q-r+1, r+1]_q$, $1 \le r \le q-1$.

All Reed-Solomon codes and the modifications described above are optimal in the sense that they meet the Singleton bound. Unfortunately they exist only for short lengths. In the above constructions we have $n \le q+1$. In the special case when q is a power of two there are two minor improvements. Let I_3 be the 3×3 identity matrix. Using $[V(3,q)|I_3]$ either as a parity check matrix or as a generator matrix we obtain linear MDS codes with parameters $[n,k,d]_q = [q+2, 3, q]_q$, $[q+2, q-1, 4]_q$, $q = 2^m$, $m = 1, 2, 3, \ldots$.

An important special case of this last construction is $(q,m) = (4,2)$. For this case we obtain a self-dual code with parameters $[n,k,d]_q =$

$[6,3,4]_4$. This code is often referred to as the *Hexacode*. The complete weight enumerator is

$$W(x,y,z,u) = x^6 + 3x^2(y^4 + z^4 + u^4) + 12x^2yzu(y+z+u)$$
$$+4(y^3z^3 + y^3u^3 + z^3u^3) + 6y^2z^2u^2.$$

The *BCH* codes can be defined as *subfield subcodes* of Reed–Solomon codes. Let $\mathbb{F}_q$ be a subfield in $\mathbb{F}_{q^m}$ and let $\mathcal{B}$ be any code in $\mathbb{F}_{q^m}^n$. The code

$$\mathcal{C}_\mathcal{B} \triangleq \{c = (c_1, c_2, \dots, c_n) \in \mathcal{B} : c_i \in \mathbb{F}_q,\ 1 \le i \le n\}$$

is said to be the *subfield subcode* in $\mathbb{F}_q^n$ generated by $\mathcal{B} \subseteq \mathbb{F}_{q^m}^n$. If $\mathcal{B}$ is a Reed–Solomon code the code $\mathcal{C}_\mathcal{B}$ is said to be a *BCH* code. The binary *BCH* codes are of particular interest. In many cases they are the best codes known.

For *BCH* codes and the shortened versions of them the parameters can—for the binary case—be estimated as follows:

BCH-bound:

$$n \le 2^m - 1, \quad k \ge 2^m - 1 - tm, \quad d \ge 2t+1, \quad m = 1,2,3,\dots ,$$

where t is any number less than half the length n. Often the real distance exceeds this bound. There are many improved bounds and numerous special techniques for determining the exact minimum distance in various special cases, but the general problem is still open.

n	$d = 4$	$d = 6$	$d = 8$
5	$\underline{2}$ $\mathcal{A}_2 + \mathcal{B}_2$		
6	$\underline{4}$ $\mathcal{A}_4$	$\underline{2}$ $3\mathcal{A}_2$	
7	$\underline{8}$ $\mathcal{B}_4$	$\underline{2}$ $2\mathcal{A}_2 + \mathcal{B}_2$	
8	$\underline{16}$ $\mathcal{C}_4$	$\underline{2}$ $4\mathcal{A}_2 + 2\mathcal{B}_2$	$\underline{2}$ $4\mathcal{A}_2$
9	$\underline{20}$ CM	$\underline{4}$ $3\mathcal{B}_2$	$\underline{2}$ $3\mathcal{A}_2 + \mathcal{B}_2$
10	$\underline{40}$ $Best$	$\underline{6}$ $\mathcal{A}_6$	$\underline{2}$ $2\mathcal{A}_2 + 2\mathcal{B}_2$
11	$\underline{72}$ $\mathcal{C}_{11}^*$	$\underline{12}$ $\mathcal{B}_6$	$\underline{2}$ $\mathcal{A}_2 + 3\mathcal{B}_2$
12	$\underline{144}$ $\mathcal{C}_{12}^*$	$\underline{24}$ $\mathcal{C}_6$	$\underline{4}$ $4\mathcal{B}_2$
13	$\underline{256}$	$\underline{32}$	$\underline{4}$ $3\mathcal{B}_2 + \mathcal{C}_2$
14	$\underline{512}$	$\underline{64}$	$\underline{8}$ $\mathcal{A}_8$
15	$\underline{1024}$	$\underline{128}$	$\underline{16}$ $\mathcal{B}_8$
16	$\underline{2048}$ $\mathcal{H}$	$\underline{256}$ $\mathcal{NR}$	$\underline{32}$ $\mathcal{C}_8$
17	2720	256	36 CM
18	5248	512	64
19	10496	1024	128
20	20480	2048 $\mathcal{H}/\mathcal{NR}$	256
21	36864	2560	$\underline{512}$
22	73728	4096	$\underline{1024}$
23	147456	8192	$\underline{2048}$
24	294912	16384	$\underline{4096}$ $Golay$

Table A.1: Best binary codes known, (Brower et al., IT-36, p. 1338, [51]). The table gives sizes M of best known codes of lengths $n = 5, 6, \ldots, 24$ and minimum distances $d = 4, 6, 8$. In most cases we also give an indication of the construction. Underlining indicates that the size is known to be largest possible. For the notation used for the codes we refer to the text.

n	$d = 10$	$d = 12$	$d = 14$
10	$\underline{2}$ $5\mathcal{A}_2$		
11	$\underline{2}$ $4\mathcal{A}_2 + \mathcal{B}_2$		
12	$\underline{2}$ $3\mathcal{A}_2 + 2\mathcal{B}_2$	$\underline{2}$ $6\mathcal{A}_2$	
13	$\underline{2}$ $2\mathcal{A}_2 + 3\mathcal{B}_2$	$\underline{2}$ $5\mathcal{A}_2 + \mathcal{B}_2$	
14	$\underline{2}$ $\mathcal{A}_2 + 4\mathcal{B}_2$	$\underline{2}$ $4\mathcal{A}_2 + \mathcal{B}_2$	$\underline{2}$ $7\mathcal{A}_2$
15	$\underline{4}$ $5\mathcal{B}_2$	$\underline{2}$ $3\mathcal{A}_2 + 3\mathcal{B}_2$	$\underline{2}$ $6\mathcal{A}_2 + \mathcal{B}_2$
16	$\underline{4}$ $4\mathcal{B}_2 + 4\mathcal{C}_2$	$\underline{2}$ $2\mathcal{A}_2 + 3\mathcal{B}_2$	$\underline{2}$ $5\mathcal{A}_2 + 2\mathcal{B}_2$
17	$\underline{6}$ $\mathcal{A}_6 + \mathcal{B}_4$	$\underline{2}$ $\mathcal{A}_2 + 5\mathcal{B}_2$	$\underline{2}$ $4\mathcal{A}_2 + 3\mathcal{B}_2$
18	$\underline{10}$ $\mathcal{A}_{10}$	$\underline{4}$ $6\mathcal{B}_2$	$\underline{2}$ $3\mathcal{A}0_2 + 2\mathcal{B}_2$
19	$\underline{20}$ $\mathcal{B}_{10}$	$\underline{4}$ $5\mathcal{B}_2 + \mathcal{C}_2$	$\underline{2}$ $2\mathcal{A}_2 + 3\mathcal{B}_2$
20	$\underline{40}$ $\mathcal{C}_{10}$	$\underline{6}$ $2\mathcal{A}_6$	$\underline{2}$ $\mathcal{A}_2 + 6\mathcal{B}_2$
21	42	$\underline{8}$ $\mathcal{A}_8 + \mathcal{B}_4$	$\underline{4}$ $7\mathcal{B}_2$
22	48 $\mathcal{C}_{12}$-2	$\underline{12}$ $\mathcal{A}_{12}$	$\underline{4}$ $6\mathcal{B}_2 + \mathcal{C}_2$
23	68	$\underline{24}$ $\mathcal{B}_{12}$	$\underline{4}$ $5\mathcal{B}_2 + 2\mathcal{C}_2$
24	128	$\underline{48}$ $\mathcal{C}_{12}$	$\underline{6}$ $\mathcal{A}_6 + 2\mathcal{B}_4$

Table A.2: Best binary codes of lengths $n = 10, 11, \ldots, 24$ and minimum distances $d = 10, 12, 14$. The notation $\mathcal{C}_{12} - 2$ ($(n, d) = (22, 10)$) indicates that two coordinates have been deleted from the code $\mathcal{C}_{12}$, ($(n, d) = (24, 12)$).

$S(37,19,18)\nexists$ $\vdots$? $S(24,6,5)$ $S(23,5,4)$ $S(22,4,3)$ $S(21,3,2)$	$S(24,8,5)$ $S(23,7,4)$ $S(22,6,3)$ $S(21,5,2)$	$S(17,5,3)$ $S(16,4,2)$	$S(25,13,12)\nexists$ $\vdots$? $S(16,4,3)$ $S(15,3,2)$
$S(33,17,16)\nexists$ $\vdots$? $S(20,4,3)$ $S(19,3,2)$		$S(14,4,3)$ $S(13,3,2)$	$S(13,4,2)$
		$S(12,6,5)$ $S(11,5,4)$ $S(10,4,3)$ $S(9,3,2)$	
		$S(8,4,3)$ $S(7,3,2)$	

Table A.3: Steiner systems $S(v,k,t)$ in the range $v \leq 24$. The bars indicate—when known—the beginning and end of a series.

n	w	d	T		n	w	d	T	
24	8	8	759		16	4	4	140	
23	7	8	253		15	3	4	35	
22	6	8	77		14	4	4	91	
21	5	8	21		13	3	4	26	
24	6	4	7084		13	4	6	13	
23	5	4	1771		12	6	4	132	
22	4	4	385		11	5	4	66	
21	3	4	70		10	4	4	30	
20	4	4	285		9	3	4	12	
19	3	4	57		8	4	4	14	
17	5	6	68		7	3	4	7	
16	4	6	20						

Table A.4: Constant weight binary codes generated from *Steiner systems*; $n \leq 24$.

Appendix B

Spheres in $\mathbb{R}^n$

For $n = 2, 3, \ldots$ denote by $Y_n(r)$ the surface area of a sphere with radius r in Euclidean n-space $\mathbb{R}^n$. Denote by $V_n(r)$ the total volume of the same sphere. We have

$$Y_n(r) = k_n r^{n-1},$$

$$V_n(r) = \int_0^r Y_n(x)dx = k_n r^n/n,$$

$$n = 2, 3, \ldots .$$

From elementary geometry we have

$$k_2 = 2\pi; \quad k_3 = 4\pi.$$

We proceed to determine k_n for $n \geq 4$.

For any pair of vectors, $x = (x_1, x_2, \ldots, x_n)$; $y = (y_1, y_2, \ldots, y_n) \in \mathbb{R}^n$ we denote by $\langle x, y \rangle$ the usual Euclidean inner product:

$$\langle x, y \rangle = x_1 y_1 + x_2 y_2 + \cdots + x_n y_n.$$

The corresponding norm is defined by

$$\|x\|^2 = \langle x, x \rangle = x_1^2 + x_2^2 + \cdots + x_n^2.$$

We denote by Ω_n the set of all unit norm vectors in $\mathbb{R}^n$:

$$\Omega_n = \{x \in \mathbb{R}^n : \|x\| = 1\}.$$

For any x in Ω_n the spherical cap $Cap(x, \theta)$ is defined as

$$Cap(x, \theta) = \{y \in \Omega_n : \langle x, y\rangle \geq \cos\theta\}.$$

Equivalently:

$$Cap(x, \theta) = \{y \in \mathbb{R}^n : \|x - y\|^2 = 2 - 2\langle x, y\rangle \leq 2 - 2\cos\theta\}.$$

Clearly the area corresponding to the set $Cap(x, \theta)$ doesn't depend on x and is given by

$$C_n(\theta) = \int_0^\theta Y_{n-1}(\sin\beta)d\beta.$$

We define

$$T_n(\theta) = \int_0^\theta (\sin\beta)^n d\beta; \quad 0 \leq \theta \leq \pi; \quad n = 0, 1, 2, \ldots \; .$$

By partial integration the following recursion is easily obtained:

$$T_n(\theta) = \frac{n-1}{n} T_{n-2}(\theta) - \frac{1}{n}\cos\theta(\sin\theta)^{n-1},$$
$$n = 2, 3, \ldots \; ,$$

with

$$T_0(\theta) = \theta, \quad T_1(\theta) = 1 - \cos\theta.$$

The case $\theta = \pi$ is of special interest. We define

$$t_n = T_n(\pi), \quad n = 0, 1, \ldots \; ,$$

and obtain

$$t_n = \frac{n-1}{n} t_{n-2}, \quad n = 2, 3, \ldots \; ,$$

$$t_0 = \pi, \quad t_1 = 2.$$

We have two separate sequences starting from t_0 and t_1 respectively. The first few terms are

n	0	2	4
t_n	π	$\pi/2$	$3\pi/8$

n	1	3	5
t_n	2	4/3	16/15

In general we have

$$t_{2m} = \pi \cdot \frac{(2m-1)!!}{(2m)!!}, \quad t_{2m+1} = 2 \cdot \frac{(2m)!!}{(2m+1)!!}, \quad m = 1, 2, 3, \ldots .$$

Now for k_n we have

$$k_n = Y_n(1) = C_n(\pi) = \int_0^\pi Y_{n-1}(\sin\beta)d\beta = k_{n-1}t_{n-2},$$

$$n = 3, 4, \ldots .$$

Depending on whether n is even or odd we obtain for $m = 2, 3, \ldots$

$$k_{2m} = k_2 \prod_{j=1}^{m-1} t_{2j-1}t_{2j} = k_2 \prod_{j=1}^{m-1} 2\pi \frac{(2j-2)!!}{(2j-1)!!} \frac{(2j-1)!!}{(2j)!!} =$$

$$= k_2 \prod_{j=1}^{m-1} \frac{2\pi}{2j} = k_2 \frac{(2\pi)^{m-1}}{(2m-2)!!} = \frac{(2\pi)^m}{(2m-2)!!},$$

$$k_{2m+1} = k_3 \prod_{j=1}^{m-1} t_{2j}t_{2j+1} = k_3 \prod_{j=1}^{m-1} 2\pi \frac{(2j-1)!!}{(2j)!!} \frac{(2j)!!}{(2j+1)!!} =$$

$$= k_3 \prod_{j=1}^{m-1} \frac{2\pi}{2j+1} = k_3 \frac{(2\pi)^{m-1}}{(2m-1)!!} = 2\frac{(2\pi)^m}{(2m-1)!!}.$$

Alternatively we have

$$k_n = \begin{cases} \dfrac{(2\pi)^{n/2}}{(n-2)!!}, & n = 2, 4, \ldots , \\ \\ 2 \cdot \dfrac{(2\pi)^{(n-1)/2}}{(n-2)!!}, & n = 3, 5, \ldots , \end{cases}$$

where we define $0!! \triangleq 1$.

We finally observe that the area of the spherical cap is given by

$$C_n(\theta) = \int_0^\theta Y_{n-1}(\sin\beta)d\beta = \int_0^\theta k_{n-1}(\sin\beta)^{n-2}d\beta = k_{n-1}T_{n-2}(\theta),$$

where $n = 3, 4, \ldots$.

Appendix C

Spherical geometry

We derive a few basic facts from spherical geometry. The results are needed in our derivation of the *Fejes Tóth bound*, but are otherwise of no importance for the theory of spherical codes. All of them could be found in any book on elementary spherical geometry. They are included here just for completeness.

In any geometry the basic elements are *points* and *lines*. Spherical geometry deals with geometrical objects on the surface of the unit sphere Ω_3. Our first objective is to define points and lines in this geometry.

A *point* is any triple (x, y, z) of real-valued coordinates such that $x^2 + y^2 + z^2 = 1$, i.e. such that $(x, y, z) \in \Omega_3$. The point $-(x, y, z) = (-x, -y, -z) \in \Omega_3$ is said to be *antipodal* to the point $(x, y, z) \in \Omega_3$. A *line* is a set of points $(x, y, z) \in \Omega_3$ orthogonal to a given fixed point $(a, b, c) \in \Omega_3$. Thus, the line $L(a, b, c)$ generated by the point $(a, b, c) \in \Omega_3$ has the following characterization:

$$L(a, b, c) = \{(x, y, z) \in \Omega_3 : ax + by + cz = 0\}.$$

Lines in Ω_3 are also called great-circles. The points $\pm(a, b, c)$ are referred to as the *poles* of the line $L(a, b, c)$.

For any pair of distinct, non-antipodal points there is always a unique line containing them. Two antipodal points are contained in an infinite number of intersecting lines.

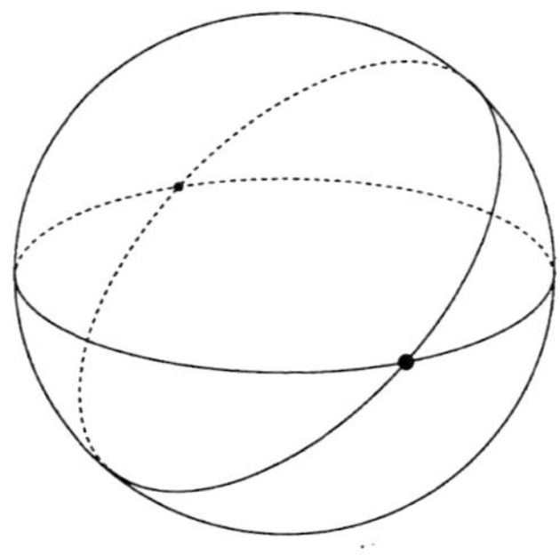

Figure C.1: A pair of distinct lines on the sphere.

Any *pair* of distinct lines intersect in precisely two antipodal points, see Figure C.1. A *triple* of lines in general position intersect in six pairwise antipodal points, see Figure C.2. In general, n lines in general position intersect in $n(n-1)$ pairwise antipodal points. In particular: there are no parallel lines in Ω_3.

A *single* line divides the sphere Ω_3 in two equal parts; a *pair* of distinct lines divide Ω_3 in four parts, (Figure C.1). A *triple* of lines in general position divides Ω_3 in $2^3 = 8$ parts, (Figure C.2). In general, n lines in general position divide Ω_3 in 2^n parts.

The eight parts into which Ω_3 is subdivided by three lines in general position are called *triangles* (Figure C.2). We call the four parts into which Ω_3 is subdivided by two distinct lines *biangles* (Figure C.1). A biangle contains two lines and two vertices. The two vertices are necessarily antipodal and the angles are always equal.

The area of a biangle equals the sum of its angles. In other words, the area is 2α for a biangle with angle α at both of its two vertices. This little result follows by simply observing that the area must be proportional to α and that for the maximal possible angle ($\alpha = \pi$) the area is that of half the sphere (2π).

It is an interesting fact that the area of a triangle can be obtained by a very similar formula. More precisely we have the following theorem.

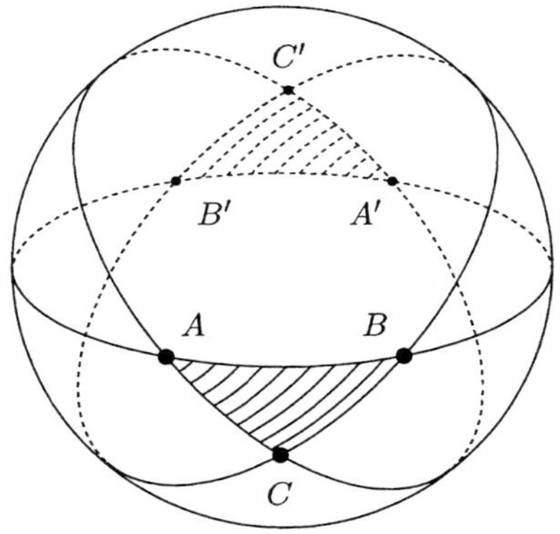

Figure C.2: A triple of distinct lines on the sphere.

Theorem C1 (Excess Theorem) *The area T of a spherical triangle with angles α, β, γ is given by*

$$T = \alpha + \beta + \gamma - \pi.$$

Remark C1: Recalling that the sum of the angles in a *plane* triangle equals π we also refer to the expression $\alpha + \beta + \gamma - \pi$ as the *excess*. The theorem says that the area equals the excess. We conclude that the excess of a spherical triangle is always positive.

Proof: Let A, B, C be the vertices of a spherical triangle and let A', B', C' be the corresponding antipodes. For each one of the vertices A, B, C there is a biangle containing the triangle and having its antipodal vertex in A', B', C' respectively (Figure C.2). The union of the three biangles has a total area of 2π (half the sphere). Denoting by T the area of the triangle we therefore have

$$2\pi = 2\alpha + 2\beta + 2\gamma - 2T,$$

taking into account the fact that the area of the triangle is counted three times when adding the areas of the biangles. Dividing by two and reordering the terms gives us the result. □

In general, let A, B, C denote the vertices of a spherical triangle, let the corresponding angles be α, β, γ and let the lengths of the

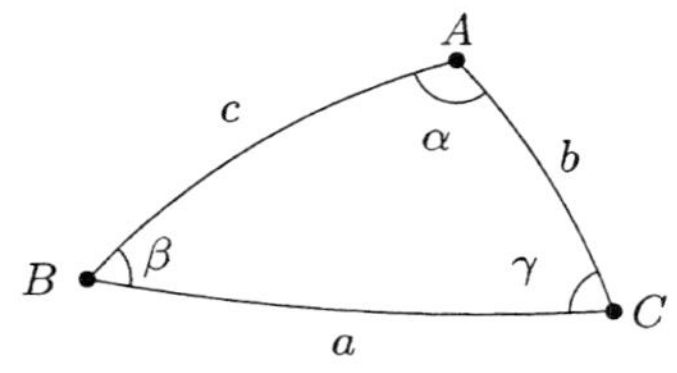

Figure C.3: A spherical triangle with vertices A, B, C, angles α,β,γ and edges a,b,c.

opposite edges be a, b, c (Figure C.3). By the *length* of an edge we understand—in this context—the angle as seen from the center of Ω_3. The well known "Cosine Theorem" of plane geometry has the following counterpart in spherical geometry.

Theorem C2 (Cosine Theorem) *In a spherical triangle angles α, β, γ and lengths a, b, c of opposite edges are related as follows:*

$$\cos\gamma = \frac{\cos c - \cos a \cos b}{\sin a \sin b}.$$

Remark C2: It is easy to check that for small triangles $(a, b, c \ll 1)$ this formula reduces to the well known "cosine theorem" of plane Euclidean geometry.

Proof: We prove the theorem by invoking some well known results from plane geometry. Consider Figure C.4. It indicates a spherical triangle, where lines connecting each one of the vertices A, B, C with the center O of the sphere Ω_3 have been indicated. Let us extend two tangents from the vertex C along the two edges AC and BC, and let A' and B' be the points where these tangents intersect the lines OA and OB respectively. Applying the usual "Cosine Theorem" of plane geometry to each one of the plane triangles $OA'B'$ and $CA'B'$ we obtain

$$(A'B')^2 = \frac{1}{\cos^2 a} + \frac{1}{\cos^2 b} - \frac{2\cos c}{\cos a \cos b} =$$

$$= \tan^2 a + \tan^2 b - 2\tan a \tan b \cos\gamma.$$

Solving with respect to $\cos\gamma$ gives us the result. □

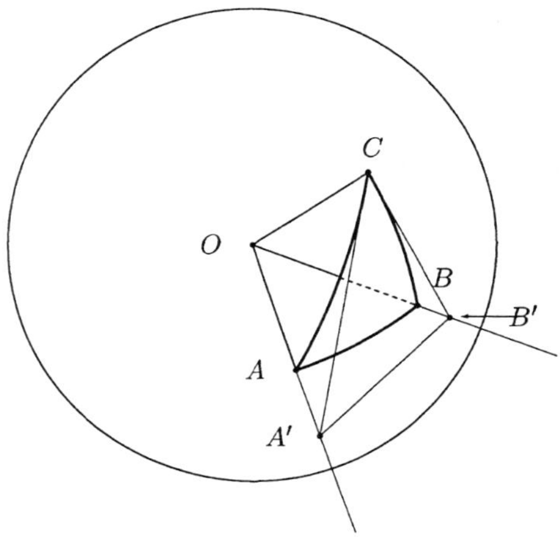

Figure C.4: Illustration to the cosine theorem: $OC = 1$, $A'O = 1/\cos b$, $B'O = 1/\cos a$, $A'C = \tan b$, $B'C = \tan a$.

There is also a counterpart to the so called "Sinus Theorem". Actually, as with the corresponding result in plane geometry, the sinus theorem for spherical geometry is logically equivalent to the cosine theorem and can be derived from it. The result is as follows.

Theorem C3 (Sinus Theorem) *In any spherical triangle angles α, β, γ and lengths of edges a, b, c are related as follows:*

$$\frac{\sin\alpha}{\sin a} = \frac{\sin\beta}{\sin b} = \frac{\sin\gamma}{\sin c} .$$

Proof: Squaring the cosine formula we get

$$\cos^2\gamma = 1 - \sin^2\gamma = \frac{\cos^2 c + \cos^2 a \cos^2 b - 2\cos a\cos b\cos c}{\sin^2 a \sin^2 b} .$$

After some elementary operations this expression can be brought to the following form:

$$\sin^2 a \sin^2 b \sin^2\gamma =$$

$$= 1 - \cos^2 a - \cos^2 b - \cos^2 c + 2\cos a\cos b\cos c .$$

We notice that the right hand member is symmetric in the three variables a, b, c. From this observation we conclude that we are allowed to perform the corresponding permutations in the left hand member of the equation, without changing the value of the expression. Thus we have

$$\sin^2 a \sin^2 b \sin^2 \gamma = \sin^2 a \sin^2 \beta \sin^2 c = \sin^2 \alpha \sin^2 b \sin^2 c.$$

Dividing and taking square roots produces the desired formula. □

A number of additional formulas can be derived from the above results. For instance, inserting $\gamma = \pi/2$ in the cosine theorem we obtain

$$\cos c = \cos a \cos b.$$

This little formula is the counterpart in spherical geometry to the *Pythagorean Theorem* in plane geometry. For small triangles it approaches the usual Euclidean formula.

Consider a spherical triangle with $\gamma = \pi/2$. By the Excess Theorem we have the following formula for the area:

$$T = \alpha - (\pi/2 - \beta).$$

Utilizing first the cosine theorem, then the Pythagorean formula, and finally the sinus theorem we obtain

$$\sin(\pi/2 - \beta) = \cos\beta = \frac{\cos b - \cos a \cos c}{\sin a \sin c} =$$

$$= \frac{\cos b - \cos^2 a \cos b}{\sin a \sin c} = \frac{\sin a \cos b}{\sin c} = \sin\alpha \cos b.$$

Inserting into the formula for the area we get

$$T = \alpha - \arcsin(\cos b \sin\alpha).$$

This formula will be used in our derivation of the Fejes Tóth bound. We remark that for fixed b the area T is a convex function of the angle α. This fact, which is easily established by differentiating twice with respect to α, is crucial in the derivation of the Fejes Tóth bound.

The final thing we need is a classical result known as *Euler's formula*. We start with some definitions.

Following Coxeter ([75], p. 6) we define a *map* as a partition of an unbounded surface into polygonal regions by means of simple curves joining pairs of points. Unbounded in this context means that the surface should not be a subset of any larger surface. The sphere Ω_3 is a valid example of such a surface. The Euclidean plane is another valid example. Both of these are examples of so called *simply connected* surfaces. A torus is an example of a *multiply-connected* surface. In its original form Euler's formula applies to simply connected surfaces only. For our purposes that restricted form is sufficient.

A map consists of three kinds of elements: *faces*, *edges* and *vertices*. The faces are the polygonal regions, the edges are the boundaries between them and the vertices are the crossing points between the boundaries. Euler's formula expresses a general relation between the cardinalities of these sets of entities.

Theorem C4 (Euler's formula) *For any map on a simply connected surface the number v of vertices, the number e of edges and the number f of faces are related by the formula*

$$v - e + f = 2.$$

Proof: A map in which each vertex is connected to precisely two edges is called a *cycle*. For a cycle the number v of vertices and the number e of edges coincide. A cycle divides a simply connected surface into two disjoint parts. Therefore, if the map is a cycle the relation obviously holds. If not we can generate a cycle by successively removing appropriate edges. In each such step the number of vertices remains unchanged, while both the number of edges and the number of faces decreases by one. It is clear that the sum $v - e + f$ remains constant throughout the process and thus equals 2 for any map on a simply connected surface. □

Remark C3: The argument fails on a multiply-connected surface. For instance, on a torus we can have a cycle which does not divide the surface in two disjoint regions.

Appendix D

Tables

D.1 Introduction

Tables of codes in dimensions $3 \leq n \leq 24$ are given. The codes have been carefully selected so as to provide illustrative examples of the constructions discussed in the book. They have also been chosen so as to provide a fair indication of the current state of art regarding codes in Euclidean spaces. All entries represent currently best known codes. Each table contains seven columns with the following labels:

N: Identification number.

M: Cardinality of the code.

ρ: Exact value of the squared minimum distance, usually in the form of an algebraic expression.

ρ_L: Decimal representation of the squared minimum distance, always rounded *downwards*.

Constr.: Brief indication of *one* possible construction. We emphasize that often alternative constructions are available.

ρ_U: Best upper bound currently known for the code parameters in question, decimal expressions rounded *upwards*.

Ref.: Reference for the upper bound used in the previous column.

D.2 Notation

Each table is accompanied with brief information regarding the various constructions. Whenever possible one of the constructions discussed in the text is employed. References are given, mainly to appropriate locations in the text, but sometimes also to external sources.

For obvious reasons the text in the explaining notes is short. The following general remarks might be helpful.

(1) By a *trivial extension* of a code $\mathcal{X}$ with parameters (n, ρ, M) we understand a code $\mathcal{Z}$ with parameters $(n+1, \rho, M)$ obtained from $\mathcal{X}$ by adding a zero in a position to the *right* in each codeword: $\mathcal{Z} = (\mathcal{X}, 0)$. The code obtained by adding i zeros is referred to as a *trivial i-extension.* Whenever we like to emphasize that the zeros are added to the *right* of the original codeword we refer to the extension as a *right extension.* If zeros are added to the *left* the resulting code is referred to as a *left extension.*

(2) Whenever a code appearing in the table is described anywhere in the book, a reference is given to the appropriate chapter.

(3) Any non-marked entry in the table is obtained as a union between a trivial extension of a code of smaller dimension and an appropriate representation of the binary antipodal code. More precisely, let $\mathcal{X}$ be any code with parameters (n, ρ, M), with $\rho \leq 2$. Choose $\mathcal{Y} = (0^n, \pm 1)$. The union $\mathcal{Z} = \mathcal{X} \cup \mathcal{Y}$ is a spherical code with parameters $(n+1, \rho, M+2)$. All non-marked entries in the table are obtained in this way.

(4) For the *Biorthogonal* code with parameters $(n, 2, 2n)$ we use the notation $\mathcal{B}_n$. Whenever a specific construction is needed this fact is explicitly indicated. For instance, whenever we write "the code $\mathcal{B}_n$ (by Y_3)" the usual Y_3-construction is assumed: $\mathcal{B}_n = (0^{n-1}, \pm 1)^C$.

(5) For the *Simplex* code with parameters $(n, 2+2/n, n+1)$ we use the notation $\mathcal{S}_n$. Again, whenever a specific construction is needed this fact is explicitly indicated. We write "$\mathcal{S}_n$ (by CW_2)"

to denote the usual representation by the CW_2-construction: $\mathcal{S}_n = (1^n, -n)^C/\sqrt{n(n+1)}$.

(6) A *polyphase* code with parameters $(n, \rho, M) = (2, 4\sin^2(\pi/q), q)$ formed by the points in the set $\mathcal{L} = \{\theta^i : 0 \le i \le q-1\}$ will be denoted $\mathcal{P}_q$.

(7) Let $\mathcal{X} : (n+1, \rho, M)$ be the extension of $\mathcal{X}_1$ and $\mathcal{X}_2$ with coefficients b_1, b_2 and with tails a_1 and $-a_2$, where

$$a_1 = \sqrt{1 - b_1^2}, \quad a_2 = \sqrt{1 - b_2^2}.$$

The maximal correlation in $\mathcal{X}$ is

$$s = b_1 b_2 s(\mathcal{X}_1, \mathcal{X}_2) - a_1 a_2.$$

By "adding one point" we understand joining the single point $(0^n, 1)$ to $\mathcal{X}$. If $a_1 \le s$ this can be done without decreasing the minimum distance. "Adding two points" means joining $(0^n, \pm 1)$. If $\max\{a_1, a_2\} \le s$ this can also be done without decreasing the minimum distance.

(8) For any spherical code $\mathcal{X} \subseteq \Omega_n$ we denote by $-\mathcal{X}$ the code obtained by replacing each codeword x in $\mathcal{X}$ with its additive inverse: $-\mathcal{X} = \{x \in \Omega_n : -x \in \mathcal{X}\}$. Clearly the codes $\mathcal{X}$ and $-\mathcal{X}$ have the same parameters (n, ρ, M).

(9) Let $\mathcal{L}$ be an arbitrary alphabet and let $\mathcal{A}$ be an arbitrary code in $\mathcal{L}^n$. We say that

$$\mathcal{A}' \triangleq \{(a_2, a_3, \ldots, a_n) \in \mathcal{L}^{(n-1)} : (i_1, a_2, \ldots, a_n) \in \mathcal{A}\}$$

is obtained by (i_1)-*reduction* of $\mathcal{A}$, $i_1 \in \mathcal{L}$. Similarly, for any k, $1 \le k < n$, and for any vector $(i_1, \ldots, i_k)$ over $\mathcal{L}$ we say that $\mathcal{A}'' \subset \mathcal{L}^{n-k}$ is obtained by $(i_1, \ldots, i_k)$-reduction of $\mathcal{A}$ if the following holds:

$$\mathcal{A}'' = \{(a_{k+1}, \ldots, a_n) : (i_1, \ldots, i_k, a_{k+1}, \ldots, a_n) \in \mathcal{A}\}.$$

(10) We use $E(\mathcal{X}_1, \mathcal{X}_2, \cdots, \mathcal{X}_s)$ to denote *any* extended union (as defined in Chapter 5.10) of the spherical codes $\mathcal{X}_1$, $\mathcal{X}_2, \ldots, \mathcal{X}_s$:

$$E(\mathcal{X}_1, \mathcal{X}_2, \ldots, \mathcal{X}_s) \triangleq \bigcup_{i=1}^{s} (b_i \mathcal{X}_i, a_i).$$

The coefficients b_i and the tails $a_i = \pm\sqrt{1 - b_i^2}$ are always explicitly indicated. Notice that in some constructions these parameters are not necessarily chosen so as to maximize the squared minimum distance in $E(\mathcal{X}_1, \mathcal{X}_2, \ldots, \mathcal{X}_s)$: if the extended union is just a part in a more complicated construction a different choice might lead to a better final result.

(11) A code $\mathcal{X}$ obtained as a union of shells with squared norms a_1, a_2,...,a_s of a sphere packing P will be denoted $\mathcal{X} = US(P; \{a_1, a_2, \ldots, a_s\})$. For such codes we give either a reference for the packing or an explicit construction. For packings obtained from the constructions $P_A(\mathcal{A})$ and $P_B^*(\mathcal{A})$ (Chapter 10.7) we describe the corresponding codes by their weight distributions.

(12) We use $\mathcal{G}_{23}$ and $\mathcal{G}_{24}$ to denote the binary *Golay* code and its extension, respectively. We use $\mathcal{NR}$ to denote the *Nordstrom–Robinson* code with parameters $(n, d, M) = (16, 6, 256)$. We always assume that these codes are represented in a form so as to contain the zero vector. In particular this implies that $\mathcal{G}_{24}$ contains the celebrated *Steiner system* $S(24, 8, 5)$ and that $\mathcal{G}_{23}$ contains the related system $S(23, 7, 4)$. Alternatively this means that the binary unrestricted code $\mathcal{G}_{24}$ contains a binary constant weight code with parameters $(N, w, d, T) = (24, 8, 8, 759)$ as a subcode and that the code $\mathcal{G}_{23}$ contains a binary constant weight code with parameters $(N, w, d, T) = (23, 7, 8, 253)$. It also means that the Nordstrom–Robinson code $\mathcal{NR}$ contains the *Reed–Muller* code $\mathcal{R}(1, 4)$ as well as two optimal constant weight codes with parameters $(N, w, d, T) = (16, 6, 6, 112), (16, 10, 6, 112)$. To these there are two important partitions (see Appendix A). Finally it implies that $\mathcal{NR}$ can be obtained as a subcode of the Hamming code $\mathcal{H}_4^*$ and that the important partition described in Appendix A holds.

(13) We say that $\mathcal{A}$ is an *even weight* code if all its codewords have *even* weight; we say that $\mathcal{A}$ is an *odd weight* code if all its codewords have *odd* weight.

N	M	ρ	ρ_L	*Constr.*	ρ_U	*Ref.*
1	7	ρ_1	1.57972	*Ex.* 5.10.4	ρ_1	[193]
2	8	$4(4-\sqrt{2})/7$	1.47759	*Ex.* 5.10.1	$4(4-\sqrt{2})/7$	[193]
3	9	$4/3$	1.33333	$UE(CW_2)$	$4/3$	[193]
4	10	ρ_2	1.19121	*Sec.* 3.2	ρ_2	[77]
5	12	$2-2/\sqrt{5}$	1.10557	$E(\mathcal{P}_5)$	$2-2/\sqrt{5}$	[106]
6	16	ρ_3	0.77541	$UE(Y_2, Y_3)$	0.83901	[43]
7	24	ρ_4	0.55384	*Sec.* 3.2	ρ_4	[185]
8	27	ρ_5	0.48322	$UE(\mathcal{P}_5)$	0.51599	[43]

Table D.3: Spherical codes of dimension $n = 3$.

D.3 Spherical codes of dimension $n = 3$

(3) A union of $\mathcal{X}_1 : (3, 3, 3)$ (the trivial extension of $\mathcal{S}_2$) and $\mathcal{X}_2 : (3, 4/3, 6)$ (the extension of two copies of $-\mathcal{S}_2$ with $b = 2/3$).

(5) The extension of $\mathcal{P}_5$ and $-\mathcal{P}_5$ with $b = 2/\sqrt{5}$, plus two points.

(6) The union of $\mathcal{X}_1$ and $\mathcal{X}_2$. The code $\mathcal{X}_1 : (3, \rho_3, 8)$ is the extension of two copies of $\mathcal{B}_2$ (by Y_2 from $\mathcal{A}_1 : (2, 1, 4)$) with b_1 and b_2, where $b_1 < b_2$, and with the tails $z_1 = a_1$ and $z_2 = -a_2$. The code $\mathcal{X}_2 : (3, \rho_3, 8)$ is the extension of $\mathcal{B}_2$ (by Y_3) with the same b_1 and b_2, and with the tails $z_1 = -a_1$ and $z_2 = a_2$. We have $\rho_3 = 2b_1^2$ and b_1 and b_2 can be found from the equations:

$$\begin{aligned} 2b_1^2 &= 2 - 2(b_1 b_2/\sqrt{2} + a_1 a_2) \\ &= 2 - 2(b_2^2/\sqrt{2} - a_2^2). \end{aligned}$$

(8) The union of $\mathcal{X}_1 : (3, \rho_5, 17)$ and $\mathcal{X}_2 : (3, \rho_5, 10)$. The code $\mathcal{X}_1$ is the extension of three copies of $\mathcal{P}_5$ with

$$b_1 = 1 - (1/t)(1 - (1/2)\sqrt{4-k}),$$

where $t = \sin^2(\pi/5)$ and $k = \sin^2(2\pi/5)$, and with the tails $z_{1,1} = a_1$, $z_{1,2} = 0$, and $z_{1,3} = -a_1$, plus two points ($\rho_5 = 2 - 2a_1$). The code $\mathcal{X}_2$ is the extension of two copies of $-\mathcal{P}_5$ with b_2 and with the tails $z_{2,1} = a_2$ and $z_{2,2} = -a_2$, where $b_2 = 1 - \rho_5/2\cos(\pi/5)$.

N	M	ρ	ρ_L	*Constr.*	ρ_U	*Ref.*
1	9	ρ_1	1.67596	$Ex.\,5.10.4$	1.84639	[43]
2	10	5/3	1.66666	CW_2	1.72804	[43]
3	12	3/2	1.50000	$UE(Y_2, Y_3)$	1.54405	[43]
4	20	$(7-\sqrt{13})/3$	1.13148	$UE(Y_2, Y_3)$	1.18351	[157]
5	24	1	1.00000	Y_3	1.04292	[43]
6	120	$(3-\sqrt{5})/2$	0.38196	$Ex.\,1.2.8$	$(3-\sqrt{5})/2$	[53]

Table D.4: Spherical codes of length $n = 4$.

D.4 Spherical codes of dimension $n = 4$

(2) $CW_2\,(\mathcal{C} : (5, 2, 2, 10))$.

(3) The union of $\mathcal{X}_1$ and $\mathcal{X}_2$. The code $\mathcal{X}_1 : (4, 3/2, 6)$ is the extension of two copies of $\mathcal{S}_2$ with the tails $z_1 = (a|0)$ and $z_2 = (-a|0)$ and with $b^2 = 1/2$. The code $\mathcal{X}_2 : (4, 3/2, 6)$ is the extension of two copies of $-\mathcal{S}_2$ with the tails $z_3 = (0|a)$ and $z_4 = (0|-a)$ and the same b.

(4) The union of $\mathcal{X}_1$ and $\mathcal{X}_2$. The code $\mathcal{X}_1 : (4, 4/3, 8)$ is the trivial extension of $\mathcal{V}_1 : (3, 4/3, 8)$ (by Y_2 from $\mathcal{A}_1 : (3, 1, 8)$). The code $\mathcal{X}_2$ is the extension of two copies of $\mathcal{B}_3$ (by Y_3) with $b = (\sqrt{13}-1)/2\sqrt{3}$.

(5) $Y_3\,(\mathcal{C} : (4, 2, 2, 6),\ \mathcal{A} : (2, 1, 4))$.

D.5 Spherical codes of dimension $n = 5$

(2) $Y_2\,(\mathcal{A} : (5, 2, 16))$.

(3) $Y_3\,(\mathcal{C} : (5, 2, 2, 10),\ \mathcal{A} : (2, 1, 4))$.

(4) The extension of $\mathcal{X}_1 : (4, 1, 24)$ and $\mathcal{X}_2 : (4, 1, 24)$ with

$$b = 2\sqrt{(1/7)(3-\sqrt{2})},$$

N	M	ρ	ρ_L	*Constr.*	ρ_U	*Ref.*
1	11	ρ_1	1.73429	*Ex.* 5.10.10	1.90558	[157]
2	16	8/5	1.60000	Y_2	8/5	[157]
3	40	1	1.00000	Y_3	1.07096	[43]
4	50	$4(3-\sqrt{2})/7$	0.90616	$EU(Y_2, Y_3)$	0.96947	[43]
5	74	$3-\sqrt{5}$	0.76393	$UE(Y_2, Y_3)$	0.81751	[43]
6	90	2/3	0.66666	$U(Y_3)$	0.74057	[43]

Table D.5: Spherical codes of length $n = 5$.

plus two points. The code $\mathcal{X}_1$ is obtained by Y_3 from $\mathcal{C}_1$: $(4,2,2,6)$ and $\mathcal{A}_1 : (2,1,4)$. The code $\mathcal{X}_2$ is the union of $\mathcal{V}_1 : (4,1,16)$ (by Y_2 from $\mathcal{A}_2 : (4,1,16)$) and $\mathcal{B}_4$ (by Y_3).

(5) The union of $\mathcal{X}_1$ and $\mathcal{X}_2$, plus two points. The code $\mathcal{X}_1 : (5,1,24)$ is the trivial extension of $\mathcal{V}_1 : (4,1,24)$ (see **(4)**). The code $\mathcal{X}_2$: $(5, 3-\sqrt{5}, 48)$ is the extension of two copies of $\mathcal{V}_2 : (4,1,24)$ (see again **(4)**) with $b = (\sqrt{5}-1)/\sqrt{2}$.

(6) The union of $\mathcal{X}_1 : (5, 2/3, 80)$ (by Y_3 from $\mathcal{C} : (5,3,2,10)$ and $\mathcal{A} : (3,1,8)$) and $\mathcal{B}_5$ (by Y_3).

D.6 Spherical codes of dimension $n = 6$

(2) The extension of $\mathcal{X}_1 : (5, 12/5, 4)$ (by Y_2 from $\mathcal{A} : (5,3,4)$) and $\mathcal{B}_5$ (by Y_3) with $b_1 = (11-\sqrt{6})/12$.

(4) The union of $\mathcal{X}_i : (6, 20/13, 10)$, $i = 1, 2$. The code $\mathcal{X}_1$ is the extension of two copies of $\mathcal{S}_4 : (4, 5/2, 5)$ with $b = 2\sqrt{2/13}$ and with the tails $z_1 = (a|0)$ and $z_2 = (-a|0)$. The code $\mathcal{X}_2$ is the extension of two copies of $-\mathcal{S}_4 : (4, 5/2, 5)$ with the same b and with the tails $z_3 = (0|a)$ and $z_4 = (0|-a)$.

(5) The union of $\mathcal{X}_1 : (6, 3/2, 15)$ and $\mathcal{X}_2 : (6, 3/2, 12)$. The code $\mathcal{X}_1$ is the trivial extension of $\mathcal{V}_1 : (5, 3/2, 15)$ (by CW_2 from $\mathcal{C}_1$: $(6,2,2,15)$). The code $\mathcal{X}_2$ is the extension of two copies of $\mathcal{S}_5$ (by CW_2) with $b = \sqrt{5/2}/2$.

N	M	ρ	ρ_L	$Constr.$	ρ_U	$Ref.$
1	13	ρ_1	1.77384	$Ex.$ 5.10.4	1.93581	[157]
2	14	$(11-\sqrt{6})/5$	1.71010	$E(Y_2, Y_3)$	1.87722	[157]
3	18	8/5	1.60000		1.69940	[157]
4	20	20/13	1.53846	$UE(\mathcal{S}_4)$	1.63750	[157]
5	27	3/2	1.50000	$UE(CW_2)$	3/2	[157]
6	28	7/5	1.40000	$U(CW_2)$	1.47806	[43]
7	32	4/3	1.33333	Y_2	1.40579	[43]
8	36	$(11-\sqrt{21})/5$	1.28348	$UE(Y_2, Y_3)$	1.35184	[43]
9	42	$(9-\sqrt{17})/4$	1.21922	$UE(Y_2, Y_3)$	1.29290	[157]
10	44	$2-2/\sqrt{6}$	1.18350	$U(Y_2, Y_3)$	1.26863	[43]
11	49	$2-2\sqrt{2}/3$	1.05719	$U(CW_2)$	1.21315	[43]
12	72	1	1.00000	$UE(Y_2, Y_3)$	1.04298	[43]
13	74	$2-2\sqrt{2/7}$	0.93095	$UE(Y_2, Y_3)$	1.03391	[43]
14	92	$2-2/\sqrt{3}$	0.84529	$U(Y_2, Y_3)$	0.96081	[43]
15	114	4/5	0.80000	$UE(Y_2, Y_3)$	0.89333	[43]
16	172	2/3	0.66666	$U(Y_3)$	0.75863	[43]
17	243	3/5	0.60000	$US(E_6+[1])$	0.67051	[43]
18	342	1/2	0.50000	$US(E_6)$	0.58752	[43]
19	558	2/5	0.40000	$US(E_6^*)$	0.48958	[43]
20	990	1/3	0.33333	$US(E_6)$	0.39295	[43]

Table D.6: Spherical codes of length $n = 6$.

(6) The union of $\mathcal{X}_1 : (6, 7/5, 21)$ (by CW_2 from $\mathcal{C} : (7, 2, 2, 21)$) and $-\mathcal{S}_6$ (by CW_2).

(7) $Y_2\,(\mathcal{A} : (6, 2, 32))$.

(8) The union of $\mathcal{X}_1 : (6, 8/5, 16)$ and $\mathcal{X}_2 : (6, (11 - \sqrt{21})/5, 20)$. The code $\mathcal{X}_1$ is the trivial extension of $\mathcal{V}_1 : (5, 8/5, 16)$ (by Y_2 from $\mathcal{A}_1 : (5, 2, 16)$). The code X_2 is the extension of two copies of $\mathcal{B}_5$ (by Y_3) with

$$b = \sqrt{(1/10)(11 - \sqrt{21})}\,.$$

(9) The union of $\mathcal{X}_1 : (6, (9 - \sqrt{17})/4, 32)$ and $\mathcal{X}_2 : (6, 2, 10)$ (the trivial extension of $\mathcal{B}_5$ (by Y_3)). The code $\mathcal{X}_1$ is the extension with

$$b = \sqrt{(5/32)(9 - \sqrt{17})}\,.$$

of $\mathcal{V}_i : (5, 8/5, 16)$, $i = 1, 2$ (by Y_2 from $\mathcal{A}_i : (5, 2, 16)$, respectively), where $d(\mathcal{A}_1, \mathcal{A}_2) = 1$, implying $\rho(\mathcal{V}_1, \mathcal{V}_2) = 4/5$.

(10) The union of $\mathcal{X}_1 : (6, 4/3, 32)$ (by Y_2 from $\mathcal{A} : (6, 2, 32)$) and $\mathcal{B}_6$ (by Y_3).

(11) The union of $\mathcal{X}_1 : (6, 7/6, 35)$ (by CW_2 from $\mathcal{C} : (7, 3, 2, 35)$), $\mathcal{S}_6$ (by CW_2), and $-\mathcal{S}_6$.

(12) The union of $\mathcal{X}_1 : (6, 1, 40)$ (the trivial extension of $\mathcal{V}_1 : (5, 1, 40)$ (by Y_3 from $\mathcal{C}_1 : (5, 2, 2, 10)$ and $\mathcal{A}_1 : (2, 1, 4)$)) and $\mathcal{X}_2 : (6, 1, 32)$ (the extension of $\mathcal{V}_i : (5, 8/5, 16)$, $i = 2, 3$, with $b = \sqrt{5/2}/2$). Both $\mathcal{V}_i$ are obtained by Y_2 from $\mathcal{A}_i : (5, 2, 16)$, $i = 2, 3$, respectively, where $d(\mathcal{A}_2, \mathcal{A}_3) = 1$, implying $\rho(\mathcal{V}_2, \mathcal{V}_3) = 4/5$.

(13) It can be obtained by the construction in **(12)**, if we add two points and use for the construction of $\mathcal{X}_2$ the parameter $b = \sqrt{5/7}$.

(14) The union of $\mathcal{X}_1 : (6, 1, 60)$ (by Y_3 from $\mathcal{C}_1 : (6, 2, 2, 15)$ and $\mathcal{A}_1 : (2, 1, 4)$) and $\mathcal{X}_2 : (6, 4/3, 32)$ (by Y_2 from $\mathcal{A}_2 : (6, 2, 32)$).

(15) The union of $\mathcal{X}_1$ and $\mathcal{X}_2$. The code $\mathcal{X}_1 : (6, 4/5, 32)$ is the trivial extension of $\mathcal{V}_1 : (5, 4/5, 32)$ (by Y_2 from $\mathcal{A}_1 : (5, 1, 32)$). The code

$\mathcal{X}_2 : (6, 4/5, 80)$ is the extension of two copies of $\mathcal{V}_2 : (5, 1, 40)$ (see **(12)** above) with $b = 2/\sqrt{5}$, plus two points.

(16) The union of $\mathcal{X}_1 : (6, 2/3, 160)$ (by Y_3 from $\mathcal{C}_1 : (6, 3, 2, 20)$ and $\mathcal{A}_1 : (3, 1, 8)$) and $\mathcal{B}_6$ (by Y_3).

(17) $US(E_6 + [1];\ \{4/3, 10/3\})$, the shift of the lattice E_6: $E_6 + [1]$ (see [71], p. 127). If E_6 is defined from E_8 as follows

$$E_6 \;=\; \{(x_1, \ldots, x_8) \in E_8 :\ x_1 + x_8 = x_2 + \cdots + x_7 = 0\},$$

then $[1] = (0, -2/3, -2/3, 1/3, 1/3, 1/3, 1/3, 0)$.

(18) $US(E_6;\ \{2, 4\})$.

(19) $US(E_6^*;\ \{4/3, 2, 10/3\})$, the lattice E_6^* (see [71], p. 127).

(20) $US(E_6;\ \{4, 6\})$.

D.7 Spherical codes of dimension $n = 7$

(2) The extension of $\mathcal{X}_1 : (6, 8/3, 4)$ (by Y_2 from $\mathcal{A} : (6, 4, 4)$) and $\mathcal{B}_6$ (by Y_3) with

$$b_1 = \sqrt{(3/46)(13 - 2\sqrt{2})}.$$

(3) The extension of $\mathcal{X}_1 : (6, 2, 8)$ (by Y_2 from $\mathcal{A} : (6, 3, 8)$) and $\mathcal{B}_6$ (by Y_3) with

$$b = \sqrt{(2/23)(12 - \sqrt{6})}.$$

(4) The extension of $\mathcal{X}_i : (6, 2, 12)$, $i = 1, 2$, with $b = 2/\sqrt{5}$. The codes $\mathcal{X}_i$ are obtained by Y_3 from $\mathcal{C}_i : (6, 2, 4, 3)$ respectively and $\mathcal{A} : (2, 1, 4)$, where $d(\mathcal{C}_1, \mathcal{C}_2) = 2$, implying $\rho(\mathcal{X}_1, \mathcal{X}_2) = 1$.

(5) The union of $\mathcal{X}_1$ and $\mathcal{X}_2$. The code $\mathcal{X}_1 : (7, 3/2, 24)$ is the trivial extension of $\mathcal{V}_1 : (6, 3/2, 24)$ (the extension of $\mathcal{W}_i : (4, 2, 8)$, $i = 1, 2, 3$, with the tail code $\mathcal{S}_2$ and with $b_1 = \sqrt{3}/2$). All $\mathcal{W}_i$, $i = 1, 2, 3$ are obtained by Y_3 from $\mathcal{C}_i : (4, 2, 4, 2)$, respectively, and from $\mathcal{A}_1$:

N	M	ρ	ρ_L	*Constr.*	ρ_U	*Ref.*
1	15	ρ_1	1.80259	*Ex.* 5.10.4	1.95364	[157]
2	16	$4(13-2\sqrt{2})/23$	1.76896	$E(Y_2, Y_3)$	1.91015	[157]
3	20	$4(12-\sqrt{6})/23$	1.66095	$E(Y_2, Y_3)$	1.76808	[157]
4	24	8/5	1.60000	$E(Y_3)$	1.66948	[157]
5	30	3/2	1.50000	$UE(Y_3)$	1.57276	[157]
6	35	7/5	1.40000	$UE(CW_2)$	1.51950	[157]
7	56	4/3	1.33333	Y_3	4/3	[157]
8	63	7/6	1.16666	$UE(CW_2)$	1.28052	[43]
9	80	8/7	1.14285	$UE(Y_2, Y_3)$	1.17935	[43]
10	126	1	1.00000	$U(Y_3)$	1.03292	[43]
11	148	$2-2\sqrt{2/7}$	0.93095	$U(Y_2, Y_3)$	0.98491	[43]
12	182	$3-\sqrt{5}$	0.76393	$UE(Y_3)$	0.92642	[43]
13	358	2/3	0.66666	$U(Y_2, Y_3)$	0.74666	[43]
14	632	4/7	0.57142	$US(P_A)$	0.62368	[43]
15	882	1/2	0.50000	$US(E_7)$	0.56311	[43]
16	1264	2/5	0.40000	$US(P_A)$	0.50185	[43]
17	2088	4/11	0.36363	$US(P_A)$	0.42843	[43]
18	2828	1/3	0.33333	$US(E_7)$	0.39108	[157]
19	3624	2/7	0.28571	$US(P_A)$	0.36116	[157]

Table D.7: Spherical codes of dimension $n = 7$.

$(2,1,4)$, where $d(\mathcal{C}_i, \mathcal{C}_j) = 2$ for $i \neq j$, implying $\rho(\mathcal{W}_i, \mathcal{W}_j) = 1$. The code $\mathcal{X}_2 : (7, 3/2, 6)$ is the trivial left 4-extension of the code $\mathcal{V}_2 : (3, 3/2, 6)$ (the extension of two copies of $-\mathcal{S}_2$ with $b_2 = \sqrt{5/8}$).

(6) The union of $\mathcal{X}_1 : (7, 7/5, 21)$ and $\mathcal{X}_2 : (7, 7/5, 14)$. The code $\mathcal{X}_1$ is the trivial extension of $\mathcal{V}_1 : (6, 7/5, 21)$ (by CW_2 from $\mathcal{C} : (7, 2, 2, 21)$). The code $\mathcal{X}_2$ is the extension of two copies of $-\mathcal{S}_6$ (by CW_2) with any b from the interval: $3/5 \leq b^2 \leq 13/20$.

(7) $Y_3\,(\mathcal{C} : (7, 3, 4, 7),\ \mathcal{A} : (3, 1, 8))$.

(8) The union of $\mathcal{X}_1$ and $\mathcal{X}_2$. The code $\mathcal{X}_1 : (7, 7/6, 35)$ is the trivial extension of $\mathcal{V}_1 : (6, 7/6, 35)$ (by CW_2 from $\mathcal{C} : (7, 3, 2, 35)$). The code $\mathcal{X}_2 : (7, 7/6, 28)$ is the extension of two copies of $\mathcal{V}_2 : (6, 5/3, 14)$ with $b = \sqrt{7/10}$, where $\mathcal{V}_2$ is the union of $\mathcal{S}_6$ (by CW_2) and $-\mathcal{S}_6$.

(9) The union of $\mathcal{X}_1$ and $\mathcal{B}_7$ (by Y_3). The code $\mathcal{X}_1 : (7, 8/7, 64)$ is the extension with $b = \sqrt{6/7}$ of $\mathcal{V}_i : (6, 4/3, 32)$, $i = 1, 2$ (by Y_2 from $\mathcal{A}_i : (6, 2, 32)$, respectively), where $d(\mathcal{A}_1, \mathcal{A}_2) = 1$, implying $\rho(\mathcal{V}_1, \mathcal{V}_2) = 2/3$.

(10) The union of $\mathcal{X} : (7, 1, 112)$ (by Y_3 from $\mathcal{C} : (7, 4, 4, 7)$ and $\mathcal{A} : (4, 1, 16)$) and $\mathcal{B}_7$ (by Y_3).

(11) The union of $\mathcal{X}_1 : (7, 1, 84)$ (by Y_3 from $\mathcal{C}_1 : (7, 2, 2, 21)$ and $\mathcal{A}_1 : (2, 1, 4)$) and $\mathcal{X}_2 : (7, 8/7, 64)$ (by Y_2 from $\mathcal{A}_2 : (7, 2, 64)$).

(12) The union of $\mathcal{X}_1, \mathcal{X}_2$, and $\mathcal{B}_7$ (by Y_3). The code $\mathcal{X}_1 : (7, 3 - \sqrt{5}, 120)$ is the extension with $b = \sqrt{3 - \sqrt{5}}$ of two copies of $\mathcal{V}_1 : (6, 1, 60)$ (by Y_3 from $\mathcal{C}_1 : (6, 2, 2, 15)$ and $\mathcal{A}_1 : (2, 1, 4)$). The code $\mathcal{X}_2 : (7, 1, 48)$ is the trivial extension of $\mathcal{V}_2 : (6, 1, 48)$ (by Y_3 from $\mathcal{C}_2 : (6, 4, 4, 3)$ and $\mathcal{A}_2 : (4, 1, 16)$).

(13) The union of $\mathcal{X}_1 : (7, 8/7, 64)$ (by Y_2 from $\mathcal{A}_1 : (7, 2, 64)$), $\mathcal{X}_2 : (7, 2/3, 280)$ (by Y_3 from $\mathcal{C}_2 : (7, 3, 2, 35)$ and $\mathcal{A}_2 : (3, 1, 8)$) and $\mathcal{B}_7$ (by Y_3).

(14) $US(P_A(\mathcal{A});\ \{3, 7\}),\ \ \mathcal{A} : (7, 4, 8),\ \ 7y^3 + y^7$.

(15) $US(E_7;\ \{2, 4\})$, the lattice E_7.

N	M	ρ	ρ_L	$Constr.$	ρ_U	$Ref.$
1	17	ρ_1	1.82453	$Ex.5.10.4$	1.96499	[157]
2	18	9/5	1.80000	CW_2	1.93159	[157]
3	22	$(15-2\sqrt{2})/7$	1.73879	$E(Y_2, Y_3)$	1.81686	[157]
4	30	$2-1/\sqrt{6}$	1.59175	$E(Y_2, Y_3)$	5/3	[157]
5	39	3/2	1.50000	$UE(Y_3)$	1.57169	[157]
6	64	4/3	1.33333	Y_3	7/5	[157]
7	71	$2-1/\sqrt{2}$	1.29289	$E(Y_3)$	1.37118	[157]
8	72	9/7	1.28571	$U(CW_2)$	1.36755	[157]
9	114	8/7	1.14285	$E(Y_3)$	1.19528	[43]
10	136	$(31-9\sqrt{2})/17$	1.07482	$E(Y_2, Y_3)$	1.14051	[43]
11	240	1	1.00000	$U(Y_3)$	1	[157]
12	310	4/5	0.80000	$US(P_A)$	0.92434	[43]
13	592	2/3	0.66666	$U(Y_2, Y_3)$	0.77795	[43]
14	996	4/7	0.57142	$US(P_A)$	0.67459	[43]
15	2400	1/2	0.50000	$US(E_8)$	0.53296	[43]
16	3616	2/5	0.40000	$US(P_A)$	0.47758	[43]
17	5118	4/11	0.36363	$US(P_A)$	0.43408	[43]
18	8880	1/3	0.33333	$US(E_8)$	0.37458	[43]
19	12128	2/7	0.28571	$US(P_A)$	0.34450	[157]

Table D.8: Spherical codes of dimension $n = 8$.

(16) $US(P_A(\mathcal{A});\ \{4,6,8,10\}),\ \ \mathcal{A}:(7,4,8),\ \ 3y^2+4y^4+y^6.$

(17) $US(P_A;\ \{7,11\})$, P_A from **(14)**).

(18) $US(E_7;\ \{4,6\})$.

(19) $US(P_A;\ \{8,10,12,14\})$, P_A from **(16)**.

D.8 Spherical codes of dimension $n = 8$

(2) $CW_2\,(\mathcal{C}:(9,4,4,18))$.

(3) The extension of $\mathcal{X}_1 : (7, 16/7, 8)$ (by Y_2 from $\mathcal{A} : (7,4,8)$) and $\mathcal{B}_7$ (by Y_3) with

$$b_1 = (1/4)\sqrt{15-2\sqrt{2}}\,.$$

(4) The extension of $\mathcal{X}_1 : (7, 12/7, 16)$ (by Y_2 from $\mathcal{A}_1 : (7, 3, 16)$) and $\mathcal{B}_7$ (by Y_3) with

$$b_1 = (1/6)\sqrt{(7/2)(12 - \sqrt{6})}\,.$$

(5) The union of $\mathcal{X}_1$ and $\mathcal{X}_2 : (8, 3, 3)$ (the trivial left 6-extension of $-\mathcal{S}_2$. The code $\mathcal{X}_1 : (8, 3/2, 36)$ is the extension of $\mathcal{V}_i : (6, 2, 12),\ i = 1, 2, 3$, with the tail code $\mathcal{S}_2$ and with $b = \sqrt{3}/2$. All $\mathcal{V}_i$ are obtained by Y_3 from $\mathcal{C}_i : (6, 2, 4, 3),\ i = 1, 2, 3$, respectively and from $\mathcal{A}_1 : (2, 1, 4)$.

(6) $Y_3\,(\mathcal{C} : (8, 3, 4, 8),\ \mathcal{A} : (3, 1, 8))$.

(7) The extension of $\mathcal{X}_1 : (7, 4/3, 56)$ (by Y_3 from $\mathcal{C}_1 : (7, 3, 4, 7)$ and $\mathcal{A}_1 : (3, 1, 8)$) and $\mathcal{B}_7$ (by Y_3), plus one point, with

$$b_1 = (1/2)\sqrt{(3/2)(4 - \sqrt{2})}\,.$$

(8) The union of $\mathcal{X} : (8, 9/7, 36)$ (by CW_2 from $\mathcal{C} : (9, 2, 2, 36)$) and $-\mathcal{X}$.

(9) The extension of $\mathcal{X}_i : (7, 4/3, 56),\ i = 1, 2$, with $b = \sqrt{6/7}$, plus two points. The code $\mathcal{X}_i$ is obtained by Y_3 from $\mathcal{C}_i : (7, 3, 4, 7),\ i = 1, 2$, respectively and $\mathcal{A} : (3, 1, 8)$, where $d(\mathcal{C}_1, \mathcal{C}_2) = 2$ implying $\rho(\mathcal{X}_1, \mathcal{X}_2) = 2/3$.

(10) The extension of $\mathcal{X}_1 : (7, 4/3, 56)$ (by Y_3 from $\mathcal{C}_1 : (7, 3, 4, 7)$ and $\mathcal{A}_1 : (3, 1, 8)$), $\mathcal{X}_2$ (by Y_2 from $\mathcal{A}_2 : (7, 2, 64)$) and $\mathcal{B}_7$ (by Y_3) with

$$b_1 = (1/2)\sqrt{3\rho}, \quad b_2 = (1/2)\sqrt{7\rho/2},$$

$$b_3 = (1/2)\sqrt{\rho(4 - \rho)}, \quad \rho = (31 - 9\sqrt{2})/17\,.$$

and with the tails $z_1 = a_1,\ z_2 = -a_2$, and $z_3 = -a_3$, plus two points.

(11) The union of $\mathcal{X} : (8, 1, 224)$ (by Y_3 from $\mathcal{C} : (8, 4, 4, 14)$ and $\mathcal{A} : (4, 1, 16)$) and $\mathcal{B}_8$ (by Y_3).

(12) $US(P_A(\mathcal{A});\ \{1, 3, 5\}),\ \mathcal{A} : (8, 4, 16),\ y + 7y^3 + 7y^5 + y^7$.

(13) The union of $X_1 : (8, 2/3, 448)$ (by Y_3 from $\mathcal{C}_1 : (8, 3, 2, 56)$ and $\mathcal{A}_1 : (3, 1, 8)$), $\mathcal{X}_2 : (8, 1, 128)$ (by Y_2 from $\mathcal{A}_2 : (8, 2, 128)$) and $\mathcal{B}_8$ (by Y_3).

(14) $US(P_A;\ \{3, 5, 7\})$, P_A from **(12)**.

(15) $US(E_8;\ \{2, 4\})$, the lattice E_8.

(16) $US(P_A(\mathcal{A});\ \{4, 6, 8\}),\ \ \mathcal{A} : (8, 4, 16),\ \ 4y^2 + 8y^4 + 4y^6$.

(17) $US(P_A;\ \{5, 7, 9, 11\})$, P_A from **(12)**.

(18) $US(E_8;\ \{4, 6\})$.

(19) $US(P_A;\ \{8, 10, 12, 14\})$, P_A from **(16)**.

D.9 Spherical codes of dimension $n = 9$

(5) $CW_2\,(\mathcal{C} : (10, 5, 4, 36))$.

(6) The union of $\mathcal{X}_1 : (9, 3/2, 36)$ (the trivial extension of $\mathcal{V}_1 : (8, 3/2, 36)$ (see **(5)** for $n = 8$)) and $\mathcal{X}_2 : (9, 3/2, 6)$ (the trivial left 6-extension of $\mathcal{V}_2 : (3, 3/2, 6)$), where $\mathcal{V}_2$ is the extension of two copies of $-\,\mathcal{S}_2$ with $b_2 = \sqrt{5/8}$.

(7) The union of $\mathcal{X}_1$ and $\mathcal{X}_2$. The code $\mathcal{X}_1 : (9, 16/11, 48)$ is the extension of $\mathcal{V}_i : (6, 2, 12),\ i = 1, 2, 3, 4$, with the tail code $\mathcal{S}_3$ and with $b = \sqrt{8/11}$. The code $\mathcal{X}_2 : (9, 8/3, 4)$ is the trivial left 6-extension of $-\,\mathcal{S}_3$. Each $\mathcal{V}_i$ is obtained by Y_3 from $\mathcal{C}_i : (6, 2, 4, 3),\ i = 1, 2, 3, 4$, respectively and from $\mathcal{A}_1 : (2, 1, 4)$, where $d(\mathcal{C}_i, \mathcal{C}_j) = 2$, implying $\rho(\mathcal{V}_i, \mathcal{V}_j) = 1$.

(6) $Y_3\,(\mathcal{C} : (9, 3, 4, 12),\ \mathcal{A} : (3, 1, 8))$.

(9) The extension of $\mathcal{X}_1$ and $\mathcal{X}_2$ with $b = \sqrt{6/7}$, plus two points. Each $\mathcal{X}_i : (8, 4/3, 64),\ i = 1, 2$, is obtained by Y_3 from $\mathcal{C}_i : (8, 3, 4, 8)$, respectively and from $\mathcal{A} : (3, 1, 8)$, where $d(\mathcal{C}_1, \mathcal{C}_2) = 2$.

N	M	ρ	ρ_L	*Constr.*	ρ_U	*Ref.*
1	19	ρ_1	1.84186	*Ex.* 5.10.4	1.97265	[157]
2	20	9/5	1.80000		1.94626	[157]
3	24	$(15-2\sqrt{2})/7$	1.73879		1.85240	[157]
4	32	$8(8-\sqrt{2})/31$	1.69955	*Ex.* 5.10.1	1.71939	[157]
5	36	8/5	1.60000	CW_2	1.67337	[157]
6	42	3/2	1.50000	$UE(Y_3)$	1.62085	[157]
7	52	16/11	1.45454	$UE(Y_3)$	1.56114	[157]
8	96	4/3	1.33333	Y_3	1.37437	[157]
9	130	8/7	1.14285	$E(Y_3)$	1.27202	[157]
10	165	12/11	1.09090	$UE(Y_3)$	1.19926	[43]
11	306	1	1.00000	$U(Y_3)$	1.05405	[43]
12	482	$4(3-\sqrt{2})/7$	0.90616	$EU(Y_2, Y_3)$	0.93851	[43]
13	610	4/5	0.80000	$US(P_A)$	0.88758	[43]
14	722	$3-\sqrt{5}$	0.76393	$UE(Y_2, Y_3)$	0.85559	[43]
15	1008	2/3	0.66666	$US(P_A)$	0.79414	[43]
16	1760	4/7	0.57142	$US(P_A)$	0.69184	[43]
17	3586	1/2	0.50000	$US(P_A)$	0.58756	[43]
18	5954	4/9	0.44444	$US(P_A)$	0.52143	[43]
19	7552	2/5	0.40000	$U(Y_2, \cdots, Y_5)$	0.49288	[43]
20	10288	4/11	0.36363	$US(P_A)$	0.45848	[43]
21	17136	1/3	0.33333	$US(P_A)$	0.40752	[157]
22	22978	4/13	0.30769	$US(P_A)$	0.37981	[157]
23	26848	2/7	0.28571	$US(P_A)$	0.36600	[157]

Table D.9: Spherical codes of dimension $n = 9$.

(10) The union of $\mathcal{X}_1$ and $\mathcal{X}_2$: $(9, 4\sin^2(\pi/5), 5)$ (the trivial left 7-extension of $\mathcal{P}_5$). The code $\mathcal{X}_1 : (9, 12/11, 160)$ is the extension of $\mathcal{V}_i : (7, 4/3, M_i)$, $i = 1, 2, 3$, with the tail code $\mathcal{S}_2$ and with $b = 3/\sqrt{11}$, where $M_1 = M_2 = 56$ and $M_3 = 48$. Each $\mathcal{V}_i$ is obtained by Y_3 from $\mathcal{C}_i : (7, 3, 4, N_i)$, $i = 1, 2, 3$, respectively and from $\mathcal{A}_1 : (3, 1, 8)$, where $d(\mathcal{C}_i, \mathcal{C}_j) = 2$ for $i \neq j$ and $N_1 = N_2 = 7$ and $N_3 = 6$.

(11) The union of $\mathcal{X} : (9, 1, 288)$ (by Y_3 from $\mathcal{C} : (9, 4, 4, 18)$ and $\mathcal{A} : (4, 1, 16)$) and $\mathcal{B}_9$ (by Y_3).

(12) The extension of $\mathcal{X}_i : (8, 1, 240)$, $i = 1, 2$, with

$$b = 2\sqrt{(1/7)(3 - \sqrt{2})}.$$

The code $\mathcal{X}_1$ is the union of $\mathcal{V}_1 : (8, 1, 224)$ (by Y_3 from $\mathcal{C}_1 : (8, 4, 4, 14)$ and $\mathcal{A}_1 : (4, 1, 16)$) and $\mathcal{B}_8$ (by Y_3), and $\mathcal{X}_2$ is the union of $\mathcal{V}_2 : (8, 1, 112)$ (by Y_3 from $\mathcal{C}_2 : (8, 2, 2, 28)$ and $\mathcal{A}_2 : (2, 1, 4)$) and $\mathcal{V}_3 : (8, 1, 128)$ (by Y_2 from $\mathcal{A}_3 : (8, 2, 128)$). This implies that $\rho(\mathcal{X}_1, \mathcal{X}_2) = 2 - \sqrt{2}$.

(13) $US(P_A(\mathcal{A});\ \{1, 5\})$, $\mathcal{A} : (9, 4, 20)$, $y + 18y^5 + y^9$.

(14) The union of $\mathcal{X}_1$ and $\mathcal{X}_2$, plus two points. The code $\mathcal{X}_1 : (9, 3 - \sqrt{5}, 480)$ is the extension of two copies of $\mathcal{V}_1 : (8, 1, 240)$ with $b = \sqrt{3 - \sqrt{5}}$, where $\mathcal{V}_1$ coincides with $\mathcal{X}_1$ in **(12)**. The code $\mathcal{X}_2 : (9, 1, 240)$ is the trivial extension of $\mathcal{V}_2 : (8, 1, 240)$, where $\mathcal{V}_2$ coincides with $\mathcal{X}_2$ in **(12)**. We have that $\rho(\mathcal{V}_1, \mathcal{V}_2) = 2 - \sqrt{2}$.

(15) $US(P_A(\mathcal{A});\ \{2, 6\})$, $\mathcal{A} : (9, 4, 16)$, $4y^2 + 12y^6$.

(16) $US(P_A(\mathcal{A});\ \{3, 7\})$, $\mathcal{A} : (9, 4, 16)$, $12y^3 + 4y^7$.

(17) $US(P_A(\mathcal{A});\ \{4, 8\})$, $\mathcal{A} : (9, 4, 20)$, $1 + 18y^4 + y^8$.

(18) $US(P_A;\ \{5, 9\})$, P_A from **(13)**.

(19) The union of $\mathcal{X}_1 : (9, 8/9, 256)$ (by Y_2 from $\mathcal{A}_1 : (9, 2, 256)$ which is even weight), $\mathcal{X}_2 : (9, 8/17, 2304)$ (by Y_4 from $\mathcal{A}_1$ and $\mathcal{C}_2 : (9, 1, 2, 9)$), $\mathcal{X}_3 : (9, 2/3, 672)$ (by Y_3 from $\mathcal{C}_3 : (9, 3, 2, 84)$ and $\mathcal{A}_3 : (3, 1, 8)$), $\mathcal{X}_4 : (9, 2/5, 4032)$ (by Y_3 from $\mathcal{C}_4 : (9, 5, 2, 126)$ and $\mathcal{A}_4$:

$(5,1,32))$, and $\mathcal{X}_5 : (9,2/5,288)$ (by Y_5 from $\mathcal{C}_5 : (9,1,2,9)$, $\mathcal{A}_{5,1} : (1,1,2)$, $\mathcal{C}_{5,0} : (8,1,2,8)$ and $\mathcal{A}_{5,01} = \mathcal{A}_{5,1}$). The code $\mathcal{X}_5$ consists of all permutations of ± 2, ± 4 and seven zeros, with the proper normalization.

(20) $US(P_A(\mathcal{A});\ \{5,7,9,11\})$, $\mathcal{A} : (9,4,20)$, $10y^3 + 6y^5 + 4y^7$.

(21) $US(P_A;\ \{8,12\})$, P_A from **(17)**.

(22) $US(P_A;\ \{9,13\})$, P_A from **(13)**.

(23) $US(P_A(\mathcal{A});\ \{8,10,12,14\})$, $\mathcal{A} : (9,4,20)$, $4y^2 + 6y^4 + 10y^6$.

D.10 Spherical codes of dimension $n = 10$

(2) The extension of $\mathcal{X}_1 : (9,8/3,4)$ (by Y_2 from $\mathcal{A} : (9,6,4)$) and $\mathcal{B}_9$ (by Y_3) with

$$b = (1/4)\sqrt{(3/2)(9 - \sqrt{3})}.$$

(3) The extension of $\mathcal{X}_1 : (9,40/21,13)$ (by CW_2 from $\mathcal{C} : (10,3,4,13)$) and $-\mathcal{S}_9$ (by CW_2) with

$$b_1 = (1/2)\sqrt{(21/170)(35 - 3\sqrt{2})}\,.$$

(4) The extension of $\mathcal{X}_1 : (9,20/9,6)$ (by Y_2 from $\mathcal{A} : (9,5,6)$) and $\mathcal{B}_9$ (by Y_3) with

$$b = (1/2)\sqrt{(1/5)(19 - \sqrt{10})}.$$

(6) The extension of $\mathcal{X}_1 : (9,5/3,30)$ (by CW_2 from $\mathcal{C} : (10,4,4,30)$) and $-\mathcal{S}_9$ (by CW_2) with

$$b_1 = (1/2)\sqrt{(3/5)(8 - \sqrt{2})}.$$

N	M	ρ	ρ_L	$Constr.$	ρ_U	$Ref.$
1	21	ρ_1	1.85593	$Ex.5.10.4$	1.97805	[157]
2	22	$(9-\sqrt{3})/4$	1.81698	$E(Y_2, Y_3)$	1.95671	[157]
3	23	$(35-3\sqrt{2})/17$	1.80925	$E(CW_2)$	1.93606	[157]
4	24	$(19-\sqrt{10})/9$	1.75974	$E(Y_2, Y_3)$	1.91619	[157]
5	34	$8(8-\sqrt{2})/31$	1.69955		1.76155	[157]
6	40	$2-\sqrt{2}/4$	1.64644	$E(CW_2)$	1.70029	[157]
7	50	$2(73-12\sqrt{2})/71$	1.57829	$E(CW_2)$	1.62999	[157]
8	57	$(17-\sqrt{10})/9$	1.53752	$E(CW_2)$	1.59564	[157]
9	66	$22/15$	1.46666	CW_2	1.56047	[157]
10	77	$2-1/\sqrt{3}$	1.42264	$U(CW_2)$	1.51114	[157]
11	104	$4/3$	1.33333	Y_3	1.43444	[157]
12	115	$2-1/\sqrt{2}$	1.29289	$E(Y_3)$	1.40855	[157]
13	156	$4(5-\sqrt{3})/11$	1.18834	$E(Y_2, Y_3)$	1.31340	[157]
14	194	$8/7$	1.14285	$E(Y_3)$	1.25316	[157]
15	500	1	1.00000	$U(Y_3)$	1.04261	[43]
16	596	$8/9$	0.88888	$UE(Y_3)$	1.00319	[43]
17	740	$(37-12\sqrt{2})/23$	0.87084	$UE(Y_2, Y_3)$	0.95038	[43]
18	1190	$4/5$	0.80000	$US(P_A)$	0.86582	[43]
19	2260	$2/3$	0.66666	$US(P_A)$	0.75490	[43]
20	3488	$4/7$	0.57142	$US(P_A)$	0.69106	[43]
21	7720	$1/2$	0.50000	$US(P_A)$	0.58292	[43]
22	13510	$4/9$	0.44444	$US(P_A)$	0.52036	[43]
23	20904	$2/5$	0.40000	$US(P_A)$	0.47782	[157]
24	26080	$4/11$	0.36363	$US(P_A)$	0.45575	[157]
25	44020	$1/3$	0.33333	$US(P_A)$	0.40740	[157]
26	66566	$4/13$	0.30769	$US(P_A)$	0.37263	[157]
27	85864	$2/7$	0.28571	$US(P_A)$	0.35284	[157]

Table D.10: Spherical codes of dimension $n = 10$.

(7) The extension of $\mathcal{X}_1 : (9, 5/3, 30)$ (from **(6)**), $\mathcal{S}_9$ (by CW_2) and $-\mathcal{S}_9$ with

$$b_1 = (6/5)(12 - \sqrt{2})/(12 + \sqrt{2}), \quad b_3 = b_1(\sqrt{3}/2).$$

and with the tails $z_1 = -a_1$, $z_2 = a_2$, $z_3 = a_3$, where b_2 can be found from the expression

$$(5/3)b_1^2 = 2 - 2(b_2 b_3/9 + a_2 a_3).$$

(8) The extension of $\mathcal{X}_1 : (9, 8/5, 36)$ (by CW_2 from $\mathcal{C} : (10, 5, 4, 36)$), $\mathcal{S}_9$ (by CW_2) and $-\mathcal{S}_9$ and with

$$b_1 = (1/6)\sqrt{(5/2)(17 - \sqrt{10})}, \quad b_2 = b_3 = 3b_1/\sqrt{10}$$

and with the tails $z_1 = -a_1$, $z_2 = a_2$, and $z_3 = a_3$, plus one point.

(9) $CW_2\,(\mathcal{C} : (11, 5, 4, 66))$.

(10) The union of $\mathcal{X} : (10, 22/15, 66)$ (by CW_2 from $C : (11, 5, 4, 66)$) and $-\mathcal{S}_{10}$ (by CW_2).

(11) $Y_3\,(\mathcal{C} : (10, 3, 4, 13),\ \mathcal{A} : (3, 1, 8))$.

(12) The extension of $\mathcal{X}_1 : (9, 4/3, 96)$ (see **(8)** for $n = 9$) and $\mathcal{B}_9$ (by Y_3), plus one point, with

$$b_1 = (1/2)\sqrt{(3/4)(4 - \sqrt{2})}.$$

(13) The extension of $\mathcal{X}_1 : (9, 4/3, 96)$ (from **(12)**), $\mathcal{X}_2 : (9, 4/3, 40)$ (by Y_2 from $\mathcal{A}_2 : (9, 3, 40)$), and $\mathcal{B}_9$ (by Y_3) with

$$b = \sqrt{(3/11)(5 - \sqrt{3})}$$

and with the tails $z_1 = a$ and $z_2 = z_3 = -a$, plus two points.

(14) The extension of $\mathcal{X}_i : (9, 4/3, 96)$, $i = 1, 2$, with $b = \sqrt{6/7}$, plus two points. The code $\mathcal{X}_i$ is obtained by Y_3 from $\mathcal{C}_i : (9, 3, 4, 12)$, $i = 1, 2$, respectively, and from $\mathcal{A} : (3, 1, 8)$, where $d(\mathcal{C}_1, \mathcal{C}_2) = 2$.

(15) The union of $\mathcal{X} : (10, 1, 480)$ (by Y_3 from $\mathcal{C} : (10, 4, 4, 30)$ and $\mathcal{A} : (4, 1, 16)$) and $\mathcal{B}_{10}$ (by Y_3).

(16) The union of $\mathcal{X}_1$ and $\mathcal{B}_{10}$ (by Y_3). The code $\mathcal{X}_1 : (10, 8/9, 576)$ is the extension with $b = 2\sqrt{2}/3$ of $\mathcal{V}_i : (9, 1, 288)$, $i = 1, 2$ (by Y_3 from $\mathcal{C}_i(9, 4, 4, 18)$, respectively, and from $\mathcal{A} : (4, 1, 16)$), where $d(\mathcal{C}_1, \mathcal{C}_2) = 2$.

(17) The union of $\mathcal{X}_i$, $i = 1, 2, 3$. The code $\mathcal{X}_1 : (10, \rho_1, 688)$, $\rho_1 = (37 - 12\sqrt{2})/23$, is the extension of $\mathcal{V}_j$, $j = 1, 2, 3$, with $b_1 = \sqrt{\rho_1}$ and with the tails

$$z_1 = (c|h), \quad z_2 = (c| - h), \quad z_3 = (-a|0),$$

where $a_1 = \sqrt{1 - b_1^2} = \sqrt{c^2 + h^2}$ and $h = b_1/2\sqrt{2}$. The codes $\mathcal{V}_j : (8, 1, 224)$, $j = 1, 2$, are obtained by Y_3 from $C_j : (8, 4, 4, 14)$, $j = 1, 2$, respectively, and from $\mathcal{A}_1 : (4, 1, 16)$, where $d(\mathcal{C}_1, \mathcal{C}_2) = 2$ implying $\rho(\mathcal{V}_1, \mathcal{V}_2) = 1/2$. The code $\mathcal{V}_3 : (8, 1, 240)$ is the union of $\mathcal{W}_1 : (8, 1, 112)$ (by Y_3 from $\mathcal{C}_3 : (8, 2, 2, 28)$ and $\mathcal{A}_2 : (2, 1, 4)$) and $\mathcal{W}_2 : (8, 1, 128)$ (by Y_2 from $\mathcal{A}_3 : (8, 2, 128)$), implying $\rho(\mathcal{V}_j, \mathcal{V}_3) = 2 - \sqrt{2}$ for $j = 1, 2$. The code $\mathcal{X}_2 : (10, \rho_2, 48)$ (where $\rho_2 = 2b_2^2$ and $b_2 = 33/50$) is the extension of three copies of $\mathcal{B}_8$ (by Y_3) with b_2 and with the tails

$$z_4 = (a_2|0), \quad z_5 = (0|a_2), \quad z_6 = (0| - a_2).$$

The code $\mathcal{X}_3 : (10, 2, 4)$ is the trivial left 8-extension of $\mathcal{B}_2$ (by Y_2 from $\mathcal{A}_2$).

(18) $US(P_A(\mathcal{A});\ \{1, 5\})$, $\mathcal{A} : (10, 4, 38)$, $y + 36y^5 + y^9$.

(19) $US(P_A(\mathcal{A});\ \{2, 6\})$, $\mathcal{A} : (10, 4, 36)$, $5y^2 + 30y^6 + y^{10}$.

(20) $US(P_A(\mathcal{A});\ \{3, 5, 7\})$, $\mathcal{A} : (10, 4, 40)$, $12y^3 + 16y^5 + 12y^7$.

(21) $US(P_A(\mathcal{A});\ \{4, 8\})$, $\mathcal{A} : (10, 4, 36)$, $1 + 30y^4 + 5y^8$.

(22) $US(P_A;\ \{5, 9\})$, P_A from **(18)**.

(23) $US(P_A;\ \{6, 10\})$, P_A from **(19)**.

(24) $US(P_A;\ \{5, 7, 9, 11\})$, P_A from **(20)**.

(25) $US(P_A;\ \{8, 12\})$, P_A from **(21)**.

(26) $US(P_A;\ \{9, 13\})$, P_A from **(18)**.

(27) $US(P_A;\ \{10, 14\})$, P_A from **(19)**.

D.11 Spherical codes of dimension $n = 11$

(2) $Y_2\,(\mathcal{A} : (11, 5, 24)$.

(4) The extension of $\mathcal{X}_1 : (10, 12/5, 6)$ (by Y_2 from $\mathcal{A} : (10, 6, 6)$) and $\mathcal{B}_{10}$ (by Y_3) with

$$b_1 = (1/3)\sqrt{5(21 - 2\sqrt{3})/13}.$$

(5) The extension of $\mathcal{X}_1 : (10, 11/6, 17)$ (by CW_2 from $C : (11, 3, 4, 17)$) and $-\mathcal{S}_{10}$ (by CW_2) with

$$b_1 = 6\sqrt{2(12 - \sqrt{2})/781}.$$

(6) The extension of $\mathcal{X}_1 : (10, 2, 12)$ (by Y_2 from $\mathcal{A} : (10, 5, 12)$) and $\mathcal{B}_{10}$ (by Y_3) with

$$b = \sqrt{2(20 - \sqrt{10})/39}\,.$$

(8) The union of $\mathcal{X}_1$ and $\mathcal{X}_2$. The code $\mathcal{X}_1 : (11, 216/131, 36)$ is the extension of $\mathcal{V}_i : (8, 27/13, 9)$, $i = 1, 2, 3, 4$, with the tail code $\mathcal{S}_3$ and with $b = 2\sqrt{26/131}$. The code $\mathcal{V}_i$ is obtained by CW_3 from $\mathcal{C}_i : (9, 3, 6, 3)$, $i = 1, 2, 3, 4$, respectively and $\mathcal{C}_5 : (3, 1, 2, 3)$ with $\beta = 1/8$, $\gamma = 5/4$. The codes $\mathcal{C}_i$, $i = 1, 2, 3, 4$, formed by the parallel classes of the resolvable Steiner triple system $S(9, 3, 2)$, implying $d(\mathcal{C}_i, \mathcal{C}_j) = 4$ for any $i \neq j$ and, therefore, $\rho(\mathcal{V}_i, \mathcal{V}_j) = 18/13$. The code $\mathcal{X}_2 : (11, 8/3, 4)$ is the trivial left 8-extension of $-\mathcal{S}_3$.

(11) The extension of $\mathcal{X}_1 : (10, 8/5, 40)$ (by Y_2 from $\mathcal{A} : (10, 4, 40)$) and $\mathcal{B}_{10}$ (by Y_3) with $b = 5(15 - 2\sqrt{2})/62$, plus one point.

(13) $Y_2\,(\mathcal{A} : (11, 4, 72)$.

(15) The union of $\mathcal{X}_1$ and $\mathcal{X}_2$. The code $\mathcal{X}_1 : (11, \rho, 72)$, (where $\rho = (17 - \sqrt{33})\,/8)$, is the extension of $\mathcal{V}_i : (10, 8/5, N_i)$, $i = 1, 2$, with $b = \sqrt{5\rho/8}$. The code $\mathcal{V}_i$ is obtained by Y_2 from $\mathcal{A}_i : (10, 4, N_i)$, $i = 1, 2$, respectively, where $\mathcal{A}_1$ and $\mathcal{A}_2$ are the (0) and (1)-reductions, respectively, of $\mathcal{A} : (11, 4, 72)$ such that $d(\mathcal{A}_1, \mathcal{A}_2) = 3$ (implying

N	M	ρ	ρ_L	*Constr.*	ρ_U	*Ref.*
1	23	ρ_1	1.86759	*Ex.* 5.10.4	1.98200	[157]
2	24	20/11	1.81818	Y_2	1.96440	[157]
3	25	$(35-3\sqrt{2})/17$	1.80925		1.94727	[157]
4	26	$4(21-2\sqrt{3})/39$	1.79855	$E(Y_2, Y_3)$	1.93065	[157]
5	28	$12(12-\sqrt{2})/71$	1.78914	$E(CW_2)$	1.89907	[157]
6	32	$4(20-\sqrt{10})/39$	1.72694	$E(Y_2, Y_3)$	1.84288	[157]
7	36	$8(8-\sqrt{2})/31$	1.69955		1.79560	[157]
8	40	216/131	1.64885	$UE(CW_2, CW_3)$	1.75605	[157]
9	42	$2-\sqrt{2}/4$	1.64644		1.73875	[157]
10	52	$2(73-12\sqrt{2})/71$	1.57829		1.67102	[157]
11	61	$4(15-2\sqrt{2})/31$	1.57052	$E(Y_2, Y_3)$	1.62870	[157]
12	68	22/15	1.46666		1.60361	[157]
13	72	16/11	1.45454	Y_2	1.59151	[157]
14	79	$2-1/\sqrt{3}$	1.42264		1.57021	[157]
15	92	$(17-\sqrt{33})/8$	1.40692	$UE(Y_2, Y_3)$	1.52629	[157]
16	94	$2-2/\sqrt{11}$	1.39697	$U(Y_2, Y_3)$	1.52059	[157]
17	136	4/3	1.33333	Y_3	1.43665	[157]
18	210	8/7	1.14285	$E(Y_3)$	1.31342	[157]
19	293	12/11	1.09090	$UE(Y_3)$	1.23656	[157]
20	582	1	1.00000	$U(Y_3)$	1.10174	[157]
21	982	8/9	0.88888	$UE(Y_3)$	0.98596	[43]
22	2154	4/5	0.80000	$US(P_A)$	0.85796	[43]
23	4604	2/3	0.66666	$US(P_A)$	0.73915	[43]
24	6792	4/7	0.57142	$US(P_A)$	0.68843	[43]
25	12994	1/2	0.50000	$US(P_A)$	0.61452	[157]
26	30418	4/9	0.44444	$US(P_A)$	0.52268	[157]
27	50768	2/5	0.40000	$US(P_A)$	0.47351	[157]
28	60184	4/11	0.36363	$US(P_A)$	0.45854	[157]
29	89124	1/3	0.33333	$US(P_A)$	0.42555	[157]
30	177746	4/13	0.30769	$US(P_A)$	0.37221	[157]
31	256696	2/7	0.28571	$US(P_A)$	0.34639	[157]

Table D.11: Spherical codes of dimension $n = 11$.

$\rho(\mathcal{V}_1, \mathcal{V}_2) = 6/5)$ and $N_1 + N_2 = 72$. The code $\mathcal{X}_2 : (11, 2, 20)$ is the trivial extension of $\mathcal{B}_{10}$ (by Y_3).

(16) The union of $\mathcal{X} : (11, 16/11, 72)$ (by Y_2 from $\mathcal{A} : (11, 4, 72)$) and $\mathcal{B}_{11}$ (by Y_3).

(17) $Y_3\,(\mathcal{C} : (11, 3, 4, 17),\ \mathcal{A} : (3, 1, 8))$.

(18) The extension of $\mathcal{X}_i : (10, 4/3, 104),\ i = 1, 2$ with $b = \sqrt{6/7}$, plus two points. Each $\mathcal{X}_i$ is obtained by Y_3 from $\mathcal{C}_i : (10, 3, 4, 13),\ i = 1, 2$, respectively, and from $\mathcal{A} : (3, 1, 8)$, where $d(\mathcal{C}_1, \mathcal{C}_2) = 2$.

(19) The union of $\mathcal{X}_1$ and $\mathcal{X}_2$. The code $\mathcal{X}_1 : (11, 12/11, 288)$ is the extension of $\mathcal{V}_i,\ i = 1, 2, 3$, with the tail code $\mathcal{S}_2$ and with $b = 3/\sqrt{11}$. Each $\mathcal{V}_i : (9, 4/3, 96)$ is obtained by Y_3 from $\mathcal{C}_i : (9, 3, 4, 12),\ i = 1, 2, 3$, respectively, and from $\mathcal{A} : (3, 1, 8)$, where $d(\mathcal{C}_i, \mathcal{C}_j) = 2$ for $i \neq j$. The code $\mathcal{X}_2 : (11, 4\sin^2(\pi/5), 5)$ is the trivial left 9-extension of $\mathcal{P}_5$.

(20) The union of $\mathcal{X} : (11, 1, 560)$ (by Y_3 from $\mathcal{C} : (11, 4, 4, 35)$ and $\mathcal{A} : (4, 1, 16)$) and $\mathcal{B}_{11}$ (by Y_3).

(21) The union of $\mathcal{X}$ and $\mathcal{B}_{11}$ (by Y_3). The code $\mathcal{X} : (11, 8/9, 960)$ is the extension of $\mathcal{V}_i : (10, 1, 480),\ i = 1, 2$, with $b = (2\sqrt{2}/3$. Each $\mathcal{V}_i$ is obtained by Y_3 from $\mathcal{C}_i(10, 4, 4, 30),\ i = 1, 2$, respectively, and from $\mathcal{A} : (4, 1, 16)$, where $d(\mathcal{C}_1, \mathcal{C}_2) = 2$.

(22) $US(P_A(\mathcal{A});\ \{1, 5\}),\ \mathcal{A} : (11, 4, 72),\ y + 66y^5 + 5y^9$.

(23) $US(P_A(\mathcal{A});\ \{2, 6\}),\ \mathcal{A} : (11, 4, 72),\ 5y^2 + 66y^6 + y^{10}$.

(24) $US(P_A(\mathcal{A});\ \{3, 7\}),\ \mathcal{A} : (11, 4, 53),\ 17y^3 + 35y^7 + y^{11}$.

(25) $US(P_A(\mathcal{A});\ \{4, 8\}),\ \mathcal{A} : (11, 4, 53),\ 1 + 35y^4 + 17y^8$.

(26) $US(P_A;\ \{5, 9\})$, P_A from **(22)**.

(27) $US(P_A;\ \{6, 10\})$, P_A from **(23)**.

(28) $US(P_A;\ \{7, 11\})$, P_A from **(24)**.

(29) $US(P_A;\ \{8, 12\})$, P_A from **(25)**.

(30) $US(P_A;\ \{9, 13\})$, P_A from **(22)**.

(31) $US(P_A;\ \{10, 14\})$, P_A from **(23)**.

N	M	ρ	ρ_L	*Constr.*	ρ_U	*Ref.*
1	25	ρ_1	1.87743	*Ex.*5.10.4	1.98498	[157]
2	26	13/7	1.85714	CW_2	1.97023	[157]
3	30	$12(12-\sqrt{2})/71$	1.78914		1.91472	[157]
4	34	$(23-\sqrt{12})/11$	1.77599	$E(Y_2, Y_3)$	1.86571	[157]
5	38	$8(8-\sqrt{2})/31$	1.69955		1.82335	[157]
6	46	$2-1/\sqrt{(10)}$	1.68377	$E(Y_2, Y_3)$	1.75599	[157]
7	52	13/8	1.62500	CW_3	1.71737	[157]
8	63	$4(15-2\sqrt{2})/31$	1.57052		1.66467	[157]
9	78	3/2	1.50000	CW_3	1.61646	[157]
10	95	$(13-2\sqrt{2})/7$	1.45308	$E(Y_2, Y_3)$	1.57431	[157]
11	168	4/3	1.33333	$U(Y_2, Y_3)$	1.44556	[157]
12	209	5/4	1.25000	$E(Y_2, Y_3)$	1.38654	[157]
13	288	6/5	1.20000	Y_4	1.30730	[157]
14	396	16/15	1.06666	$UE(Y_3)$	1.24434	[157]
15	840	1	1.00000	$U(Y_3)$	1.10412	[43]
16	1144	8/9	0.88888	$E(Y_3)$	1.03923	[43]
17	2605	4/5	0.80000	$US(P_A)$	0.90855	[43]
18	4512	8/11	0.72727	$UE(Y_2, Y_3)$	0.82171	[43]
19	8952	2/3	0.66666	$US(P_A)$	0.73277	[43]
20	13280	4/7	0.57142	$US(P_A)$	0.68422	[43]
21	27216	1/2	0.50000	$US(P_A)$	0.60528	[43]
22	49126	4/9	0.44444	$US(P_A)$	0.55130	[157]
23	120816	2/5	0.40000	$US(P_A)$	0.47132	[157]
24	141088	4/11	0.36363	$US(P_A)$	0.45889	[157]
25	231336	1/3	0.33333	$US(P_A)$	0.42065	[157]
26	1179360	1/4	0.25000	$US(K_{12})$	0.31568	[157]

Table D.12: Spherical codes of dimension $n = 12$.

D.12 Spherical codes of dimension $n = 12$

(2) CW_2 ($\mathcal{C} : (13, 6, 6, 26)$).

(3) The extension of $\mathcal{X}_1 : (11, 24/11, 12)$ (by Y_2 from $\mathcal{A} : (11, 6, 12)$) and $\mathcal{B}_{11}$ (by Y_3) with

$$b_1 = (1/2)\sqrt{(23-2\sqrt{3})/6}\,.$$

(6) The extension of $\mathcal{X}_1 : (11, 20/11, 24)$ (by Y_2 from $\mathcal{A} : (11, 5, 24)$)

and $\mathcal{B}_{11}$ (by Y_3) with

$$b_1 = (1/2)\sqrt{11(2 - 1/\sqrt{10})/5}\,.$$

(7) $CW_3\,(\mathcal{C} : (13, 4, 6, 13),\ \mathcal{C}_1 : (4, 1, 2, 4),\ \beta = 2/11,\ \gamma = 15/11)$.

(9) $CW_3\,(\mathcal{C} : (13, 4, 6, 13),\ \mathcal{C}_1 : (4, 2, 2, 6),\ \beta = 0,\ \gamma = 1)$.

(10) The extension of $\mathcal{X}_1 : (11, 16/11, 72)$ (by Y_2 from $\mathcal{A} : (11, 4, 72)$) and $\mathcal{B}_{11}$ (by Y_3), plus one point, with

$$b_1 = (1/4)\sqrt{11(13 - 2\sqrt{2})/7}\,.$$

(11) The union of $\mathcal{X}_1 : (12, 4/3, 144)$ (by Y_2 from $\mathcal{A} : (12, 4, 144)$) and $\mathcal{B}_{12}$ (by Y_3).

(12) The extension of $\mathcal{X}_1 : (11, 16/11, 72)$ (by Y_2 from $\mathcal{A} : (11, 4, 72)$) and $\mathcal{X}_2 : (11, 4/3, 136)$ (by Y_3 from $\mathcal{C} : (11, 3, 4, 17)$ and $\mathcal{A} : (3, 1, 8)$) with $b_1 = \sqrt{55}/8$, plus one point.

(13) $Y_4\,(\mathcal{A} : (12, 6, 24),\ \mathcal{C} : (12, 1, 2, 12))$.

(14) The union of $\mathcal{X}_1$ and $\mathcal{X}_2$. The code $\mathcal{X}_1$ is the extension with $b = 2/\sqrt{5}$ and with the tail code $\mathcal{S}_3$ of $\mathcal{V}_i : (9, 4/3, 96),\ i = 1, 2, 3, 4$ (by Y_3 from $\mathcal{C}_i : (9, 3, 4, 12)$, respectively, and from $\mathcal{A} : (3, 1, 8)$), where $d(\mathcal{C}_i, \mathcal{C}_j) = 2$ for $i \neq j$. The code $\mathcal{X}_2 : (12, 2 - 2/\sqrt{5}, 12)$ is the trivial left 9-extension of $\mathcal{V} : (3, 2 - 2/\sqrt{5}, 12)$ (see **(5)** for $n = 3$).

(15) The union of $\mathcal{X}_1 : (12, 1, 816)$ (by Y_3 from $\mathcal{C} : (12, 4, 4, 51)$ and $\mathcal{A} : (4, 1, 16)$) and $\mathcal{B}_{12}$ (by Y_3).

(16) The union of $\mathcal{X}_1 : (12, 8/9, 1120)$ and $\mathcal{B}_{12}$ (by Y_3). The code $\mathcal{X}_1$ is the extension of $\mathcal{V}_i : (11, 1, 560),\ i = 1, 2$, with $b = 2\sqrt{2}/3$ (by Y_3 from $\mathcal{C}_i : (11, 4, 4, 35),\ i = 1, 2$, respectively, and from $\mathcal{A} : (4, 1, 16)$), where $d(\mathcal{C}_i, \mathcal{C}_j) = 2$.

(17) $US(P_{\mathcal{A}}(\mathcal{A});\ \{1, 5\}),\ \ \mathcal{A} : (12, 4, 101),\ \ y^1 + 80y^5 + 20y^9$.

(18) The union of $\mathcal{X}_1, \mathcal{X}_2, \mathcal{X}_3$, and $\mathcal{X}_4$. The code $\mathcal{X}_1 : (12, 1, 220)$ is the trivial extension of $\mathcal{V}_1 : (11, 1, 220)$ (by Y_3 from $\mathcal{C}_1 : (11, 2, 2, 55)$ and $\mathcal{A}_1 : (2, 1, 4)$). The code $\mathcal{X}_2 : (12, 12/11, 44)$ is the extension of two copies of $\mathcal{B}_{11}$ (by Y_3) with $b_2 = 2\sqrt{2/11}$. The code $\mathcal{X}_3 : (12, 8/11, 24)$ is the extension of two copies of $\mathcal{V}_3 : (11, 24/11, 12)$ (by Y_2 from $\mathcal{A}_3 : (11, 6, 12)$) with $b_3 = 1/\sqrt{3}$. The code $\mathcal{X}_4 : (12, 8/11, 4224)$ is the extension of $\mathcal{V}_i : (11, 4/5, 2112)$, $i = 4, 5$, with $b_4 = \sqrt{10/11}$, where $\mathcal{V}_i$ are obtained by Y_3 from $\mathcal{C}_i : (11, 5, 4, 66)$, $i = 4, 5$, respectively, and from $\mathcal{A}_4 : (5, 1, 32)$ where $d(\mathcal{C}_4, \mathcal{C}_5) = 2$.

(19) $US(P_A(\mathcal{A});\ \{2, 6\})$, $\mathcal{A} : (12, 4, 144)$, $6y^2 + 132y^6 + 6y^{10}$.

(20) $US(P_A(\mathcal{A});\ \{3, 7\})$, $\mathcal{A} : (12, 4, 101)$, $20y^3 + 80y^7 + y^{11}$.

(21) $US(P_A(\mathcal{A});\ \{4, 8\})$, $\mathcal{A} : (12, 4, 104)$, $1 + 51y^4 + 51y^8 + y^{12}$.

(22) $US(P_A;\ \{5, 9\})$, P_A from **(17)**.

(23) $US(P_A;\ \{6, 10\})$, P_A from **(19)**.

(24) $US(P_A;\ \{7, 11\})$, P_A from **(20)**.

(25) $US(P_A;\ \{8, 12\})$, P_A from **(21)**.

(26) $US(K_{12};\ \{10, 12, 14, 16\})$, the Coxeter–Todd lattice K_{12}.

D.13 Spherical codes of dimension $n = 13$

(2) $CW_2\,(\mathcal{C} : (14, 5, 6, 28))$.

(3) $Y_2\,(\mathcal{A} : (13, 6, 32))$.

(4) $CW_2\,(\mathcal{C} : (14, 6, 6, 42))$.

(6) The extension of $\mathcal{X}_1 : (12, 13/7, 26)$ (by CW_2 from $\mathcal{C} : (13, 6, 6, 26)$), $\mathcal{S}_{12}$ (by CW_2) and $-\mathcal{S}_{12}$, with the tails a_1, a_2, and $-a_3$, where $\rho_2 = (13/7)b_1^2$. The coefficients b_1, b_2, b_3 are determined by the following equations:

$$(1/6)\sqrt{7/2}\; b_1 b_2 - a_1 a_2 = (1/2)\sqrt{2/7}\; b_1 b_3 - a_1 a_3 =$$

N	M	ρ	ρ_L	$Constr.$	ρ_U	$Ref.$
1	27	ρ_1	1.88585	$Ex.5.10.4$	1.98727	[157]
2	28	28/15	1.86666	CW_2	1.97474	[157]
3	32	24/13	1.84615	Y_2	1.92707	[157]
4	42	7/4	1.75000	CW_2	1.82900	[157]
5	48	$8(12-\sqrt{3})/47$	1.74773	$Ex.5.10.xx$	1.78403	[157]
6	52	ρ_2	1.67035	$E(CW_2)$	1.75879	[157]
7	56	$4(19-\sqrt{10})/39$	1.62438	$E(Y_2, Y_3)$	1.73671	[157]
8	64	20/13	1.53846	Y_2	1.70015	[157]
9	104	3/2	1.50000	Y_3	8/5	[157]
10	140	84/61	1.37704	CW_3	1.53040	[157]
11	208	4/3	1.33333	Y_3	1.45006	[157]
12	252	21/17	1.23529	CW_3	1.40167	[157]
13	306	16/13	1.23076	$E(Y_2, Y_3)$	1.35532	[157]
14	416	8/7	1.14285	Y_4	1.29213	[157]
15	1154	1	1.00000	$UE(Y_2, Y_3)$	1.11334	[43]
16	1490	$4(3-\sqrt{2})/7$	0.90616	$UE(Y_2, Y_3)$	1.06355	[43]
17	3986	4/5	0.80000	$US(P_A)$	0.91863	[43]
18	11176	2/3	0.66666	$US(P_A)$	0.77670	[43]
19	25616	4/7	0.57142	$US(P_A)$	0.67904	[43]
20	51586	1/2	0.50000	$US(P_A)$	0.61550	[157]
21	100770	4/9	0.44444	$US(P_A)$	0.55475	[157]
22	191840	2/5	0.40000	$US(P_A)$	0.49995	[157]
23	330736	4/11	0.36363	$US(P_A)$	0.45793	[157]
24	525704	1/3	0.33333	$US(P_A)$	0.42533	[157]

Table D.13: Spherical codes of dimension $n = 13$.

$$= b_2 b_3/12 + a_2 a_3 = 1 - (13/14)\ b_1^2.$$

(7) The extension of $\mathcal{X}_1 : (12, 20/12, 32)$ (by Y_2 from $\mathcal{A} : (12, 5, 32)$) and $\mathcal{B}_{12}$ (by Y_3) with

$$b_1 = 2\sqrt{(19 - \sqrt{10})/65}.$$

(8) $Y_2\,(\mathcal{A} : (13, 5, 64))$.

(9) $Y_3\,(\mathcal{C} : (13, 4, 6, 13),\ \mathcal{A} : (4, 2, 8))$.

(10) $CW_3\,(\mathcal{C} : (14, 5, 6, 28),\ \mathcal{C}_1 : (5, 1, 2, 5),\ \beta = 3/11,\ \gamma = 17/11)$.

(11) $Y_3\,(\mathcal{C} : (13, 3, 4, 26),\ \mathcal{A} : (3, 1, 8))$.

(12) $CW_3\,(\mathcal{C} : (14, 6, 6, 42),\ \mathcal{C}_1 : (6, 1, 2, 6)\ ,\ \beta = 2/5,\ \gamma = 9/5)$.

(13) The extension of $\mathcal{X}_1 : (12, 4/3, 160)$ (by Y_3 from $\mathcal{C} : (12, 3, 4, 20)$ and $\mathcal{A} : (3, 1, 8)$) and $\mathcal{X}_2 : (12, 4/3, 144)$ (by Y_2 from $\mathcal{A} : (12, 4, 144)$) with $b = 2\sqrt{3/13}$, plus two points.

(14) $Y_4\,(\mathcal{A} : (13, 6, 32),\ \mathcal{C} : (13, 1, 2, 13))$.

(15) The union of $\mathcal{X}_1$, $\mathcal{X}_2$ and $\mathcal{X}_3$, plus two points. The code $\mathcal{X}_1 : (13, 1, 816)$ is the trivial extension of $\mathcal{V}_1 : (12, 1, 816)$ (by Y_3 from $\mathcal{C} : (12, 4, 4, 51)$ and $\mathcal{A}_1 : (4, 1, 16)$). The code $\mathcal{X}_2 : (13, 1, 288)$ is the extension of two copies of $\mathcal{V}_2 : (12, 4/3, 144)$ (by Y_2 from $\mathcal{A}_2 : (12, 4, 144)$) with $b_2 = \sqrt{3}/2$. The code $\mathcal{X}_3 : (13, 1, 48)$ is the extension of two copies of $\mathcal{B}_{12}$ (by Y_3) with $b_3 = b_2$.

(16) The extension of $\mathcal{X}_1 : (12, 1, 840)$ (see **(12)** for $n = 12$) and $\mathcal{X}_2 : (12, 1, 648)$ with

$$b = 2\sqrt{(3 - \sqrt{2})/7},$$

plus two points. The code $\mathcal{X}_2$ is the union of $\mathcal{V}_1 : (12, 1, 264)$ (by Y_3 from $\mathcal{C}_1 : (12, 2, 2, 66)$ and $\mathcal{A}_1 : (2, 1, 4)$) and $\mathcal{V}_2 : (12, 1, 384)$ (by Y_3 from $\mathcal{C}_2 : (12, 8, 8, 3)$ and $\mathcal{A}_2 : (8, 2, 128)$), implying $\rho(\mathcal{X}_1, \mathcal{X}_2) = 2 - \sqrt{2}$.

(17) $US(P_{\mathcal{A}}(\mathcal{A});\ \{1,5\}),\ \ \mathcal{A}:(13,4,190),\ \ y+123y^5+65y^9+y^{13}.$

(18) $US(P_{\mathcal{A}}(\mathcal{A});\ \{2,6\}),\ \ \mathcal{A}:(13,4,198),\ \ 6y^2+166y^6+26y^{10}.$

(19) $US(P_{\mathcal{A}}(\mathcal{A});\ \{3,7\}),\ \ \mathcal{A}:(13,4,198),\ \ 26y^3+166y^7+6y^{11}.$

(20) $US(P_{\mathcal{A}}(\mathcal{A});\ \{4,8\}),\ \ \mathcal{A}:(13,4,190),\ \ 1+65y^4+123y^8+y^{12}.$

(21) $US(P_A;\ \{5,9\})$, P_A from **(17)**.

(22) $US(P_A;\ \{6,10\})$, P_A from **(18)**.

(23) $US(P_A;\ \{7,11\})$, P_A from **(19)**.

(24) $US(P_A;\ \{8,12\})$, P_A from **(20)**.

D.14 Spherical codes of dimension $n = 14$

(4) $CW_2\,(\mathcal{C}:(15,5,6,42))$.

(7) $CW_2\,(\mathcal{C}:(15,6,6,70))$.

(8) The union of $\mathcal{X}_1 : (14,5/3,70)$ (by CW_2 from $\mathcal{C} : (15,6,6,70)$) and $-\mathcal{S}_{14}$ (by CW_2).

(9) The extension of $\mathcal{X}_1$, $\mathcal{X}_2$, and $\mathcal{X}_3$, with

$$b_1 = b_2 = \sqrt{15}\,(\sqrt{57}-1)/28, \quad b_3 = 1$$

and with the tails $z_1 = a_1$, $z_2 = -a_1$, and $z_3 = 0$. Both $\mathcal{X}_i : (13,28/15,28)$, $i = 1,2$ are obtained by CW_2 from $\mathcal{C}_i : (14,5,6,28)$, $i = 1,2$, respectively, which are the (01) and (10)-reductions, respectively, of $\mathcal{C} : (16,6,6,112)$ such that $d(\mathcal{C}_1,\mathcal{C}_2) = 4$, implying $\rho(\mathcal{X}_1,\mathcal{X}_2) = 56/45$. The code $\mathcal{X}_3 : (13,14/8,42)$ is obtained by CW_2 from $\mathcal{C}_3 : (14,6,6,42)$, which is the (00)-reduction of $\mathcal{C}$ such that $d(\mathcal{C}_i,\mathcal{C}_3) = 5$, $i = 1,2$ and, therefore, $\rho(\mathcal{X}_i,\ \mathcal{X}_3) = 2-2/\sqrt{15}$, $i = 1,2$.

(10) $Y_3\,(\mathcal{C}:(14,4,6,14),\ \mathcal{A}:(4,2,8))$.

N	M	ρ	ρ_L	*Constr.*	ρ_U	*Ref.*
1	29	ρ_1	1.89314	*Ex.* 5.10.4	1.98908	[157]
2	30	28/15	1.86666		1.97830	[157]
3	34	24/13	1.84615		1.93697	[157]
4	42	9/5	1.80000	CW_2	1.86507	[157]
5	50	$8(12-\sqrt{3})/47$	1.74773		1.80792	[157]
6	64	12/7	1.71428	Y_2	1.73588	[157]
7	70	5/3	1.66666	CW_2	1.71287	[157]
8	85	$2-2/\sqrt{21}$	1.56356	$U(CW_2)$	1.66886	[157]
9	98	$(29-\sqrt{57})/14$	1.53215	$E(CW_2)$	1.64137	[157]
10	112	3/2	1.50000	Y_3	1.61887	[157]
11	156	10/7	1.42857	$U(Y_2, Y_3)$	1.54997	[157]
12	210	15/11	1.36363	CW_3	3/2	[157]
13	224	4/3	1.33333	Y_3	1.48496	[157]
14	466	6/5	1.20000	$E(Y_2, Y_3)$	1.32340	[157]
15	896	12/11	1.09090	Y_4	1.21520	[157]
16	1606	1	1.00000	$UE(Y_2, Y_3)$	1.11788	[157]
17	1996	$2-2\sqrt{2/7}$	0.93095	$U(Y_2, Y_3)$	1.07838	[43]
18	6460	4/5	0.80000	$U(Y_2, Y_3)$	0.91407	[43]
19	19180	2/3	0.66666	$U(Y_2, Y_3)$	0.77990	[43]
20	52732	4/7	0.57142	$U(Y_2, Y_3)$	0.67244	[43]
21	102136	1/2	0.50000	$US(P_A)$	0.61617	[157]
22	189958	4/9	0.44444	$US(P_A)$	0.56241	[157]
23	403768	2/5	0.40000	$US(P_A)$	0.50393	[157]
24	736224	4/11	0.36363	$US(P_A)$	0.46073	[157]
25	1254156	1/3	0.33333	$US(P_A)$	0.42552	[157]
26	7784588	1/4	0.25000	$US(P_B)$	0.32430	[157]

Table D.14: Spherical codes of dimension $n = 14$.

(11) The union of $\mathcal{X}_1 : (14, 20/14, 128)$ (by Y_2 from $\mathcal{A} : (14, 5, 128)$) and $\mathcal{B}_{14}$ (by Y_3).

(12) $CW_3\,(\mathcal{C} : (15, 5, 6, 42),\ \mathcal{C}_1 : (5, 1, 2, 5),\ \beta = 1/4,\ \gamma = 3/2)$.

(13) $Y_3\,(\mathcal{C} : (14, 3, 4, 28),\ \mathcal{A} : (3, 1, 8))$.

(14) The extension of $\mathcal{X}_1 : (13, 4/3, 208)$ (by Y_3 from $\mathcal{C}_1 : (13, 3, 4, 26)$ and $\mathcal{A}_1 : (3, 1, 8)$) and $\mathcal{X}_2 : (13, 16/13, 256)$ (by Y_2 from $\mathcal{A}_2 : (13, 4, 256)$) with $b_1 = 3/\sqrt{10}$, plus two points.

(15) $Y_4\,(\mathcal{A} : (14, 6, 64),\ \mathcal{C} : (14, 1, 2, 14))$.

(16) The union of $\mathcal{X}_1$, $\mathcal{X}_2$ and $\mathcal{X}_3$, plus two points. The code $\mathcal{X}_1 : (14, 1, 1040)$ is the trivial extension of $\mathcal{V}_1 : (13, 1, 1040)$ (by Y_3 from $\mathcal{C}_1 : (13, 4, 4, 65)$ and $\mathcal{A}_1 : (4, 1, 16)$). The code $\mathcal{X}_2 : (14, 1, 512)$ is the extension with $b_2 = \sqrt{13}/4$ of $\mathcal{V}_i : (13, 16/13, 256)$, $i = 2, 3$, (by Y_2 from $\mathcal{A}_i : (13, 4, 256)$, $i = 2, 3$, respectively), where $\mathcal{A}_2$ and $\mathcal{A}_3$ are the (1) and (0)-reductions, respectively, of $\mathcal{A} : (14, 4, 512)$ such that $d(\mathcal{A}_2, \mathcal{A}_3) = 3$, implying $\rho(\mathcal{V}_2,\ \mathcal{V}_3) = 12/13$. The code $\mathcal{X}_3 : (14, 1, 52)$ is the extension of two copies of $\mathcal{B}_{13}$ (by Y_3) with $b_3 = \sqrt{3}/2$.

(17) The union of $\mathcal{X}_1 : (14, 16/14, 512)$ (by Y_2 from $\mathcal{A}_1 : (14, 4, 512)$), $\mathcal{X}_2 : (14, 1, 1456)$ (by Y_3 from $\mathcal{C}_2 : (14, 4, 4, 91)$ and $\mathcal{A}_2 : (4, 1, 16)$) and $\mathcal{B}_{14}$ (by Y_3).

(18) The union of $\mathcal{X}_1 : (14, 4/5, 5408)$ (by Y_3 from $\mathcal{C}_1 : (14, 5, 4, 169)$ and $\mathcal{A}_1 : (5, 1, 32)$), $\mathcal{X}_2 : (14, 6/7, 1024)$ (by Y_2 from $\mathcal{A}_2 : (14, 3, 1024)$) and $\mathcal{B}_{14}$ (by Y_3).

(19) The union of $\mathcal{X}_1 : (14, 2/3, 17792)$ (by Y_3 from $\mathcal{C}_1 : (14, 6, 4, 278)$ and $\mathcal{A}_1 : (6, 1, 64)$), $\mathcal{X}_2 : (14, 6/7, 1024)$ (see **(18)** above) and $\mathcal{X}_3 : (14, 1, 364)$ (by Y_3 from $\mathcal{C}_3 : (14, 2, 2, 91)$ and $\mathcal{A}_3 : (2, 1, 4)$).

(20) The union of $\mathcal{X}_1 : (14, 4/7, 41600)$ (by Y_3 from $\mathcal{C}_1 : (14, 7, 4, 325)$ and $\mathcal{A}_1 : (7, 1, 128)$), $\mathcal{X}_2 : (14, 4/7, 2^{13})$ (by Y_2 from $\mathcal{A}_2 : (14, 2, 2^{13})$, $\mathcal{X}_3 : (14, 2/3, 2912)$ (by Y_3 from $\mathcal{C}_3 : (14, 3, 2, 364)$ and $\mathcal{A}_3; (3, 1, 8)$) and $\mathcal{B}_{14}$ (by Y_3).

(21) $US(P_A(\mathcal{A});\ \{4, 8\}),\ \ \mathcal{A} : (14, 4, 377),\ \ 1 + 91y^4 + 278y^8 + 7y^{12}$.

(22) $US(P_A(\mathcal{A});\ \{5,9\})$, $\mathcal{A}:(14,4,340)$, $y+169y^5+169y^9+y^{13}$.

(23) $US(P_A(\mathcal{A});\ \{6,10\})$, $\mathcal{A}:(14,4,377)$, $7y^2+278y^6+91y^{10}+y^{14}$.

(24) $US(P_A(\mathcal{A});\ \{7,11\})$, $\mathcal{A}:(14,4,381)$, $28y^3+325y^7+28y^{11}$.

(25) $US(P_A;\ \{8,12\})$, P_A from **(21)**.

(26) $US(P_B(\mathcal{A});\ \{20,24,28,32\})$, $\mathcal{A}:(14,8,8)$, $1+7y^8$.

D.15 Spherical codes of dimension $n = 15$

(2) $Y_2\,(\mathcal{A}:(15,7,32))$.

(7) The extension of $\mathcal{X}_1(14,12/7,64)$ (by Y_2 from $\mathcal{A}:(14,6,64)$) and $\mathcal{B}_{14}$ (by Y_3).

(8) $Y_2\,(\mathcal{A}:(15,6,128))$

(9) The union of $\mathcal{X}_1:(15,3/2,128)$ and $\mathcal{X}_2:(15,2,28)$ which is the trivial extension of $\mathcal{B}_{14}$ (by Y_3). The code $\mathcal{X}_1$ is the extension of $\mathcal{V}_i:(14,12/7,64)$, $i=1,2$, with $b=\sqrt{7/8}$, where $\mathcal{V}_i$ is obtained by Y_2 from $\mathcal{A}_i:(14,6,64)$, respectively, and where $\mathcal{A}_i$, $i=1,2$, are the (0) and (1)-reductions, respectively, of $\mathcal{A}:(15,6,128)$ such that $d(\mathcal{A}_1,\ \mathcal{A}_2)=5$ and $\rho(\mathcal{V}_1,\ \mathcal{V}_2)=10/7$.

(10) $Y_4\,(\mathcal{A}:(15,8,16),\ \mathcal{C}:(15,1,2,15))$.

(11) $Y_3\,(\mathcal{C}:(15,6,8,10),\ \mathcal{A}:(6,2,32))$.

(12) $CW_3\,(\mathcal{C}:(16,6,6,112),\ \mathcal{C}_1:(6,1,2,6),\ \beta=1/3,\ \gamma=5/3)$.

(13) The union of $\mathcal{X}:(15,8/7,960)$ (by Y_3 from $\mathcal{C}:(15,7,8,15)$ and $\mathcal{A}:(7,2,64)$) and $\mathcal{B}_{15}$ (by Y_3).

(14) $Y_4\,(\mathcal{A}:(15,6,128),\ \mathcal{C}:(15,1,2,15))$.

(15) The union of $\mathcal{X}_1$, $\mathcal{X}_2$, $\mathcal{X}_3$ and $\mathcal{X}_4$. The code $\mathcal{X}_1:(15,1,1456)$ is the trivial extension $\mathcal{V}_1:(14,1,1456)$ (by Y_3 from $\mathcal{C}_1:(14,4,4,91)$

N	M	ρ	ρ_L	*Constr.*	ρ_U	*Ref.*
1	31	ρ_1	1.89952	*Ex.* 5.10.4	1.99053	[157]
2	32	28/15	1.86666	Y_2	1.98116	[157]
3	36	24/13	1.84615		1.94502	[157]
4	44	9/5	1.80000		1.88087	[157]
5	52	$8(12-\sqrt{3})/47$	1.74773		1.82836	[157]
6	66	12/7	1.71428		44/25	[157]
7	92	$4(23-\sqrt{12})/47$	1.66262	$E(Y_2,Y_3)$	1.68274	[157]
8	128	8/5	1.60000	Y_2	1.62605	[157]
9	156	3/2	1.50000	$UE(Y_2,Y_3)$	1.58707	[157]
10	240	32/23	1.39130	Y_4	1.51493	[157]
11	320	4/3	1.33333	Y_3	1.45078	[157]
12	672	6/5	1.20000	CW_3	1.30935	[157]
13	990	8/7	1.14285	$U(Y_3)$	1.24936	[157]
14	1920	24/23	1.04347	Y_4	1.14549	[157]
15	2564	1	1.00000	$UE(Y_2,Y_3)$	1.09718	[43]
16	2734	$2-4/\sqrt{15}$	0.96720	$U(Y_2,Y_3)$	1.08735	[43]
17	9662	4/5	0.80000	$U(Y_2,Y_3)$	0.91907	[43]
18	29220	2/3	0.66666	$US(P_B)$	0.79318	[43]
19	87872	4/7	0.57142	$US(P_B)$	0.68232	[43]
20	227730	1/2	0.50000	$US(P_B)$	0.60623	[157]
21	491072	4/9	0.44444	$US(P_B)$	0.54529	[157]
22	980370	2/5	0.40000	$US(P_B)$	0.49578	[157]
23	1975872	4/11	0.36363	$US(P_B)$	0.45049	[157]
24	3755350	1/3	0.33333	$US(P_B)$	0.41255	[157]
25	28252780	1/4	0.25000	$US(P_B)$	0.31174	[157]

Table D.15: Spherical codes of dimension $n = 15$.

and $\mathcal{A}_1 : (4, 1, 16)$). The code $\mathcal{X}_2$ is the extension with $b_2 = \sqrt{7/8}$ of $\mathcal{V}_i : (14, 8/7, 512)$, $i = 2, 3$, (by Y_2 from $\mathcal{A}_i : (14, 4, 512)$, respectively) where $\mathcal{A}_2$ and $\mathcal{A}_3$ are the (1) and (0)-reductions of $\mathcal{A} : (15, 4, 1024)$, respectively, such that $d(\mathcal{A}_2, \mathcal{A}_3) = 3$, implying $\rho(\mathcal{V}_2, \ \mathcal{V}_3) = 6/7$. The code $\mathcal{X}_3 : (15, 2, 28)$ is the trivial extension of $\mathcal{B}_{14}$ (by Y_3 from $\mathcal{C}_3 : (14, 1, 2, 14)$ and $\mathcal{A}_3' : (1, 1, 2)$). The code $\mathcal{X}_4 : (15, 1, 56)$ is the extension of two copies of $\mathcal{V}_4 = \mathcal{B}_{14}'$ (by Y_3 from $\mathcal{C}_4 : (14, 2, 4, 7)$ and $\mathcal{A}_4 : (2, 1, 4)$) with $b_4 = 1/\sqrt{2}$.

(16) The union of $\mathcal{X}_1 : (15, 16/15, 1024)$ (by Y_2 from $\mathcal{A}_1 : (15, 4, 1024)$), $\mathcal{X}_2 : (15, 1, 1680)$ (by Y_3 from $\mathcal{C}_2 : (15, 4, 4, 105)$ and $\mathcal{A}_2 : (4, 1, 16)$) and $\mathcal{B}_{15}$ (by Y_3).

(17) The union of $\mathcal{X}_1 : (15, 4/5, 7584)$ (by Y_3 from $\mathcal{C}_1 : (15, 5, 4, 237)$ and $\mathcal{A}_1 : (5, 1, 32)$), $\mathcal{X}_2 : (15, 4/5, 2048)$ (by Y_2 from $\mathcal{A}_2 : (15, 3, 2048)$) and $\mathcal{B}_{15}$ (by Y_3).

(18) $US(P_B(\mathcal{A}); \ \{8, 12\}), \ \ \mathcal{A} : (15, 8, 16), \ \ 1 + 15y^8$.

(19) $US(P_B(\mathcal{A}); \ \{6, 10, 14\}), \ \ \mathcal{A} : (15, 8, 16), \ \ 10y^6 + 6y^{10}$.

(20) $US(P_B; \ \{8, 12, 16\})$, P_B from **(18)**.

(21) $US(P_B; \ \{6, 10, 14, 18\})$, P_B from **(19)**.

(22) $US(P_B; \ \{8, 12, 16, 20\})$, P_B from **(18)**.

(23) $US(P_B; \ \{10, 14, 18, 22\})$, P_B from **(19)**.

(24) $US(P_B; \ \{12, 16, 20, 24\})$, P_B from **(18)**.

(25) $US(P_B; \ \{20, 24, 28, 32\})$, P_B from **(18)**.

D.16 Spherical codes of dimension $n = 16$

(2) $CW_2\,(\mathcal{C} : (17, 8, 8, 34))$.

N	M	ρ	ρ_L	$Constr.$	ρ_U	$Ref.$
1	33	ρ_1	1.90515	$Ex.$ 5.10.4	1.99171	[157]
2	34	17/9	1.88888	CW_2	1.98349	[157]
3	38	24/13	1.84615		1.95164	[157]
4	46	9/5	1.80000		1.89417	[157]
5	62	$2-1/\sqrt{14}$	1.73273	$E(Y_2, Y_3)$	1.80629	[157]
6	68	12/7	1.71428		1.78131	[157]
7	94	$4(23-\sqrt{12})/47$	1.66262		1.70568	[157]
8	130	8/5	1.60000		1.64856	[157]
9	159	$(21-2\sqrt{3})/11$	1.59417	$E(Y_2, Y_3)$	1.61721	[157]
10	288	3/2	1.50000	$U(Y_2, Y_3)$	1.51683	[157]
11	512	4/3	1.33333	Y_3	1.39762	[157]
12	768	6/5	1.20000	Y_3	1.33033	[157]
13	1024	$2-\sqrt{2/3}$	1.18350	$U(Y_3, Y_4)$	1.29023	[157]
14	1586	16/15	1.06666	$UE(Y_2, Y_3)$	1.22869	[157]
15	4320	1	1.00000	$U(Y_2, Y_3)$	1.07816	[43]
16	4992	$2-3/2\sqrt{2}$	0.93933	$U(Y_3, Y_4)$	1.06248	[43]
17	5408	8/9	0.88888	$US(P_{B^*})$	1.05175	[43]
18	18944	4/5	0.80000	$US(P_B)$	0.89304	[43]
19	34592	3/4	0.75000	$U(Y_3, Y_4)$	0.83057	[43]
20	72608	2/3	0.66666	$U(Y_3, Y_4)$	0.75264	[43]
21	147680	3/5	0.60000	$U(Y_3, Y_4)$	0.69639	[157]
22	212480	4/7	0.57142	$US(P_B)$	0.66544	[157]
23	588480	1/2	0.50000	$US(\Lambda_{16})$	0.58459	[157]
24	1338880	4/9	0.44444	$US(P_B)$	0.52646	[157]
25	2915904	2/5	0.40000	$US(P_B)$	0.47701	[157]
26	5923328	4/11	0.36363	$US(P_B)$	0.43541	[157]
27	11756640	1/3	0.33333	$US(\Lambda_{16})$	0.39860	[157]
28	101551200	1/4	0.25000	$US(\Lambda_{16})$	0.30137	[157]

Table D.16: Spherical codes of dimension $n = 16$.

(5) The extension of $\mathcal{X}_1 : (15, 28/15, 32)$ (by Y_2 from $\mathcal{A} : (15, 7, 32)$) and $\mathcal{B}_{15}$ (by Y_3) with

$$b_1 = (1/2)\sqrt{15(2 - 1/\sqrt{11})/7}\,.$$

(9) The extension of $\mathcal{X}_1 : (15, 8/5, 128)$ (by Y_2 from $\mathcal{A} : (15, 6, 128)$) and $\mathcal{B}_{15}$ (by Y_3), plus one point, with

$$b_1 = (1/2)\sqrt{5(21 - 2\sqrt{3})/22}\,.$$

(10) The union of $\mathcal{X}_1 : (16, 3/2, 256)$ (by Y_2 from $\mathcal{N}R$) and $\mathcal{B}_{16}$ (by Y_3).

(11) $Y_3\,(\mathcal{C} : (16, 6, 8, 16),\ \mathcal{A} : (6, 2, 32))$.

(12) $Y_3\,(\mathcal{C} : (16, 5, 6, 48),\ \mathcal{A} : (5, 2, 16))$.

(13) The union of $\mathcal{X}_1 : (16, 4/3, 512)$ (by Y_4 from $\mathcal{R}(1, 4)$ and $\mathcal{C}_1 : (16, 1, 2, 16)$) and $\mathcal{X}_2 : (16, 4/3, 512)$ (by Y_3 from $\mathcal{C}_2 : (16, 6, 8, 16)$ and $\mathcal{A}_2 : (6, 2, 32)$). The code $\mathcal{C}_2$ is a subcode of $\mathcal{N}R$, and $\mathcal{A}_2$ is the even weight code.

(14) The union of $\mathcal{X}_1$ (from **(13)** above) and $\mathcal{X}_2$, plus two points. The code $\mathcal{X}_2 : (16, 16/15, 1024)$ is the trivial extension of $\mathcal{V} : (15, 16/15, 1024)$ (by Y_2 from $\mathcal{A} : (15, 4, 1024)$).

(15) The union of $\mathcal{X}_1 : (16, 1, 2240)$ (by Y_3 from $\mathcal{C}_1 : (16, 4, 4, 140)$ and $\mathcal{A}_1 : (4, 1, 16)$), $\mathcal{X}_2 : (16, 1, 2048)$ (by Y_2 from $\mathcal{A}_2 : (16, 4, 2048)$) and $\mathcal{B}_{16}$ (by Y_3).

(16) The union of $\mathcal{X}_1 : (16, 1, 3840)$ (by Y_4 from $\mathcal{R}(1, 4)$ and $\mathcal{C}_1 : (16, 2, 2, 120)$), $\mathcal{X}_2 : (16, 1, 1120)$ (by Y_3 from $\mathcal{C}_2 : (16, 4, 4, 140)$ (the subcode of $\mathcal{H}_4^*$) and $\mathcal{A}_2 : (4, 2, 8)$ which is odd weight), and $\mathcal{B}_{16}$ (by Y_3).

(17) $US(P_{B^*}(\mathcal{A});\ \{5, 7, 9\}),\ \mathcal{A} : (16, 8, 32),\ 6y^5 + 10y^7 + 10y^9 + 6y^{11}$.

(18) $US(P_B(\mathcal{A});\ \{6, 10\}),\ \mathcal{A} : (16, 8, 32),\ 16y^6 + 16y^{10}$.

(19) The union of $\mathcal{X}_1 : (16, 3/4, 30720)$ (by Y_4 from $\mathcal{NR}$ and $\mathcal{C}_1 : (16, 2, 2, 120)$), $\mathcal{X}_2 : (16, 1, 3840)$ (by Y_3 from $\mathcal{C}_2 : (16, 8, 8, 30)$ (the subcode of $\mathcal{NR}$) and $\mathcal{A}_2 : (8, 2, 128)$ which is odd weight) and $\mathcal{B}_{16}$ (by Y_3).

(20) The union of $\mathcal{X}_1 : (16, 2/3, 39360)$ (by Y_3 from $\mathcal{C}_1 : (16, 6, 4, 615)$ and $\mathcal{A}_1 : (6, 1, 64)$), $\mathcal{X}_2 : (16, 2/3, 32768)$ (by Y_4 from $\mathcal{A}_2 : (16, 4, 2048)$ and $\mathcal{C}_2 : (16, 1, 2, 16)$) and $\mathcal{X}_3 : (16, 1, 480)$ (by Y_3 from $\mathcal{C}_3 : (16, 2, 2, 120)$ and $\mathcal{A}_3 : (2, 1, 4)$).

(21) The union of $\mathcal{X}_1 : (16, 3/5, 143360)$ (by Y_4 from $\mathcal{NR}$ and $\mathcal{C}_1 : (16, 3, 2, 560)$), $\mathcal{X}_2 : (16, 1, 3840)$ (by Y_3 from $\mathcal{C}_2 : (16, 8, 8, 30)$ (the subcode of $\mathcal{NR}$) and $\mathcal{A}_2 : (8, 2, 128)$ which is even weight) and $\mathcal{X}_3 : (16, 1, 480)$ (from **(20)** above).

(22) $US(P_B;\ \{6, 10, 14\}),\ \ P_B$ from **(18)**.

(23) $US(\Lambda_{16};\ \{4, 6, 8\})$, the Barnes–Wall lattice Λ_{16}.

(24) $US(P_B;\ \{6, 10, 14, 18\}),\ \ P_B$ from **(18)**.

(25) $US(P_B(\mathcal{A});\ \{8, 12, 16, 20\}),\ \ \mathcal{A} : (16, 8, 32),\ \ 4y^4 + 24y^8 + 4y^{12}$.

(26) $US(P_B;\ \{10, 14, 18, 22\}),\ \ P_B$ from **(18)**.

(27) $US(\Lambda_{16};\ \{6, 8, 10, 12\})$.

(28) $US(\Lambda_{16};\ \{10, 12, 14, 16\})$.

D.17 Spherical codes of dimension $n = 17$

(4) The extension of $\mathcal{X}_i : (15, 32/15, 16)$, $i = 1, 2, 3$, with the tail code $\mathcal{S}_2$ and with $b = 3\sqrt{5/53}$. All $\mathcal{X}_i$ are obtained by Y_2 from $\mathcal{A}_i : (15, 8, 16)$, $i = 1, 2, 3$, respectively, which are subcodes of $\mathcal{A} : (15, 6, 128)$, implying $\rho(\mathcal{X}_i,\ \mathcal{X}_j) = 8/5$ for $i \neq j$.

(11) $CW_3(C : (18, 6, 8, 21),\ \mathcal{C}_1 : (6, 2, 2, 15),\ \beta = 1/8,\ \gamma = 5/4)$.

N	M	ρ	ρ_L	$Constr.$	ρ_U	$Ref.$
1	35	ρ_1	1.91016	$Ex.$ 5.10.4	1.99268	[157]
2	36	17/9	1.88888		1.98542	[157]
3	40	24/13	1.84615		1.95715	[157]
4	48	96/53	1.81132	$E(Y_2)$	1.90545	[157]
5	64	16/9	1.77777	$Ex.$ 5.10.1	1.82397	[157]
6	70	12/7	1.71428		1.80018	[157]
7	96	$4(23-\sqrt{12})/47$	1.66262		1.72648	[157]
8	132	8/5	1.60000		1.66922	[157]
9	161	$(21-2\sqrt{3})/11$	1.59417		1.64127	[157]
10	290	3/2	1.50000		1.54771	[157]
11	315	18/13	1.38461	CW_3	1.53556	[157]
12	546	4/3	1.33333	$UE(Y_2, Y_3)$	1.42605	[157]
13	553	$8(47-5\sqrt{6})/213$	1.30526	$E(Y_2, Y_3)$	1.42366	[157]
14	1026	$8(10-\sqrt{6})/47$	1.28519	$E(Y_3, Y_4)$	1.32849	[157]
15	1538	$(9-\sqrt{17})/4$	1.21922	$UE(Y_3, Y_4)$	1.27154	[157]
16	2050	$(19-\sqrt{105})/8$	1.09413	$UE(Y_3, Y_4)$	1.23528	[157]
17	5346	1	1.00000	$UE(Y_3, Y_4)$	1.09871	[43]
18	7906	$4(9-2\sqrt{3})/23$	0.96276	$E(Y_3)$	1.05278	[43]
19	11298	$2(4-\sqrt{7})/3$	0.9028	$UE(Y_3, Y_4)$	1.01056	[43]
20	35906	16/21	0.76190	$UE(Y_3, Y_4)$	0.87680	[43]
21	83232	2/3	0.66666	$US(P_B)$	0.79247	[43]
22	117474	16/25	0.64000	$UE(Y_3, Y_4)$	0.76124	[43]
23	204928	4/7	0.57142	$US(P_B)$	0.71903	[157]
24	815618	1/2	0.50000	$US(P_B)$	0.61026	[157]
25	1520256	4/9	0.44444	$US(P_B)$	0.56771	[157]
26	4432386	2/5	0.40000	$US(P_B)$	0.49922	[157]
27	7688064	4/11	0.36363	$US(P_B)$	0.46707	[157]
28	19747490	1/3	0.33333	$US(P_B)$	0.41682	[157]
29	196398944	1/4	0.25000	$US(P_B)$	0.31538	[157]

Table D.17: Spherical codes of dimension $n = 17$.

(12) The union of $\mathcal{X}_1$, $\mathcal{X}_2$, plus two points. The code $\mathcal{X}_1 : (17, 4/3, 512)$ is the extension with $b = 2\sqrt{2}/3$ of $\mathcal{V}_i : (16, 3/2, 256)$, $i = 1, 2$, (by Y_2 from $\mathcal{A}_i : (16, 6, 256)$, respectively), which are subcodes of $\mathcal{H}_4^*$, implying $\rho(\mathcal{V}_1,\ \mathcal{V}_2) = 1$. The code $\mathcal{X}_2 : (17, 2, 32)$ is the trivial extension of $\mathcal{B}_{16}$ (by Y_3).

(13) The extension of $\mathcal{X}_1 : (16, 3/2, 256)$ (by Y_2 from $\mathcal{NR}$) and $\mathcal{X}_2 : (16, 4/3, 296)$ (by Y_3 from $\mathcal{C}_2 : (16, 3, 4, 37)$ and $\mathcal{A}_2 : (3, 1, 8)$), plus one point

$$b_1 = (4/3)\sqrt{(47 - 5\sqrt{6})/71}\,.$$

(14) The extension of $\mathcal{X}_i : (16, 4/3, 512)$, $i = 1, 2$ (see **(13)** for $n = 16$), plus two points, with

$$b = \sqrt{6(10 - \sqrt{6})/47}\,.$$

(15) The union of $\mathcal{X}_1$ and $\mathcal{X}_2$, plus two points. The code $\mathcal{X}_1 : (17, 4/3, 512)$ is the trivial extension of $\mathcal{V}_1 : (16, 4/3, 512)$ (by Y_4 from $\mathcal{NR}$ and $\mathcal{C}_1 : (16, 1, 2, 16)$). The code $\mathcal{X}_2 : (17, \rho, 1024)$ is the extension with

$$b = (1/4)\sqrt{3(9 - \sqrt{17})}\,, \quad \rho = 4b^2/3.$$

of $\mathcal{V}_i : (16, 4/3, 512)$, $i = 2, 3$ (by Y_3 from $\mathcal{C}_i : (16, 6, 8, 16)$, $i = 2, 3$, respectively (which are subcodes of $\mathcal{NR}$), and from $\mathcal{A}_2 : (6, 2, 32)$ (which is even weight)). This implies that $\rho(\mathcal{V}_1,\ \mathcal{V}_i) = 2 - 2/\sqrt{6}$, for $i = 2, 3$, and $\rho(\mathcal{V}_2,\ \mathcal{V}_3) = 1$.

(16) The union of $\mathcal{X}_1 : (17, 4/3, 512)$ (from **(15)** above), $\mathcal{X}_2$ and $\mathcal{X}_3$, plus two points. The code $\mathcal{X}_2$ is the extension of $\mathcal{V}_2$ and $\mathcal{V}_3$ (from **(15)** again) with

$$b = (1/4)\sqrt{3(19 - \sqrt{105})/2}\,.$$

The code $\mathcal{X}_3 : (17, 4/3, 512)$ is the trivial extension of $\mathcal{V}_4 : (16, 4/3, 512)$, obtained in the same way as $\mathcal{V}_2$ and $\mathcal{V}_3$ from **(15)**, using $\mathcal{C}_4 : (16, 6, 8, 16)$ (the subcode of $\mathcal{NR}$).

(17) The union of $\mathcal{X}_1$, $\mathcal{X}_2$ and $\mathcal{X}_3$, plus two points. The code $\mathcal{X}_1; (17, 1, 1024)$ is the extension of two copies of $\mathcal{V}_1 : (16, 4/3, 512)$

(by Y_4 from $\mathcal{R}(1,4)$ and $\mathcal{C}_1 : (16,1,2,16)$) with $b = \sqrt{3}/2$. The code $\mathcal{X}_2 : (17,1,3840)$ is the trivial extension of $\mathcal{V}_2 : (16,1,3840)$ (by Y_3 from $\mathcal{C}_2 : (16,8,8,30)$ (the subcode of $\mathcal{R}(1,4)$) and from $\mathcal{A}_2 : (8,2,128)$ (which is even weight)). The codes $\mathcal{A}_1$ and $\mathcal{C}_2$ are subcodes of $\mathcal{N}R$. The code $\mathcal{X}_3 : (17,1,480)$ is the trivial extension of $\mathcal{V}_3 : (16,1,480)$ (by Y_3 from $\mathcal{C}_3 : (16,2,2,120)$ and $\mathcal{A}_3 : (2,1,4)$). This implies $\rho(\mathcal{V}_1,\ \mathcal{V}_i) = 2 - 2/\sqrt{3}$, for $i = 2,3$ and $\rho(\mathcal{V}_2,\ \mathcal{V}_3) = 1$.

(18) The extension of $\mathcal{X}_1 : (16,1,3584)$ (by Y_3 from $\mathcal{C}_1 : (16,6,6,112)$ and $\mathcal{A}_1 : (6,2,32)$) and $\mathcal{X}_2$, plus two points, with

$$b = 2\sqrt{(9 - 2\sqrt{3})/23}\,.$$

The code $\mathcal{X}_2 : (16,1,4320)$ is the union of $\mathcal{V}_2 : (16,1,3840)$ (by Y_3 from $\mathcal{C}_2 : (16,8,8,30)$ and $\mathcal{A}_2 : (8,2,128)$) and $\mathcal{V}_3 : (16,1,480)$ (by Y_3 from $\mathcal{C}_3 : (16,2,2,120)$ and $\mathcal{A}_3 : (2,1,4)$). Both $\mathcal{C}_1$ and $\mathcal{C}_2$ are subcodes of $\mathcal{N}R$, implying $\rho(\mathcal{X}_1,\ \mathcal{X}_2) = 2 - 2/\sqrt{3}$.

(19) The union of $\mathcal{X}_1$, $\mathcal{X}_2$ and $\mathcal{B}_{17}$ (by Y_3). The code $\mathcal{X}_1$ is the extension of $\mathcal{V}_1 : (16,1,3840)$ (by Y_4 from $\mathcal{R}(1,4)$ and $\mathcal{C}_1 : (16,2,2,120)$) and $\mathcal{V}_2 : (16,1,3840)$ (by Y_3 from $\mathcal{C}_2 : (16,8,8,30)$ (the subcode of $\mathcal{R}(1,4)$) and from $\mathcal{A}_2 : (8,2,128)$) with $b = (\sqrt{7}-1)/\sqrt{3}$. The code $\mathcal{X}_2 : (17,1,3584)$ is the trivial extension of $\mathcal{V}_3; (16,1,3584)$ (by Y_3 from $\mathcal{C}_3 : (16,6,6,112)$ (the subcode of $\mathcal{N}R$) and from $\mathcal{A}_3 : (6,2,32)$). Both $\mathcal{A}_2$ and $\mathcal{A}_3$ are the odd weight codes, implying $\rho(\mathcal{V}_1,\ \mathcal{V}_2) = 3/4$, and $\rho(\mathcal{V}_i,\ \mathcal{V}_3) = 2 - 2/\sqrt{3}$ for $i = 1,2$.

(20) The union of $\mathcal{X}_1$ and $\mathcal{X}_2$, plus two points. The code $\mathcal{X}_1$: $(17,16/21,35840)$ is the extension with $b_1 = 2\sqrt{5/21}$ of $\mathcal{V}_i$: $(16,4/5,17920)$, $i = 1,2$ (by Y_4 from $\mathcal{A}_i : (16,8,32)$, $i = 1,2$, respectively, and from $\mathcal{C} : (16,3,2,560)$), where $\mathcal{A}_1$ and $\mathcal{A}_2$ are subcodes of $\mathcal{N}R$ implying $\rho(\mathcal{V}_1,\ \mathcal{V}_2) = 3/5$. The code $\mathcal{X}_2$ is the extension of two copies of $\mathcal{B}_{16}$ (by Y_3) with $b_2 = \sqrt{17/21}$.

(21) $US(P_B(\mathcal{A});\ \{8,12\}),\ \ \mathcal{A} : (17,8,36),\ \ 1 + 34y^8 + y^{16}$.

(22) The union of $\mathcal{X}_1$, $\mathcal{X}_2$ and $\mathcal{B}_{17}$ (by Y_3). The code $\mathcal{X}_1$: $(17,16/25,116480)$ is the extension with $b_1 = 2\sqrt{6}/5$ of $\mathcal{V}_i$:

$(16, 2/3, 58240)$, $i = 1, 2$ (by Y_4 from $\mathcal{A}_i : (16, 8, 32)$, $i = 1, 2$, respectively, and from $\mathcal{C} : (16, 4, 2, 1820)$, where $\mathcal{A}_1$ and $\mathcal{A}_2$ are subcodes of $\mathcal{N}R$ implying $\rho(\mathcal{V}_1,\ \mathcal{V}_2) = 1/2$. The code $\mathcal{X}_2$ is the extension of two copies of $\mathcal{V}_3 : (16, 1, 480)$ (by Y_3 from $\mathcal{C}_3 : (16, 2, 2, 120)$ and $\mathcal{A}_3 : (2, 1, 4)$) with $b_2 = 4/5$.

(23) $US(P_B(\mathcal{A});\ \{6, 8, 10, 12, 14\})$, $\mathcal{A} : (17, 8, 36)$, $12y^6 + 10y^8 + 10y^{10} + 4y^{12}$.

(24) $US(P_B(\mathcal{A});\ \{8, 12, 16\})$, P_B from **(21)**.

(25) $US(P_B(\mathcal{A});\ \{6, 8, 10, 12, 14, 16, 18\})$, P_B from **(23)**.

(26) $US(P_B(\mathcal{A});\ \{8, 12, 16, 20\})$, P_B from **(21)**.

(27) $US(P_B(\mathcal{A});\ \{10, 12, 14, 16, 18, 20, 22\})$, P_B from **(23)**.

(28) $US(P_B(\mathcal{A});\ \{12, 16, 20, 24\})$, P_B from **(21)**.

(29) $US(P_B(\mathcal{A});\ \{20, 24, 28, 32\})$, P_B from **(21)**.

D.18 Spherical codes of dimension $n = 18$

(3) The extension of $\mathcal{X}_i : (14, 16/7, 8)$, $i = 1, \cdots, 5$, with the tail code $\mathcal{S}_4$ and with $b = \sqrt{35/43}$. The codes $\mathcal{X}_i$ are obtained by Y_2 from the codes $\mathcal{A}_i : (14, 8, 8)$, $i = 1, \cdots, 5$, respectively, which are subcodes of $\mathcal{A} : (14, 6, 64)$, such that $d(\mathcal{A}_i, \mathcal{A}_j) = 6$, implying $\rho(\mathcal{X}_i,\ \mathcal{X}_j) = 12/7$ for all $i \neq j$.

(5) The extension of $\mathcal{X}_1$ and $\mathcal{X}_2$ with $b_1 = \sqrt{319/5}/8$, plus one point. The code $\mathcal{X}_1 : (17, 160/87, 40)$ is the extension of $\mathcal{V}_i : (14, 20/9, 10)$, $i = 1, 2, 3, 4$, with the tail code $\mathcal{S}_3$ and with $b = (3/4)\sqrt{3/2}$. The codes $\mathcal{V}_i$ are obtained by CW_2 from $\mathcal{C}_i : (15, 6, 8, 10)$, $i = 1, 2, 3, 4$, respectively, where $d(\mathcal{C}_i, \mathcal{C}_j) = 6$ for $i \neq j$, implying $\rho(\mathcal{V}_i,\ \mathcal{V}_j) = 5/3$. The code $\mathcal{X}_2 : (17, 8/3, 4)$ is the trivial left 14-extension of $-\mathcal{S}_3$.

(6) The extension of $\mathcal{V}_i : (14, 20/9, 10)$, $i = 1, \cdots, 5$ (see **(5)** above) with the tail code $\mathcal{S}_4$ and with $b = 3/\sqrt{11}$.

N	M	ρ	ρ_L	$Constr.$	ρ_U	$Ref.$
1	37	ρ_1	1.91465	$Ex.5.10.4$	1.99349	[157]
2	38	17/9	1.88888		1.98703	[157]
3	40	80/43	1.86046	$E(Y_2)$	1.97427	[157]
4	42	24/13	1.84615		1.96178	[157]
5	45	11/6	1.83333	$E(CW_2)$	1.94361	[157]
6	50	20/11	1.81818	$E(CW_2)$	1.91508	[157]
7	52	38/21	1.80952	CW_2	1.90432	[157]
8	55	9/5	1.80000	$E(CW_2)$	17/9	[157]
9	66	16/9	1.77777		1.83953	[157]
10	70	7/4	1.75000	$E(Y_2, Y_3)$	1.82416	[157]
11	78	19/11	1.72727	CW_2	1.79700	[157]
12	88	76/45	1.68888	CW_2	1.76871	[157]
13	96	12/7	1.71428	$E(Y_3)$	1.74972	[157]
14	120	76/47	1.61702	CW_3	1.70664	[157]
15	144	8/5	1.60000	Y_3	1.67721	[157]
16	168	76/49	1.55102	CW_3	1.65597	[157]
17	292	3/2	1.50000		1.57190	[157]
18	364	38/27	1.40740	CW_3	1.54126	[157]
19	672	4/3	1.33333	Y_3	1.42559	[157]
20	1570	16/13	1.23076	$UE(Y_3, Y_4)$	1.30611	[157]
21	2053	6/5	1.20000	$UE(Y_3, Y_4)$	1.27262	[157]
22	2661	8/7	1.14285	$UE(Y_3, Y_4)$	1.23814	[157]
23	3141	$4t/(1+3t)$	1.07422	$UE(Y_3, Y_4)$	1.21289	[157]
24	7398	1	1.00000	$UE(Y_3, Y_4)$	1.10474	[43]
25	11302	ρ_2	0.93906	$UE(Y_3, Y_4)$	1.05726	[43]
26	15744	8/9	0.88888	$UE(Y_2, Y_3, Y_4)$	1.01653	[43]
27	36480	4/5	0.80000	$US(P_B)$	0.92210	[43]
28	161184	2/3	0.66666	$US(P_B)$	0.77949	[43]
29	569728	4/7	0.57142	$US(P_B)$	0.68461	[157]
30	1738952	1/2	0.50000	$US(P_B)$	0.60428	[157]
31	4548992	4/9	0.44444	$US(P_B)$	0.54248	[157]
32	10955712	2/5	0.40000	$US(P_B)$	0.49118	[157]
33	24352896	4/11	0.36363	$US(P_B)$	0.44852	[157]
34	51186916	1/3	0.33333	$US(P_B)$	0.41242	[157]
35	586862848	1/4	0.25000	$US(P_B)$	0.31219	[157]

Table D.18: Spherical codes of dimension $n = 18$.

(7) $CW_2\,(\mathcal{C} : (19,7,8,52))$.

(8) The union of $\mathcal{X}_1$ and $\mathcal{X}_2$. The code $\mathcal{X}_1 : (18, 9/5, 50)$ is the extension of five codes $\mathcal{V}_i : (14, 20/9, 10)$ (see **(5)** above) with the tail code $\mathcal{S}_4$ and with $b = \sqrt{21}/5$. The code $\mathcal{X}_2 : (18, 5/2, 5)$ is the trivial left 14-extension of $-\,\mathcal{S}_4$.

(10) The extension of $\mathcal{X} : (17, 32/17, 36)$ (by Y_2 from $\mathcal{A} : (17, 8, 36)$) and $\mathcal{B}_{17}$ (by Y_3) with $b_1 = \sqrt{119/2}\,/8$.

(11) $CW_2\,(\mathcal{C} : (19,8,8,78))$.

(12) $CW_2\,(\mathcal{C} : (19,9,8,88))$.

(13) The extension of $\mathcal{X}_i : (16, 2, 32)$, $i = 1, 2, 3$, with the tail code $\mathcal{S}_2$ and $b = \sqrt{6/7}$. All $\mathcal{X}_i$, $i = 1, 2, 3$, are obtained by Y_3 from $\mathcal{C}_i : (16, 4, 8, 4)$, respectively, and $\mathcal{A} : (4, 2, 8)$, where $d(\mathcal{C}_i,\ \mathcal{C}_j) = 6$ ($\mathcal{C}_i$ are the different parallel classes of a resolvable Steiner system $S(16, 4, 2)$) which implies $\rho(\mathcal{X}_i,\ \mathcal{X}_j) = 3/2$.

(14) $CW_3\,(\mathcal{C} : (19,5,8,12),\ \mathcal{C}_1 : (5,2,2,10),\ \beta = 1/18,\ \gamma = 10/9)$.

(15) $Y_3\,(\mathcal{C} : (18,5,8,9),\ \mathcal{A} : (5,2,16))$.

(16) $CW_3\,(\mathcal{C} : (19,6,8,28),\ \mathcal{C}_1 : (6,1,2,6),\ \beta = 4/15,\ \gamma = 23/15)$.

(18) $CW_3\,(\mathcal{C} : (19,7,8,52),\ \mathcal{C}_1 : (7,1,2,7),\ \beta = 5/14,\ c = 12/7)$.

(19) $Y_3\,(\mathcal{C} : (18,6,8,21),\ \mathcal{A} : (6,2,32))$.

(20) The union of $\mathcal{X}_1$, $\mathcal{X}_2$, and $\mathcal{X}_3$, plus two points. The code $\mathcal{X}_1$: $(18, 16/13, 512)$ is the extension of $\mathcal{V}_1$: $(16, 4/3, 512)$ (by Y_4 from $\mathcal{R}(1,4)$ and $\mathcal{C}_1$: $(16,1,2,16)$) with $b_1 = 2\sqrt{3/13}$ and with the tail $z_1 = (a_1|0)$. The code $\mathcal{X}_2$: $(18, 16/13, 1024)$ is the extension of $\mathcal{V}_i$: $(16, 4/3, 512)$, $i = 2, 3$ (by Y_3 from $\mathcal{C}_i$: $(16, 6, 8, 16)$, $i = 2, 3$ respectively, (the subcodes of $\mathcal{NR}$) and $\mathcal{A}_2$: $(6, 2, 32)$ (which is even weight) with $b_2 = b_1$ and with the tails $z_{2,1} = (0|a_2)$ and $z_{2,2} = (0|-a_2)$. The code $\mathcal{X}_3 : (18, 16/13, 32)$ is the extension of $\mathcal{B}_{16}$ (by Y_3) with $b_3 = 2\sqrt{2/13}$ and with the tail $z_3 = (-h|g)$ where $h = 2/\sqrt{13}$ and $g = 1/\sqrt{13}$. This implies that $\rho(\mathcal{V}_1, \mathcal{V}_i) = 2 - 2/\sqrt{6}$ for $i = 2, 3$ and $\rho(\mathcal{V}_2, \mathcal{V}_3) = 1$. The additional points are $(0, 0, \ldots, 0|0| \pm 1)$.

(21) The union of $\mathcal{X}_1$, $\mathcal{X}_2$ and $\mathcal{X}_3$. the code $\mathcal{X}_1 : (18, 4/3, 512)$ is the trivial extension of $\mathcal{V}_1$ (see **(20)**). The code $\mathcal{X}_2 : (18, 6/5, 1536)$ is the extension of $\mathcal{V}_i : (16, 4/3, 512)$, $i = 2, 3, 4$, (by Y_3 from $\mathcal{C}_i$: $(16, 6, 8, 16)$, $i = 2, 3, 4$, respectively, and from $\mathcal{A}_2 : (6, 2, 32)$; see **(20)** again) with $b = 3/\sqrt{10}$ and with the tail code $\mathcal{S}_2$. The code $\mathcal{X}_3 : (18, 4\sin^2(\pi/5), 5)$ is the trivial left 16-extension of $\mathcal{P}_5$.

(22) The union of $\mathcal{X}_1$ (from **(21)** above), $\mathcal{X}_2$, $\mathcal{X}_3$ and $\mathcal{X}_4$. The code $\mathcal{X}_2$ is the extension of $\mathcal{V}_i : (16, 4/3, 512)$, $i = 2, 3, 4, 5$, (by Y_3 from $\mathcal{C}_i : (16, 6, 8, 16)$, $i = 2, 3, 4, 5$, respectively, and $\mathcal{A}_2; (6, 2, 32)$; see **(20)**) with $b = \sqrt{6/7}$ and with the tail code $\mathcal{B}_2$. The code $\mathcal{X}_3$: $(18, 3/2, 96)$ is the trivial extension of $\mathcal{V}_6 : (16, 3/2, 96)$ (by Y_2 from $\mathcal{A}_6 : (16, 6, 96)$), where

$$\mathcal{A}_6 = \mathcal{NR} \setminus \{\{ \bigcup_{i=2,3,4,5} \mathcal{C}_i\} \bigcup \{ \bigcup_{i=2,3,4,5} \bar{\mathcal{C}}_i\} \bigcup \mathcal{R}(1,4)\},$$

where $\bar{\mathcal{C}}_i$ denotes the binary complement of $\mathcal{C}_i$, i.e. the code $\bar{\mathcal{C}}_i$: $(16, 10, 8, 16)$. The code $\mathcal{X}_4$ is $\mathcal{X}_3$ from **(21)** above.

(23) Here $t = \rho(\mathcal{P}_5) = 4\sin^2(\pi/5)$. The union of $\mathcal{X}_1$ (from **(21)**), $\mathcal{X}_2$, $\mathcal{X}_3$ and $\mathcal{X}_4$ (the code $\mathcal{X}_3$ from **(21)**). The code $\mathcal{X}_2 : (18, 4t/(1 + 3t), 2560)$ is the extension of $\mathcal{V}_i : (16, 4/3, 512)$, $i = 2, \cdots, 6$, (by Y_3 from $\mathcal{C}_i : (16, 6, 8, 16)$, $i = 2, \cdots, 6$, respectively, and from $\mathcal{A}_2; (6, 2, 32)$; see **(20)**) with $b = \sqrt{3t/(1 + 3t)}$ and with the tail code $\mathcal{P}_5$. The code $\mathcal{X}_3 : (18, 3/2, 64)$ is the trivial extension of $\mathcal{V}_7 : (16, 3/2, 64)$ (by Y_2 from $\mathcal{A}_7 : (16, 6, 64)$), where

$$\mathcal{A}_7 = \mathcal{NR} \setminus \{\{ \bigcup_{i=2,\cdots,6} \mathcal{C}_i\} \bigcup \{ \bigcup_{i=2,\cdots,6} \bar{\mathcal{C}}_i\} \bigcup \mathcal{R}(1,4)\}.$$

(24) The union of $\mathcal{X}_i$, $i = 1, \cdots, 6$. The code $\mathcal{X}_1; (18, 1, 1024)$ is the extension of two copies of $\mathcal{V}_1 : (16, 4/3, 512)$ (by Y_4 from $\mathcal{R}(1,4)$ and $\mathcal{C}_1 : (16, 1, 2, 16)$) with $b = \sqrt{3}/2$ and with the tails $z_{1,1} = (a, a)/\sqrt{2}$ and $z_{1,2} = (-a, -a)/\sqrt{2}$. The code $\mathcal{X}_2 : (18, 1, 3840)$ is the trivial extension of $\mathcal{V}_2 : (16, 1, 3840)$ (by Y_3 from $\mathcal{C}_2 : (16, 8, 8, 30)$ (the subcode of $\mathcal{R}(1,4)$) and from $\mathcal{A}_2 : (8, 2, 128)$ which is even weight), implying $\rho(\mathcal{V}_1, \mathcal{V}_2) = 2 - 2/\sqrt{3}$. The code $\mathcal{X}_3 : (18, 1, 1024)$ is the extension of two copies of $\mathcal{V}_3 : (16, 4/3, 512)$ (the same as $\mathcal{V}_3$ in **(20)** above) with the same b and with the tails $z_{3,1} = (a|0)$ and $z_{3,2} = (-a|0)$. The code

$\mathcal{X}_4 : (18, 1, 1024)$ is the extension of two copies of $\mathcal{V}_4 : (16, 4/3, 512)$ (see again **(20)** above) with the same b and with the tails $z_{4,1} = (0|a)$ and $z_{4,2} = (0|-a)$. Therefore, $\rho(\mathcal{V}_1, \mathcal{V}_i) = 2-2/\sqrt{6}$ for $i = 3, 4$. The code $\mathcal{X}_5 : (18, 1, 480)$ is the trivial extension of $\mathcal{V}_5 : (16, 1, 480)$ (by Y_3 from $\mathcal{C}_5 : (16, 2, 2, 120)$ and $\mathcal{A}_5 : (2, 1, 4)$). The code $\mathcal{X}_6 : (18, 1, 6)$ is the trivial left 16-extension of $\mathcal{P}_6$.

(25) The union of $\mathcal{X}_1$, $\mathcal{X}_2$, and $\mathcal{X}_3$ (which coincides with $\mathcal{X}_6$ in **(24)** above). The code $\mathcal{X}_1 : (18, \rho_2, 11264)$ is the extension of $\mathcal{V}_1$, $\mathcal{V}_2$ and $\mathcal{V}_3$ (see **(19)** for $n = 17$) with b and with the tails $z_1 = (c|h)$, $z_2 = (c|-h)$ and $z_3 = (-a|0)$ where $h = b/4$ and $c^2 = 1 - 17b^2/16$. Here $b^2 = \rho_2$ is the root of the quadratic equation $x^2k_2 + xk_1 - 3 = 0$, where

$$k_2 = t^2 - 17/16, \quad k_1 = 33/16 - 4t, \quad t = 1 + 2/\sqrt{3}.$$

The code $\mathcal{X}_2 : (18, 2, 32)$ is the trivial extension of $\mathcal{B}_{16}$ (by Y_3).

(26) The union of $\mathcal{X}_1$, $\mathcal{X}_2$, $\mathcal{X}_3$ and $\mathcal{X}_4$. The code $\mathcal{X}_1 : (18, 8/9, 15360)$ is the extension of $\mathcal{V}_i$, $i = 1, 2, 3, 4$, with $b_1 = 2\sqrt{2}/3$ and with the tail code $\mathcal{B}_2$ (by Y_2 from $\mathcal{A} : (2, 1, 4)$). The codes $\mathcal{V}_i : (16, 1, 3840)$, $i = 1, 2, 3, 4$, are obtained by Y_4 from $\mathcal{A}_i : (16, 8, 32)$, respectively, and $\mathcal{C}_1 : (16, 2, 2, 120)$, where all $\mathcal{A}_i$ are subcodes of $\mathcal{N}R$, and therefore $\rho(\mathcal{V}_i, \mathcal{V}_j) = 3/4$ for $i \neq j$. The code $\mathcal{X}_i : (18, 8/9, 128)$, $i = 2, 3$ is the extension of two copies of $\mathcal{V}_j : (16, 3/2, 64)$, $j = 5, 6$, respectively with $b_2 = 4/3\sqrt{3}$ and with the tails $z_{2,1} = (a_2|0)$, $z_{2,2} = (-a_2|0)$ for $\mathcal{V}_5$ and $z_{3,1} = (0|a_2)$, $z_{3,2} = (0|-a_2)$ for $\mathcal{V}_6$. Both $\mathcal{V}_j$, $j = 5, 6$ are obtained by Y_2 from $\mathcal{A}_j : (16, 6, 64)$, $j = 5, 6$, respectively, where $\cup_{i=1,\cdots,6} \mathcal{A}_i = \mathcal{N}R$. This implies $\rho(\mathcal{V}_i, \mathcal{V}_j) = 2 - 3/\sqrt{2}$ for all $i = 1, 2, 3, 4$, and $j = 5, 6$. The code $\mathcal{X}_4 : (18, 8/9, 128)$ is the extension of four copies of $\mathcal{B}_{16}$ (by Y_3) with $b_3 = 2/3$ and with the tail code $\mathcal{B}_2$ (by Y_2 from $\mathcal{A} : (2, 1, 4)$).

(27) $US(P_B(\mathcal{A});\ \{6, 10\})$, $\mathcal{A} : (18, 8, 64)$, $20y^6 + 40y^{10} + 4y^{14}$.

(28) $US(P_B(\mathcal{A});\ \{4, 8, 12\})$, $\mathcal{A} : (18, 8, 64)$, $4y^4 + 40y^8 + 20y^{12}$.

(29) $US(P_B(\mathcal{A});\ \{6, 10, 14\})$, P_B from **(27)**.

(30) $US(P_B(\mathcal{A});\ \{8, 12, 16\})$, $\mathcal{A} : (18, 8, 64)$, $1 + 46y^8 + 16y^{12} + y^{16}$.

(31) $US(P_B(\mathcal{A});\ \{6, 10, 14, 18\})$, P_B from **(27)**.

(32) $US(P_B(\mathcal{A});\ \{8, 12, 16, 20\}),\ \ P_B$ from **(28)**.

(33) $US(P_B(\mathcal{A});\ \{10, 14, 18, 22\}),\ \ P_B$ from **(27)**.

(34) $US(P_B(\mathcal{A});\ \{12, 16, 20, 24\}),\ \ P_B$ from **(30)**.

(35) $US(P_B(\mathcal{A});\ \{20, 24, 28, 32\}),\ \ P_B$ from **(30)**.

D.19 Spherical codes of dimension $n = 19$

(2) $CW_2\,(\mathcal{C} : (20, 6, 8, 40))$.

(3) The extension of $\mathcal{X}_1$ and $\mathcal{X}_2$ with $b_1 = \sqrt{895/14}\,/8$, plus one point. The code $\mathcal{X}_1 : (18, 336/179, 36)$ is the extension of six codes $\mathcal{V}_i : (13, 7/3, 6),\ i = 1, \ \cdots\ , 6$, with the tail code $\mathcal{S}_5$ and with $b = 12/\sqrt{179}$. The codes $\mathcal{V}_i,\ i = 1, \ \cdots\ , 6$ are obtained by CW_2 from $\mathcal{C}_i : (14, 6, 8, 6)$, respectively. Each $\mathcal{C}_i$ is the (00)-reduction of $\mathcal{C}'_i : (16, 6, 8, 16)$, where all $\mathcal{C}'_i$ are the subcodes of $\mathcal{C}' : (16, 6, 6, 112)$. This implies that $\rho(\mathcal{V}_i,\ \mathcal{V}_j) = 7/4$ for $i \neq j$. The code $\mathcal{X}_2 : (18, 12/5, 6)$ is the trivial left 13-extension of $-\,\mathcal{S}_5$, implying $\rho(\mathcal{X}_1,\ \mathcal{X}_2) = 2 - 2a/5$ where $a = \sqrt{35/179}$.

(4) The union of $\mathcal{X}_1$ and $\mathcal{X}_2$. The code $\mathcal{X}_1 : (19, 13/7, 42)$ is the extension of $\mathcal{V}_i : (13, 7/3, 6),\ i = 1, \ \cdots\ , 7$ (see **(3)**) with the tail code $\mathcal{S}_6$ and with $b = 2\sqrt{10}/7$. The code $\mathcal{X}_2 : (19, 7/3, 7)$ is the trivial left 13-extension of $-\,\mathcal{S}_6$.

(5) The union of $\mathcal{X}_1$ and $\mathcal{X}_2$. The code $\mathcal{X}_1 : (19, 11/6, 48)$ is the extension of $\mathcal{V}_i,\ i = 1, \ \cdots\ , 6$, with the tail code $\mathcal{S}_5$ and with $b = \sqrt{119}/12$. The codes $\mathcal{V}_i : (14, 16/7, 8),\ i = 1, \ \cdots\ , 6$ are obtained by Y_2 from $\mathcal{A}_i(14, 8, 8)$, respectively, which are subcodes of $\mathcal{A} : (14, 6, 64)$ such that $\rho(\mathcal{V}_i,\ \mathcal{V}_j) = 12/7$ for $i \neq j$. The code $\mathcal{X}_2 : (19, 12/5, 6)$ is the trivial left 14-extension of $-\,\mathcal{S}_5$.

(6) The extension of $\mathcal{X}_1$ and $\mathcal{X}_2$ with $b_1 = 1$, plus one point. The code $\mathcal{X}_1 : (18, 20/11, 50)$ is the extension of $\mathcal{V}_i : (14, 20/9, 10),\ i = 1, \ \cdots\ , 5$, where $\rho(\mathcal{V}_i,\ \mathcal{V}_j) = 5/3$, (see **(5)** for $n = 18$) with the tail

N	M	ρ	ρ_L	$Constr.$	ρ_U	$Ref.$
1	39	ρ_1	1.91870	$Ex.5.10.4$	1.99417	[157]
2	40	40/21	1.90476	CW_2	1.98838	[157]
3	43	15/8	1.87500	$E(CW_2)$	1.97129	[157]
4	49	13/7	1.85714	$UE(CW_2)$	1.93870	[157]
5	54	11/6	1.83333	$UE(Y_2)$	1.91350	[157]
6	56	20/11	1.81818	$E(CW_2)$	1.90397	[157]
7	66	240/133	1.80451	$E(CW_2)$	1.86096	[157]
8	68	16/9	1.77777		1.85327	[157]
9	85	160/91	1.75824	$UE(Y_2)$	1.79848	[157]
10	100	12/7	1.71428	$E(Y_2, Y_3)$	1.76264	[157]
11	132	32/19	1.68421	$UE(Y_2)$	1.71070	[157]
12	160	160/99	1.61616	CW_2	1.68137	[157]
13	192	8/5	1.60000	Y_3	1.65806	[157]
14	240	20/13	1.53846	CW_3	1.62461	[157]
15	294	3/2	1.50000		1.59439	[157]
16	560	32/23	1.39130	CW_3	1.49269	[157]
17	896	4/3	1.33333	Y_3	1.41231	[157]
18	2666	32/27	1.18518	$UE(Y_2, Y_3, Y_4)$	1.27737	[157]
19	3626	8/7	1.14285	$UE(Y_2, Y_3, Y_4)$	1.23261	[157]
20	10668	1	1.00000	$U(Y_3)$	1.10207	[157]
21	22144	32/35	0.91428	$U(Y_2, Y_4)$	1.01963	[43]
22	28096	8/9	0.88888	$US(P_{B^*})$	0.99066	[43]
23	69248	4/5	0.80000	$US(P_B)$	0.90578	[43]
24	158736	8/11	0.72727	$US(P_{B^*})$	0.82458	[43]
25	333024	2/3	0.66666	$US(P_B)$	0.77150	[157]
26	1263488	4/7	0.57142	$US(P_B)$	0.67023	[157]
27	4076050	1/2	0.50000	$US(P_B)$	0.59216	[157]
28	11374976	4/9	0.44444	$US(P_B)$	0.53116	[157]
29	28810048	2/5	0.40000	$US(P_B)$	0.48123	[157]
30	67034496	4/11	0.36363	$US(P_B)$	0.43966	[157]
31	146139198	1/3	0.33333	$US(P_B)$	0.40415	[157]
32	577390080	2/7	0.28571	$US(P_B)$	0.34845	[157]
33	1926769668	1/4	0.25000	$US(P_B)$	0.30599	[157]

Table D.19: Spherical codes of dimension $n = 19$.

code $\mathcal{S}_4$ and with $b = 3/\sqrt{11}$. The code $\mathcal{X}_2 : (18, 5/2, 5)$ is the trivial left 14-extension of $-\mathcal{S}_4$.

(7) The union of $\mathcal{X}_1$ and $\mathcal{X}_2$. The code $\mathcal{X}_1 : (19, 240/133, 60)$ is the extension of $\mathcal{V}_i : (14, 20/9, 10)$, $i = 1, \cdots, 6$, (see **(6)**) with the tail code $\mathcal{S}_5$ and $b = 6\sqrt{3/133}$. The code $\mathcal{X}_2 : (19, 12/5, 6)$ is the trivial left 14-extension of $-\mathcal{S}_5$.

(9) The union of $\mathcal{X}_1$ and $\mathcal{X}_2$. The code $\mathcal{X}_1 : (19, 160/91, 80)$ is the extension of $\mathcal{V}_i$, $i = 1, \cdots, 5$, with the tail code $\mathcal{S}_4$ and with $b = 5\sqrt{3/91}$. Each $\mathcal{V}_i : (15, 32/15, 16)$, $i = 1, \cdots, 5$ is obtained by Y_2 from $\mathcal{A}_i : (15, 8, 16)$, respectively, which are subcodes of $\mathcal{A} : (15, 6, 128)$, such that $\rho(\mathcal{V}_i, \mathcal{V}_j) = 8/5$. The code $\mathcal{X}_2 : (19, 5/2, 5)$ is the trivial left 15-extension of $-\mathcal{S}_4$.

(10) The extension of $\mathcal{X}_1 : (18, 16/9, 64)$ (by Y_2 from $\mathcal{A} : (18, 8, 64)$) and $\mathcal{B}_{18}$ (by Y_3) with $b_1 = (3/2)\sqrt{3/7}$.

(11) The union of $\mathcal{X}_1$ and $\mathcal{X}_2$. The code $\mathcal{X}_1 : (19, 32/19, 128)$ is the extension of $\mathcal{V}_i$, $i = 1, 2, 3, 4$, with the tail code $\mathcal{S}_3$ and with $b = 4/\sqrt{19}$. The codes $\mathcal{V}_i : (16, 2, 32)$ are obtained by Y_2 from $\mathcal{A}_i : (16, 8, 32)$ which are subcodes of $\mathcal{N}R$, such that $\rho(\mathcal{V}_i, \mathcal{V}_j) = 3/2$. The code $\mathcal{X}_2 : (19, 8/3, 4)$ is the trivial left 16-extension of $-\mathcal{S}_3$.

(12) $CW_2\,(\mathcal{C} : (20, 9, 8, 160))$.

(13) $Y_3\,(\mathcal{C} : (19, 5, 8, 12),\ \mathcal{A} : (5, 2, 16))$.

(14) $CW_3\,(\mathcal{C} : (20, 6, 8, 40),\ \mathcal{C}_1 : (6, 1, 2, 6),\ \beta = 1/4,\ \gamma = 3/2)$.

(16) $CW_3\,(\mathcal{C} : (20, 7, 8, 80),\ \mathcal{C}_1 : (7, 1, 2, 7),\ \beta = 1/3,\ \gamma = 5/3)$.

(17) $Y_3\,(\mathcal{C} : (19, 6, 8, 28),\ \mathcal{A} : (6, 2, 32))$.

(18) The union of $\mathcal{X}_1$, $\mathcal{X}_2$, $\mathcal{X}_3$ and $\mathcal{X}_4$. The code $\mathcal{X}_1 : (19, 32/27, 2048)$ is the extension of $\mathcal{V}_i : (16, 4/3, 512)$, $i = 1, 2, 3, 4$, with $b = 2\sqrt{2}/3$ and with the tail code $\mathcal{S}_3$. The codes $\mathcal{V}_i$ are obtained by Y_3 from $\mathcal{C}_i : (16, 6, 8, 16)$ and $\mathcal{A}_1; (6, 2, 32)$ (see $\mathcal{V}_i$ in **(20)** for $n = 18$). The code $\mathcal{X}_2 : (19, 4/3, 512)$ is the trivial extension of the code $\mathcal{V}_5 : (16, 4/3, 512)$ (by Y_4 from $\mathcal{R}(1, 4)$ and $\mathcal{C}_5 : (16, 1, 2, 16)$ (see again **(20)** for $n = 18$)). The code $\mathcal{X}_3 : (19, 4/3, 96)$ is the extension of $\mathcal{V}_6 : (16, 3/2, 96)$ with the same b and with the tail $z_3 = ay$ where $y \in -\mathcal{S}_3$. The code $\mathcal{V}_6$ is obtained by Y_2 from $\mathcal{A}_6 : (16, 6, 96)$ where

$$\mathcal{A}_6 = \mathcal{NR} \setminus \{\{ \bigcup_{i=1,2,3,4} \mathcal{C}_i\} \bigcup \{ \bigcup_{i=1,2,3,4} \bar{\mathcal{C}}_i\} \bigcup \mathcal{R}(1,4)\}.$$

The code $\mathcal{X}_4 : (19, \rho_2, 10)$ is the trivial left 16-extension of $\mathcal{W} : (3, \rho_2, 10)$ (see **(4)** for $n = 3$).

(19) The union of $\mathcal{X}_1$, $\mathcal{X}_2$, $\mathcal{X}_3$ and $\mathcal{X}_4$ (which is $\mathcal{X}_4$ from **(18)**). The code $\mathcal{X}_1 : (19, 8/7, 3072)$ is the extension of $\mathcal{V}_i : (16, 4/3, 512)$, $i = 1, \cdots, 6$, with $b = \sqrt{6/7}$ and with the tail code $\mathcal{B}_3$ (by Y_3). All $\mathcal{V}_i$ are the same as in **(18)**. The code $\mathcal{X}_2 : (19, 4/3, 512)$ is the trivial extension of $\mathcal{V}_7 : (16, 4/3, 512)$, obtained by Y_4 (see again **(18)**). The code $\mathcal{X}_3 : (19, 2, 32)$ is the trivial extension of $\mathcal{B}_{16}$, obtained by Y_2 from $\mathcal{A}_8 : (16, 8, 32)$ (the last subcode of $\mathcal{NR}$, which has not been used for the construction of $\mathcal{V}_i$, $i = 1, \cdots, 7$).

(20) The union of $\mathcal{X}_1 : (19, 1, 9984)$ (by Y_3 from $\mathcal{C}_1 : (19, 8, 8, 78)$ and $\mathcal{A}_1 : (8, 2, 128)$) and $\mathcal{X}_2 : (19, 1, 684)$ (by Y_3 from $\mathcal{C}_2 : (19, 2, 2, 171)$ and $\mathcal{A}_2 : (2, 1, 4)$).

(21) The union of $\mathcal{X}_1 : (19, 32/35, 21888)$ (by Y_4 from $\mathcal{A}_1 : (19, 8, 128)$ and $\mathcal{C}_1 : (19, 2, 2, 171)$) and $\mathcal{X}_2 : (19, 28/19, 256)$ (by Y_2 from $\mathcal{A}_2 : (19, 7, 256)$), where $\mathcal{A}_1$ is the subcode of $\mathcal{A}_2$, implying $\rho(\mathcal{X}_1, \mathcal{X}_2) = 2 - 26/\sqrt{665}$.

(22) $US(P_{B^*}(\mathcal{A});\ \{5, 9\}),\ \mathcal{A} : (19, 8, 128),\ 12y^5 + 88y^9 + 28y^{13}$.

(23) $US(P_B(\mathcal{A});\ \{6, 10\}),\ \mathcal{A} : (19, 8, 128),\ 28y^6 + 88y^{10} + 12y^{14}$.

(24) $US(P_{B^*}(\mathcal{A});\ \{3, 7, 11\}),\ \mathcal{A}$: $(19, 8, 128),\ y^3 + 48y^7 + 78y^{11} + y^{19}$.

(25) $US(P_B(\mathcal{A});\ \{4, 8, 12\}),\ \mathcal{A} : (19, 8, 128),\ 4y^4 + 72y^8 + 52y^{12}$.

(26) $US(P_B;\ \{6, 10, 14\}),\ P_B$ from **(23)**.

(27) $US(P_B(\mathcal{A});\ \{8, 12, 16\}),\ \mathcal{A} : (19, 8, 128),\ 1 + 78y^8 + 48y^{12} + y^{16}$.

(28) $US(P_B;\ \{6, 10, 14, 18\})$, P_B from **(23)**.

(29) $US(P_B;\ \{8, 12, 16, 20\})$, P_B from **(25)**.

(30) $US(P_B;\ \{10, 14, 18, 22\})$, P_B from **(23)**.

(31) $US(P_B;\ \{12, 16, 20, 24\})$, P_B from **(27)**.

(32) $US(P_B;\ \{16, 20, 24, 28\})$, P_B from **(25)**.

(31) $US(P_B;\ \{20, 24, 28, 32\})$, P_B from **(27)**.

D.20 Spherical codes of dimension $n = 20$

(2) The extension of $\mathcal{X}_i : (14, 12/5, 6)$, $i = 1, \cdots, 7$, with the tail code $\mathcal{S}_6$ and with $b = \sqrt{35/11}\,/2$. The codes $\mathcal{X}_i$, $i = 1, \cdots, 7$, are obtained by CW_2 from $\mathcal{C}_i : (15, 5, 8, 6)$, respectively, where $d(\mathcal{C}_i,\ \mathcal{C}_j) = 6$ implying $\rho(\mathcal{X}_i,\ \mathcal{X}_j) = 9/5$ for all $i \neq j$.

(4) $CW_2\,(C : (21, 6, 8, 56))$. The code with such parameters can be obtained also from strongly regular Gewirtz graph $(56, 10, 0, 2)$ (see Section 9.4).

(5) The union of $\mathcal{X}_1$ and $\mathcal{X}_2$. The code $\mathcal{X}_1 : (20, 112/61, 56)$ is the extension of $\mathcal{V}_i$, $i = 1, \cdots, 7$, with the tail code $\mathcal{S}_6$ and with $b = 7/\sqrt{61}$. The codes $\mathcal{V}_i : (14, 16/7, 8)$ are obtained by Y_2 from $\mathcal{A}_i : (14, 8, 8)$ which are subcodes of $\mathcal{A} : (14, 6, 64)$, such that $\rho(\mathcal{V}_i,\ \mathcal{V}_j) = 12/7$. The code $\mathcal{X}_2 : (20, 7/3, 7)$ is the trivial left 14-extension of $-\mathcal{S}_6$.

(7) The union of $\mathcal{X}_1$ and $\mathcal{X}_2$. The code $\mathcal{X}_1 : (20, 70/39, 70)$ is the extension of $\mathcal{V}_i : (14, 20/9, 10)$, $i = 1, \cdots, 7$, where $\rho(\mathcal{V}_i,\ \mathcal{V}_j) = 5/3$ for $i \neq j$ (see **(5)** for $n = 18$) with the tail code $\mathcal{S}_6$ and with $b = \sqrt{21/26}$. The code $\mathcal{X}_2 : (20, 7/3, 7)$ is the trivial left 14-extension of $-\mathcal{S}_6$.

(8) The extension of $\mathcal{X}_1 : (19, 36/19, 40)$ (by Y_2 from $\mathcal{A} : (19, 9, 40)$) and $\mathcal{B}_{19}$ (by Y_3) with

$$b_1 = (1/6)\sqrt{19(12 - \sqrt{2})/6}\,.$$

N	M	ρ	ρ_L	$Constr.$	ρ_U	$Ref.$
1	41	ρ_1	1.92237	$Ex.5.10.4$	1.99476	[157]
2	42	21/11	1.90909	$E(CW_2)$	1.98954	[157]
3	45	15/8	1.87500		1.97411	[157]
4	56	28/15	1.86666	CW_2	1.92147	[157]
5	63	112/61	1.83606	$UE(Y_2)$	1.89188	[157]
6	68	240/133	1.80451		1.87269	[157]
7	77	70/39	1.79487	$E(CW_2)$	1.84204	[157]
8	78	$(12-\sqrt{2})/6$	1.76429	$E(Y_2, Y_3)$	1.83893	[157]
9	87	160/91	1.75824		1.81323	[157]
10	105	7/4	1.75000	CW_3	1.77241	[157]
11	120	12/7	1.71428	CW_2	1.74645	[157]
12	141	$2-1/\sqrt{10}$	1.68377	$U(CW_2)$	1.71856	[157]
13	170	5/3	1.66666	$E(Y_2, Y_3)$	1.69076	[157]
14	231	21/13	1.61538	$U(CW_2)$	1.65417	[157]
15	256	8/5	1.60000	Y_2	1.63707	[157]
16	294	$(33-\sqrt{65})/16$	1.55860	$UE(Y_2, Y_3)$	1.61632	[157]
17	301	14/9	1.55555	$U(CW_2)$	1.61304	[157]
18	378	84/55	1.52727	$U(CW_2)$	1.58475	[157]
19	896	84/61	1.37704	$U(CW_2, CW_3)$	1.44305	[157]
20	1280	4/3	1.33333	Y_3	1.39304	[157]
21	1701	14/11	1.27272	$U(CW_2, CW_3)$	1.36186	[157]
22	5416	8/7	1.14285	$U(Y_2, Y_3)$	1.21465	[157]
23	17400	1	1.00000	$U(Y_3)$	1.09020	[157]
24	49152	8/9	0.88888	$U(Y_2, Y_4)$	0.97681	[43]
25	127232	4/5	0.80000	$US(P_B)$	0.88926	[43]
26	302080	8/11	0.72727	$US(P_{B^*})$	0.82107	[157]
27	670920	2/3	0.66666	$US(P_B)$	0.75734	[157]
28	2730752	4/7	0.57142	$US(P_B)$	0.65925	[157]
29	9362160	1/2	0.50000	$US(P_B)$	0.58288	[157]
30	27726592	4/9	0.44444	$US(P_B)$	0.52247	[157]
31	74056848	2/5	0.40000	$US(P_B)$	0.47306	[157]
32	179854592	4/11	0.36363	$US(P_B)$	0.43212	[157]
33	407756760	1/3	0.33333	$US(P_B)$	0.39751	[157]
34	1743159088	2/7	0.28571	$US(P_B)$	0.34292	[157]
35	6177187800	1/4	0.25000	$US(P_B)$	0.30115	[157]

Table D.20: Spherical codes of dimension $n = 20$.

(10) $CW_3\,(\mathcal{C} : (21, 5, 8, 21),\ \mathcal{C}_1 : (5, 1, 2, 5),\ \beta = 1/6,\ \gamma = 4/3)$.

(11) $CW_2\,(\mathcal{C} : (21, 7, 8, 120))$. The code with such parameters can be obtained also from the strongly regular graph $(120, 42, 8, 18)$ (see Section 9.4).

(12) The union of the code from **(11)** above and $-\mathcal{S}_{20}$ (by CW_2).

(13) The extension of $\mathcal{X}_1$, $\mathcal{X}_2$ and $\mathcal{B}_{19}$ with

$$b_1 = (1/4)\sqrt{95/6}, \quad b_2 = (1/2)\sqrt{5/2}, \quad b_3 = \sqrt{5/6},$$

and with the tails $z_1 = a_1$, $z_2 = a_2$, and $z_3 = -a_3$. The code $\mathcal{X}_1 : (19, 32/19, 128)$ is obtained by Y_2 from $\mathcal{A}_1 : (19, 8, 128)$ which is the (11111)-reduction of $\mathcal{G}_{24}$. Assume that $(11100\cdots0) \in \mathcal{A}_1$. Then $\mathcal{X}_2 : (19, 8/3, 4)$ is the trivial 16-extension of $\mathcal{S}_3$ (by Y_2 from $\mathcal{A}_2 : (3, 2, 4)$). By construction $\rho(\mathcal{X}_1,\ \mathcal{X}_2) = 2 - 2/\sqrt{57}$.

(14) The union of $\mathcal{X}_1 : (20, 21/13, 210)$ (by CW_2 from $\mathcal{C} : (21, 8, 8, 210)$) and $-\mathcal{S}_{20}$ (by CW_2).

(15) $Y_2\,(\mathcal{A} : (20, 8, 256))$.

(16) The union of $\mathcal{X}_1$ and $\mathcal{X}_2$. The code $\mathcal{X}_1 : (20, \rho, 256)$, $\rho = (33 - \sqrt{65})/16$, is the extension with $b = (1/4)\sqrt{19\,\rho/2}$ of $\mathcal{V}_i : (19, 32/19, 128)$, $i = 1, 2$, (by Y_2 from $\mathcal{A}_i : (19, 8, 128)$, respectively) where $d(\mathcal{A}_1,\ \mathcal{A}_2) = 7$, and therefore $\rho(\mathcal{V}_1,\ \mathcal{V}_2) = 28/19$. The code $\mathcal{X}_2 : (20, 2, 38)$ is the trivial extension of $\mathcal{B}_{19}$ (by Y_3).

(17) The union of $\mathcal{X}_1 : (20, 14/9, 280)$ (by CW_2 from $\mathcal{C} : (21, 8, 8, 280)$) and $-\mathcal{S}_{20}$ (by CW_2).

(18) The union of $\mathcal{X}_1 : (20, 84/55, 336)$ (by CW_2 from $\mathcal{C} : (21, 10, 8, 336)$), $\mathcal{S}_{20}$ (by CW_2) and $-\mathcal{S}_{20}$.

(19) The union of $\mathcal{X}_1 : (20, 84/61, 840)$ (by CW_3 from $\mathcal{C} : (21, 6, 8, 56)$ and $\mathcal{C}_1 : (6, 2, 2, 15)$ $(\beta = 2/19,\ \gamma = 23/19)$) and $\mathcal{X}_2 : (20, 28/15, 56)$, where the code $-\mathcal{X}_2$ is obtained by CW_2 from $\mathcal{C}$.

(20) $Y_3\,(\mathcal{C} : (20, 6, 8, 40),\ \mathcal{A} : (6, 2, 32))$.

(21) The union of $\mathcal{X}_1 : (20, 14/11, 1680)$ (by CW_3 from $\mathcal{C} : (21, 8, 8, 210)$ and $\mathcal{C}_1 : (8, 1, 2, 8)$ ($\beta = 2/5,\ \gamma = 9/5$)) and $-\mathcal{S}_{20}$ (by CW_2).

(22) The union of $\mathcal{X}_1 : (20, 8/7, 5120)$ (by Y_3 from $\mathcal{C}_1 : (20, 7, 8, 80)$ and $\mathcal{A}_1 : (7, 2, 64)$), $\mathcal{X}_2 : (20, 8/5, 256)$ (by Y_2 from $\mathcal{A}_2 : (20, 8, 256)$) and $\mathcal{B}_{20}$ (by Y_3). The code $\mathcal{A}_2$ is the (0000)-reduction of $\mathcal{G}_{24}$ and $\mathcal{C}_1$ is the (1000)-reduction of $\mathcal{C} : (24, 8, 8, 759)$, implying $\rho(\mathcal{X}_1,\ \mathcal{X}_2) = 2 - \sqrt{5/7}$.

(23) The union of $\mathcal{X}_1 : (20, 1, 16640)$ (by Y_3 from $\mathcal{C}_1 : (20, 8, 8, 130)$ and $\mathcal{A}_1 : (8, 2, 128)$) and $\mathcal{X}_2 : (20, 1, 760)$ (by Y_3 from $\mathcal{C}_2 : (20, 2, 2, 190)$ and $\mathcal{A}_2 : (2, 1, 4)$).

(24) The union of $\mathcal{X}_1 : (20, 8/9, 48640)$ (by Y_4 from $\mathcal{A}_1 : (20, 8, 256)$ and $\mathcal{C}_1 : (20, 2, 2, 190)$) and $\mathcal{X}_2 : (20, 7/5, 512)$ (by Y_2 from $\mathcal{A}_2 : (20, 7, 512)$) such that $\mathcal{A}_1$ is the subcode of $\mathcal{A}_2$, implying $\rho(\mathcal{X}_1,\ \mathcal{X}_2) = 2 - 7/3\sqrt{5}$.

(25) $US(P_B(\mathcal{A});\ \{6, 10\}),\ \ \mathcal{A} : (20, 8, 256),\ \ 40y^6 + 176y^{10} + 40y^{14}$.

(26) $US(P_{B^*}(\mathcal{A});\ \{7, 11\}),\ \ \mathcal{A} : (20, 8, 256),\ \ 80y^7 + 160y^{11} + 16y^{15}$.

(27) $US(P_B(\mathcal{A});\ \{4, 8, 12\}),\ \ \mathcal{A} : (20, 8, 256),\ \ 5y^4 + 120y^8 + 130y^{12} + y^{20}$.

(28) $US(P_B;\ \{6, 10, 14\}),\ \ P_B$ from **(25)**.

(29) $US(P_B(\mathcal{A});\ \{4, 8, 12\}),\ \ \mathcal{A} : (20, 8, 256),\ \ x^{20} + 130y^8 + 120y^{12} + 5y^{16}$.

(30) $US(P_B;\ \{6, 10, 14, 18\}),\ \ P_B$ from **(25)**.

(31) $US(P_B;\ \{8, 12, 16, 20\}),\ \ P_B$ from **(27)**.

(32) $US(P_B;\ \{10, 14, 18, 22\}),\ \ P_B$ from **(25)**.

(33) $US(P_B;\ \{12, 16, 20, 24\}),\ \ P_B$ from **(29)**.

(34) $US(P_B;\ \{16, 20, 24, 28\}),\ \ P_B$ from **(27)**.

(35) $US(P_B;\ \{20, 24, 28, 32\}),\ \ P_B$ from **(29)**.

N	M	ρ	ρ_L	*Constr.*	ρ_U	*Ref.*
1	43	ρ_1	1.92570	$Ex.5.10.4$	1.99525	[157]
2	44	21/11	1.90909		1.99053	[157]
3	50	40/21	1.90476	$E(CW_2)$	1.96287	[157]
4	58	28/15	1.86666		1.92843	[157]
5	65	112/61	1.83606		1.90102	[157]
6	77	11/6	1.83333	CW_2	1.86021	[157]
7	80	$8(20-\sqrt{5})/79$	1.79887	$Ex.5.10.1$	1.85117	[157]
8	112	16/9	1.77777	$E(CW_2)$	16/9	[157]
9	119	224/129	1.73643	$UE(Y_2)$	1.76606	[157]
10	162	12/7	1.71428	$E(CW_2, CW_3)$	12/7	[157]
11	198	176/105	1.67619	$U(CW_2)$	1.68733	[157]
12	233	21/13	1.61538		1.66887	[157]
13	336	8/5	1.60000	Y_3	1.61858	[157]
14	554	32/21	1.52380	$U(Y_2, Y_3)$	1.55315	[157]
15	660	22/15	1.46666	$U(CW_2)$	1.52320	[157]
16	716	16/11	1.45454	$U(CW_2)$	1.50924	[157]
17	1232	11/8	1.37500	$U(CW_2, CW_3)$	1.42513	[157]
18	1792	4/3	1.33333	Y_3	1.38165	[157]
19	8234	8/7	1.14285	$U(Y_2, Y_3)$	1.19805	[157]
20	27720	1	1.00000	$U(Y_3)$	1.07533	[43]
21	82768	8/9	0.88888	$US(P_{B^*})$	0.96739	[43]
22	227584	4/5	0.80000	$US(P_B)$	0.87551	[43]
23	566784	8/11	0.72727	$US(P_{B^*})$	0.80466	[43]
24	1300040	2/3	0.66666	$US(P_B)$	0.74933	[157]
25	2826208	8/13	0.61538	$US(P_{B^*})$	0.69556	[157]
26	5758720	4/7	0.57142	$US(P_B)$	0.65120	[157]
27	11214848	8/15	0.53333	$US(P_{B^*})$	0.61034	[157]
28	21007490	1/2	0.50000	$US(P_B)$	0.57576	[157]
29	65934080	4/9	0.44444	$US(P_B)$	0.51583	[157]
30	184437890	2/5	0.40000	$US(P_B)$	0.46721	[157]
31	470933248	4/11	0.36363	$US(P_B)$	0.42685	[157]
32	1111716074	1/3	0.33333	$US(P_B)$	0.39276	[157]

Table D.21: Spherical codes of dimension $n = 21$.

D.21 Spherical codes of dimension $n = 21$

(3) The extension of $\mathcal{X}_1 : (20, 21/11, 42)$ (see **(2)** for $n = 20$) and $\mathcal{X}_2$ with $b_1 = \sqrt{440/441}$, plus one point. The code $\mathcal{X}_2 : (20, 7/3, 7)$ is the trivial left 14-extension of $-\mathcal{S}_6$. A code with these parameters can also be obtained from the Hoffman–Singleton strongly regular graph with parameters $(50, 7, 0, 1)$ (see Section 9.4).

(6) $CW_2\,(\mathcal{C} : (22, 6, 8, 77))$. The code with such parameters can be obtained also from strongly regular graph $(77, 16, 0, 4)$ (see Section 9.4).

(8) The extension of $\mathcal{X}_1$ and $-\mathcal{X}_2$ with $b = \sqrt{20/21}$. Both $\mathcal{X}_i$: $(20, 28/15, 56)$, $i = 1, 2$, are obtained by CW_2 from $\mathcal{C}_i : (21, 6, 8, 56)$, respectively, where $\mathcal{C}_1$ and $\mathcal{C}_2$ are the (110) and (101)-reductions, respectively, of $\mathcal{C} : (24, 8, 8, 759)$. By construction $\lambda(c_1, c_2) \in \{1, 3\}$ for any $c_1 \in \mathcal{C}_1$ and any $c_2 \in \mathcal{C}_2$, implying $\rho(\mathcal{X}_1, -\mathcal{X}_2) = 5/3$. This code $\mathcal{X}$ is the first member of the infinite family of optimal spherical codes (see Section 9.4 for the construction of all this family).

(9) The union of $\mathcal{X}_1$ and $\mathcal{X}_2$. The code $\mathcal{X}_1 : (21, 224/129, 112)$ is the extension of $\mathcal{V}_i$, $i = 1, \cdots, 7$, with the tail code $\mathcal{S}_6$ and with $b = \sqrt{35/43}$. The codes $\mathcal{V}_i : (15, 32/15, 16)$ are obtained by Y_2 from $\mathcal{A}_i : (15, 8, 16)$, respectively, which are subcodes of $\mathcal{A} : (15, 6, 128)$, such that $\rho(\mathcal{X}_i, \mathcal{X}_j) = 8/5$ for $i \neq j$. The code $\mathcal{V}_2 : (21, 7/3, 7)$ is the trivial left 15-extension of $-\mathcal{S}_6$.

(10) The extension of $\mathcal{X}_1 : (20, 7/4, 105)$ (by CW_3 from $\mathcal{C}_1$: $(21, 5, 8, 21)$ and $\mathcal{C}_{1,1} : (5, 1, 2, 5)$, with parameters $\beta = 1/6$, $\gamma = 4/3$) and $\mathcal{X}_2 : (20, 28/15, 56)$ (by CW_2 from $\mathcal{C}_2 : (21, 6, 8, 56)$), plus one point, with

$$b_1 = 4\sqrt{3}/7, \quad b_2 = 3\sqrt{5}/7.$$

Here $\mathcal{C}_1$ and $\mathcal{C}_2$ are the (111) and (110)-reductions, respectively, of $\mathcal{C} : (24, 8, 8, 759)$, which implies that $\rho(\mathcal{X}_1, \mathcal{X}_2) = 2 - \sqrt{3/5}\,/2$ and, therefore

$$\rho = 7b_1^2/4 = 28b_2^2/15 = 2 - 2(\sqrt{3/5}\,b_1 b_2/4 - a_1 a_2) = 2 - 2a_1 = 12/7.$$

The code with such parameters can be obtained also from strongly regular McLaughlin graph $(162, 56, 10, 24)$ (see Section 9.4).

(11) The union of $\mathcal{X}_1; (21, 176/105, 176)$ (by CW_2 from $\mathcal{C}$: $(22, 7, 8, 176)$) and $-\mathcal{S}_{21}$. The code with the same parameters as $\mathcal{X}_1$ can be obtained also from strongly regular graph $(176, 70, 18, 34)$ (see Section 9.4).

(13) $Y_3\,(\mathcal{C} : (21, 5, 8, 21),\ \mathcal{A} : (5, 2, 16))$.

(14) The union of $\mathcal{X} : (21, 32/21, 512)$ (by Y_2 from $\mathcal{A} : (21, 8, 512)$) and $\mathcal{B}_{21}$ (by Y_3).

(15) The union of $\mathcal{X}$: $(21, 12/15, 616)$ (by CW_2 from $\mathcal{C}$: $(22, 10, 8, 616)$), $\mathcal{S}_{21}$ (by CW_2), and $-\mathcal{S}_{21}$.

(16) The union of $\mathcal{X}_1$: $(21, 16/11, 672)$ (by CW_2 from $\mathcal{C}$: $(22, 11, 8, 672)$), $\mathcal{S}_{21}$ (by CW_2), and $-\mathcal{S}_{21}$.

(17) The union of $\mathcal{X}_1 : (21, 11/8, 1155)$ (by CW_3 from $\mathcal{C} : (22, 6, 8, 77)$ and $\mathcal{C}_1 : (6, 2, 2, 15)$, $\beta = 1/10$, $\gamma = 6/5$) and $-\mathcal{X}_2$, where $\mathcal{X}_2$: $(21, 11/6, 77)$ is obtained by CW_2 from $\mathcal{C}$.

(18) $Y_3\,(\mathcal{C} : (21, 6, 8, 56),\ \mathcal{A} : (6, 2, 32))$.

(19) The union of $\mathcal{X}_1$, $-\mathcal{X}_2$ and $\mathcal{B}_{21}$. The code $\mathcal{X}_1 : (21, 8/7, 7680)$ is obtained by Y_3 from $\mathcal{C}_1 : (21, 7, 8, 120)$ (the (100)-reduction of $\mathcal{C}$: $(24, 8, 8, 759)$) and from $\mathcal{A}_1 : (7, 2, 64)$ (which is even weight). The code $\mathcal{X}_2 : (21, 32/21, 512)$ is obtained by Y_2 from $\mathcal{A}_2 : (21, 8, 512)$ (the (000)-reduction of $\mathcal{G}_{24}$), implying $\rho(\mathcal{X}_1,\ \mathcal{X}_2) = 2 - \sqrt{5/7}$.

(20) The union of $\mathcal{X}_1 : (21, 1, 26880)$ (by Y_3 from $\mathcal{C}_1 : (21, 8, 8, 210)$ and $\mathcal{A}_1$: $(8, 2, 128)$) and $\mathcal{X}_2$: $(21, 1, 840)$ (by Y_3 from $\mathcal{C}_2$: $(21, 2, 2, 210)$ and $\mathcal{A}_2 : (2, 1, 4)$).

(21) $US(P_{B^*}(\mathcal{A}); \{5, 9\})$, $\mathcal{A} : (21, 8, 512)$, $21y^5 + 280y^9 + 210y^{13} + y^{21}$.

(22) $US(P_B(\mathcal{A});\ \{6, 10\})$, $\mathcal{A} : (21, 8, 512)$, $56y^6 + 336y^{10} + 120y^{14}$.

(23) $US(P_{B^*}(\mathcal{A});\ \{7, 11\})$, $\mathcal{A} : (21, 8, 512)$, $120y^7 + 336y^{11} + 56y^{15}$.

(24) $US(P_B(\mathcal{A});\ \{8, 12\})$, $\mathcal{A} : (21, 8, 512)$, $1 + 210y^8 + 280y^{12} + 21y^{16}$.

(25) $US(P_{B^*};\ \{5, 9, 13\})$, P_B from **(21)**.

(26) $US(P_B;\ \{6,10,14\})$, P_B from **(22)**.

(27) $US(P_{B^*};\ \{7,11,15\})$, P_B from **(23)**.

(28) $US(P_B;\ \{8,12,16\})$, P_B from **(24)**.

(29) $US(P_B;\ \{6,10,14,18\})$, P_B from **(22)**.

(30) $US(P_B;\ \{8,12,16,20\})$, P_B from **(24)**.

(31) $US(P_B;\ \{10,14,18,22\})$, P_B from **(22)**.

(32) $US(P_B;\ \{12,16,20,24\})$, P_B from **(24)**.

D.22 Spherical codes of dimension $n = 22$

(2) The union of $\mathcal{S}_{22}$ and $-\mathcal{S}_{22}$.

(7) The extension of $\mathcal{X}_1$: $(21, 11/6, 77)$ (by CW_2 from $\mathcal{C}$: $(22,6,8,77)$) and $-\mathcal{S}_{21}$, plus one point, where $b_1 = 2\sqrt{30}/11$ and $b_2 = \sqrt{105}/11$. We have $\rho(\mathcal{X}_1,\ \mathcal{X}_2) = 2 - 1/\sqrt{14}$.

(10) The union of $\mathcal{X}_1$ and $\mathcal{X}_2$. The code $\mathcal{X}_1$: $(22, 64/37, 128)$ is the extension of $\mathcal{V}_i$: $(15, 32/15, 16)$, $i = 1, \cdots, 8$, with the tail code $\mathcal{S}_7$ and with $b = \sqrt{30/37}$. The codes $\mathcal{V}_i$, $i = 1, \cdots, 8$, are obtained by Y_2 from $\mathcal{A}_i : (15,8,16)$, respectively, which are subcodes of $\mathcal{A}$: $(15,6,128)$, such that $\rho(\mathcal{V}_i,\ \mathcal{V}_j) = 8/5$ for $i \neq j$. The code $\mathcal{X}_2 : (22, 16/7, 8)$ is the trivial left 15-extension of $-\mathcal{S}_7$.

(13) The extension of $\mathcal{X}_1$, $-\mathcal{X}_2$, and $-\mathcal{S}_{21}$ (by CW_2) with

$$b_1 = (5/4)\sqrt{7/11}, \quad b_2 = \sqrt{10/11}, \quad b_3 = (1/2)\sqrt{35/11}$$

and with the tails $z_1 = a_1$, $z_2 = -a_2$, $z_3 = a_3$, plus one point. The code $\mathcal{X}_1$: $(21, 176/105, 176)$ is obtained by CW_2 from $\mathcal{C}_1$: $(22,7,8,176)$. The code $\mathcal{X}_2$: $(21, 11/6, 77)$ is obtained by CW_2 from $\mathcal{C}_2$: $(22,6,8,77)$. The codes $\mathcal{C}_1$ and $\mathcal{C}_2$ are the (10) and (11)-reductions, respectively, of $\mathcal{C}$: $(24,8,8,759)$. We have that $\lambda(c_1, c_2) \in$

N	M	ρ	ρ_L	*Constr.*	ρ_U	*Ref.*
1	45	ρ_1	1.92876	*Ex.*5.10.4	1.99569	[157]
2	46	21/11	1.90909	$U(CW_2)$	1.99139	[157]
3	52	40/21	1.90476		1.96616	[157]
4	60	28/15	1.86666		1.93452	[157]
5	67	112/61	1.83606		1.90910	[157]
6	79	11/6	1.83333		1.87076	[157]
7	100	20/11	1.81818	$E(CW_2)$	20/11	[157]
8	114	16/9	1.77777		1.79133	[157]
9	121	224/129	1.73643		1.77981	[157]
10	136	64/37	1.72972	$UE(Y_2)$	1.75857	[157]
11	164	12/7	1.71428		1.72832	[157]
12	200	176/105	1.67619		1.70113	[157]
13	275	5/3	1.66666	$E(CW_2)$	5/3	[157]
14	338	8/5	1.60000		1.63661	[157]
15	506	43/28	1.53571	$U(CW_2)$	1.59176	[157]
16	529	23/15	1.53333	$E(CW_2)$	1.58495	[157]
17	891	3/2	1.50000	$E(Y_2, Y_3)$	3/2	[157]
18	1068	16/11	1.45454	$U(Y_2, Y_3)$	1.47180	[157]
19	1334	46/33	1.39393	$U(CW_2)$	1.44044	[157]
20	1771	23/17	1.35294	CW_3	1.40667	[157]
21	2464	4/3	1.33333	Y_3	1.37236	[157]
22	12332	8/7	1.14285	$U(Y_2, Y_3)$	1.18775	[157]
23	49896	1	1.00000	$UE(Y_3, Y_4)$	1.04723	[43]
24	87024	8/9	0.88888	$US(P_{B^*})$	0.99947	[43]
25	396704	4/5	0.80000	$US(P_B)$	0.86704	[43]
26	1037312	8/11	0.72727	$US(P_{B^*})$	0.79613	[43]
27	2487452	2/3	0.66666	$US(P_B)$	0.74301	[157]
28	3446912	8/13	0.61538	$US(P_{B^*})$	0.71997	[157]
29	11866976	4/7	0.57142	$US(P_B)$	0.64399	[157]
30	23869440	8/15	0.53333	$US(P_{B^*})$	0.60471	[157]
31	46072312	1/2	0.50000	$US(P_B)$	0.56916	[157]
32	153049952	4/9	0.44444	$US(P_B)$	0.51025	[157]
33	451035640	2/5	0.40000	$US(P_B)$	0.46226	[157]
34	2963328764	1/3	0.33333	$US(P_B)$	0.38870	[157]

Table D.22: Spherical codes of dimension $n = 22$.

$\{1,3\}$ for any $c_1 \in \mathcal{C}_1$ and any $c_2 \in \mathcal{C}_2$, implying $\rho(\mathcal{X}_1, -\mathcal{X}_2) = 2 - (1/3)\sqrt{10/7}$. We have

$$\rho(\mathcal{X}_1, -\mathcal{S}_{21}) = 2 - 2/(3\sqrt{5}), \quad \rho(\mathcal{X}_2, \mathcal{S}_{21}) = 2 - (4/3)\sqrt{2/7}.$$

This code corresponds to the strongly regular McLaughlin graph $(275, 112, 30, 56)$ (see Section 9.4).

(15) The union of $\mathcal{X}$: $(22, 23/14, 253)$ (by CW_2 from $\mathcal{C}$: $(23, 7, 8, 253)$) and $-\mathcal{X}$. Any two distinct words of $\mathcal{C}$ have the correlations 1, or 3, implying $\rho(\mathcal{X}, -\mathcal{X}) = 43/28$.

(16) The extension of $\mathcal{X}_1 : (21, 11/7, 330)$, $\mathcal{X}_2 : (21, 176/105, 176)$, and $-\mathcal{S}_{21}$ (by CW_2) with

$$b_1 = \sqrt{161/165}, \quad b_2 = (1/4)\sqrt{161/11}, \quad b_3 = 1,$$

with the tails $z_1 = a_1$, $z_2 = -a_2$, $z_3 = 0$, plus one point. The codes $\mathcal{X}_1$ and $\mathcal{X}_2$ are obtained by CW_2 from $\mathcal{C}_1 : (22, 8, 8, 330)$ and $\mathcal{C}_2 : (22, 7, 8, 176)$, respectively, which are the (0)- and (1)-reductions of $\mathcal{C} : (23, 8, 8, 506)$, respectively. This gives $\rho(\mathcal{X}_1, \mathcal{X}_2) = 2 - 16/7\sqrt{15}$ and

$$\rho(\mathcal{X}_1, -S) = 2 - (4/7)(1/\sqrt{3}), \quad \rho(\mathcal{X}_2, -S) = 2 - (2/3)(1/\sqrt{5}).$$

(17) The extension of $\mathcal{X}_1$, $\mathcal{X}_2$, and $\mathcal{B}_{21}$ (by Y_3) with

$$b_1 = 3\sqrt{7}/8, \quad b_2 = \sqrt{15}/4, \quad b_3 = \sqrt{3}/2,$$

and with the tails $z_1 = a_1$, $z_2 = -a_2$, $z_3 = a_3$, plus one point. The code $\mathcal{X}_1 : (21, 32/21, 512)$ is obtained by Y_2 from $\mathcal{A}_1 : (21, 8, 512)$. The code $\mathcal{X}_2 : (21, 8/5, 336)$ is obtained by Y_3 from $\mathcal{C}_2 : (21, 5, 8, 21)$ and $\mathcal{A}_2 : (5, 2, 16)$, which is even weight. The code $\mathcal{A}_1$ is the (100)-reduction of $\mathcal{G}_{24}$ and $\mathcal{C}_2$ is the (111)-reduction of $\mathcal{C} : (24, 8, 8, 759)$, implying $\rho(\mathcal{X}_1, \mathcal{X}_2) = 2 - 2\sqrt{3/35}$.

(18) The union of $\mathcal{X}; (22, 16/11, 1024)$ (by Y_2 from $\mathcal{A} : (22, 8, 1024)$) and $\mathcal{B}_{22}$ (by Y_3).

(19) The union of $\mathcal{X}_1$: $(22, 46/33, 1288)$ (by CW_2 from $\mathcal{C}_1$: $(23, 11, 8, 1288)$), $\mathcal{S}_{22}$ (by CW_2) and $-\mathcal{S}_{22}$.

(20) $CW_3\,(\mathcal{C}:(23,7,8,253),\ \mathcal{C}_1:(7,1,2,7),\ \beta = 5/18, \gamma = 14/9)$.

(21) $Y_3\,(\mathcal{C}:(22,6,8,77),\ \mathcal{A}:(6,2,32))$.

(22) The union of $\mathcal{X}_1; (22, 8/7, 11264)$ (by Y_3 from $\mathcal{C}_1; (22,7,8,176)$ and $\mathcal{A}_1; (7,2,64)$, which is even weight), $\mathcal{X}_2 : (22, 16/11, 1024)$ (by Y_2 from $\mathcal{A}_2 : (22,8,1024)$) and $\mathcal{B}_{22}$ (by Y_3). The code $\mathcal{C}_1$ is the (01)-reduction of $\mathcal{C} : (24,8,8,759)$, and $\mathcal{A}_2$ is the (00)-reduction of $\mathcal{G}_{24}$, implying $\rho(\mathcal{X}_1,\ \mathcal{X}_2) = 2 - 5\sqrt{2/77}$.

(23) The union of $\mathcal{X}_i$, $i = 1,2,3,4$. The code $\mathcal{X}_1 : (22,1,21504)$ is the extension of $\mathcal{V}_i : (21, 32/29, 10752)$, $i = 1,2$, with $b_1 = (1/4)\sqrt{29/2}$ where $\mathcal{V}_i$ are obtained by Y_4 from $\mathcal{A}_i : (21,8,512)$, $i = 1,2$, respectively, and $\mathcal{C}_1 : (21,1,2,21)$. The codes $\mathcal{A}_1$ and $\mathcal{A}_2$ are the (000) and (110)-reductions, respectively, of $\mathcal{G}_{24}$, implying $\rho(\mathcal{V}_1, \mathcal{V}_2) = 24/29$. The code $\mathcal{X}_2 : (22,1,26880)$ is the trivial extension of $\mathcal{V}_3 : (21,1,26880)$ (by Y_3 from $\mathcal{C}_3 : (21,8,8,210)$ and $\mathcal{A}_3 : (8,2,128)$ which is even weight). The code $\mathcal{C}_3$ is the (000)-reduction of $\mathcal{C} : (24,8,8,759)$, implying

$$\rho(\mathcal{V}_i, \mathcal{V}_3) = 2 - 4\sqrt{2/29}, \quad i = 1,2.$$

The code $\mathcal{X}_3 : (22,1,672)$ is the extension with $b_2 = (1/2)\sqrt{5/2}$ of two copies of $\mathcal{V}_4 : (21, 8/5, 336)$ (by Y_3 from $\mathcal{C}_4 : (21,5,8,21)$ and $\mathcal{A}_4 : (5,2,16)$ which is even weight). The code $\mathcal{C}_4$ is the (111)-reduction of $\mathcal{C}$, which implies

$$\rho(\mathcal{V}_1, \mathcal{V}_4) = \rho(\mathcal{V}_2, \mathcal{V}_4) = 2 - 2\sqrt{5/29}, \quad \rho(\mathcal{V}_3, \mathcal{V}_4) = 2 - 4/\sqrt{10}.$$

The code $\mathcal{X}_4 : (22,1,880)$ is the trivial extension of $\mathcal{V}_5 : (21,1,880)$ (by Y_3 from $\mathcal{C}_5 : (21,2,2,210)$ and $\mathcal{A}_5 : (2,1,4)$). This implies

$$\rho(\mathcal{V}_4, \mathcal{V}_5) = 2 - 2\sqrt{2/5}.$$

(24) $US(P_{B^*}(\mathcal{A});\ \{5,7,9\}),\ \mathcal{A}:(22,8,1024)$,

$$21y^5 + 56y^7 + 280y^9 + 336y^{11} + 210y^{13} + 120y^{15} + y^{21}.$$

(25) $US(P_B(\mathcal{A});\ \{6,10\}),\ \mathcal{A}:(22,8,1024),\ 77y^6 + 616y^{10} + 330y^{14} + y^{22}$.

(26) $US(P_{B^*}(\mathcal{A});\ \{7,11\}),\ \mathcal{A}: (22,8,1024),\ 176y^7+672y^{11}+176y^{15}$.

(27) $US(P_B(\mathcal{A});\{8,12\}),\mathcal{A} : (22,8,1024), 1+330y^8+616y^{12}+77y^{16}$.

(28) $US(P_{B^*};\ \{5,7,9,11\}),\ P_B$ from **(24)**.

(29) $US(P_B;\ \{6,10,14\}),\ P_B$ from **(25)**.

(30) $US(P_{B^*};\ \{7,11,15\}),\ P_B$ from **(26)**.

(31) $US(P_B;\ \{8,12,16\}),\ P_B$ from **(27)**.

(32) $US(P_B;\ \{6,10,14,18\}),\ P_B$ from **(25)**.

(33) $US(P_B;\ \{8,12,16,20\}),\ P_B$ from **(27)**.

(34) $US(P_B;\ \{12,16,20,24\}),\ P_B$ from **(27)**.

D.23 Spherical codes of dimension $n = 23$

(2) The union of $\mathcal{S}_{23}$ and $-\mathcal{S}_{23}$.

(14) The union of $\mathcal{X}_1$ and $\mathcal{X}_2$. The code $\mathcal{X}_1 : (23, 8/5, 506)$ is the extension of $\mathcal{V}_1$ and $\mathcal{V}_2$ with $b_1 = b_2 = 4\sqrt{7/115}$ and with the tails $z_1 = a_1$ and $z_2 = -a_2$. The code $\mathcal{V}_1 : (22, 23/14, 253)$ is obtained by CW_2 from $\mathcal{C}_1 : (23, 7, 8, 253)$, and $\mathcal{V}_2 = -\mathcal{V}_1$, implying $\rho(\mathcal{V}_1,\ \mathcal{V}_2) = 43/28$. The code $\mathcal{X}_2 : (23, 8/5, 46)$ is the extension of $\mathcal{S}_{22}$ (by CW_2) and $-\mathcal{S}_{22}$ with $b_3 = b_4 = 2\sqrt{22/115}$ and with the tails $z_3 = -a_3$ and $z_4 = a_4$. We have

$$\rho(\mathcal{V}_1, \mathcal{S}_{22}) = 2 - 4\sqrt{2/77}, \quad \rho(\mathcal{V}_2, \mathcal{S}_{22}) = 2 - \sqrt{7/22}/2.$$

(15) The union of $\mathcal{X}_1 : (23, 3/2, 759)$ (by CW_2 from $\mathcal{C} : (24, 8, 8, 759)$) and $-\mathcal{S}_{23}$ (by CW_2).

(16) The union of $\mathcal{X}_1$ and $\mathcal{X}_2$. The code $\mathcal{X}_1 : (23, 16/11, 1024)$ is the trivial extension of $\mathcal{V}_1 : (22, 16/11, 1024)$ (by Y_2 from $\mathcal{A}_1; (22, 8, 1024)$). The code $\mathcal{X}_2 : (23, 16/11, 88)$ is the extension of

N	M	ρ	ρ_L	$Constr.$	ρ_U	$Ref.$
1	47	ρ_1	1.93156	$Ex.5.10.4$	1.99606	[157]
2	48	44/23	1.91304	$U(CW_2)$	1.99213	[157]
3	54	40/21	1.90476		1.96904	[157]
4	62	28/15	1.86666		1.93989	[157]
5	69	112/61	1.83606		1.91627	[157]
6	81	11/6	1.83333		1.88025	[157]
7	102	20/11	1.81818		1.82994	[157]
8	116	16/9	1.77777		1.80381	[157]
9	123	224/129	1.73643		1.79252	[157]
10	138	64/37	1.72972		1.77157	[157]
11	166	12/7	1.71428		1.74145	[157]
12	202	176/105	1.67619		1.71410	[157]
13	277	5/3	1.66666		1.67910	[157]
14	552	8/5	1.60000	$UE(CW_2)$	8/5	[157]
15	783	3/2	1.50000	$U(CW_2)$	1.54585	[157]
16	1112	16/11	1.45454	$UE(Y_2, Y_3)$	1.49091	[157]
17	4600	4/3	1.33333	$UE(Y_2, Y_3)$	4/3	[157]
18	18286	8/7	1.14285	$U(Y_2, Y_3)$	1.18059	[157]
19	45058	16/15	1.06666	$UE(Y_2, Y_3)$	1.09340	[157]
20	93150	1	1.00000	$UE(Y_3, Y_4)$	1.02591	[43]
21	122000	8/9	0.88888	$US(P_{B^*})$	1.00394	[43]
22	582958	32/39	0.82051	$U(Y_3, Y_4)$	0.87325	[43]
23	1853248	8/11	0.72727	$US(P_{B^*})$	0.79105	[43]
24	4646644	2/3	0.66666	$US(P_B)$	0.73602	[157]
25	98831506	1/2	0.50000	$US(P_B)$	0.56535	[157]
26	1078170770	2/5	0.40000	$US(P_B)$	0.45915	[157]
27	7727861862	1/3	0.33333	$US(P_B)$	0.38613	[157]

Table D.23: Spherical codes of dimension $n = 23$.

$\mathcal{V}_2$ and $\mathcal{V}_3$ with $b = 2\sqrt{2/11}$, where $\mathcal{V}_i : (22, 2, 44)$, $i = 2, 3$, are obtained by Y_3 from $\mathcal{C}_i : (22, 2, 4, 11)$, $i = 2, 3$, respectively, and $\mathcal{A}_2 : (2, 1, 4)$, where $d(\mathcal{C}_2, \mathcal{C}_3) = 2$, implying $\rho(\mathcal{V}_2, \mathcal{V}_3) = 1$, and $\rho(\mathcal{X}_1, \mathcal{X}_2) = 2 - 4\sqrt{2}/11$.

(17) The union of $\mathcal{X}_1$, $\mathcal{X}_2$ and $\mathcal{X}_3$. The code $\mathcal{X}_1 : (23, 4/3, 2464)$ is the trivial extension of $\mathcal{V}_1 : (22, 4/3, 2464)$ (by Y_3 from $\mathcal{C}_1 : (22, 6, 8, 77)$ and $\mathcal{A}_1 : (6, 2, 32)$ which is odd weight). The code $\mathcal{X}_2 : (23, 4/3, 2048)$ is the extension of $\mathcal{V}_i : (22, 16/11, 1024)$, $i = 2, 3$, with $b_2 = \sqrt{11/12}$, where $\mathcal{V}_i$, $i = 2, 3$ are obtained by Y_2 from $\mathcal{A}_i : (22, 8, 1024)$, $i = 2, 3$, respectively. The code $\mathcal{X}_3 : (23, 4/3, 88)$ is the extension of two copies of $\mathcal{B}_{22}$ (by Y_3) with $b_3 = \sqrt{2/3}$. The code $\mathcal{C}_1$ is the (11)-reduction of $\mathcal{C} : (24, 8, 8, 759)$, and $\mathcal{A}_2$ and $\mathcal{A}_3$ are the (11) and (00)-reductions, respectively, of code $\mathcal{G}_{24}$. For any $c \in \mathcal{C}_1$ and for any $v \in \mathcal{A}_2 \cup \mathcal{A}_3$ the correlation $\lambda(c, v)$ is even, implying $\rho(\mathcal{V}_1, \mathcal{V}_i) = 2 - 4/\sqrt{33}$, $i = 2, 3$. Using that $\rho(\mathcal{V}_1, \mathcal{B}_{22}) = 2 - 2/\sqrt{6}$, and $\rho(\mathcal{V}_i, \mathcal{B}_{22}) = 2 - 2/\sqrt{22}$, $i = 2, 3$, we obtain $\rho(\mathcal{X}_i, \mathcal{X}_j) = 4/3$ for $i \neq j$.

(18) The union of $\mathcal{X}_1; (23, 8/7, 16192)$ (by Y_3 from $\mathcal{C}_1 : (23, 7, 8, 253)$ and $\mathcal{A}_1 : (7, 2, 64)$ which is even weight), $\mathcal{X}_2 : (23, 32/23, 2048)$ (by Y_2 from $\mathcal{A}_2 : (23, 8, 2048)$) and $\mathcal{B}_{23}$ (by Y_3). The code $\mathcal{C}_1$ is the (1)-reduction of $\mathcal{C} : (24, 8, 8, 759)$, and $\mathcal{A}_2$ is the (0)-reduction of $\mathcal{G}_{24}$, such that $\mathcal{C}_1$ and $\mathcal{A}_2$ have only even correlations, implying $\rho(\mathcal{X}_1, \mathcal{X}_2) = 2 - 10/\sqrt{161}$.

(19) The union of $\mathcal{X}_1$ and $\mathcal{X}_2$. The code $\mathcal{X}_1 : (23, 16/15, 22528)$ is the trivial extension of $\mathcal{V}_1 : (22, 16/15, 22528)$ (by Y_4 from $\mathcal{A}_1 : (22, 8, 1024)$ and $\mathcal{C}_1 : (22, 1, 2, 22)$). The code $\mathcal{X}_2 : (23, 16/15, 22530)$ is the extension of $\mathcal{V}_i : (22, 8/7, 11264)$, $i = 2, 3$, with $b = \sqrt{14/15}$, plus two points. The codes $\mathcal{V}_i$, $i = 2, 3$, are obtained by Y_3 from $\mathcal{C}_i : (22, 7, 8, 176)$, $i = 2, 3$, respectively, and $\mathcal{A}_2 : (7, 2, 64)$ which is even weight. The code $\mathcal{A}_1$ is the (00)-reduction of $\mathcal{G}_{24}$, and $\mathcal{C}_2$ and $\mathcal{C}_3$ are the (10) and (01)-reductions of $\mathcal{C} : (24, 8, 8, 759)$, respectively. We have

$$\rho(\mathcal{V}_1, \mathcal{V}_2) = \rho(\mathcal{V}_1, \mathcal{V}_3) = 2 - 2\sqrt{7/30}, \quad \rho(\mathcal{V}_2, \mathcal{V}_3) = 6/7.$$

(20) The union of $\mathcal{X}_i$, $i = 1, 2, 3, 4$. The code $\mathcal{X}_1 : (23, 1, 45056)$ is the extension with $b_1 = \sqrt{15}/4$ of $\mathcal{V}_i : (22, 16/15, 22528)$, $i = 1, 2$, (by Y_4

from $\mathcal{A}_i : (22, 8, 1024)$, $i = 1, 2$, respectively, and $\mathcal{C}_1 : (22, 1, 2, 22)$). The codes $\mathcal{A}_1$ and $\mathcal{A}_2$ are (00) and (11)-reductions, respectively, of $\mathcal{G}_{24}$, implying $\rho(\mathcal{V}_1, \mathcal{V}_2) = 4/5$. The code $\mathcal{X}_2 : (23, 1, 42240)$ is the trivial extension of $\mathcal{V}_3 : (22, 1, 42240)$ (by Y_3 from $\mathcal{C}_3 : (22, 8, 8, 330)$ and $\mathcal{A}_3 : (8, 2, 128)$ which is even weight). The code $\mathcal{C}_3$ is the (00)-reduction of $\mathcal{C} : (24, 8, 8, 759)$, implying $\rho(\mathcal{V}_i, \mathcal{V}_3) = 2 - 4/\sqrt{15}$, $i = 1, 2$. The code $\mathcal{X}_3 : (23, 1, 4928)$ is the extension of two copies of $\mathcal{V}_4$ with $b_2 = \sqrt{3}/2$, plus two points, where $\mathcal{V}_4 : (22, 4/3, 2464)$ is obtained by Y_3 from $\mathcal{C}_4 : (22, 6, 8, 77)$ and $\mathcal{A}_4 : (6, 2, 32)$ which is even weight. The code $\mathcal{C}_4$ is the (11)-reduction of $\mathcal{C}$, implying $\rho(\mathcal{V}_i, \mathcal{V}_4) = 2 - 2/\sqrt{5}$, $i = 1, 2$. and $\rho(\mathcal{V}_3, \mathcal{V}_4) = 2 - 2/\sqrt{3}$. The code $\mathcal{X}_4 : (23, 1, 924)$ is the trivial extension of $\mathcal{V}_5 : (22, 1, 924)$ (by Y_3 from $\mathcal{C}_5 : (22, 2, 2, 231)$ and $\mathcal{A}_5 : (2, 1, 4)$). We have $\rho(\mathcal{V}_i, \mathcal{V}_5) = 2 - 4/\sqrt{15}$, $i = 1, 2$, $\rho(\mathcal{V}_3, \mathcal{V}_5) = 1$, and $\rho(\mathcal{V}_4, \mathcal{V}_5) = 2 - 2/\sqrt{3}$.

(21) $US(P_{B^*}(\mathcal{A});\ \{5, 7, 9\})$, $\mathcal{A} : (23, 8, 2048)$,

$$21y^5 + 112y^7 + 400y^9 + 672y^{11} + 546y^{13} + 240y^{15} + 56y^{17} + y^{21}.$$

(22) The union of $\mathcal{X}_1 : (23, 32/39, 518144)$ (by Y_4 from $\mathcal{A}_1$; $(23, 8, 2048)$ and $\mathcal{C}_1 : (23, 2, 2, 253)$), $\mathcal{X}_2 : (23, 1, 64768)$ (by Y_3 from $\mathcal{C}_2 : (23, 8, 8, 506)$ and $\mathcal{A}_2 : (8, 2, 128)$ which is odd weight) and $\mathcal{B}_{23}$ (by Y_3). The code $\mathcal{A}_1$ is the (0)-reduction of $\mathcal{G}_{24}$ and $\mathcal{C}_2$ is the (0)-reduction of $\mathcal{C} : (24, 8, 8, 759)$.

(23) $US(P_{B^*}(\mathcal{A});\ \{7, 11\})$, $\mathcal{A} : (23, 8, 2048)$, $253y^7 + 1288y^{11} + 506y^{15} + y^{23}$.

(24) $US(P_B(\mathcal{A});\ \{8, 12\})$, $\mathcal{A} : (23, 8, 2048)$, $1 + 506y^8 + 1288y^{12} + 253y^{16}$.

(25) $US(P_B;\ \{8, 12, 16\})$, P_B from **(24)**.

(26) $US(P_B;\ \{8, 12, 16, 20\})$, P_B from **(24)**.

(27) $US(P_B;\ \{12, 16, 20, 24\})$, P_B from **(24)**.

N	M	ρ	ρ_L	$Constr.$	ρ_U	$Ref.$
1	49	ρ_1	1.93415	$Ex.$5.10.4	1.99639	[157]
2	50	25/13	1.92307	CW_2	1.99279	[157]
3	56	40/21	1.90476		1.97156	[157]
4	64	28/15	1.86666		1.94463	[157]
5	71	112/61	1.83606		1.92265	[157]
6	83	11/6	1.83333		1.88879	[157]
7	104	20/11	1.81818		1.84069	[157]
8	118	16/9	1.77777		1.81533	[157]
9	125	100/57	1.75438	$U(CW_2)$	1.80429	[157]
10	140	64/37	1.72972		1.78367	[157]
11	150	50/29	1.72413	CW_3	1.77187	[157]
12	168	12/7	1.71428		1.75375	[157]
13	204	176/105	1.67619		1.72630	[157]
14	279	5/3	1.66666		1.69084	[157]
15	297	18/11	1.63636	$UE(Y_2, Y_3)$	1.68491	[157]
16	330	50/31	1.61290	$U(CW_2, CW_3)$	1.67457	[157]
17	554	8/5	1.60000		1.61470	[157]
18	807	3/2	1.50000	$UE(CW_2)$	1.56380	[157]
19	2138	32/23	1.39130	$UE(Y_2, Y_3)$	1.43095	[157]
20	4604	4/3	1.33333	$UE(Y_2, Y_3)$	1.35532	[157]
21	18332	8/7	1.14285	$UE(Y_2, Y_3)$	1.20705	[157]
22	196560	1	1.00000	$U(Y_2, Y_3, Y_4)$	1	[157]
23	1322460	4/5	0.80000	$US(P_1)$	0.85339	[43]
24	3150820	8/11	0.72727	$US(P_2)$	0.81778	[43]
25	16969680	2/3	0.66666	$US(\Lambda_{24})$	0.69288	[157]
26	58859348	4/7	0.57142	$US(P_1)$	0.62508	[157]
27	415003680	1/2	0.50000	$US(\Lambda_{24})$	0.53087	[157]
28	1028092956	4/9	0.44444	$US(P_1)$	0.49200	[157]
29	5044384800	2/5	0.40000	$US(\Lambda_{24})$	0.43081	[157]
30	10273188232	4/11	0.36363	$US(P_1)$	0.40565	[157]
31	39461844240	1/3	0.33333	$US(\Lambda_{24})$	0.36208	[157]

Table D.24: Spherical codes of dimension $n = 24$.

D.24 Spherical codes of dimension $n = 24$

(2) $CW_2\,(\mathcal{C} : (25, 12, 12, 50))$.

(9) The union of $\mathcal{X}_1; (24, 100/57, 100)$ (by CW_2 from $\mathcal{C}$: $(25, 6, 8, 100)$) and $-\mathcal{S}_{24}$ (by CW_2).

(11) $CW_3\,(\mathcal{C} : (25, 5, 8, 30),\ \mathcal{C}_1 : (5, 1, 2, 5),\ \beta = 3/22,\ \gamma = 14/11)$.

(15) The union of $\mathcal{X}_1$ and $\mathcal{X}_2$. The code $\mathcal{X}_1 : (24, 18/11, 288)$ is the extension of $\mathcal{V}_i : (16, 2, 32)$, $i = 1, 2, \cdots, 9$, with $b = 3/\sqrt{11}$ and with the tail code $\mathcal{S}_8$. The codes $\mathcal{V}_i$, $i = 1, 2, \cdots, 8$, are obtained by Y_2 from $\mathcal{A}_i : (16, 8, 32)$, respectively, which are all eight subcodes of $\mathcal{NR}$, and $\mathcal{V}_9 = \mathcal{B}_{16}$ (by Y_3), implying $\rho(\mathcal{V}_i,\ \mathcal{V}_j) = 3/2$ for all $i \neq j$ where $i, j \in \{1, \cdots, 9\}$. The code $\mathcal{X}_2 : (24, 9/4, 9)$ is the trivial left 16-extension of $-\mathcal{S}_8$.

(16) The union of $\mathcal{X}_1 : (24, 50/31, 300)$ (by CW_3 from $\mathcal{C} : (25, 5, 8, 30)$ and $\mathcal{C}_1 : (5, 2, 2, 10)$, $\beta = 1/24$, $\gamma = 13/12$) and $\mathcal{X}_2 : (24, 2, 30)$ (by CW_2 from $\mathcal{C}$).

(18) The union of $\mathcal{X}_1$ and $\mathcal{X}_2$. The code $\mathcal{X}_1 : (24, 3/2, 759)$ is the trivial extension of $\mathcal{V}_1 : (23, 3/2, 759)$ (by CW_2 from $\mathcal{C}; (24, 8, 8, 759)$). The code $\mathcal{X}_2 : (24, 3/2, 48)$ is the extension of $\mathcal{S}_{23}$ (by CW_2) and $-\mathcal{S}_{23}$ with $b = (1/4)\sqrt{23/2}$.

(19) The union of $\mathcal{X}_1$ and $\mathcal{X}_2$, plus two points. The code $\mathcal{X}_1$: $(24, 32/23, 2048)$ is the trivial extension of $\mathcal{V}_1 : (23, 32/23, 2048)$ (by Y_2 from $\mathcal{A}_1 : (23, 8, 2048)$). The code $\mathcal{X}_2 : (24, 32/23, 88)$ is the extension with $b = 4/\sqrt{23}$ of $\mathcal{V}_i : (23, 2, 44)$, $i = 2, 3$ (by Y_3 from $\mathcal{C}_i$: $(23, 2, 4, 11)$, respectively, and from $\mathcal{A}_2 : (2, 1, 4)$), where $d(\mathcal{C}_2,\ \mathcal{C}_3) = 2$, implying $\rho(\mathcal{V}_2,\ \mathcal{V}_3) = 1$. Both $\mathcal{C}_i$ can be chosen such, that one position, say, the last 23-th position, will be zero for both codes. So, we add two points $(00 \cdots 0| \pm 1|0)$.

(20) The union of $\mathcal{X}_1 : (24, 4/3, 4600)$ (the trivial extension of $\mathcal{V}_1$: $(23, 4/3, 4600)$ (see **(17)** for $n = 23$)) and $\mathcal{X}_2 : (24, 4/3, 4)$ (the trivial left 22-extension of the following code $\mathcal{V}_2 : (2, 4/3, 4) = \{(\pm a_1|\pm a_2)\}$, where $a_1 = 1/\sqrt{3}$, $a_2 = \sqrt{2/3}$.

(21) The only difference from the code $\mathcal{X} : (23, 8/7, 18286)$ (see **(18)** for $n = 23$) is that instead of $\mathcal{B}_{23}$ we take the code $\mathcal{X}_3 : (24, 8/7, 92)$ (the extension of two copies of $\mathcal{B}_{23}$ (by Y_3) with $b = \sqrt{5/7}$. The other codes $\mathcal{X}_1; (24, 8/7, 16192)$ and $\mathcal{X}_2 : (24, 32/23, 2048)$ are the trivial extensions of the corresponding codes of length 23.

(22) The union of $\mathcal{X}_i$, $i = 1, 2, 3$. The code $\mathcal{X}_1 : (24, 1, 98304)$ is obtained by Y_4 from $\mathcal{G}_{24}$ and $\mathcal{C}_1 : (24, 1, 2, 24)$. The code $\mathcal{X}_2 : (24, 1, 97150)$ is obtained by Y_3 from $\mathcal{C}_2 : (24, 8, 8, 759)$ and $\mathcal{A}_2; (8, 2, 128)$ which is even weight. For any $c \in \mathcal{C}_2$ and for any $v \in \mathcal{G}_{24}$ the correlation $\lambda(c, v)$ is even, implying $\rho(\mathcal{X}_1, \mathcal{X}_2) = 1$. The code $\mathcal{X}_3 : (24, 1, 1104)$ is obtained by Y_3 from $\mathcal{C}_3; (24, 2, 2, 276)$ and $\mathcal{A}_3 : (2, 1, 4)$.

(23) $US(P_1);\ \{3, 5, 6, 7, 9, 10\})$, $P_1 = P_{B^*}(\mathcal{A}) \cup \{P_{B^*}(\mathcal{A}) + \mathbf{v}\}$. The code $\mathcal{A}$ is a coset of $\mathcal{G}_{24}$,

$$y^3 + 21y^5 + 168y^7 + 640y^9 + 1218y^{11}$$
$$+1218y^{13} + 640y^{15} + 168y^{17} + 21y^{19} + y^{21},$$

and $\mathbf{v}$ is given by $\mathbf{v} = (-3/2, (1/2)^{23})$.

(24) $US(P_2);\ \{1, 5, 6, 7, 9, 10, 11\})$, $P_2 = P_{B^*}(\mathcal{A}) \cup \{P_{B^*}(\mathcal{A}) + \mathbf{v}\}$. The code $\mathcal{A}$ is the coset of $\mathcal{G}_{24}$,

$$y + 253y^7 + 506y^9 + 1288y^{11} + 1288y^{13} + 506y^{15} + 253y^{17} + y^{23},$$

and $\mathbf{v}$ is from **(23)**.

(25) $US(\Lambda_{24};\ \{8, 12\})$, the Leech lattice Λ_{24}.

(26) $US(P_1;\ \{3, 5, 6, 7, 9, 10, 11, 13, 14\})$.

(27) $US(\Lambda_{24};\ \{8, 12, 16\})$.

(28) $US(P_1;\ \{6, 7, 9, 10, 11, 13, 14, 15, 17, 18\})$.

(29) $US(\Lambda_{24};\ \{8, 12, 16, 20\})$.

(30) $US(P_1;\ \{9, 10, 11, 13, 14, 15, 17, 18, 19, 21, 22\})$.

(31) $US(\Lambda_{24};\ \{12, 16, 20, 24\})$.

Bibliography

[1] M. Abramowitz and I.A, Stegun, *Handbook of Mathematical functions*, Dover, New York, 1965.

[2] J. Aczél and T. Szele, Aufgabe 35, *Matematikai Lapok*, Vol. 3, No. 1, pp. 94-95, 1952.

[3] Eric Agrell, Alexander Vardy and Kenneth Zeger, Upper Bounds for Constant-Weight Codes, *IEEE Trans. on Inform. Theory*, 2000, to appear.

[4] N.N. Andreev, One spherical code, *Russian Survey in Math.*, Vol. 54, No. 1, pp. 251-253, 1999.

[5] N.N. Andreev, Minimal spherical 11-design on three dimensional sphere, *Mathematical Notes*, Vol. 67, No. 4, pp. 417-424, 2000.

[6] N.N. Andreev, V.A. Yudin, Problems of Approximation Theory in Discrete Geometry, Math. Research, (Advances in Multivariate Approximation), Vol. 107, pp. 19-32, 1999.

[7] R. Askey, Orthogonal polynomials and special functions, *Regional Conference Lectures in Applied Mathematics*, SIAM, 21, 1975.

[8] J.T. Astola, The Tietäväinen bound for spherical codes, *Discrete Appl. Math.*, Vol. 7, pp. 17-21, 1984.

[9] E. Bannai and T. Ito, *Algebraic Combinatorics I: Association Schemes*, Benjamin, New York, 1984.

[10] E. Bannai and N.J.A. Sloane, Uniqueness of certain spherical codes, *Canad. J. Math.*, Vol. 33, pp. 437 - 449, 1981.

[11] E. Bannai, Spherical designs and group representations, *Contemp. Math.*, Vol. 34, pp. 95-107, 1984.

[12] E. Bannai, Spherical t-designs which are orbits of finite groups, *J. Math. Soc. Japan*, Vol. 36, pp. 341-354, 1984.

[13] E. Bannai, R.M. Damerell, Tight spherical designs I, *J. Math. Soc. Japan*, Vol. 31, pp. 199-207, 1979.

[14] E. Bannai, R.M. Damerell, Tight spherical designs II, *J. London Math. Soc.*, Vol. 21, pp. 13-30, 1980.

[15] E.S. Barnes and G.E. Wall, Some extreme forms defined in terms of Abelian groups, *J. Amer. Math. Soc.*, Vol. 1, pp. 47 - 63, 1959.

[16] L.A. Bassalygo, New upper bounds for error-correcting codes, *Problems of Information Transmission*, Vol.1, No.4, pp. 41-44, 1965.

[17] H. Bateman and A. Erdély, Higher tanscendental functions, Vol. 2, New York/Toronto/London, MacGraw–Hill, 1953.

[18] R. Bellman, *Introduction to Matrix Analysis*, McGraw-Hill, New York, 1960.

[19] C. Bender, Bestimmungen der grössten Anzahl gleicher Kugeln, welche sich aut eine Kugel von demselben Radius, wie die übrigen, auflegen lassen, Archiv Math. Physik (Grunert), Vol. 56, pp. 302-306, 1874.

[20] M.R. Best, Binary Codes with a minimum distance of four, PGIT, Vol. 26, pp. 738-742, 1980.

[21] K. Bezdek, Kepler's conjecture and the dodecahedral conjecture, Mitteilungen der DMV, Vol. 4, pp. 52-54, 1996.

[22] T. Beth, D. Jungnickel and H. Lenz, *Design Theory*, B.I.Wissenschaftsverlag, Bibliographisches Institut, Zurich, 1985.

[23] N. Biggs, *Finite Groups of Automorphisms*, Cambridge AT The University Press, 1971.

[24] E. Biglieri and M. Elia, On the existence of group codes for the Gaussian channel, *IEEE Trans. on Inform. Theory*, Vol. IT-18, pp. 399-402, May 1972.

[25] E. Biglieri and M. Elia, Cyclic-Group Codes for the Gaussian Channel, *IEEE Trans. on Inform. Theory*, Vol. IT-22, No. 5, pp. 624-629, Sep. 1976.

[26] E. Biglieri and M. Elia, Optimum permutation modulation codes and their asymptotic performance, *IEEE Trans. on Inform. Theory*, Vol. IT-22, No. 6, pp. 751-753, Nov. 1976.

[27] E. Biglieri and M. Elia, Signal sets generated by groups, in: Longo (ed.), *The information theory approach to communications*, pp. 263-306, Berin, Heidelberg, New York: Springer, 1977.

[28] I.F. Blake, Codes over certain rings, *Information and Control*, Vol. 20, pp. 396-404, May 1972.

[29] I.F. Blake, Distance properties of group codes for the Gaussian channel, *SIAM J. Appl. Math.*, Vol. 23, No. 3, 1972.

[30] I.F. Blake, Configuration Matrices for Group Codes, *IEEE Trans. on Inform. Theory*, Vol. IT-20, pp. 95-100, Jan. 1974.

[31] I.F. Blake, Permutation codes for discrete channels, *IEEE Trans. on Inform. Theory*, Vol. IT-20, pp. 138-140, Jan. 1974.

[32] I.F. Blake, Codes over Integers Residue Rings, *Information and Control*, Vol. 29, pp. 295-300, 1975.

[33] I.F. Blake and R.I. Mullin, *The mathematical theory of coding*, New York, Academic Press, 1975.

[34] H.F. Blichfeldt, The minimum values of positive quadratic forms in six, seven and eight variables, *Math.Zeit*, Vol.39, pp. 1-15, 1935.

[35] E.L. Blokh and V.V. Zyablov, *Generalized Concatenated Codes* ,Moscow: Svyaz, 1976, (in Russian).

[36] E.L. Blokh and V.V. Zyablov, *Linear concatenated codes*, Moscow: Nauka, 1982, (in Russian).

[37] S. Bochner, Hilbert distances and positive definite functions, Ann. Math., Vol. 42, pp. 647-656, 1941.

[38] A. Bos, Code over groups with arbitrary metrics, Technological University Eindhoven, Department of Mathematics, The Netherlands, *TH-Report 80-WSK-06*, July 1980.

[39] R.C. Bose, Strongly regular graphs, partial geometries, and partially balanced designs, *Pacific J. Math.*, Vol. 13, pp. 389-419, 1963.

[40] R.C. Bose and D.M. Mesner, On linear associative algebras corresponding to association schemes of partially balanced designs, *Annals Math. Stat.*, Vol. 30, pp. 21-38, 1959.

[41] R.C. Bose and T. Shimamoto, Classification and anaysis of partially balanced incomplete block designs with two associate classes, *J. Amer. Statist. Soc.*, Vol. 47, pp. 151-184, 1952.

[42] P.J. Boyvalenkov and S.M. Dodunekov, Some new bounds on the kissing numbers, *Proc. of the Sixth Joint Swedish-Russian Workshop on Information Theory*, Mölle, Sweden, pp. 389-393, 1993.

[43] P.G. Boyvalenkov, I. Landgev, On maximal spherical codes I, *Lecture Notes in Computer Science*, Vol. 0948, pp. 158-168, 1995.

[44] P.G. Boyvalenkov, D.P. Danev and I. Landgev, On maximal spherical codes II, *Journal of Combinatorial Designs*, Vol. 7, No. 5, pp. 316-326, September 1999.

[45] P.G. Boyvalenkov, D.P. Danev and S.P. Bumova (Dimcheva), Upper bounds on the minimum distance of spherical codes, *IEEE Trans. on Inform. Theory*, Vol. IT-42, pp. 1576-1581, September 1996.

[46] P.G. Boyvalenkov and D.P. Danev, On maximal codes in polynomial metric spaces, *Lecture Notes in Computer Science*, Vol. 1255, pp. 29-38, 1997.

[47] P.G. Boyvalenkov and D.P. Danev, On linear programming bounds for codes in polynomial metric spaces, *Problems of Information Transmissions*, Vol. 33, pp. 108-120, 1998.

[48] P. Boyvalenkov, Computing distance distribution of spherical designs, *Lin. Alg. Appl.*, Vol. 226/228, pp. 277-286, 1995.

[49] P. Boyvalenkov and D. Danev, Uniqueness of the 120-point spherical 11-design in four dimensions, Archiv der Mathematik, to appear.

[50] Peter Boyvalenkov, Danyo Danev and Peter Kazakov, Indexes of spherical codes, in "Codes and Association schemes", Proceedings of AMS-DIMACS series, to appear.

[51] A.E. Brouwer, J.B. Shearer, N.J.A. Sloane and W.D. Smith, A new table of constant weight codes, *IEEE Trans. on Inform. Theory*, Vol. IT-36, No. 6, pp. 1334-1380, Nov. 1990.

[52] K.A. Bush, Orthogonal arrays of index unity, *Ann. Math. Stat.*, Vol. 23, No. 3, pp. 426-434, 1952.

[53] K. Böröczky, Packing of spheres in spaces of constant curvature, *Acta Mathematica Academia Scientiarum Hungariacae*, Vol. 32, pp. 243-261, 1978.

[54] K. Böröczky, The problem of Tammes for n=11, *Studia Scientiarium Mathematicarum Hungaria*, Vol. 18, pp. 165-171, 1983.

[55] A.R. Calderbank, Multilevel Codes and Multistage Decoding, *IEEE Trans. on Communications*, Vol. 37, No. 3, pp. 222-229, March, 1989.

[56] A.R. Calderbank, P.J. Cameron, M.Kantor and J.J. Seidel, Orthogonal Symplectic Geometry, Kerdock Codes over $\mathbb{Z}_2$ and $\mathbb{Z}_4$ and Extraspecial 2-Groups, *preprint*, 1994.

[57] A.R. Calderbank, P.J. Cameron, M.Kantor and J.J. Seidel, $\mathbb{Z}_4$-Kerdock Codes, orthogonal spreads and extremal Euclidean line sets, *Proc. London Math. Soc.*, Vol. 75, pp. 436-480, 1997.

[58] A.R. Calderbank and N.J.A. Sloane, New Trellis Codes Based on Lattices and Cosets, *IEEE Trans. on Inform. Theory*, Vol. IT-33, No. 2, pp. 177-195, 1987.

[59] A.R. Calderbank, The art of Signalling: Fifty Years of Coding Theory, *IEEE Trans. on Inform. Theory*, Vol. 44, No. 6, pp. 2561-2595, 1998.

[60] P.J. Cameron, J.-M. Goethals and J.J. Seidel, Strongly regular graphs having strongly regular subconstituents, *J. of Algebra*, Vol. 43, pp. 257-280, 1978.

[61] P.J. Cameron and J.H. van Lint, Designs, Graphs, Codes and their links, *London Mathematical Society Student Texts*, Vol. 22, Cambridge University Press, Cambridge, 1991.

[62] P.J. Cameron, J.-M. Goethals and J.J. Seidel, The Krein condition, spherical designs, Norton algebras and permutation groups, *Proc. Konink. Nederl. Akad. Wetensch.* A, 81, (=*Indag. Math.*, Vol. 40), pp. 196-206, 1978.

[63] P.J. Cameron, J.-M. Goethals and J.J. Seidel, Line Graphs, Root Systems, and Elliptic Geometry, *J. of Algebra*, Vol. 43, pp. 305-327, 1976.

[64] P. Camion, Association schemes, in V.Pless, W.C. Huffman and R.A.Brualdo Ed, *Handbook on coding theory*, Amsterdam, Elsevier 1998.

[65] J.W.S. Cassels, *An introduction to the geometry of numbers*, Springer-Verlag 1959.

[66] C. Chabauty, Resultat sur l'empilement de calottes egalessur une perisphere de $\mathbb{R}^n$ et correction a un travail anterieur, *C.R.A.S*, Vol. 236, pp. 1462-1464, 1953.

[67] D. Chase, A class of algorithms for decoding block codes with channel measurement information, *IEEE Trans. on Inform. Theory*, IT-18, pp. 170-182, Jan. 1972.

[68] B.W. Clare and D. L. Kepert, The closest packing of equal circles on a sphere, *Proc. R. Soc. Lond.*, Vol. 405, pp. 329-344, 1986.

[69] P.M. Cohn, Algebra Vol. 1, Wiley, London 1974.

[70] C.J. Colbourn and J.H. Dinitz, *The CRC Handbook of Combinatorial Designs*, CRC Press, Boca Raton, New York, London, Tokyo, 1996.

[71] J.H. Conway and N.J.A. Sloane, *Sphere Packings, Lattices and Groups*, Sec. Ed., Springer-Verlag, 1993.

[72] H.S.M. Coxeter, Discrete groups generated by reflections, *Annal. Math.*, Vol. 35, pp. 588-621, 1934.

[73] H. S. M. Coxeter, The problem of packing a number of equal nonoverlapping circles on a sphere, *Trans. N.Y. Acad. Sci.* ser. II, Vol. 24, pp. 320-331, 1962.

[74] H.S.M. Coxeter, An upper bound for the number of equal nonoverlapping spheres that can touch another of the same size, *Proc. Symp. in Pure Mathematics*, Vol.VII, pp. 53-72, 1963. Reprinted in H.S.M. Coxeter, Twelve geometric essays, Southern Illinois University Press, Carbondale Il, 1968.

[75] H.S.M. Coxeter, Regular polytopes, *Dover publications*, Third ed., New York 1973.

[76] H.S.M. Coxeter and J.A. Todd, An extreme duodenary form, *Canad. J. Math.*, Vol. 5, pp. 384-392, 1953.

[77] L. Danzer, Endliche Punktmengen auf der 2-Sphäre mit möglichst gro em Minimalabstand, *Habilitationsschrift*, Universität Göttingen, 1965.

[78] H. Davenport and G. Hajos, Aufgabe 35, *Matematikai Lapok*, Vol. 58, No. 2, 1951.

[79] P. Delsarte, Bounds for unrestricted codes by linear programming, *Philips Res. Reports*, Vol. 27, pp. 272-289, 1972.

[80] P. Delsarte, An algebraic approach to the association schemes of coding theory, *Philips Res. Repts.* Suppl. No. 10, 1973.

[81] P. Delsarte, J.M. Goethals and J.J. Seidel, Bounds for Systems of Lines and Jacobi Polynomials, *Philips Res. Repts.*, 30, pp. 91-105, 1975.

[82] P. Delsarte, J.M. Goethals and J.J. Seidel, Spherical codes and designs, *Geometriae Dedicata*, Vol. 6, pp. 363-388, 1977.

[83] P. Delsarte and J.J. Seidel, Fisher type inequalities for Euclidean t-designs, *Lin. Alg. Appl.*, Vol. 114/115, pp. 213-230, 1989.

[84] D. Djokovic and I. F. Blake, An optimization problem for unitary and orthogonal representations of finite groups, *Trans. Amer. Math. Soc.*, Vol. 164, 1972.

[85] S.M. Dodunekov, V.A. Zinoviev and Th. Ericson, Concatenated method of construction of spherical codes in n-dimensional Euclidean space, *Probl. Inform. Transm.*, Vol. 27, No. 1, pp. 34-38, 1991.

[86] C.P. Downey and J.K. Karlof, On the existence of [M,n] group codes for the Gaussian channel with M and n odd, *IEEE Trans. on Inform. Theory*, Vol. IT-23, pp. 500-503, July 1977.

[87] C.P. Downey and J.K. Karlof, Optimal [M,3] group codes for the Gaussian channel, *IEEE Trans. on Inform. Theory*, Vol. IT-24, pp. 760-761, Nov. 1978.

[88] I.I. Dumer, V.A. Zinoviev and V.V. Zyablov, Cascade decoding with respect to minimal generalized distance, *Problems of Control and Inform. Theory*, Vol. 10, No. 1, pp.1-17, 1982.

[89] G. Einarsson, Polyphase Coding for a Gaussian Channel, *Ericsson Technics*, No. 2, pp. 75-130, 1967.

[90] G. Einarsson, Performance of Polyphase Signals on a Gaussian Channel, *Ericsson Technics*, No. 4, pp. 411-433, 1967.

[91] M. Elia and E. Biglieri, The generating groups of geometrically uniform spherical signal sets, *Applicable Algebra in Engineering, Communication and Computing*, Vol. 3, pp. 163-181, 1992.

[92] P. Erdös, Aufgabe 20, *Matematikai Lapok*, Vol. 1, No. 3, p. 226, 1950.

[93] Th. Ericson, A simple analysis of the Blokh-Zyablov decoding algorithm, in: Th. Beth and M. Clausen ed, *Lecture Notes in Computer Science*, 307, Springer, 1986.

[94] Th. Ericson, Bounds on the size of a code, in: *Lecture Notes in Control and Inform. Sciences*, Vol. 128, Ed. M. Thoma and A. Wyner, Springer, Verlag, pp. 45-72, 1989.

[95] Th. Ericson, Permutation Codes, *INRIA - Institut National de Recherce en Informatique et en Automatique*, No. 2109, Novembre, 1993.

[96] Th. Ericson and V.A. Zinoviev, Spherical codes and binary partitions of finite pointsets on the real line, in: *Proc. of the Fifth Joint Soviet-Swedish Workshop on Information Theory*, Moscow, USSR, 1991, pp. 57-60, 1990.

[97] Th. Ericson and V.A. Zinoviev, Concatenated spherical codes, in: "EUROCODE 90, Intern. Symp. on Coding Theory and Applications", Udine, Italy, "Lecture Notes in Computer Science", Springer, Verlag, Vol. 514, pp. 2-5, 1991.

[98] Th. Ericson and V.A. Zinoviev, Binary partitions of symmetric pointsets, in: *Fifth Tirrenia Intern. Conf. on Inform. theory*, Tirrenia, Italy, 1991, *Coded Modulation and Bandwidth-Efficient Transmission*, Ed. E. Biglieri and M. Luise, Elsevier Science Publishers B.V., pp. 249- 259, 1992.

[99] Th. Ericson and V.A. Zinoviev, Concatenated spherical codes in Euclidean space, in: *Fifth Tirrenia Intern. Conf. on Inform. theory*, Tirrenia, Italy, 1991, *Coded Modulation and Bandwidth-Efficient Transmission*, Ed. E. Biglieri and M. Luise, Elsevier Science Publishers B.V., pp. 261-268, 1992.

[100] Th. Ericson and V.A. Zinoviev, On spherical codes generating the kissing number in dimensions 8 and 24, *Discrete Math.*, Vol. 106/107, pp. 199-207, 1992.

[101] Th. Ericson and V.A. Zinoviev, New packings on Euclidean sphere for finite dimensions, *Probl. Inform. Transmi.*, Vol. 28, No. 2, pp. 47- 53, 1992.

[102] Th. Ericson and V.A. Zinoviev, Ternary spherical codes by balanced mappings of constant weight binary codes, in: *Proc. of the Sixth Joint Swedish-Russian Workshop on Information Theory*, Molle, Sweden, 1993, pp. 140-144, 1993.

[103] Th. Ericson and V.A. Zinoviev, Sperical codes from non-symmetric alphabets, *Proc. of the 15th Benelux Symposium on Inform. Theory*, Louvain-la-neuve, May 30-31, 1994.

[104] Th. Ericson and V.A. Zinoviev, Spherical codes generated by binary partitions of symmetric pointsets, *IEEE Trans. Inform. Theory*, Vol. IT-41, pp 107-129, Jan., 1995.

[105] L. Fejes Tóth, Über die Abschatzung des kürzesten Abstandes zweier Punkte eines auf einer Kugelfläche liegenden Punktsystemes, *Jber. Deut. Math. Verein.*, Vol. 53, pp. 66-68, 1943.

[106] L. Fejes Tóth, *Lagerungen in der Ebene, auf der Kugel und im Raum*, Springer, Berlin, 1953.

[107] G.D. Forney, Generalized minimum distance decoding, *IEEE Trans. Inform. Theory*, IT-12, pp. 125-131, 1966.

[108] G.D. Forney, Jr., Concatenated codes, *MIT Research Monograph*, No. 37, MIT Press, Cambridge, Mass., 1966.

[109] G.D. Forney, Jr., R.G. Gallager, G.R. Lang, F.M. Longstaff and S.U. Quershi, Efficient modulation for band-limited channels, *IEEE J. Selected Areas Comm.*, Vol. SAC-2, pp. 632-647, 1984.

[110] G.D. Forney, Jr., and M.D. Trott, The Dynamics of Group Codes: Dual Group Codes and Systems, 1994.

[111] G.D. Forney, Jr. and G. Ungerboeck, Modulation and coding for linear Gaussian channels, *IEEE Trans. on Inform. Theory*, Vol. 44, No. 6, pp. 2384-2415, 1998.

[112] G.D. Forney, Geometrically uniform codes, *IEEE Trans. on Inform. Theory*, Vol. 37, No. 5, pp. 1241-1260, 1991.

[113] C.F. Gauss, Besprechung des Buchs von L.A. Seeber, *Untersuchungen über die Eigenschaften der positiven ternären quadratischen Formen usw. Göttingsche Gelehrte Anzeigen*, July 9, 1831 = Werke, II, pp. 188-196, 1876.

[114] A. Gewirtz, Graphs of maximal even girth, Canad. J. Math., Vol. 21, pp. 915-934, 1969.

[115] V. V. Ginzburg, Multidimensional signals for a continuous channel, *Probl. Inform. Transm.*, Vol. 20, No. 1, pp. 28-46, 1984.

[116] C.D. Godsil, *Algebraic Combinatorics*, Chapman and Hal, New York-London, 1993.

[117] J.-M. Goethals and J.J. Seidel, Strongly regular graphs derived from combinatorial designs, *Canad. J. Math.*, Vol. 22, pp. 597-614, 1970.

[118] J.M. Goethals and J.J. Seidel, Cubature formulae, polytopes and spherical designs, in "The Geometric Vein, Coxeter Festschrift", (B. Grünbaum, C. Davis and F.A. Sherk, ed.), Springer, pp. 203-218, 1981.

[119] T. Gosset, On the regular and semi-regular figures in space of n dimensions, *Messenger Math.*, Vol. 29, pp. 43-48, 1900.

[120] P.J. Grabner and R.F. Tichy, Spherical designs, discrepancy and numerical integration, *Math. Comp.*, Vol. 60, pp. 327-336, 1991.

[121] L.C. Grove, C.T. Benson, Finite Reflection Groups, Springer-Verlag, New York - Berlin - Heidelberg - Tokyo, 1971.

[122] M. Hall Jr, Combinatorial theory, *Wiley Interscience Series in discrete Mathematics*, Sec. Ed., Wiley 1986.

[123] A.R. Hammons Jr, P.V. Kumar, A.R. Calderbank, N.J.A. Sloane and P. Sole, The $\mathbb{Z}_4$ -linearity of Kerdock, Preparata, Goethals and related codes, *IEEE Trans. on Inform. Theory*, Vol. IT-40, pp. 301-319, 1994.

[124] A.A. Hassan and W.E. Stark, On decoding concatenated codes, *IEEE Trans. on Inform. Theory*, Vol. IT-36, No 3, pp. 677-683, May 1990.

[125] R.H. Hardin, N.J.A. Sloane, New Spherical 4-Designs, *Discrete Math.*, Vol. 106/107, pp. 255-264, 1992.

[126] J.N. Herstein, *Topics in Algebra*, Blaisdell Publ. Company, New York - Toronto - London, 1964.

[127] D.G. Higman, C.C. Sims, A simple group of order 44 352 000, Math.Z., Vol. 105, pp. 110-113, 1968.

[128] E. Hlawka, Ausfüllung und überdeckung konvexer Körper durch konvexer Körper, Monatsh. Math. Phys., Vol. 53, pp. 81-131, 1949.

[129] E. Hlawka, Zur Geometrie der Zahlen, Math. Z., Vol. 49, pp. 285-312, 1943.

[130] A.J. Hoffman, R.R. Singleton, On Moore graphs of diameter 2 and 3, IBM J. Res. Development, Vol. 4, pp. 497-504, 1960

[131] R. Hoppe, Bemerkung der Reduction, *Archiv Math. Physic*, (Grunert), Vol. 56, pp. 307-312, 1874.

[132] Hsiang, Wu-Yi, On the sphere packing problem and the proof of Kepler's conjecture. Internat. J. Math., Vol., No. 5, pp. 739-831, 1993.

[133] J.E. Humphreys, Reflection Groups and Coxeter Groups, Cambridge University Press, 1990.

[134] H. Imai and S. Hirakawa, A new multilevel coding method using error-correcting codes, *IEEE Trans. on Inform. Theory*, Vol. IT-23, pp. 371-377, 1977.

[135] I. Ingemarsson, Commutative Group Codes for the Gaussian Channel, *IEEE Trans. on Inform. Theory*, Vol. IT-19, pp. 215-219, March 1973.

[136] I. Ingemarsson, Group Codes for the Gaussian Channel, in: *Lecture Notes in Control and Inform. Sciences*, Vol. 128, Ed. M. Thoma and A. Wyner, Springer, Verlag, pp. 73-108, 1989.

[137] I. Ingemarsson, Optimized permutation modulation, *IEEE Trans. on Inform. Theory*, Vol. IT-36, No. 5, pp. 1098-1100, Sept. 1990.

[138] S. M. Johnson, A new upper bound for error-correcting codes, *IEEE Trans. on Inform. Theory*, Vol. IT-8, pp. 203-207, 1962.

[139] G.A. Kabatianskii, On decoding concatenated codes in certain spaces, in: *Proceedings of Fifth Joint Soviet-Swedish Intern. Workshop on Information Theory*, January 13-19, 1991, Moscow, USSR, pp. 86-90, 1990.

[140] G.A. Kabatianskii and V.I. Levenshtein, Bounds for packings on a sphere and in space, Problems of Information Transmission, Vol. 14, No 1, pp. 1-17, 1978.

[141] J. K. Karlof, Permutation codes for the Gaussian channel, *IEEE Trans. on Inform. Theory*, Vol. IT-35, No. 4, pp. 726-732, July 1989.

[142] J. K. Karlof and C.P. Downey, Odd group codes for the Gaussian channel, *SIAM J. Appl. Math.*, Vol. 34, pp. 715-716, 1978.

[143] A.M. Kerdock, A class of low-rate nonlinear codes, *Information and Control*, Vol. 20, pp. 182-187, 1972.

[144] A. Korkine and G. Zolotareff, Sur les formes quadratique positive quaternaires, *Math. Ann.*, Vol. 5, pp. 581-583, 1872.

[145] A. Korkine and G. Zolotareff, Sur les formes quadratique positives, *Math. Ann.*, Vol. 11, pp. 242-292, 1877.

[146] T. Kornwinder, The addition formula for Jacobi polynomials and spherical harmonics, SIAM J. on Applied Math., Vol. 25, pp. 236-246, 1973.

[147] G. Lachaud and J. Stern, Polynomial-Time Construction of Codes II: Spherical Codes and the Kissing Number of Spheres, *IEEE Trans. on Inform. Theory*, Vol. IT-40, No. 4, pp. 1140-1146, July, 1994.

[148] H.J. Landau, How does a porcupine separate its quills?, *IEEE Trans. on Inform. Theory*, Vol. IT-17, No 2, pp. 157-161, March 1971.

[149] D.G. Larman, C.A. Rogers and J.J. Seidel, On two-distance sets in Euclidean space, *Bull. London Math. Soc.*, Vol. 9, pp. 261-267, 1977.

[150] J. Leech, The problem of the thirteen spheres, *Math. Gaz.* Vol. 40, pp. 22-23, 1956.

[151] J. Leech, Some sphere packings in higher space, *Canad. J. Math.*, Vol. 16, pp. 657-682, 1964.

[152] J. Leech, Notes on sphere packings, *Canad. J. Math.*, Vol. 19, pp. 251-267, 1967.

[153] J. Leech and N.J.A. Sloane, Sphere packings and error-correcting codes, *Canad. J. Math.*, Vol. 23, No 4, pp. 718-745, 1971.

[154] P.W.H. Lemmens and J.J. Seidel, Equiangular lines, *J. Algebra*, Vol. 24, pp. 494-512, 1973.

[155] V. I. Levenshtein, On choosing polynomials to obtain bounds in packing problems, In: *Proc. Seventh All-Union Conf. on Coding Theory and Information Transmission*, Part II, Moscow-Vilnius, pp. 103-108, 1978 (in Russian).

[156] V.I. Levenshtein, On bounds for packings in n-dimensional Euclidean space, *Soviet Math. Doklady*, Vol. 20, No 2, pp. 417-421, 1979.

[157] V.I. Levenshtein, Bounds for packings of metric spaces and some of their applications, *Problemy Kibernetiki*, Vol. 40, Moscow, Nauka, pp. 43-110, 1983 (in Russian).

[158] V.I. Levenshtein, Designs as maximum codes in polynomial metric spaces, *Acta Applicandae Mathematicae*, Vol. 29, pp. 1-82, 1992.

[159] V.I. Levenshtein, Universal bounds for codes and designs, in *Handbook of Coding Theory*, V. Pless, W.C. Huffman and R.A. Brualdi, Ed. Amsterdam, Elsevier, 1998.

[160] J.H. Lindsey, Sphere-packing in $\mathbb{R}^3$, *Math.*, Vol. 33, pp. 137-147, 1986.

[161] J.H. van Lint and J.J. Seidel, Equilateral Point Sets in Elliptic Geometry, *Proc. Konink. Nederl. Akad. Wetensch.* A, 69, (=*Indag. Math.*, Vol. 28), pp. 335-348, 1966.

[162] J.H. van Lint, Introduction to Coding Theory, Graduate Texts in Math. 86, Springer-Verlag, Berlin, 1982.

[163] S.P. Lloyd, Hamming association schemes and codes on spheres, *SIAM J. Math. Anal.*, Vol. 11, No 3, pp. 488-505, May 1980.

[164] H.-A. Loeliger, Signal sets matched to groups, *IEEE Trans. on Inform. Theory*, Vol. 37, No. 6, pp. 1675-1682, 1991.

[165] Y.I. Lyubich and L.N. Vaserstein, Isometric embeddings between classical Banach spaces, cubature formulas and spherical designs, *Geom. Dedic.*, Vol. 47, pp. 327-362, 1993.

[166] A.L. MacKay, The packing of three-dimensional spheres on the surface of a four-dimensional hypersphere, *J. Phys.*, Vol. A13, pp. 3373-3379, 1980.

[167] F.J. MacWilliams and N.J.A. Sloane, *The theory of error-correcting codes*, North-Holland, 1977.

[168] R.J. McEliece, E.R. Rodemich, H. Rumsey Jr, and L.R. Welch, New upper bounds on the rate of a code via the Delsarte-MacWilliams inequalities, *IEEE Trans. on Inform. Theory*, IT-23, pp. 157-166, 1977.

[169] H. Minkowski, Diskontinuitätsbereich für arithmetische Aequivalenz, *J reine angew. Math.*, Vol. 129, pp. 220-274, 1905.

[170] H. Minkowski, Dichteste gitterförmige Lagerungen kongruenter Körper, Nachr. Ges. Wiss. Göttingen (1904), pp 311-355, Ges. Abh. II, pp. 3-42, 1911.

[171] T. Mittelholzer and J. Lahtonen, Group codes generated by finite reflection groups, *IEEE Trans. on Inform. Theory*, Vol. IT-42, No 2, pp. 519-528, 1996.

[172] A.A. Nechaev, Kerdock code in cyclic form, *Discrete Math. and Appl. (USSR)*, Vol. 1, No. 4, pp. 365-384, 1989.

[173] A.A. Nechaev, Linear codes over finite rings and QF modules, In: *Fourth Intern. Workshop on Algebraic and Combinatorial Coding Theory*, ACCT4 Proceedings, Novgorod, Russia, September 11-17, pp. 154-157, 1994.

[174] M. Nilsson, Linear block codes over rings for phase shift keying, *Linköping Studies in Science and Technology, Dissertations, No. 321*, Linköping, Sweden, 1993.

[175] A.W. Nordstrom, J.P. Robinson, An optimum nonlinear code, Info. and Control, Vol. 11, pp. 613-616, 1967.

[176] A.M. Odlyzko and N.J.A. Sloane, New upper bounds on the number of unit spheres that can touch a unit sphere in n dimensions, *J. Combin. Th.*, Vol. A 26, pp. 210-214, 1979.

[177] P.R.J. Östergård, T. Baicheva and E. Kolev, Optimal binary one-error-correcting codes of length 10 having 72 codewords, *IEEE Trans. on Inform. Theory*, Vol IT-45, No 4, pp. 1229-1231, May 1999.

[178] R. Ottosson, Performance of phase- and amplitude-modulated signals on a Gaussian channel, *Ericsson Technics*, No 3, pp. 153-198, 1969.

[179] R. Ottosson, Group codes for phase- and amplitude-modulated signals on a Gaussian channel, *IEEE Trans. on Inform. Theory*, Vol. IT-17, No 3, pp. 315-321, May 1971.

[180] S.E. Payne and J.A. Thas, Finite generalized quadrangles, Pitman, New York, 1985.

[181] P. Piret, Bounds for codes over the unit circle, *IEEE Trans. on Inform. Theory*, Vol. IT-32, No 6, pp. 760-767, 1986.

[182] S. L. Portnoi, Characterizations of modulation and encoding systems as concatenated codes, *Probl. Inform. Transm.*, Vol. 21, No 3, pp. 14-27, 1985.

[183] R.A. Rankin, On the closest packing of spheres in n dimensions, *Ann. Math.* Vol. 48, pp. 1062-1081, 1947.

[184] R. A. Rankin, The closest packing of spherical caps in n dimensions, *Proc. Glascow Math. Assoc.*, Vol. 2, pp. 139-144, 1955.

[185] R. M. Robinson, Arrangements of 24 points on a sphere, *Math. Ann.*, Vol. 144, pp. 17-48, 1961.

[186] C.A. Rogers, The packing of equal spheres, *Proc. London Math. Soc,*, Vol. 8, pp. 609-620, 1958.

[187] S.S. Ryskov and E.P. Baranovskii, Classical methods in the theory of lattice packings, Russian Math. Surveys, Vol. 34, No 4, pp. 3-63, 1979.

[188] K. Sarkadi and T. Szele, Aufgabe 20, *Matematikai Lapok*, Vol. 2, No. 1, p. 76-77, 1951.

[189] S.I. Sayegh, A class of optimum block codes in signal space, *IEEE Trans. on Commun.*, Vol. COM-34, No. 10, pp. 1043-1045, 1986.

[190] L. Schläfi, Réduction d'une intégrale multiple, qui comprend l'arc de cercle et l'aire du triangle sphérique comme cas particuliers, *Gesammelte mathematische Abhandlungen*, Bd. 2, pp. 164-190.

[191] I.L. Schoenberg, Positive definite functions on spheres, Duke Math. J. Vol. 9, pp. 96-107, 1942.

[192] P.H. Schoute, Analytical treatment of the polytopes regularly derived from the regular polytopes, *Ver. der Koninklije Akad. van Wetenschappen te Amsterdam* (eerstie sectie), Vol. 11, No 5, 1913.

[193] K. Schütte and B.L. van der Waerden, Auf welcher Kugel haben 5, 6, 7, oder 8 Punkte mit Mindestabstand Eins Platz?, *Math. Ann.* Vol. 123, pp. 96-124, 1951.

[194] K. Schütte and B.L. van der Waerden, Das Problem der dreizehn Kuheln, Math. Annalen 53, (1953), pp 325-334.

[195] G. Segö, Orthogonal polynomials, *Am. Math. Soc*, Coll. Publications, Vol. XXIII, 1939.

[196] J.J. Seidel, Spherical Designs and Tensors, *Recent Progress in Algebraic Combin. (E. Bannai and A. Munemasa, Eds.), Advanced Studies in Pure Math., Math. Soc. Japan*, 1996, Vol. 24, pp. 309-321.

[197] J.J. Seidel, D.E. Taylor, Two-graphs, a second survey, *Coll. Math. Soc. János Bolyai. 25. Algebraic Methods in Graph Theory*, Szeged (Hungary), pp. 689-711, 1978

[198] N.V. Semakov and V.A. Zinoviev, Complete and quasi-complete constant-weight codes, *Probl. Inform. Transm.*, Vol. 5, No.2, pp. 11-13, 1969.

[199] N.V. Semakov and V.A. Zinoviev, Constant weight codes and tactical configurations, *Probl. Inform. Transmiss.*, Vol. 5, No 3, pp. 38-50, 1969.

[200] N.V. Semakov, V.A. Zinoviev and G.V. Zaitsev, On duality of Preparata and Kerdock codes, in: *Fifth All-Union Conference on Coding Theory*, Moscow-Gorkii, Part 2, pp. 55-58, 1972, (in Russian).

[201] J.-P. Serre, Linear Representations of Finite Groups, Springer-Verlag, New York - Heidelberg - Berlin, 1977.

[202] P.D. Seymour and T. Zaslavsky, Averaging sets, *Adv. Math.*, Vol. 52, pp. 213-240, 1984.

[203] C.E. Shannon, Probability of error for optimal codes in a Gaussian channel, *Bell Syst. Techn. J.*, Vol. 38, pp. 611-656, 1959.

[204] V.M. Sidelnikov, On the densest packing of balls on the surface of an n-dimensional Eucliqean sphere and the number of binary code vectors with a given code distance, *Soviet Math. Doklady*, Vol. 14, pp. 1851-1855, 1973.

[205] V.M. Sidelnikov, On extremal polynomials used to estimate the size of codes, Problems of Inform. Transmission, Vol. 16, No 3, pp. 174-186, 1980.

[206] V.M. Sidelnikov, On one finite matrix group and codes on Euclidean spheres, Problems of Information Transmission, Vol. 33, No. 1, pp. 29-44, 1997.

[207] V.M. Sidelnikov, On one finite matrix group and orbit codes on Euclidean sphere, Doklady Akademii Nauk, Vol. 359, No. 3, pp. 313-315, 1998.

[208] V.M. Sidelnikov, Spherical 7-designs in 2^n-dimensional Euclidean space, J. of Algebraic Combinatorics, Vol. 10, pp. 279-288, 1999.

[209] V.M. Sidelnikov, Orbital spherical 11-designs whose initial point is a root of an invariant polynomial, St. Petersburg Math. J., Vol. 11, No. 4, 2000.

[210] D. Slepian, Bounds on communication, *Bell Syst. Techn. J.*, Vol. 42, pp. 681-707, May 1963.

[211] D. Slepian, Permutation modulation, *Proc. IEEE*, Vol. 53, No 3, pp. 228-236, March 1965.

[212] D. Slepian, Group codes for the Gaussian channel, *Bell Syst. Techn. J.*, Vol. 47, No 4, pp. 575-602, April, 1968.

[213] D. Slepian, On neighbor distances and symmetry in group codes, *IEEE Trans. on Inform.Theory*, Vol. IT-17, No 5, pp. 630-632, September 1971.

[214] N.J.A. Sloane, Tables of sphere packings and spherical codes, *IEEE Trans on Inform. Theory*, Vol. IT-27, pp. 327-338, 1981.

[215] N.J.A. Sloane, Recent bounds for codes, sphere packing and related problems obtained by linear programming and other methods, *Contemp. Math.*, Vol. 9, pp. 153-185, 1982.

[216] M.S. Smith, On rank 3 permutation groups, *J. Algebra*, Vol. 33, pp. 22-42, 1975.

[217] S.L. Sobolev, On formulas of mechanical cubature on a sphere invariant under action of finite rotation groups, Soviet Math. Doklady, Vol. 146, No 3, pp 1307-1310, 1962.

[218] S.L. Sobolev, Introduction to the Theory of Cubature Formules, Nauka, Moscow, 1974.

[219] A.B. Stott, Geometrical deduction of semiregular from regular polytopes and space fillings, *Ver. der Koninklijke Akademie van Wetenschappen te Amsterdam* (eerstie sectie), Vol. 11, No 1, 1910.

[220] P. M. L. Tammes, On the origin of number and arrangement of the places of exit on the surface of pollen-grains, *Recueil des travaux botaniques néerlandais* (Nederlandsche Botanische Vereeniging), Vol. 27, pp. 1-84, 1930.

[221] D.E. Taylor, Regular 2-Graphs, *Proc. London Math. Soc.*, Vol. 35, pp. 257-274, 1977.

[222] J. Tits, *Sur la trialité et certains groupes qui s'en déduisent*, Institut des Hautes Études Scientifiques, Publications Mathématiques, vol. 2, pp. 14-60, 1959.

[223] J. Tits, *Buildings and BN-Pairs of Spherical Type*, Lecture Notes in Math., No. 386, Springer Verlag, Berlin, 1974.

[224] R.Turyn, Hadamard matrices, Baumert–Hall units, four-symbol sequences, pulse compression, and surface wave encodings, Journal of Combinatorial Theory A, Vol. 16, pp. 313-333, 1974.

[225] G. Ungerböck, Channel coding with multilevel/phase signals, *IEEE Trans. on Inform. Theory*, Vol. IT- 28, No 1, pp. 55-67, Jan. 1982.

[226] Z.-X. Wan, Geometry of classical groups over finite fields, *Studentlitteratur*, Chartwell-Bratt, Lund, Sweden, 1993.

[227] E. Witt, Über Steinerische Systeme, *Abh. Math. Sem. Hamb.*, Vol. 12, pp. 265–275, 1938.

[228] A. D. Wyner, Capabilities of bounded discrepancy decoding, *Bell Syst. Tech. J.*, Vol. 44, pp. 1061-1122, 1965.

[229] I.M. Yaglom, Some results concerning distributions in n-dimensional space, *Appendix to the Russian edition of [Fej2]*, Moscow, 1958.

[230] V.A. Yudin, On the Variations of Designs, *Problems of Information Transmission*, Vol. 35, No. 4, pp. 68-73, 1999.

[231] V.A. Yudin, Rotations of Spherical Designs, *Problems of Information Transmission*, Vol. 36, No 3, pp. 39-45, 2000.

[232] G.V. Zaitsev, V.A. Zinoviev and N.V. Semakov, Interrelation of Preparata and Hamming codes and extension of Hamming codes to new double-error-correcting codes, *Proc. of Second Intern. Symp. on Inform. Theory* (Tsahkadsor, Armenia, 1971), Budapest, Acad. Kiado, pp. 257-263, 1973.

[233] L.H. Zetterberg, A Class of Codes for Polyphase Signals on a Bandlimited Gaussian Channel, *IEEE Trans. on Inform. Theory*, Vol. IT-11, pp. 385-395, July 1965.

[234] L.H. Zetterberg, Detection of a Class of Coded and Phase-Modulated Signals , *IEEE Trans. on Inform. Theory*, Vol. IT-12, pp. 153-161, April 1966.

[235] L.H. Zetterberg and H. Brändström, Codes for the combined phase and amplitude modulated signals in a four-dimensional space, *IEEE Trans. on Comm.*, Vol. COM-25, pp. 943-950, Sept. 1977.

[236] V.A. Zinoviev, On generalized concatenated codes, *Colloq. Math. Soc. Janos Bol. Vol. 16. Topics in Information Theory*, (Keszthely, Hungary, 1975), pp. 587-592, 1975.

[237] V.A. Zinoviev, Generalized concatenated codes, *Problems of Information Transmission*, Vol. 12, No. 1, pp. 5-15, 1976.

[238] V.A. Zinoviev, Generalized concatenated codes for channels with bursts of errors and independent errors, *Problems of Information Transmission*, Vol. 17, No. 4, pp. 53-62, 1981.

[239] V.A. Zinoviev and T. Ericson, New lower bounds for contact number in small dimensions, *Problems of Information Transmission*, Vol. 35, No. 4, pp. 3-11, 1999.

[240] V.A. Zinoviev, S.N. Litsyn and S.L. Portnoi, Cascade codes in Euclidean space, *Problems of Information Transmission*, Vol. 25, No. 3, pp. 62-75, 1989.

[241] V.A. Zinoviev and V.V. Zyablov, Decoding of nonlinear generalized concatenated codes, *Problems of Information Transmission*, Vol. 14, No. 2, pp. 46-52, 1978.

[242] C. Zong, Sphere packigns, Springer-Verlag New York, 1999.

[243] V.V. Zyablov, Optimization of cascaded decoding algorithms, *Problems of Information Transmission*, Vol. 9, No. 1, pp. 26-32, 1973.

[244] V.V. Zyablov and S.L. Portnoi, On one modulation coding system for a Gaussian channel, *Problems of Information Transmission*, Vol. 23, No 3, pp. 18-26, 1987.

Index